Introduction to
Pascal
with Applications in
Science and Engineering

Introduction to
Pascal
with Applications in Science and Engineering

SUSAN FINGER
Boston University

ELLEN FINGER

D. C. HEATH AND COMPANY
Lexington, Massachusetts   Toronto

to the memory of our grandmothers:
    Ethel Simmons Finger
    Mildred Trachier Donovan

Cover: Graphic design based on an engraving of Pascal's adding machine.

Copyright © 1986 by D. C. Heath and Company.

All rights reserved. No part of this publication may be reproduced or transmitted in any form or by any means, electronic or mechanical, including photocopy, recording, or any information storage or retrieval system, without permission in writing from the publisher.

Published simultaneously in Canada.

Printed in the United States of America.

International Standard Book Number: 0-669-08609-6

Library of Congress Catalog Card Number: 85-60979

Preface

This *Introduction to Pascal with Applications in Science and Engineering* is written for students taking a first course in programming. The text provides an introduction to programming and problem solving using the language Pascal and problems from science and engineering. The only prerequisites for the use of this text are high school algebra and trigonometry and an aptitude for and interest in science and mathematics. We assume that individuals with mathematical ability are likely to enjoy programming and to use computers in their work and recreation. We also assume that students in science and engineering want to learn a computer language to facilitate solving problems. This text will both interest and challenge such students.

Features

Most chapters are in two parts: a **topic in Pascal** and an **application topic in science or engineering** that illustrates the use of the Pascal topic covered in the first part of the chapter. The design of the book is modular; the Pascal topics can be taught independently of the application topics and most of the application topics can be taught independently of each other.

Each Pascal topic and each application topic includes a **case study** relevant to a science or engineering problem. The case studies in the Pascal sections generally use less rigorous mathematics than do the case studies in the application sections, and thus are suitable for a more general audience. For example, in Chapter 8 the case study for the Pascal section involves developing a program for determining population growth in discrete time intervals, whereas the case study in the application section requires the use of the exponential function (which is explained in the section) to determine population growth. Some problems are introduced at a simple level early in the book and expanded in later chapters so that students can develop and build on programs throughout the course.

Top-down design, **modular programming**, and **problem solving** are stressed throughout the book. An algorithm is developed following each case study. No formal pseudocode is used for the algorithms, and, in fact, the format of the algorithms varies so that students see the development of the algorithm as a process rather than a formula. In many cases, the original problem statement is quite general, allowing students to see how solutions to problems evolve. As the topic is presented, Pascal code is developed to implement the algorithm. Each section ends with a solution to the case study in which the code that has been developed throughout the section is brought together in a complete program. Sample output from the program is shown.

The **application topics** have been chosen from a wide range of engineering fields and from the physical and life sciences. Some applications, such as dimensional analysis and computational errors, are important topics in all sciences; other applications, such as the exponential function and complex numbers, are more relevant to engineering.

Debugging and **style hints** are presented throughout the text. These and **reminders** and **warnings** are boxed. Examples of both correct and incorrect code are included with the output (often from more than one compiler) generated by the code. Students are made aware of mistakes that are frequently made and of the error messages that result from these mistakes.

Numerous exercises and **programming problems** are presented at the end of each chapter. Exercises and problems are divided into two sections: those that require understanding and use of the Pascal topic, and those that require understanding and use of both the Pascal topic and the application topic. Solutions to the odd-numbered exercises can be found at the back of the book.

Organization and Content

The modular design of the text allows its use for either a one- or two-semester course. The instructor may omit the application topics altogether or choose the application topics most suitable to a particular course. The Pascal topics in Chapters 2 through 14 cover the material usually presented in a one-semester introductory Pascal course.

Chapter 1 provides a general introduction to computers. In Chapter 2, problem solving and algorithm development are discussed, and topics in Pascal that allow the student to begin to write simple programs are introduced; Chapter 3 introduces more elementary Pascal syntax. Since students will be familiar with the basic components of a program and with input and output by this time, Chapter 4 is devoted to procedures, program development, and program design. This early introduction of procedures and emphasis on top-down design and modular programming encourages good problem solving and programming habits. In Chapter 5, boolean and conditional statements are introduced, and the process of developing programs using procedures is stressed. The use of parameters within procedures is introduced in Chapter 6. Chapter 6 also presents functions and functions with parameters. After Chapter 6, all procedures and functions developed in the text use parameters. In addition, general-purpose functions and procedures are developed that are used throughout the remainder of the text. The looping constructions are covered in Chapters 7 and 8. Chapters 9 through 14 cover structured data types. Chapter 9 introduces arrays as the first structured data type, followed by textfiles in Chapter 10. These two structures are introduced first because they are necessary for handling the data from experiments in science and engineering. Chapter 11 covers enumerated types and character arrays. The applications are searching and sorting. Chapter 12 incorporates the discussion of sets with a discussion of program design and development at a more advanced level than in Chapters 4 and 6. Chapter 12 discusses the process of solving a problem, using modular, top-down design, from an initial problem statement to a correct, working program. Chapter 13 covers records including variant records and Chapter 14 covers files.

Chapter 15 contains two optional modules that can be covered after Chapter 8. The first module covers recursion. Recursion may be introduced any time after procedures with parameters (Chapter 6) and conditional loops (Chapter 8) have been covered. The other module in Chapter 15 covers subprograms as parameters. Finally, Chapter 16 is an introduction to dynamic variables.

Instructor's Guide

An instructor's guide is available with sample chapter tests, answers to the even-numbered exercises, and additional programming problems.

Advanced Applications

A supplement of advanced applications is available to students. It includes such topics as discrete-time systems, matrix inversion, and the solution of simultaneous equations.

Acknowledgments

The need for a text that uses Pascal as a vehicle for teaching programming in a science and engineering environment became apparent while developing and teaching a required freshman course in engineering computation at

Boston University. We thank Susan's colleagues at B.U. who taught this course with her: Gail Nagle, Nancy Conrad, Michael Ruane, Richard Vidale, James Wisdom, and particularly Anne Clough, who read the entire manuscript and offered many insights and suggestions. We are also thankful to the engineering students at B.U. for whom the course was created and who have shaped its content and texture, and to the teaching assistants who created the initial versions of many of the programs in the text. Some of the material in this textbook appeared in a preliminary form in *Pascal Programming for Engineers Using VPS*, published by Kendall/Hunt.

We are indebted to Clinton Foulk of Ohio State Unviersity for his careful review and classroom testing of our manuscript. His comments and suggestions were invaluable in our final revisions of the manuscript. T. J. Mueller of Harvey Mudd College offered us many helpful (and often very clever) suggestions on the overall approach and design of the text, which were useful in the early stages of refining the manuscript.

Others who reviewed our manuscript are: Richard Albright, University of Delaware; Dick Bulterman, Brown University; David A. Carlson, University of Massachusetts; Robert M. Holloway, University of Wisconsin-Madison; Edwin J. Kay, Lehigh University; Karen Lemone, Worcester Polytechnic Institute; Thomas W. Lynch, Northwestern University; and Benjamin Wah, Purdue University. Of course, we assume responsibility for any remaining errors and urge readers to inform us of them.

Thank you to Pamela Kirshen, Antoinette Tingley Schleyer, and Ruth Thompson at D. C. Heath. Pam has been a source of encouragement and support throughout the entire process of developing our book. Antoinette and Ruth have seen to all the details involved in turning our manuscript into a textbook.

Thank you to the Laboratory for Manufacturing and Productivity at the Massachusetts Institute of Technology, and particularly to its director, Professor Nam P. Suh, for inspiration and support during this year.

The compiler used for many of the sample programs in the text is the Pascal 8000/2.0 compiler developed by the Australian Atomic Energy Commission. These programs were executed on an IBM 3081 computer running under the Boston University VPS operating system. The computer programs and output were printed on a Taleris 1200 laser printer from VAX 11/780.

Finally, we thank those people who helped us in various ways in the preparation of our manuscript: June Correia, Avi Weiss, John Bausch, Homayoon Kazerooni, John Perrotti, Stuart Clough, Carol Adkins, Jerry Balchunas, Gail Greenwood, Angela Pitter, Alyson Biggs, Larry Spazianni, and particularly Geoffrey Ward, Ronald House, and our parents, Mary Elizabeth and Jack Finger.

Susan Finger
Ellen Finger

Contents

CHAPTER 1

Introduction to Computers   1

1.1	Early calculating machines	1
1.2	Components of computers	2
1.3	Storing information	5
1.4	Representation of numbers and characters	6
1.5	Continuous versus discrete mathematics	8
1.6	High-level languages	8
1.7	Pascal	9

CHAPTER 2

Introduction to Pascal   11

	■ CASE STUDY 1: Computing the radius of a circle	11
2.1	Algorithms	12
2.2	Identifiers	15
2.3	Data types	17
2.4	Constants	20
2.5	Variables	23
2.6	Comment statements	24
2.7	Assignment statements	25

Contents

2.8	Standard functions	28
2.9	Program structure	30
2.10	Output statements	32
2.11	Program execution	39
	Solution to Case Study 1	44

Summary 46   Exercises 47   Problems 49

CHAPTER 3

Input Statements, Arithmetic Operators, and Expressions 50

■ CASE STUDY 2: Computing the current in a circuit with resistors connected in parallel — 50
Algorithm development — 51

3.1	Input statement—read and readln	53
3.2	Arithmetic operators and expressions	59
3.3	Expressions	64
	Solution to Case Study 2	71

APPLICATION: Dimensional Analysis — 74

■ CASE STUDY 3: Speed of sound in furlongs per fortnight — 74
Algorithm development — 74
Systems of measurement — 75
Conversion factors — 81
Solution to Case Study 3 — 85

Summary 89   Exercises 89   Problems 91

CHAPTER 4

Procedures and Program Development 96

■ CASE STUDY 4: Computing the current in a circuit connected in series — 96

4.1	Algorithm development	97
4.2	Walkthroughs	101
4.3	Procedures	104

4.4	Block structure	110
4.5	Program development	113
4.6	Debugging strategies	116
4.7	Documentation	125
	Solution to Case Study 4	126

Summary   132   Exercises   132   Problems   136

CHAPTER 5

Boolean Expressions and Conditional Statements   139

	CASE STUDY 5: Highest and lowest temperatures to date	139
	Algorithm development	140
5.1	Applications of boolean algebra to programming	145
5.2	The if statement	154
5.3	The if-then-else statement	157
5.4	Compound if statements	159
5.5	The case statement	163
	Solution to Case Study 5	166

APPLICATION: Computational Errors   171

	CASE STUDY 6: Errors in computing the travel time of a ball	171
	Algorithm development	172
	Solution to Case Study 6	183

Summary   187   Exercises   187   Problems   190

CHAPTER 6

Subprograms with Parameters   194

	CASE STUDY 7: Minimum and maximum temperatures	194
	Algorithm development	194
6.1	Parameters	195
6.2	Procedures with variable parameters	196
6.3	Procedures with value parameters	204

	Solution to Case Study 7	209
6.4	Functions	213
	CASE STUDY 8: Minimum and maximum temperature solved with functions	213
	Algorithm development	213
	Solution to Case Study 8	222

Summary 222   Exercises 223   Problems 224

CHAPTER 7

The For Statement 226

	CASE STUDY 9: Average particle velocity	226
	Algorithm development	227
7.1	The for statement	229
7.2	Using a for statement with a compound statement	234
7.3	Using for statements with subprograms	236
7.4	Using a for statement to perform exponentiation	239
	Solution to Case Study 9	241
APPLICATION: Sequences and Series		**247**
	CASE STUDY 10: Bouncing ball	247
	Algorithm development	247
	Solution to Case Study 10	254

Summary 259   Exercises 259   Problems 261

CHAPTER 8

Conditional Loops 268

	CASE STUDY 11: Population growth	268
	Algorithm development	269
8.1	The while-do statement	271
8.2	The repeat-until statement	274
8.3	Simple data checking	278
	Solution to Case Study 11	280

APPLICATION: The Exponential Function	285
■ CASE STUDY 12: Exponential growth	285
Algorithm development	285
Discrete time growth	286
Exponential growth	294
Exponentiation in Pascal	300
Solution to Case Study 12	302

Summary 307   Exercises 307   Problems 310

CHAPTER 9

Arrays   313

■ CASE STUDY 13: Sample averages of experimental data	313
Algorithm development	314
9.1 Structured data types	315
9.2 One-dimensional arrays	315
9.3 Arrays as parameters	321
9.4 Multidimensional arrays	324
Solution to Case Study 13	327
APPLICATION: Vectors and Matrices	334
■ CASE STUDY 14: Torque of a bicycle wheel	334
Algorithm development	334
Vector operations	335
Matrix operations	353
Solution to Case Study 14	358

Summary 363   Exercises 363   Problems 365

CHAPTER 10

Textfiles   369

■ CASE STUDY 15: Saving experimental data	369
Algorithm development	370
10.1 Declaring textfiles	370

10.2 Reading from textfiles	372
10.3 Writing textfiles	380
10.4 Assigning disk files to textfiles	381
Solution to Case Study 15	382

APPLICATION: Introduction to Probability — 386

■ CASE STUDY 16: Arrival of phone calls	386
Algorithm development	386
Random variables	387
Discrete distributions	393
Continuous distributions	398
Solution to Case Study 16	402

APPLICATION: Introduction to Statistics — 407

■ CASE STUDY 17: Variance of experimental data	407
Algorithm development	407
Statistics and experiments	407
Hypothesis testing	411
Solution to Case Study 17	413

Summary 418   Exercises 419   Problems 421

CHAPTER 11

Enumerated Types and Character Arrays 425

■ CASE STUDY 18: Fuel types for power plants	425
Algorithm development	426
11.1 Enumerated data types	426
11.2 Subrange types	430
11.3 Character arrays	432
Solution to Case Study 18	437

APPLICATION: Searching — 443

■ CASE STUDY 19: Searching for a power plant	443
Algorithm development	443
Sequential search	444
Binary search	446

Solution to Case Study 19	448
APPLICATION: Sorting	**450**
■ CASE STUDY 20: Sorting power plants	450
Algorithm development	450
Selection sort	451
Bubble sort	453
Insertion sort	457
Solution to Case Study 20	461

Summary 462   Exercises 462   Problems 463

CHAPTER 12

Sets and Program Design 470

■ CASE STUDY 21: The chemical elements	470
Algorithm development	471
12.1 Mathematical sets	472
12.2 Sets in Pascal	477
12.3 Program design and validation	488

Summary 518   Exercises 518   Problems 520

CHAPTER 13

Records 522

■ CASE STUDY 22: Tagging birds	522
Algorithm development	522
13.1 Declaring a record	523
13.2 Arrays of records	529
13.3 Records of records	531
13.4 Variant records	533
Solution to Case Study 22	535
APPLICATION: Complex Numbers	**540**
■ CASE STUDY 23: Complex number package	540

Contents

Solution to Case Study 23 550
Summary 555   Exercises 555   Problems 557

CHAPTER 14

Files 560

■ CASE STUDY 24: Constructing a stellar data base 560
Algorithm development 561
14.1 File structure 562
14.2 Internal files 564
14.3 External files 569
14.4 Standard file procedures and functions 573
Solution to Case Study 24 577

APPLICATION: Random Number Generators   579

■ CASE STUDY 25: Satellite transmission 579
Algorithm development 579
Generating random numbers using a computer 580
Solution to Case Study 25 585
Summary 591   Exercises 592   Problems 593

CHAPTER 15

Recursive Subprograms and Subprograms as Parameters 599

■ CASE STUDY 26: Computing the length of a river 599
Algorithm development 600
15.1 Recursive subprograms 602
Solution to Case Study 26 611
15.2 Subprograms as parameters 618

APPLICATION: Plotting a Function   622

■ CASE STUDY 27: Plotting the gravitational force 622
Algorithm development 623

Plotting a function	624
Solution to Case Study 27	634

Summary 636   Exercises 637   Problems 638

CHAPTER 16

Dynamic Variables 639

CASE STUDY 28: A flexible manufacturing system	639
Algorithm development	641
16.1 Pointers	642
16.2 Dynamic variables	642
16.3 Linked lists	650
Solution to Case Study 28	660

Summary 666   Exercises 666   Problems 667

APPENDIX A

Reserved and Standard Elements and Character Codes A1

APPENDIX B

Syntax Diagrams A8

APPENDIX C

Binary, Octal, Decimal, and Hexadecimal Number Systems A15

APPENDIX D

Other Pascal Constructions A17

APPENDIX E
External Subprograms in Pascal   A22

APPENDIX F
Turbo Pascal   A30

APPENDIX G
UCSD Pascal   A42

APPENDIX H
VAX-11 Pascal   A52

Answers to Odd-Numbered Exercises   B1

Index   I1

CHAPTER 1

Introduction to Computers

This chapter will provide an overview of how computers work and how they are used. If you have had previous experience using computers, you may wish to skim or skip this chapter.

1.1 Early calculating machines

Computers, as they are thought of today, came into existence only in the late 1930s and early 1940s. However, the concept of a machine that could follow instructions and perform computations is not new. Inventors from many different cultures have created devices such as abacuses, slide rules, and mechanical adding machines to help alleviate the drudgery of performing arithmetic. Blaise Pascal (1623–1662), a French mathematician, philosopher, and physicist, created one of the earliest mechanical adding machines. His machine could add, subtract, and convert money into different denominations. The computer language Pascal is named in his honor. But an adding machine, no matter how complicated, is not a computer because it is not automatic; it requires a person to enter each number and instruction and to decide on the course of action.

In the early nineteenth century, inventors began to create mechanical devices, such as clocks and weaving looms, that could automatically follow a sequence of instructions without human intervention. The two important characteristics of these machines were that they were *automatic* and that they had the ability to *conditionally* perform a sequence of instructions. For example, a clock could strike a particular sound *if* it was the quarter hour, and a loom could make a pattern in the cloth by performing a different operation *if* it was at the end of the row.

Charles Babbage (1791–1871), an English mathematician, conceived of a mechanical computer, which he called an analytical engine, that could solve equations. His machine could perform arithmetical and logical operations in what Babbage called the mill and today is called the central processing unit (CPU). The operation of the mill was based on that of the mechanical adding machines that had already been developed. The operations to be performed by the mill could be changed by the person operating the machine. The data on which the operations were to be performed could also be changed by the operator. Babbage used punched cards, based on a technology already developed for entering patterns into weaving looms, to communicate both the instructions and the data to the machine. His analytical engine also had a store, which today is called the memory, in which the input data, intermediate results, and final results were stored. The results of the calculations were found by inspecting the store, although Babbage also conceived of a card punch that could be used to transfer the results from the store to a card for permanent storage. The mechanical technology available to Babbage was not sufficiently precise for him to create a working version of his analytical engine. After Babbage's death, his son built a version of the analytical engine that computed π to 32 places before it failed. The mathematician Augusta Ada Byron (1816–1851), Countess of Lovelace and daughter of Lord Byron the poet, worked with Babbage in creating programs for the analytical engine. The programming language Ada, which was influenced by the language Pascal, is named in her honor because she was the first computer programmer.

A more complete history of computers can be found in the references at the end of the chapter. Many other major figures, including Hollerith, Torres y Quevedo, Stibitz, Turing, Von Neumann, and Dijkstra, contributed to the development of computers.

1.2 Components of computers

Babbage's analytical engine had all the physical elements of the modern computer: input/output, memory, and a central processing unit. These elements are usually referred to as the *hardware* of the computer. Babbage's analytical engine also used the concept of a program—that is, a sequence of instructions that are entered into the computer and that result in prescribed actions within the computer. The programs are referred to as the *software* of the computer. So, even though computers are usually thought of as silicon chips surrounded by video screens, printers, and disk drives, computers can and have been constructed from mechanical gears, switches, and vacuum tubes. The physical devices that make up the computer will continue to evolve, so that what you think of as being a computer today is probably quite different from the computers that will be in use 10 years from now.

Computers are often categorized as being microcomputers, minicomputers, or mainframe computers. The distinctions among the types are not sharply defined, and as computer technology advances, the capabilities of microcomputers approach the capabilities of older mainframe computers. The basic elements of computers that are described in the next sections are essentially the same for all types of computers. The major distinction as far as users are concerned is that microcomputers can be used by one person at a time, minicomputers can be used by several people at a time, and mainframes can be used by hundreds of people at a time. The major impact of the number of users is on the way that resources are allocated among users. This topic will be covered later under operating systems.

Hardware components

The four main hardware components of a computer are (1) an *input* device that is used to enter the program and data into the computer, (2) a *central processing unit (CPU)* that follows the instructions and utilizes the information, (3) *storage* or *memory* that allows programs and data to be stored, and (4) an *output* device that reports the results. The input and output (I/O) devices are called the *I/O peripherals.* Typical input devices are terminals, magnetic disk packs, card readers, and magnetic tapes. Typical output devices are terminals, magnetic disk packs, line printers, and magnetic tapes.

The CPU consists of three sections: an arithmetic logic unit (ALU), internal storage, and a control section. Data and instructions are transmitted among the sections by electronic links called *busses.* The control section coordinates the execution of instructions and the retrieval and storage of data.

The CPU is designed by its manufacturer to perform certain operations such as basic arithmetic and logical comparisons. Some instructions are permanently programmed into the memory, so that when instructed to do so, the computer will perform basic operations such as addition, subtraction, and comparison. These operations are called *hard-wired,* and all other instructions are combinations of the hard-wired operations.

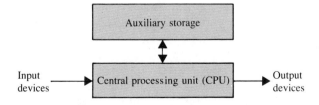

Central processing unit (CPU)	
Control section	
Arithmetic logic unit (ALU)	Internal storage

FIGURE 1.1
Overview of a computer system.

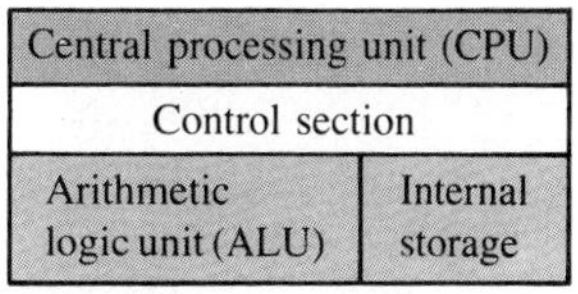

FIGURE 1.2
Overview of the central processing unit (CPU).

The CPU's memory is analogous to the memory in a calculator. Most scientific calculators have memory locations in which you can store numbers. If you want to use a number that you have stored, you must refer to it by its location; for example, Store-3 may refer to the number stored in memory location 3. The memory location of a piece of information stored in the CPU is called the *address*. Thus Store-3 can be seen as an address where data has been stored.

Information that must be quickly accessed is stored in the CPU in *registers*. Typically, there are from 8 to 16 registers that are used to store the results of the latest calculation, the address where data is to be found or stored, and other information relating to the execution and control of a program. Registers used to accumulate the results of arithmetic operations are called *accumulators*. An accumulator is similar to the display on a calculator except that the computer does not display the value in the accumulator unless instructed to do so.

A computer has more storage locations than a calculator. Also, computers have both short- and long-term memory. When a program is run on the computer, the program and its associated data are stored in the short-term memory of the computer. The computer can be directed to move information from short-term memory to the auxiliary storage, or long-term memory, so that programs and the results of programs can be saved. If programs or data have previously been stored in long-term memory, the computer can be instructed to retrieve them. Some devices used for storage of information in long-term memory are magnetic disks, drums, and tapes. In general, there is a tradeoff between speed and cost for different types of storage: the faster the information can be retrieved from a storage device, the more the device costs to purchase.

The operating system

The *operating system* is a program that serves as a bookkeeper and resource allocator for the computer. It locates programs and data in storage and makes them accessible to the user. It transfers information from short- to long-term memory. It regulates the flow of input, processing, and output.

The operating systems for the original mainframe computers operated in *batch* mode. In batch mode, the jobs to be processed by the computer are submitted in a queue, and each job must complete execution before the next job can be started. The turnaround, or the time from the submission of a job to its completion, is very uncertain in batch mode because a single large job can tie up the queue for hours. To alleviate this problem, *time-sharing* operating systems were developed. In a time-sharing system, jobs are usually worked on in rotation. The CPU will work on one job for a short time and then move on to the next job, picking up where it left off on that job. A short job is finished quickly because it takes only a few rotations through the CPU to complete.

The next advance in operating systems was the development of *interactive* systems that essentially allowed the user to have a conversation with the computer. In batch mode, all the commands are entered into the computer at the same time, traditionally through a deck of computer cards. If any one of the cards is wrong, the job will not run, and the entire deck must be resubmitted. In an interactive system, commands can be given to the computer one at a time. The computer responds after each command or set of commands. The user can take corrective actions, change the course of action, and also get immediate results when a job is run. By its nature, an interactive system requires that users have terminals that allow them to communicate directly with the computer.

Before the development of interactive systems, programs and data were punched on continuous paper tapes or on computer cards, with each line on a separate card. Once interactive systems became available, editors were developed that allowed computer users to create and edit interactively entire programs stored in a computer's long-term memory. Advances such as the introduction of interactive editors have eased the burdens of communicating with a computer and have greatly expanded the number of people who have easy access to computers.

Because input and output devices tend to be slow, the operating system has *buffers* to collect input and output information. These buffers hold information until there is a sufficient amount to send to the CPU so that the CPU is not tied up waiting for someone to type the next letter.

On microcomputers linked by networks, on minicomputers, and on mainframes, the operating systems identify users seeking access to the computer and keep track of the amount of time that each individual has used. The operating system also can control access to files, ensuring that two users do not alter a file at the same time and that files are accessed only by authorized users.

In a time-sharing system, the operating system ensures that, even if many people are working at terminals, each user has access to the CPU, one user at a time. The operating system allows a single fast CPU to service many users by switching them in and out of the CPU. Because computers operate at such high speeds, users cannot usually tell that they are sharing the CPU.

1.3 Storing information

In a digital computer, the central processing unit can be thought of as an ordered series of on-off switches. Actually, the switches are usually in the form of the presence or absence of an electrical voltage. The presence or absence of voltage can be interpreted as a binary (base 2) number, where on equals 1 and off equals 0. A computer thus stores information in the binary number system. Each 0 or 1 is called a *bit,* short for *binary digit.* In the binary system, as in the decimal system, the value of a digit is determined by its position. For example, you know that

$$1568_{10} = (1 \times 10^3) + (5 \times 10^2) + (6 \times 10^1) + (8 \times 10^0)$$
$$= 1000 + 500 + 60 + 8$$

where 1568_{10} means that the number 1568 is to be interpreted using the base 10 number system. By analogy,

$$11010_2 = (1 \times 2^4) + (1 \times 2^3) + (0 \times 2^2) + (1 \times 2^1) + (0 \times 2^0)$$
$$= 16 + 8 + 0 + 2 = 26_{10}$$

The array of bits within the computer for the above binary number would be

```
- - - + + - + -
0 0 0 1 1 0 1 0
```

where − represents off and + represents on. The +'s and −'s are used to remind you that the computer does not actually store the numbers 0 and 1.

Bits are commonly organized into groups called *bytes*. The most common size for a byte is eight bits. The largest binary number that can be represented by eight bits, or a standard byte, is 11111111 (all switches on). This number is equivalent to the decimal number 255. Two eight-bit bytes together can be used to store numbers up to 65535 (1111111111111111).

Bits are frequently combined to make *words*. A word is usually larger than a byte, and, on most computers, a word is the smallest unit in which data is stored and operated on. Common word sizes are 8, 16, 32, and 64 bits. You may also encounter 12-, 18-, 36-, 48-, and 60-bit words.

1.4 Representation of numbers and characters

Once the bits have been organized into words, the bits can be interpreted as representing binary numbers. In this section, we will discuss how the bits in memory can be organized and interpreted as integer numbers, real numbers, and characters.

For every integer number in base 10, there is a corresponding number in base 2. Consequently, unless the binary number has more bits than the computer can store, every integer can be converted and stored in its binary form.

There is another system for representing numbers called binary coded decimal (BCD), in which each digit of the number in base 10 is stored separately. For example, the number 12 would be stored as 0001 0010 ($0001_2 = 1_{10}$, $0010_2 = 2_{10}$). (You can assume that all binary numbers are given in base 2 as in the previous illustrations unless you are told otherwise.)

Real numbers and characters must also be represented, and they too must be represented using only 0s and 1s. A convention must be defined that can translate characters and real numbers into binary notation and consistently translate the binary information back into characters and real numbers for output to the user.

Characters must be represented in binary notation so that they are recognized by the computer. Every letter, number, and symbol that can be created on a keyboard or computer terminal is a character. Codes have been designed that assign a unique binary number to each letter and character. If it is known that a piece of data represents a character, the code can be looked up in a chart to determine what character the code represents. Two widely used conversion codes are ASCII (pronounced "ask-key") and EBCDIC (pronounced "ebbs-dick"). ASCII is the most common code, but EBCDIC is used on many mainframe computers. Both codes are given in Appendix A.3.

For real numbers, the computer word is divided up into fields. One field is used to represent the *mantissa* of the number. Another field is used to represent the *exponent*. You can visualize a real number as

| + | m | m | m | m | m | m | m | m | + | e | e |

This visualization is quite different from what a real number actually looks like in the computer. In the computer, the number is stored in binary, the fields may be in the opposite order, and the exponent may be stored in biased format.* The exact manner in which real numbers are stored and manipulated is the subject of a proposed Institute of Electrical and Electronics Engineers (IEEE) standard.

The number above represents the number $+0.\text{mmmmmmmm} \times 10^{+ee}$. The mantissa and the exponent are stored internally in binary, but numbers in base 10 will be used in the following examples so that they will be clearer. The number 9.876543 can be written as $.98765430 \times 10^1$. The mantissa is .98765430 and the exponent is 1. In a computer, this number might be stored in two fields: one for the mantissa, +98765430, and the other for the exponent, +01. The number 0.01234 might be stored as +12340000 and −01. Which bits represent the sign, the mantissa, and the exponent of a number are defined by the convention in use for a particular computer.

Depending on the word size for the computer, some of the trailing digits of a real number may be truncated. Therefore, when using real numbers on a computer, it is always necessary to know how many digits are *significant* or reliable. This topic is discussed in more detail in Chapter 5.

---

*When a number is in biased format, a constant, or *bias*, is added to it so that it is always greater than zero. For example, suppose that the exponent were stored in biased format with a bias of 50. If the last field were 53, the exponent would be +3. If the last field were 49, the exponent would be −1.

1.5 Continuous versus discrete mathematics

Most branches of mathematics use the real number line, which is continuous and has infinitely many numbers. On a computer, however, the number line is not continuous, and the number of numbers that can be represented in a computer is finite. The gaps in the computer's number line arise because the size of the mantissa and the exponent are fixed. If the mantissa is fixed at eight digits, then the number after 0.12345678 is 0.12345679; the infinitely many numbers that come between these two numbers on a real number line cannot be represented in the computer. For example, 0.123456788 and 0.1234567801 cannot be represented if the mantissa is fixed at eight digits.

In calculus, integration over an interval is performed by dividing the interval into subintervals and taking the limit as the length of the subintervals approaches zero. Similarly, differentiation in calculus involves taking the limit of the quotient of two differences as the divisor approaches zero. With a computer, an interval cannot approach zero. There is some smallest interval that can be represented on the computer. The largest number that can be represented on a computer is also a finite number; thus the concept of infinity cannot be used in solving mathematical problems on a computer. Because neither infinitely small nor infinitely large numbers can be represented, the operations of integration and differentiation must be approximated when one is using a computer. Some of the later chapters will deal with the approximation methods that must be used because the real number line cannot be represented on the computer.

1.6 High-level languages

We have already discussed the storage of numbers in the computer's memory. To do anything with the stored numbers, you must be able to give instructions to the computer, telling it what to do with the data. Most computers have a *machine language* that is formed from their basic hard-wired instructions. Each operation has a code, referred to as the *op-code*. Instructions to the CPU are given through op-codes. For example, the addition operation might have the op-code 01. These op-codes are decoded into basic hard-wired instructions so that addition can be performed.

Writing programs in machine language is possible, but very tedious. All instructions and data must be entered as binary numbers. The computer must be instructed about which storage locations to use. To add two numbers, you must instruct the computer to place each of them in a specific storage location. You could then instruct the computer to retrieve the contents of one of these locations and to put the number in the accumulator of the CPU. Then you could instruct the computer to get the contents of the second storage location, to add that number to the contents of the accumulator, and to store the sum in the location where the computer found the first number. Each instruction must be entered using the numerical op-code.

Assembly language is a step above machine language. Instead of each instruction having a code, each instruction has a mnemonic (that is, an alphanumeric code), and commonly executed sequences of instructions also have mnemonic codes. Assembly language is easier to use than machine language, but can also be cumbersome.

Computer languages have been devised to make it easier for people to communicate with computers. Programs written in computer languages are translated by *compilers* into machine language. A compiler is similar to a one-way bilingual dictionary, because it translates a language that people can understand into the binary code used by computers. Note that compilers are also programs: compilers' input is the code written in a high-level language and their output is computer code written in machine language.

High-level languages are computer languages that have a grammar that is closer to that of human languages. In a high-level language, most of the bookkeeping is taken care of for you. When you use a high-level computer language, you can give the storage locations names. You do not have to remember that the running total is stored in storage location 2; rather, you can give storage location 2 the name **RunningTotal**. Then, any time you want to refer to the number stored in storage location 2, you just use the name **RunningTotal** and the computer will find the number. If you want to add two numbers using the Pascal language, you can say

```
RunningTotal := RunningTotal + NewNumber
```

and the number stored in **NewNumber** will be added to the number stored in **RunningTotal,** with the result placed in **RunningTotal.** This one statement is equivalent to the whole set of instructions given at the beginning of this section. The compiler translates this instruction into a sequence of machine-language instructions for you.

Computer languages are written by a small group of people or a single individual. And, unlike the rules of the English language, the rules of a computer language must be stated precisely, because the rules must be used in the compiler program that translates from the language into machine code. Computer languages do have dialects and do evolve over time, but their evolution is usually overseen by committees.

1.7 Pascal

The first version of Pascal was written by Niklaus Wirth (1934–) in 1968. Wirth based the language Pascal on Algol-60, a powerful algorithmic language. The major differences between Algol and Pascal are in the handling of input/output and the data structures that can be created by the user. Wirth gives these reasons for developing Pascal:

> The development of the language *Pascal* is based on two principal aims. The first is to make available a language suitable to teach programming as a systematic discipline based on certain fundamental concepts clearly and naturally reflected by the language. The second is to develop implementations of this language which are both reliable and efficient on presently available computers.
>
> The desire for a new language for the purpose of teaching programming is due to . . . my conviction that the language in which the student is taught to express his ideas profoundly influences his habits of thought and invention. [reference 2]

Since the time that Wirth wrote Pascal, its usage has extended far beyond its original function as a teaching language. Because of its unified design, Pascal found wide use outside computer programming instruction. With its growing use, it became necessary to define a standard for Pascal so that programs could be transported among computers without extensive revisions.

Wirth's final version of Pascal was described in a book published in 1974 (see reference 2). His description of the language is the basis for *standard Pascal,* although there are several different standards for Pascal. In this book, the term "standard Pascal" will refer to the International Standards Organization (ISO) standard. The American National Standards Institute (ANSI)/Institute for Electrical and Electronics Engineers (IEEE) standard is equivalent except for a few rather technical differences.

References

1. Doug Cooper, *Standard Pascal, User Reference Manual,* New York: W. W. Norton and Company, 1983.

2. Kathleen Jensen and Niklaus Wirth, *Pascal, User Manual and Report,* Second Edition, New York: Springer-Verlag, 1974.

3. René Moreau, *The Computer Comes of Age,* Cambridge, Mass.: MIT Press, 1984.

4. *American National Standard Pascal, Computer Programming Language,* IEEE, New York, 1983.

5. "A Proposed Standard for Binary Floating-Point Arithmetic," Draft 8.0 of IEEE Task P754, *Computer,* March 1981, pp. 51–62.

CHAPTER 2

Introduction to Pascal

Most chapters in this book begin with a general introductory paragraph giving the topic of the chapter. Following the introduction most chapters have a case study that is used in explaining the Pascal topic. Each chapter will include complete sample programs; a listing of the program, often called the *source code,* will be provided, followed by the computer output.

When you are learning to program, you will spend much of your time editing and correcting your programs. To help you, there are programs and program fragments throughout this book that illustrate common mistakes. There are also suggestions for ways to avoid mistakes and ways to make mistakes easier to find.

In this chapter you will be introduced to the concept of algorithm development, and you will learn to write simple Pascal programs. You will learn how information is stored and referenced in a Pascal program and how information is displayed to a person running a program.

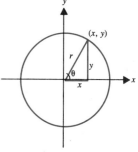

FIGURE 2.1

Circle with radius r and angle θ.

CASE STUDY 1
Computing the radius of a circle

A point on a circle can be represented by its x and y coordinates, as illustrated in Figure 2.1. Given the point (5.0, 10.0), write a program to compute and report the magnitude of the radius r in inches and the angle θ expressed in both degrees and radians.

2.1 Algorithms

In order to utilize a computer language to solve a problem, you must write an *algorithm*. An algorithm is a finite sequence of well-defined steps that unambiguously solves a problem. For example, the instructions for assembling a model airplane ought to be an algorithm. The instructions, when followed correctly, should transform a jumble of parts into an airplane that will fly. Of course, you know that you do not always get an airplane that flies the first time. There are frequently *bugs* in any set of instructions. A bug is anything that prevents the problem from being solved. In assembling the plane, you may encounter bugs because the directions were incorrect or were misunderstood.

Prior to writing a computer program, you need to write an algorithm. Learning to write algorithms—that is, to write instructions that get you from the input to the desired output—is the most difficult part of programming. Writing effective algorithms requires a lot of practice and experience. Because algorithms are usually written in English using a syntax similar to the computer language that will be used, algorithm development will be discussed more completely in Chapter 4, after you have gained some understanding of the Pascal language.

In writing a sequence of steps to solve a problem, a technique called *top-down design* is often used. With this technique, the problem is initially stated in general terms and then broken down into subproblems whose solutions are necessary to the solution of the overall problem. Each subproblem is then broken down into simpler subproblems until each subproblem can be solved directly. For the case study, the overall problem could be stated as

ALGORITHM 2.1 First iteration for an algorithm to solve Case Study 1

Compute and report the radius r in inches and the angle θ in radians and degrees, given the point (5.0, 10.0) on the circle

Algorithm 2.1 must be broken down into simpler subproblems because it cannot be solved directly as stated. Separating the compound sentence into simple sentences and arranging the sentences in order, the algorithm becomes

ALGORITHM 2.2 Second iteration for an algorithm to solve Case Study 1

1. Compute the radius of the circle in inches, given the point (5.0, 10.0)
2. Compute the angle θ in radians
3. Convert the value of the angle θ from radians to degrees

4. Report the radius r in inches and the angle θ in radians and in degrees

Most of these steps still cannot be solved directly, but now you can concentrate on each step individually and yet still see the overall problem. For Step 1, computing the magnitude of the radius, you must know the formula for the relationship between the x and y coordinates and the radius. Although it was not stated in the problem, from looking at Figure 2.1, you can assume that the center of the circle lies at the origin of the x and y axes. Using the Pythagorean theorem, you can write the formula for the magnitude of the radius:

$$\text{radius} = \sqrt{x^2 + y^2} \qquad (2.1)$$

The algorithm for Step 1 can now be stated as

ALGORITHM 2.3 Refinement of Step 1 of Algorithm 2.2

1. Compute the radius of the circle, given the point (5.0, 10.0)
   a. Set $x = 5.0$ and $y = 10.0$
   b. Use the formula radius $= \sqrt{x^2 + y^2}$ to compute the radius of the circle

The refinement of Step 1 is now essentially finished. When you first write algorithms, one way of testing whether the refinement is complete is to carry out the steps on a hand-held calculator. Step 1b could be refined further by stating how the squares, sum, and square root are calculated, but for most people the level of refinement of Step 1b is sufficient.

Step 2, computing the angle in radians, can be refined next. From Figure 2.1, you can see that θ is the angle between the x axis and the radial line connecting the origin to the point. The tangent of this angle is defined in terms of x and y:

$$\tan \theta = y/x \qquad (2.2)$$

In Equation 2.2 the only unknown variable is θ, but Equation 2.2 does not give the value of θ directly. To solve for θ, you must use the *inverse tangent* or *arctangent,* which is defined as

$$\theta = \tan^{-1}(y/x)$$

or $\qquad (2.3)$

$$\theta = \arctan(y/x)$$

Whether the angle in Equation 2.3 is in radians or degrees will depend on how the arctangent is computed. Assume for now that the angle is returned in radians. The algorithm for Step 2 now becomes

ALGORITHM 2.4 Refinement of Step 2 of Algorithm 2.2

2. Compute the angle θ in radians
   a. Given $x = 5.0$ and $y = 10.0$, compute the angle θ in radians using the relationship

$$\theta = \arctan(y/x)$$

After the value of the angle in radians has been found, the next step is to compute the angle in degrees. For this conversion, you need to know that there are 2π radians or $360°$ in a circle:

$$2\pi \text{ radians} = 360°$$

or (2.4)

$$1 = 180°/\pi \text{ radians}$$

Multiplying by $180°/\pi$ radians is the same as multiplying by 1. So to convert an angle from radians to degrees, you can multiply by $180°/\pi$ radians:

$$z \text{ radians} = z \text{ radians} \left(\frac{180 \text{ degrees}}{\pi \text{ radians}}\right) = z \left(\frac{180}{\pi} \text{ degrees}\right) \quad (2.5)$$

For the refinement of Step 3, you can state the conversion formula:

ALGORITHM 2.5 Refinement of Step 3 of Algorithm 2.2

3. Convert the value of the angle θ from radians to degrees
   a. Given the angle θ in radians, convert the angle into degrees using the relationship

$$\text{angle in degrees} = (\text{angle in radians})(180°/\pi \text{ radians})$$

The last step, that of reporting the values, does not need to be broken down further. The entire algorithm can be written as:

ALGORITHM 2.6 Refinement of Algorithm 2.2

1. Compute the radius of the circle, given the point (5.0, 10.0)
   a. Set $x = 5.0$ and $y = 10.0$
   b. Use the formula radius $= \sqrt{x^2 + y^2}$ to compute the radius of the circle
2. Compute the angle θ in radians
   a. Given $x = 5.0$ and $y = 10.0$, compute the angle θ in radians using the relationship

$$\theta = \arctan(y/x)$$

3. Convert the value of the angle θ from radians to degrees

a. Given the angle θ in radians, convert the angle into degrees using the relationship

$$\text{angle in degrees} = (\text{angle in radians})(180°/\pi \text{ radians})$$

4. Report the radius r in inches and the angle θ in radians and in degrees

Algorithm 2.6 is a series of steps that solve the problem stated in the case study. Keep in mind, however, that many algorithms and many programs can be written to solve a problem. There is never one unique solution, although all solutions should yield equivalent results.

Once you have developed your algorithm, you must be able to translate the algorithm into a language that can be used on a computer. All languages can be described by their *syntax*. Syntax is the organization and structure of a language. Thus, the syntax of a computer language is the grammar or the rules that must be followed when one is using the language. You will discover that the grammar rules for computer languages are stricter and more limited than English grammar rules. They also tend to be more logical and less ambiguous. The rest of this chapter will introduce some of the fundamental rules of Pascal syntax.

2.2 Identifiers

To begin to implement the algorithm to solve the case study, you must set up locations in which to store information within the computer's memory. Each storage location must have an *identifier* that is used to specify the location where the information is being stored. Identifiers are like the names of storage locations in a calculator. Using a calculator, you can store or retrieve data from storage location 5, for example. In a high-level language, such as Pascal, you can give storage locations names like **Radius.** Identifiers make it easier to manipulate data because you can use a name that describes the data rather than a meaningless storage location number.

In Pascal every sequence of instructions that you write must have an identifying name, and every storage location that you want to use must have an identifying name. A sequence of instructions may be either a complete program or a subprogram. Subprograms will be covered in Chapters 4 and 6 after you have learned to write simple programs.

Identifiers in Pascal must be formed using these rules:

1. Every identifier must start with a letter of the alphabet.

2. The rest of the identifier can be either letters or digits. Letters and digits are called *alphanumeric* characters. (See Exercise 3.)

3. Valid identifiers cannot contain spaces.

4. In standard Pascal an identifier can have as many characters as you want. Many compilers, however, ignore all characters after the eighth or sixteenth character. (See Exercise 10.)

The Pascal language is described by *syntax diagrams* that can help you determine if a Pascal statement follows Pascal's rules of grammar. The syntax diagram for an identifier in standard Pascal looks like

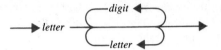

To follow a syntax diagram, enter from the left. The first thing you must have is a letter. Then you can exit to the right or you can loop down and get another letter or loop up and get a digit. You can loop as many times as you want. Suppose you decide to call the radius of the circle in the case study **Radius.** Let us run this identifier through the diagram. Start with an **R.** Loop through a letter, **a.** Loop through a letter, **d.** Loop through a letter, **i.** Loop through a letter, **u.** Loop through a letter, **s.** Exit.

The following names are valid identifiers:

`Force2   ANGLE   distance   Circumference23new   Circumference23next`

For some compilers the last two identifiers would reference the same storage location since their first sixteen characters are the same. Pascal does not distinguish between upper and lower case, so the identifiers

`RADIUS   radius   Radius`

are equivalent and reference the same storage location.

The following names are invalid identifiers:

`Cost*2,   3NAME,   NEW WORD,   force%2,   A>B,   SUM.1`

Try to run **Cost * 2** through the syntax diagram. When you get to the *, there is no place to go, because * is neither a letter nor a digit. **Cost * 2** is an invalid identifier. The second identifier is invalid because it starts with a number not a letter. The third identifier is invalid because it contains a space. The other identifiers are invalid because they include nonalphanumeric characters.

The following words are also invalid identifiers, not because they violate the rules above, but because they are *reserved words* in Pascal*:

`begin   end   program   to   var   const`

---

*In the text, all Pascal identifiers are typed in boldface. Reserved words and standard identifiers are in lower case. Identifiers that are created by the programmer are in mixed upper and lower case.

The reserved words are part of the Pascal syntax and are used by the compiler or the interpreter in translating your program into machine language. Their meaning cannot be changed within a program. A complete list of reserved words is given in Appendix A.

From the algorithm for the case study, you can see that you will need storage locations for the name of the program, the x and y coordinates, the radius, the angle in radians, the angle in degrees, and π. The identifiers that you use should describe the information that will be stored. It is much easier to see what your program is doing if the names you choose are descriptive of the data. Identifiers for the case study storage locations could be **X, Y, Radius, AngleRadians, AngleDegrees,** and **Pi.** You can give the program the identifying name **Circle.**

STYLE HINT

Give your identifiers reasonable self-descriptive names. Call your variables **Radius, Angle,** and **Distance,** not **X1, Tom,** and **Thing.**

2.3 Data types

Pascal is a *strongly typed* language. This means that each storage location must have an explicitly declared data type. The basic data types for Pascal are defined using the following words:

integer	for integer data
real	for real data
char	for character data
boolean	for logical (true-false) data

You can visualize the storage locations as boxes, each labeled with an identifier, as follows:

RealData   **IntegerData**   **CharacterData**   **BooleanData**

The identifier for each storage location is a valid identifier following the rules given above. Each identifier is also descriptive, because it implies the type of data that will be stored inside that location. Any piece of data that is stored in a Pascal program will be interpreted as one of the four basic data types. Only one piece of data can be in a storage location at any time. That is, each location can hold one real number, or one integer number, or one character, or one boolean value. This section will describe the syntax for specifying integer and real numbers and characters so that they can later be placed in storage locations. Boolean variables will be covered in Chapter 5.

Integers

An integer is any number made up of only an initial sign (+ or −) and digits. The syntax diagram for an integer number is

The following are valid integers:

$$-1050 \quad 45089 \quad +9854 \quad 0$$

The following are invalid integers:

$$-1050.0 \quad 1000e10 \quad 3456.2$$

If the initial sign is not given, the *default* is positive. A default is what you get if you do not say exactly what you want. For example, the default for pie à la mode is vanilla ice cream. If you want chocolate ice cream, you have to say so. Most computer commands have defaults because too many options exist to be able to state them all every time you want to do something. In this case, if you want a negative number, you must include the minus sign.

Real numbers

A real number can be represented in one of two basic ways. The most common way is to use a sign, then an integer, then a decimal point, and then another unsigned integer; for example,

$$0.45 \quad +4356.00 \quad 0.0 \quad -3456.2$$

The other way to represent real numbers is to use *scientific notation;* for example,

$$1.345e+02 \quad 2.058e-03$$

1.345e+02 is equivalent to 1.345×10^2 or 134.5, and 2.058e−03 is equivalent to 2.058×10^{-3} or 0.002058. For the first number the decimal point is shifted to the right by two places because the exponent is +2. For the second number the decimal point is shifted to the left by three places because the exponent is −3. If the exponent is positive, shift the decimal point that many places to the right. If the exponent is negative, shift the decimal point that many places to the left.

The syntax diagram for a real number is

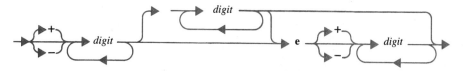

All of the following are valid real numbers. The **e** for the exponent can be a capital or a small e:

    0.45   1e10   −1e−04   0.0   1.0E+3   +456789.0

The following are invalid real numbers:

    1e0.1   12.0x10   12a+9

The first number is invalid because the exponent is not an integer. The other two are invalid because they contain alphabetic characters that are not part of the syntax for real numbers.

According to the syntax diagram, the following numbers should be invalid real numbers, but they will be accepted as real numbers in Pascal:

    .45   1234

In the first case, the compiler inserts the digit 0 to the left of the decimal place so that the number begins with an integer. In the second case, a decimal point and the digit 0 are added after the number.

Characters

Anything on your keyboard can be a character. A character is specified by enclosing a symbol in single quotation marks. **Char** refers to a piece of data that is a single character. The distinction between the identifier **A** and the character A is made by adding single quotation marks. Within a Pascal program, to refer to the letter A, you would write the letter A inside single quotation marks: 'A'. To refer to an identifier called **A,** you would write the letter A by itself: A. A syntax diagram for **char** data is not usually given because it would be just

These are valid examples of the data type **char:**

    'a'   ','   '>'   'A'   'q'   '3'

The following cannot be assigned to variables of the type **char:**

$$\text{'ab'} \qquad \text{'A} \qquad \text{q} \qquad \text{"b"}$$

The first one is invalid because because a single character must be used for the data type **char.** The second one is invalid because the right-hand quotation mark is missing. The third one is invalid because it is not enclosed in quotation marks, and the last one is invalid because it is enclosed in double quotation marks instead of single quotation marks.

In writing integers, real numbers, and characters, you do not need to worry about how the decimal number 4 is converted into the binary number 100 or how the character 'a' is converted into the ASCII code 97. You can use the decimal number system and the alphabet just as you normally would. The codes for the number system and the alphabet are part of the Pascal language. Numbers and letters as well as your instructions about what to do with them will be translated into binary codes for you.

2.4 Constants

Physical and mathematical constants such as π, e, and the speed of light in a vacuum have values that never change. Conversion factors, the first letter of the alphabet, and the number of things in a dozen also have values that do not change. In Pascal you can use identifiers called *constants* to name particular values that never change. The syntax diagram for defining constants is

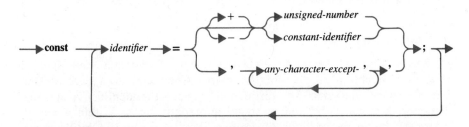

where *unsigned-number* is an integer or a real number. In the syntax diagram above, the words and punctuation that are part of the Pascal syntax are written in boldface. These parts of the syntax must be reproduced exactly. The words that are written in italics represent words and values that are chosen by you.

The following are valid constant definitions constructed using the syntax diagram above:

```
const Digit = 4;
      Number = -10.45;
      Letter = 'a';
```

Notice that, if you are defining more than one constant, the syntax diagram does not require you to repeat the word **const** before each constant is declared. It appears once at the beginning of the statement. Also notice that the constant identifier is separated from its value by an equal sign (=) and that each declaration ends in a semicolon (;).

After the **const** statement above has executed, the storage locations can be visualized as

Digit	**Number**	**Letter**
4	−10.45	a

The boxes for the storage locations have been drawn with thick lines to remind you that once the value for a constant has been defined, it cannot be changed.

In the case study, we begin with three known values: π, x, and y. Because the value of π never changes, the identifier name **Pi** can be used to refer to the number 3.14159. Since the values of x and y do not change for this problem, you could define x and y as constants. However, values along the x and y coordinates are not really constants, for x and y could take on any values on the circle's perimeter. So, for the case study there is only one constant, π, and its definition would be

```
const Pi = 3.14159;
```

Because Pascal is a strongly typed language, every constant must have a data type. The data type of a constant is given by the data type of its defined value. **Pi** is thus a constant of the data type **real.** In the previous constant definitions, **Digit** is an integer constant, **Number** is a real constant, and **Letter** is a character constant.

In the syntax diagram for constants, one of the valid choices is another constant identifier. This type of definition is called *recursive* because a construction is defined in terms of itself. So, the following constant declaration is valid:

```
const Years = 5;
      Age = Years;
```

This constant definition would result in assignment of the same value, 5, to the two identifiers **Age** and **Years.**

After defining a constant, you can use it throughout your program just as you would use a number like 10 or 1.0, except that the number is referred to by its identifier. Constants make programs easier to write because they make it unnecessary for you to keep typing the same number over and over. Whenever you type a number more than once, there is always the danger that you will mistype it. You may call π 3.142 in one place, 22/7 in another, and 3.14159 somewhere else. Constants make your program consistent. Constants also allow changes to be made consistently. If you use a constant for π and decide that you need more significant digits, you need to change the value of **Pi** only in the constant declaration. If you did not use a constant, you would have to go searching through your program for all the different places where you used a value equivalent to π and change each one.

Constants also make programs easier to read. Most physical constants are not as easily recognized as π. For example, if you were writing a program to calculate the output of a power plant, you would need the conversion factor from British Thermal Units (Btu) to kilowatt-hours. In a physics book, the conversion will be given as

$$1 \text{ Btu} = 3.412 \text{ kilowatt-hours}$$

or

$$\text{Btu/kilowatt-hour} = 3.412$$

To define the constant for the conversion factor, you can use the dimensions as the identifier name:

```
const BtusPerKwh = 3.412;
```

If your program used both π and the conversion factor for Btus to kilowatt-hours and if you did not use constants, your program would be scattered with the numbers 3.142 and 3.412. It would be difficult to tell which number was which constant and whether or not one number was a typographical error. Using constant definitions eliminates this problem.

The memory locations in a computer can store only dimensionless numbers. The number 3.412 is the same number whether it is the conversion factor for Btus to kilowatt-hours or the number of cubic feet in an aquarium. In order to give the number meaning, you must give it an identifying name that says what it is. You cannot store the dimensions as part of a number.

Pascal has several *standard identifiers* that can be used in any program. These standard identifiers have default definitions as part of the Pascal language. One of the standard identifiers, a constant called **maxint,** is equal to the largest integer that can be stored in the computer's memory. The value of **maxint** depends on the computer for which the Pascal compiler is installed. In Program 2.3 the use of **maxint** will be demonstrated in a Pascal program, and in Exercise 7 you are asked to write a program to determine **maxint** for your compiler.

2.5 Variables

In Pascal, every variable must be declared before it is used. You can think of this process as saving a place in memory for the data that you are going to store under a particular identifier name. To declare a variable, you must use the reserved Pascal word **var.** This word indicates that the identifier names that follow represent variables. When you declare a variable, you must also declare its data type. The data types that you have seen so far are **integer, char,** and **real.**

The syntax diagram for declaring a variable is*

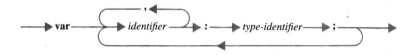

where *identifier* can be any valid Pascal identifier and *type-identifier* must be a valid data type.

For the case study, all the identifiers for variables have the data type **real.** The variable declaration can be constructed from the syntax diagram:

```
var   X, Y : real;
      Radius : real;
      AngleRadians : real;
      AngleDegrees : real;
```

Notice that the word **var** is not repeated before each variable that is declared. It appears once at the beginning of the declaration. Also, the variable name is separated from its data type by a colon (:), and each declaration ends in a semicolon (;). If several identifiers are of the same data type, they can be included on the same line with the identifiers separated by commas (,).

You should also note the identation of the examples above. Unlike many computer languages, Pascal has no requirements about which column statements must start in or how long a statement can be. The syntax for declaring variables requires only that a semicolon separate each declaration of variables of a new type. The variable declaration given above could have been written as

---

*This syntax diagram has been simplified, as have other syntax diagrams throughout the book, in order to avoid introducing constructions that have not been covered. Complete syntax diagrams are given in Appendix B.

```
var X, Y : real; Radius : real; AngleRadians : real; AngleDegrees: real;
```

However, this statement is not in good style. The statement is difficult to read and interpret. It will help you in debugging your programs if you write each statement on a separate line.

STYLE AND DEBUGGING HINT

The visual form of your program should reflect its content. Instructions that are separated into steps and grouped logically are much easier to follow than instructions that are written in a single paragraph. To save yourself debugging time, start with a program that is easy to read.

2.6 Comment statements

Comment statements allow you to include explanatory information within a program. Comment statements are ignored by the compiler, so you can use them to tell whoever is looking at your program exactly what the program does, without interfering with the running of the program. Very often a programmer will write a program with no comments because the person believes that what the program does is obvious. Then, two weeks later, looking at the program, that same person (probably you) has absolutely no recollection of what the program does or how it does it.

In Pascal comment statements can be put anywhere in a program. All that is required is that the comment start with { and end with } or start with (* and end with *). The two forms, { } and (* *), are equivalent. A comment beginning with (* and ending with } is legal, but it is confusing and should not be used.

Comments are used to clarify and explain. The exact purpose of each variable declared above would not be obvious to someone who had not read the case study. A comment statement can be added after each variable to explain its purpose; for example,

```
var  X : real;             (* x coordinate of the point on circle in *)
                           (* inches -- given. *)
     Y : real;             (* y coordinate of the point on circle in *)
                           (* inches -- given. *)
     Radius : real;        (* computed radius in inches. *)
     AngleRadians : real;  (* computed angle between x axis and radius *)
                           (* in radians. *)
     AngleDegrees : real;  (* computed angle between x axis and radius *)
                           (* in degrees. *)
```

A comment statement such as

```
const Pi = 3.14159;    (* declare constant Pi equal to 3.14159. *)
```

is not useful. It merely restates what the constant definition says and is obvious to anyone who knows Pascal. A more useful comment would be

```
const Pi = 3.14159;    (* Pi will be used to convert radians to degrees. *)
```

> *STYLE HINT*
>
> Comment statements should be used to clarify, not to restate. They should include information such as the units of a variable if they are not included in the variable name. If your program has too many comments, they will be ignored by someone reading the program. If your program does not have enough comments, it may not be understood.

2.7 Assignment statements

A variable identifier in a computer language is like the name of a storage location in a calculator. The numerical value in the storage location may change as you work toward a solution, but what it represents, such as a running total, is the same. This concept is slightly different from the concept of a variable used in math or physics. In a math problem, you may be presented with a series of equations and asked to solve for x, the unknown. The value of x is fixed by the equations, and you are trying to find out what that fixed value is. In a math problem, you tend to work with abstract variables, like x, y, and z, until the last step, when numerical values are substituted. This approach will not work with a computer because computers cannot operate on unknown data. Every variable must have a value before it can be used by a computer in a calculation. This section will show you how to assign values to variables.

So far in the solution to the case study, a statement has been written that puts a value of 3.14159 into a storage location called **Pi**. The **var** statement reserves storage locations for real numbers identified as **X, Y, Radius, AngleRadians,** and **AngleDegrees**. **X** and **Y** will be used to store the values of the x and y coordinates that were given. The other storage locations will be used to store the results of the problem. Currently, the storage locations for this problem are as follows:

Pi	X	Y
3.14159		

Radius	AngleRadians	AngleDegrees

The entries for the storage locations for the variables are empty because the storage locations do not yet have values stored in them; that is, they are *undefined*.

To compute the angle and the radius, you need to be able to assign values to the x and y coordinates. One way to give an undefined variable a value is by using the *assignment operator*. The assignment operator can be thought of as a left-pointing arrow. The value on the right side of the arrow is put into the storage location of the variable to the left of the arrow. In Pascal the assignment operator is composed of a colon (:) and an equal sign (=) written together. The statement

$$X := 2.5$$

results in the placement of a value of 2.5 in the storage location called **X.**

The assignment operator is not the same as the algebraic equal sign. In algebra, the statement

$$X = X + 1$$

does not make sense because no number is equal to itself plus 1. The Pascal statement

$$X := X + 1.0$$

makes sense. This statement takes the value that is currently stored in the location called **X,** adds one to that value, and puts the result back in the location called **X**. The old value of **X** is lost and replaced by the new value, **X** + 1.0. Suppose that the variable **X** had been assigned a value of 2.5:

X
2.5

When the statement

$$X := X + 1.0$$

is executed, the right side of the assignment operator is *evaluated*. That is, the number in the storage location called **X** is retrieved, and the value 1.0 is added to it, so the right-hand side evalutes to 3.5. Then the assignment operation is performed:

$$X \leftarrow 3.5$$

The value stored in the storage location **X** is now 3.5:

```
        X
    ┌───────┐
    │  3.5  │
    └───────┘
```

When the assignment operation is performed, all the variables on the right side of the statement are evaluated, and the resulting value is assigned to the variable on the left. If any of the variables on the right do not have values, the evaluation cannot be performed.

Notice the difference in syntax between declaring constants and using the assignment operator. The equal sign (=) is used in the constant declaration to set the value of **Pi** identically equal to 3.14159. The assignment operator (:=) is used to put a value of 2.5 into the storage location called **X**. The value of the variable **X** can be changed afterward by using another assignment statement. The value of **Pi** cannot be changed.

Because the values of constants cannot be changed, only variables can appear on the left side of the assignment operator. And only one variable can appear on the left side of an assignment operator.

As mentioned earlier, Pascal is a strongly typed language; the data type of the value produced by evaluating the right side of an assignment statement must match the data type of the variable on the left side, with only one exception: if the variable on the left side is **real,** the type on the right side may be either **integer** or **real.** Using the constant definitions and variable declarations

```
const Pi = 3.142;
      Index = 5;
      LastLetter = 'z';

var   RealVar, RealNum, ARealNumber, RealValue : real;
      IntVar, IntNum : integer;
      CharVar, Character : char;
```

the assignment statements below are examples in which the data types of the left and right sides agree or, technically, are *assignment compatible:*

```
IntVar := 101;
IntNum := Index;

RealVar := Pi;
RealNum := 6.73;
ARealNumber := 5;
RealValue := Index;

CharVar := LastLetter;
Character := '*'
```

Using the identifiers that were declared for the case study and the assignment statement, you can now begin to translate the algorithm into Pascal statements. The first step in the algorithm sets the values of *x* and *y*: x is assigned the value 5.0 and *y* is assigned the value 10.0. The following assignment statements will place these values in the storage locations **X** and **Y**:

```
X := 5.0;
Y := 10.0
```

After these statements are executed, the storage locations are as follows:

Pi	X	Y
3.14159	5.0	10.0

Radius	AngleRadians	AngleDegrees

2.8 Standard functions

Assigning a value to the variable **Radius** involves computing the sum of the squares of **X** and **Y** and then taking the square root of the sum. The square and square root functions are supplied as *standard functions* in the Pascal language. Standard functions are provided so that you do not have to write statements to compute the square root of a number or the natural log of a number if you need them in a program. When a standard function is *invoked*—that is, when its name is encountered on the right side of an assignment statement during program execution—a specific set of machine instructions is executed. A function is said to *return* a resulting value. Most standard functions operate on one *argument*. The argument appears between parentheses after the function name. For example, the argument of **sqrt(2.2)** is 2.2, the number whose square root is to be computed.

TABLE 2.1
Partial List of Standard Pascal Functions

	Argument Data Type	Data Type of Value Returned	Value Returned
abs(X)	real or integer	same as argument	absolute value of **X**
arctan(X)	real or integer	real	the angle, in radians, whose tangent is **X**
cos(X)	real or integer	real	cosine of the angle **X** (**X** must be in radians)
ln(X)	real or integer	real	natural log of **X**
round(X)	real	integer	**X** rounded to the nearest integer
sin(X)	real or integer	real	sine of the angle **X** (**X** must be in radians)
sqr(X)	real or integer	same as argument	**X** squared
sqrt(X)	real or integer	real	square root of **X**
trunc(X)	real	integer	**X** truncated (that is, with its fractional part removed)

Table 2.1 is an abbreviated list of functions that you will find useful as you are learning Pascal. The argument for each standard function is restricted to specific data types, and the result of the function must be assigned to a specific data type, as indicated. For example, the standard function called **round** rounds a real number to the nearest integer. Thus its argument must be a real number, and the value returned is an integer. Similarly, the function called **trunc** truncates a real number to its integer value. It requires a real expression as its argument, and the value returned is an integer.

Using the standard functions **sqr** and **sqrt,** Equation 2.1, which gives the radius as a function of x and y, can be translated into Pascal as follows:

```
Radius := sqrt (sqr(X) + sqr(Y))
```

Be warned that the only standard trigonometric functions in Pascal are sine, cosine, and arctangent.* The arguments for **sin** and **cos** must be in radians not degrees, and the value returned by **arctan** is in radians. A complete list of the standard Pascal functions is given in Appendix A.

---

*Note that the points (2, 2) and (−2, −2) have the same arctangent—that is, 0.79 radians or 45°. The arctangent function in Pascal returns the principal angle, which is the angle in the first or fourth quadrant. In Chapter 5, you will learn the **if** statement that will allow you to test which quadrant the point is in. (See Problems 3 and 4 in Chapter 5.)

2.9 Program structure

So far, you have seen the syntax for constant definitions, variable declarations, and assignment statements. As you might suspect, syntax rules govern the way that the pieces of a program are put together. Every Pascal program has the following structural elements:

heading

declarative part

body

The *heading* gives the name of the program. The heading has the following syntax diagram:

The first *identifier* is the name of your program. It must be a valid Pascal identifier. On most systems, the identifiers between the parentheses must include the standard identifiers, **input** and **output.** The identifier **output** must appear in the program heading if the program has any output, and **input** must appear if the program reads any data. It is a good habit to include **input** and **output** in every program heading even though they are not always necessary.

The reserved Pascal word **program** must be followed by a blank space, so that the program name is a separate word. The program heading is always followed by a semicolon. A comment statement describing what the program does should follow the heading. The beginning of the program to solve the case study is

```
program Circle (input, output);
(*      This program computes the radius of a circle in inches and the    *)
(*      angle between the radius and the x axis given the point           *)
(*      (5.0, 10.0) on the circle's perimeter.  The angle in radians      *)
(*      is then converted to degrees.                                     *)
```

The *declarative part* of the program consists of the definitions of the constants and the declarations of the variables. In Pascal every constant and variable identifier that will be used in the program must be included in the declarative part. If constants are used, their definitions are placed first and are followed by the declaration of variables. You have seen how to construct this part of a program in the previous section.

The *body* consists of the executable statements that solve the problem. The program body has the following syntax diagram:

2.9 Program structure

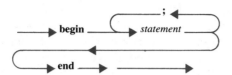

The first executable statement must be preceded by the reserved word **begin**. The last executable statement must be followed by the reserved word **end**. **Begin** and **end** are used like parentheses to set off sections of code in Pascal. The executable statements are separated by semicolons (;), and the program ends with a period (.).

In Program 2.1 the statements that have been developed so far for the solution to the case study are put together into a program called **Circle**. Each part of the program is identified as being part of the heading, the declarative part, or the body. Program 2.1 is an incomplete solution to Case Study 1.

PROGRAM 2.1 Circle

heading

```
program Circle (input, output);
(*      This program computes the radius of a circle in inches and the    *)
(*      angle between the radius and the x axis given the point           *)
(*      (5.0, 10.0) on the circle's perimeter.  The angle in radians      *)
(*      is then converted to degrees.                                     *)
```

declarative part
```
const Pi = 3.14159;                 (* used in conversion of radians to *)
                                    (* degrees. *)

var   X : real;                     (* x coordinate of the point on the *)
                                    (* circle in inches -- given. *)
      Y : real;                     (* y coordinate of the point on the *)
                                    (* circle in inches -- given. *)
      Radius : real;                (* computed radius in inches. *)
      AngleRadians : real;          (* computed angle between x axis and *)
                                    (* radius in radians. *)
      AngleDegrees : real;          (* computed angle between x axis and *)
                                    (* radius in degrees. *)
```

body
```
begin  (* main program *)
   X := 5.0;                        (* assign value to x coordinate. *)
   Y := 10.0;                       (* assign value to y coordinate. *)

   Radius := sqrt (sqr(X) + sqr(Y)) (* compute radius using Pythagorean *)
                                    (* theorem. *)
end. (* program Circle *)
```

Program 2.5 at the end of the chapter is a completed version of Program 2.1.

2.10 Output statements

Most computer programs perform a series of computations and then display the results of the computations. When Program 2.1 is executed by the computer, values will be stored in the computer's memory in the storage locations called **Radius, AngleRadians,** and **AngleDegrees.** Although the values have been computed, you cannot yet see what they are; an *output statement* that instructs the computer to report the contents of memory locations is needed. The output statement can also be used to write a message or explanatory information to the person using the program. In Pascal the output statements are the **write** and **writeln** statements. As covered in this section, the **write** and **writeln** statements will cause the output to be printed at the default output device—either at your terminal or on a line printer. The default output device depends on your computer system. In the remainder of the book, the default output device will be assumed to be the user's terminal.

The writeln *statement*

The **writeln** (pronounced "write line") statement is a *standard Pascal procedure.* That is, when you use the **writeln** statement in a program, the Pascal compiler translates the **writeln** statement into a series of machine-level instructions. This sequence of instructions has already been defined and is part of the Pascal compiler on your computer. The **writeln** statement is the first standard procedure that you will learn. A list of all the standard Pascal procedures is given in Appendix A.

The **writeln** procedure allows you to display information from your program on the terminal where you are working or on a printer. The syntax diagram for the **writeln** statement is

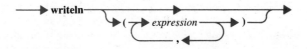

The *expressions* that you have learned so far are constants and variables. Character strings are another form of expressions that are covered later in this section. Expressions are covered in more detail in Chapter 3.

Each new **writeln** statement produces a new line of output. The effect of the **writeln** statement is that the current value of each expression is printed at the terminal. Writing the value of a variable expression does not change its value. The value stored in a memory location has the same value both before and after the **writeln** statement is executed.

2.10 Output statements

An example of a **writeln** statement is

```
writeln (X, Y)
```

=>
5.0000000000000e+00 1.0000000000000e+01

(The symbol => should be interpreted as follows: the above code when executed as part of a correct program produces the following output. You will see the symbol => both with the output of complete programs and with the output of fragments of programs used as examples. Execution of programs is covered in the next section.)

The output from the fragment above is difficult to understand, both because the numbers are printed in scientific notation and because without looking at the program's source code you cannot tell which number corresponds to which variable. The output can be made clearer by adding messages describing the output and by specifying a different way to print the numbers. A message can be used as an expression in a **writeln** statement. The message is enclosed in single quotation marks, and then the exact symbols within the quotation marks are printed.* For example,

```
writeln (' X =', X, ' Y =', Y)
```

=>
X = 5.0000000000000e+00 Y = 1.0000000000000e+01

The words enclosed within the single quotation marks are referred to as a *character string*. A character string is not a standard data type. A character string is a constant and can be formed using the syntax diagram for a constant given in Section 2.4. Because character strings do not fit neatly into the structure of Pascal, they will not be covered until Chapter 11. However, you will find it quite natural to use character strings in your **writeln** statements without worrying about their exact definition.

DEBUGGING HINT

Always include meaningful character strings in your output statements to identify what value is being printed. Do not print just the value of the variable.

---

*In the examples, each character string that starts a new line begins with a blank because the first character in a line is used for carriage control on some printers. If you are using a line printer for output, check your system documentation to find out what the carriage control characters are.

Chapter 2 □ Introduction to Pascal

The following **writeln** statement would look clear in a program:

```
Radius := sqrt (sqr(X) + sqr(Y));
writeln (' The answer is', Radius)
```

However, when the program executes, the output at the terminal would be

```
=>
 The answer is 1.1180339887498e+01
```

Only the value of the variable is printed, not its name, so the significance of the answer is lost. The message that you choose should explain what the output signifies and provide any necessary units:

```
writeln (' The radius is', Radius, ' inches.');
```

```
=>
 The radius is 1.1180339887498e+01 inches.
```

Output formatting

The process of making your output better looking and easier to read is referred to as *formatting*. Formatting allows you to put blank lines in your output, print numbers in any desired notation, and line up columns.

In Pascal the easiest way to skip lines is to use the **writeln** statement without any expressions:

```
writeln (' That''s the end of that.');
writeln;
writeln (' Here''s something new.')
```

```
=>
 That's the end of that.

 Here's something new.
```

Notice that, to get an apostrophe in your output, you put two single quotation marks in the expression, as demonstrated above.

A variable within an output statement is printed according to the default format for its data type unless you specify another valid format. The *field width* of a variable is the number of columns that the variable is allotted in the output. The default field widths depend on the version of Pascal that you are using. Typical default field widths are

12 columns for integer data

20 columns for real data

1 column for each character

Formatting integer *expressions*

Integers are printed right justified. This means that, if the integer does not take as many columns as the field width, blanks are put in front of the integer. Given the following assignment and output statements, a default field width of 12 columns will result in the output shown:

```
IntVar := 9;
writeln (' The integer printed in the default format is', IntVar)
```
=>
```
The integer printed in the default format is           9
```

There are 11 blanks between the word "is" and the number "9," for a total of 12 columns. You can specify the number of columns that a number takes up using the field width option in the **writeln** statement. The syntax diagram for formatting integer expressions is

The first *integer expression* is the integer being printed. The second *integer expression* is the field width that specifies how many columns the first integer expression should take up in the output. A colon (:) separates the identifier from the field width.

The following fragment uses field width formatting:

```
writeln (' The integer printed in a field width of 2 is', IntVar:2);
writeln (' The integer printed in a field width of 3 is', IntVar:3)
```
=>
```
The integer printed in a field width of 2 is 9
The integer printed in a field width of 3 is   9
```

The **9** is first printed in a field width of 2: one blank and one column for the number. It is then printed in a field width of 3: two blanks and one column for the number.

If the field width specified is too small, the number will be printed in a field just wide enough to contain all the digits of the number, including the sign if the number is negative. This feature can be used when the magnitude of a number is not known, as in the following fragment:

```
Index := 5463;
writeln (' In a field width of 1, Index =', Index:1)
```

=>
In a field width of 1, Index =5463

Formatting real *expressions*

The default format for real variables is *scientific notation*, which is +x.xxxxxxxxxxxxxE+mm if the default field width is 20. In scientific notation, also called *exponential notation* or *floating point notation*, there is always one digit to the left of the decimal point. The exponent tells you how many places to move the decimal point to get decimal numbers that look more familiar. For example,

$$1.0E-01 = 0.1$$
$$1.0E+00 = 1.0$$
$$1.0E+01 = 10.0$$

When a number is printed in scientific notation, a minimum of seven columns are required: one for the sign of the number, one for the digit that precedes the decimal point, one for the decimal point, one for the E, one for the sign of the exponent, and two for the exponent. So, if you specify a field width of 20, 13 decimal places will be printed.

The following fragment uses the default format for real numbers:

```
Force := -1.0;
writeln (' The force is', Force, 'newtons.')
```

=>
The force is-1.0000000000000e+00newtons.

The -1.0 in the output is in exponential notation with a field width of 20. The exponent is 00 because the decimal point is where it belongs. There are no blanks between the words and the number because the first character string in the **writeln** statement ends in the letter "s," and the numerical field begins immediately. The second character string begins immediately after the numerical field. To get a space, you can put a blank at the end of the first string or at the beginning of the second, as follows:

```
writeln (' The force is ', Force, ' newtons.')
```

=>
```
The force is -1.0000000000000e+00 newtons.
```

To specify a field width for real expressions, the syntax diagram is similar to the syntax diagram for integer expressions:

where the value of the *integer expression* specifies the number of columns for the real expression in the output. For example, changing the field width in the example to 12 gives 5 decimal places:

```
writeln (' The force is ', Force:12, ' newtons.')
```

=>
```
The force is -1.00000e+00 newtons.
```

Most people would not recognize the number above as -1.0. Therefore, Pascal allows you to specify both the field width and the number of decimal places using the following syntax:

where the first *integer expression* specifies the total number of columns for the real expression in the output and the second *integer expression* specifies the number of decimal places for the real number. If you specify both the field width and the number of decimal places for a real number, the number is written in familiar decimal notation. Adding a decimal field to the example above gives

```
writeln (' The force is ', Force:6:1, ' newtons.')
```

=>
```
The force is   -1.0 newtons.
```

In this example, the number -1.0 takes up four columns. Because the field width is 6 and the number only takes up four columns, there are two blanks before the number. The third blank appears because a blank was inserted at the end of the character string. When formatting a real number, remember that the minus sign and the decimal point each require one column. If you do not specify a large enough field width, the output that you get will depend on the compiler that you are using. The number will be printed, but it may not be in the format that you intended or express the accuracy you desired.

Formatting char *expressions*

The default field width for a variable of type **char** is one column, but the field width can be modified in the same way as for integers. The program fragment below illustrates unformatted and formatted **char** variables. Assume that **Letter** is declared as a variable of type **char**.

```
Letter := 'A';
writeln (' The first letter is', Letter);
Letter := 'Z';
writeln (' The last letter is', Letter:2)

=>
 The first letter isA
 The last letter is Z
```

Program 2.2, **Demonstrate,** shows the syntax for output statements, formatting **real, integer,** and **char** expressions. Program 2.2 also uses the standard Pascal constant **maxint.** Notice that **maxint** is not defined in the program, but its value can be printed. The value of **Maxint** is printed for two different systems: the first output is from the VAX-11 Pascal compiler and the second from an eight-bit version of the Turbo Pascal compiler.

PROGRAM 2.2 Demonstrate

```
program Demonstrate (input, output);
(*      This program shows how to declare, assign, and print variables of   *)
(*      data types integer, real, and character.  The value of maxint (the  *)
(*      standard constant representing the largest integer that can be      *)
(*      stored in the system that is used to run the program) is also       *)
(*      printed.  Output is given for different systems, allowing you to    *)
(*      see the difference in the value of maxint.                          *)

var   Force : real;        (* real variable to be used in demonstration. *)
      Largest : integer;   (* integer variable to be used in demonstration. *)
      Letter : char;       (* char variable to be used in demonstration. *)
```

```
begin (* main body *)
  writeln;                      (* write a blank line. *)

  Force := 25.6;
  writeln (' The force is', Force:12:2, ' newtons.');

  Largest := maxint;            (* the standard constant maxint does not have to *)
                                (* be defined before it is used. *)

  writeln (' The largest integer is', Largest:12);

  Letter := 'z';
  writeln(' The letter is', Letter:12)
end. (* program Demonstrate *)
```

Output 1

=>

```
 The force is          25.60 newtons.
 The largest integer is  2147483647
 The letter is             z
```

Output 2

=>

```
 The force is          25.60 newtons.
 The largest integer is       32767
 The letter is             z
```

2.11 Program execution

After you have written a program down on paper, you must enter it into the computer. Depending on the system you are using, you may type your program on cards, or you may use an editor program to enter it directly into storage on the computer. In addition to the Pascal program itself, you may need to include special statements that indicate that your program is written in Pascal. The exact requirements must be found in the documentation for your computer system. In Appendices F through H, the requirements for Turbo Pascal, UCSD Pascal, and VAX-11 Pascal are given.

When your program is submitted to the computer, the program will be either *compiled* or *interpreted*. Each statement of the program is checked against the rules for Pascal syntax. If the syntax of any of your instructions is incorrect, you will get a message stating approximately what the problem is.

Such errors are called *compile-time errors* because they are found by the compiler and prevent it from translating your program. You must go back and edit your program to correct the syntax mistakes.

The first programs that you write may have many syntax errors because you are just learning the grammar for a new language. Program 2.3, **Undeclared,** a variation on Program 2.2, demonstrates a common bug. It shows the compile-time error that results if a variable is not declared in the declarative part of the program. In this program, **Largest** is not declared. Some compilers flag only the first occurrence of an undeclared variable.

PROGRAM 2.3 Undeclared

```
program Undeclared (input, output);
(*      This program demonstrates what happens if you use a variable in    *)
(*      the body of a program that has not been declared in the var        *)
(*      statements.  The variable Largest is undeclared but is used in     *)
(*      the program.                                                       *)

var    Force : real;      (* real variable to be used in demonstration. *)
       Letter : char;     (* char variable to be used in demonstration. *)


begin (* main body *)
   writeln;               (* write a blank line *)

   Force := 25.6;
   writeln (' The force is', Force:12:2, ' newtons');

   Largest := maxint;     (* the standard constant maxint does not have to be *)
                          (* defined before it is used. *)

   writeln (' The largest integer is', Largest:12);

   Letter := 'z';
   writeln (' The letter is', Letter:12)
end. (* program Undeclared *)


=>
     Largest := maxint;
***error***  $104
        writeln (' The largest integer is', Largest:12);
***error***                                       $104

* 2 lines flagged in pascal program *

error messages :
****************

 104: identifier not declared.
```

2.11 Program execution

The following fragments illustrate other common syntax mistakes. Included with each illegal assignment statement is a typical message that would be printed by a compiler. Comment statements are included to explain why the error occurred.

```
Digit := 5.6;                  (* A real number cannot be assigned to an *)
                               (* integer variable. *)
***error***   $129

  129: type conflict of operands.



Letter := a;                   (* A char variable must be enclosed in *)
                               (* single quotes. *)
***error***   $104

  104: identifier not declared.



6 := 3 + 3;                    (* The left side of an assignment *)
                               (* statement must be a variable. *)
***error*** $167
        $5,6

    5: ":" expected.
    6: illegal symbol (possibly missing ";" in line above).
  167: undeclared label.



Number + Digit := 4.0;         (* Only one variable can appear on the *)
                               (* left side of an assignment statement. *)
***error*** $59

   59: error in variable.



Number : = 5.78;               (* The : and the = cannot be separated by *)
                               (* a blank. *)
***error*** $59      $51

   51: ":=" expected.
   59: error in variable.
```

```
    Pi := 2.3;                         (* The value of a constant cannot be *)
                                       (* changed. *)
***error***    $103,104

103: identifier is not of appropriate class.
104: identifier not declared.
```

As you can see, the messages from the compiler are not always informative. Many obscure error messages will disappear when the obvious errors are corrected. In the statements above, a label does not have to be declared as implied by error message 167. Rather, a variable must replace the constant "6" on the left side of the equation. To see the error messages generated by your compiler, you can create a program with the same illegal statements.

DEBUGGING HINT

Always fix the most obvious errors first, and then recompile the program. Most of the obscure error messages will be eliminated, and you will be able to correct the errors that remain.

Once you have corrected all the syntax mistakes, your statements must be resubmitted to the compiler for translation into machine language. Then your program can be *executed;* that is, your instructions can be followed by the computer.

At this point, your program may contain *run-time errrors*. These are errors that occur because the computer cannot follow your instructions as given. Program 2.4, **Undefined,** another variation on Program 2.2, has a run-time error. It has a variable that is undefined. In most versions of Pascal, a run-time error results if a variable is undefined, as shown in the first output block. However, in some versions of Pascal, all variables are *initialized* by the compiler; that is, they are assigned values at the beginning of program's execution. For beginning programmers, this feature can lead to problems. In the alternative output block for Program 2.4, the program compiles and runs without any error messages. A value is printed for **Largest** even though no value was assigned to it. You can see that -1863987188 is not a reasonable value for **Largest,** and you may suspect that there is a problem. In a long program, however, initialization errors may not be so obvious. If your compiler has an option to display undefined variables, you should use the option. Otherwise you may have bugs in your program of which you are unaware.

PROGRAM 2.4 Undefined

```
program Undefined (input, output);
(*      This program demonstrates what happens if you try to print a   *)
(*      variable to which you have not assigned a value.  Largest is not *)
(*      assigned a value, so an error occurs when the command to print  *)
(*      its value is encountered.                                       *)

var Force : real;            (* real variable to be used in demonstration. *)
    Largest : integer;       (* integer variable to be used in demonstration. *)
    Letter : char;           (* char variable to be used in demonstration. *)

begin (* main body *)
   writeln;                  (* write a blank line. *)

   Force :=  25.6;
   writeln (' The force is', Force:12:2, ' newtons.');

   writeln (' The largest integer is', Largest:12);

   Letter := 'z';
   writeln (' The letter is', Letter:12)
end. (* program Undefined *)
```

Output 1

```
=>

  The force is        25.60 newtons.
  The largest integer is
                                            pascal termination log
                                            ------ ----------- ---
*** error ***         undefined variable used in expression

program terminated at offset 0000f0 in main program

local vars: force   =  2.560000000000e+01 largest = <undefined>          letter
   = 'z'
```

Output 2

```
=>

  The force is        25.60 newtons.
  The largest integer is -1863987188
  The letter is              z
```

Whether your output will resemble Output 1 or Output 2 will depend on the Pascal compiler that you are using. Standard Pascal does not dictate what action should be taken when a variable is undefined.

Finally, you may have a program that compiles and runs without error messages, but that still does not give the correct results. Recognizing that you have incorrect results and correcting the program logic to yield correct results can be the most time-consuming part of programming.

Solution to Case Study 1

Program 2.1 completed only Step 1 of the algorithm to solve the case study, but the remainder of the program is straightforward now that you have learned about assignment statements, standard functions, and output statements.

The standard function **arctan** can be used to translate Step 2, computing the angle in radians given x and y, into Pascal. The assignment statement for the variable **AngleRadians** is

```
AngleRadians := arctan(Y/X)
```

Step 3 of the algorithm, converting the angle from radians to degrees, can be implemented using the statement

```
AngleDegrees := AngleRadians * 180.0 / Pi
```

In the last two assignment statements, some arithmetic operators were used that have yet to be discussed. The asterisk or star (*) stands for multiplication, and the slash (/) stands for division. These operations will be covered in Chapter 3.

The last step, reporting the values for the radius and the angle, can be implemented using a sequence of output statements. Adding the assignment statements above and the output statements to Program 2.1 completes the algorithm. Program 2.5 is the completed version of the program for the case study.

PROGRAM 2.5 Circle2

```
program Circle2 (input, output);
(*      This program computes the radius of a circle in inches and the    *)
(*      angle between the radius and the x axis given the point           *)
(*      (5.0, 10.0) on the circle's perimeter.  The angle is computed in  *)
(*      radians and then converted to degrees.                            *)

const Pi = 3.14159;              (* used in conversion of radians to *)
                                 (* degrees. *)

var   X : real;                  (* x coordinate of the point on *)
                                 (* circle in inches -- given. *)
      Y : real;                  (* y coordinate of the point on *)
                                 (* circle in inches -- given. *)

      Radius : real;             (* computed radius in inches. *)
      AngleRadians : real;       (* computed angle between x axis and *)
                                 (* radius in radians. *)

      AngleDegrees : real;       (* computed angle between x axis and *)
                                 (* radius in degrees. *)

begin   (* main program *)
  X := 5.0;
  Y := 10.0;

  Radius := sqrt (sqr(X) + sqr(Y));         (* compute radius using *)
                                            (* Pythagorean theorem. *)
  AngleRadians := arctan(Y/X);              (* compute angle in radians. *)
  AngleDegrees := AngleRadians * 180 / Pi;  (* convert angle to degrees. *)

  (* Print the values of the X and Y coordinates and the computed *)
  (* values of the radius and the angle in radians and in degrees. *)

  writeln (' For an x coordinate of', X:5:1, ' inches');
  writeln (' and a y coordinate of', Y:5:1, ' inches,');
  writeln (' the radius of the circle is', Radius:6:1,' inches,');
  writeln (' and the angle is', AngleRadians:7:2, ' radians');
  writeln (' or', AngleDegrees:7:2, ' degrees.')
end.    (* program Circle2 *)


=>
 For an x coordinate of  5.0 inches
 and a y coordinate of 10.0 inches,
 the radius of the circle is  11.2 inches,
 and the angle is   1.11 radians
 or  63.44 degrees.
```

Summary

An algorithm is a finite sequence of steps that unambiguously solves a problem. In top-down design, a problem is divided into subproblems, then each subproblem is refined until the algorithm can be stated as a sequence of steps, each of which can be solved directly. This approach is useful in developing algorithms. A program is a translation of an algorithm that can be implemented on a computer.

Identifiers are names used in a program that can be created by the programmer. In standard Pascal, the number of characters in an identifier is unlimited; the identifier must start with a letter of the alphabet, and the rest of the characters must be alphanumeric. Constant identifiers can be set to values that do not change within a program through use of the **const** statement. Variable identifiers can be declared, through use of the **var** statement, to refer to locations in which values are stored. A variable must be declared to have a particular data type, and the value stored in it must be compatible with the declared data type. Pascal is a strongly typed language in which every piece of data must have a data type. This chapter covered the data types **real, integer,** and **char.** Variables can be given values using the assignment operator (:=), which evaluates the expression on the right side of the operator and assigns the resulting value to the variable on the left side of the operator.

Programs consist of the following structural elements: the heading, the declarative part, and the body. The heading gives the name of the program. The declarative part defines the constants and declares the variables to be used in the program. The body contains all the statements that are to be executed, separated by semicolons. The body must start with the reserved word **begin** and end with **end.** Comment statements, which are ignored by the compiler, can be included within a program to explain what the program does. Comment statements are set off by the characters { } or (* *).

Standard functions, such as **sqr** and **sqrt,** are part of the Pascal language and can be used without prior declaration. Functions operate on arguments, which appear in parentheses after the function name, to return a value. Standard procedures, which can be used without definition, are also part of the Pascal language. **Writeln,** which can be used to print the values of variables or to print character strings, was introduced in this chapter. Output must be formatted if it is to be printed in a format other than its default format. For example, real values are printed in scientific notation unless another format is specified.

Exercises

1. Which of the following are valid identifiers in standard Pascal? What is wrong with the invalid identifiers?

   (a) Fiftyfive
   (b) HowMany#
   (c) Why?
   (d) yes
   (e) program
   (f) INDEX
   (g) $dump
   (h) Count3
   (i) j77
   (j) 77j

2. Can you use the character θ as an identifier in a Pascal program?

3. As an extension to standard Pascal, many versions of Pascal allow characters other than just letters and numbers to be used in identifiers. Some common nonstandard characters are the underscore (_) and the dollar size ($). The underscore is useful since it can be used to represent a space, as in the identifier **New_Word**. What nonstandard characters does your version of Pascal allow? Can a nonstandard character be used as the first character in an identifier name? You can find out either by reading the documentation or by writing a program that includes identifiers with nonstandard characters.

4. What is the data type of each of the following constants?

   (a) const Remainder = 6;
   (b) Mass = 23.1e-05;
   (c) Seven = 7;
   (d) NegNum = -Remainder;
   (e) Distance = 572.35;
   (f) Zed = 'z'

5. Find and correct the syntax mistakes in the following program.

   ```
   program FindErrors (input, output);
   (*     This program contains several mistakes.  Find and correct the
   (*     mistakes.

   const Ten := 10;
   var Number ; Integer:

   begin (* main program )
     writeln (' The number is,);
     Number = 378;
     writeln( Number)
     writeln (' Ten times the number is");
     Number := Ten * Number;
     writeln (umber);
   end (* program FindErrors *)
   ```

6. One of the first things you should do when you begin to work with a Pascal compiler is to determine the value of **maxint** for your compiler. On microcomputers **maxint** is often a number that is easy to exceed, and some

systems do not warn you when you have exceeded it. Exceeding **maxint** can make your output nonsensical. Write a program to determine **maxint** for your compiler.

7. Given the **var** declaration

    ```
    var A, B : integer;
        X, Y : real;
    ```

    what is the result of each of the following assignment statements? Make a table giving the values of **A, B, X,** and **Y** after each assignment statement is evaluated. You may need to use a calculator.

    (a) A := 3;
        B := sqr(A)

    (b) X := 1.0;
        Y := ln(X)

    (c) A := -5;
        B := abs(A);
        Y := sqrt(B)

    (d) X := 1.0;
        X := X + 0.5;
        A := round(X);
        B := trunc(X)

8. What is the output from the program **RoundAndTrunc**?

```
program RoundAndTrunc (input, output);
(*      This program compares the results of the functions round and     *)
(*      trunc.                                                           *)

var   RealVar : real;      (* variable used as argument of trunc or round. *)
      IntVar1 : integer;   (* variable to store value returned by round. *)
      IntVar2 : integer;   (* variable to store value returned by trunc. *)

begin (* main program *)
  RealVar := 5.6;
  IntVar1 := round (RealVar);
  IntVar2 := trunc (RealVar);

  writeln (' The real number is', RealVar);
  writeln (' Its value rounded is', IntVar1);
  writeln (' Its value truncated is', IntVar2)
end. (* program RoundAndTrunc *)
```

    What would be the output if **RealVar** were initialized to -10.6?

9. Why is the syntax for formatting expressions using the field width and decimal places valid only for real expressions?

10. Write a program that will allow you to determine the number of characters in an identifier that your compiler recognizes. That is, determine whether your compiler recognizes only the first eight characters, sixteen characters, or all characters.

Problems

1. Write an algorithm to be used for programming your household robot to get you a glass of water. Remember that the robot will do only what you tell it to do and will do exactly what you tell it to do.

2. Modify Algorithm 2.6 from the case study so that the diameter and the circumference of the circle are also computed.

3. Write an algorithm for taking the sum of the squares of two numbers on a calculator. Assume that the person following the algorithm has no experience with a calculator and must be told exactly what to do for each step. If possible, have someone who does not know how to use a calculator follow the steps of your algorithm. Be sure to specify the calculator for which the algorithm is written.

4. Angles are sometimes given in degrees/minutes/seconds, but for an angle to be used as the argument of a standard function, the angle must be given in radians. Write an algorithm to convert degrees/minutes/seconds to radians.

5. Write a program that prints a tree made out of stars. Center the tree in the middle of your output page, as follows:

```
   *
  ***
 *****
*******
   *
```

6. Write a program that prints a title page for your programs, with your name and the date or any other important information. For example,

>   Jane Q. Jones
>   EE 100
>   Prof. Smith
>   Assignment #1

7. The largest real number that can be created varies from compiler to compiler. However, there is no standard identifier like **maxint** for the largest real number. Determine the largest real number for your compiler.

8. Test the accuracy of your computer using the identities

$$\pi = \tan^{-1}(0)$$
$$\sin \pi = 1$$

and

or

$$\sin(\tan^{-1}(0)) = 1$$

Print the value that you get for π by taking the arctangent of 0 radians. Then print the value of the sine of π, where π has the value computed using the arctangent. Label your output. Should you format the variables in which the values for π and 1 are stored?

CHAPTER 3

Input Statements, Arithmetic Operators, and Expressions

In this chapter you will learn the input statements **read** and **readln** that are used to assign values to variables after a program begins execution. The **read** statement enables you to write a program that operates on different data each time the program is run. This chapter will also cover the basic arithmetic operations and the creation of expressions. You will learn more about the **writeln** statement along with another output statement, the **write** statement.

CASE STUDY 2
Computing the current in a circuit with resistors connected in parallel

In an electrical circuit, a current, I, is generated if a voltage, V, is applied across a resistance, R, as illustrated in Figure 3.1. The formula for the current is given by

$$I = \frac{V}{R} \tag{3.1}$$

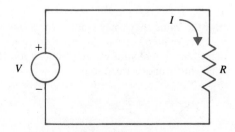

FIGURE 3.1
Electrical circuit with voltage, V, resistance, R, and current, I.

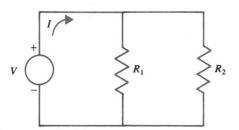

FIGURE 3.2
Circuit with two resistors connected in parallel.

where I is in amperes, V is in volts, and R is in ohms.

Resistors can be connected in parallel, as shown in Figure 3.2. For a circuit with two resistors in parallel, the current depends on the voltage and the combined resistance of the two resistors. The current in the circuit with the resistors in parallel is

$$I = \frac{V}{R_{eq}} \tag{3.2}$$

where R_{eq} is the *equivalent resistance,* which is defined as

$$\frac{1}{R_{eq}} = \frac{1}{R_1} + \frac{1}{R_2}$$

or (3.3)

$$R_{eq} = \frac{R_1 R_2}{R_1 + R_2}$$

If the two resistors in parallel are replaced by one resistor with a resistance of R_{eq}, the current in the circuit is the same.

Write a Pascal program that accepts from the program user values for the two resistors and for the voltage, and uses these values to find the current when the voltage is applied across two resistors connected in parallel.

Algorithm development

For the purpose of developing an algorithm for this problem, the overall problem stated in the last sentence above can be broken down into subproblems.

ALGORITHM 3.1 First iteration of the algorithm for Case Study 2

1. Write messages to the program user asking for values for the two resistors in ohms and for the voltage in volts.

2. Assign the values given by the user to the variable identifiers for the resistors and the voltage.
3. Compute the current in amps using the equivalent resistance.
4. Report the current.

To translate the algorithm into Pascal, you will need identifiers for the program name, the two resistors, the voltage, the equivalent resistance, and the current. You already know how to write the program heading and the declarative part of the program:

```
program ComputeCurrent (input, output);
(*      This program computes the current that flows when a voltage is    *)
(*      applied across two resistors connected in parallel.  The values   *)
(*      for the voltage and resistance are read interactively.            *)

var     Resist1 : real;        (* user supplied resistance 1 in ohms. *)
        Resist2 : real;        (* user supplied resistance 2 in ohms. *)
        Voltage : real;        (* user supplied voltage in volts. *)
        EquivResist : real;    (* equivalent resistance in ohms, *)
                               (* intermediate variable used in computation *)
                               (* of the current. *)
        Current : real;        (* computed current in amps. *)
```

You also know how to use the **writeln** statement to write the messages that will appear at the terminal to ask the user for information and to report the results. Steps 1 and 4 of the algorithm do not need further refinement because the Pascal code for these steps can be written once you have decided on the variable identifiers.

```
writeln (' Enter the voltage in volts.');
writeln (' Enter the resistance in ohms for the first resistor.');
writeln (' Enter the resistance in ohms for the second resistor.');
            .
            .
            .
writeln (' The resulting current is', Current:10:2, ' ohms.')
```

The *ellipsis*

.
.
.

is used to indicate that statements in this fragment have been omitted. Whenever you see three dots like these in a program or a fragment, you can assume that the missing statements are not relevant to the current discussion or that the missing statements are identical to statements in a previous version of the program. In this case, the statements for Steps 2 and 3 will be written later and will be placed between the **writeln** statements used to ask for input values and the statements used to report the output.

For Step 2 you must be able to assign values supplied by the user to the variables. Because this step requires the use of the standard procedure **readln,** it will be covered in the next section.

For Step 3 the equivalent resistance and the current must be computed using Equations 3.2 and 3.3. Although the value of the equivalent resistance is not a required output of the program, that value must be computed in order to compute the current. Such a variable is called an *intermediate variable.* As a matter of style, intermediate variables often make programs easier to read. With equivalent resistance used as an intermediate variable, Step 3 of the algorithm can be refined to the following.

ALGORITHM 3.2   Refinement of Step 3 of Algorithm 3.1

3. Compute the current in amps using the equivalent resistance.
   a. Compute the equivalent resistance in ohms using the user-supplied values for the resistors and Equation 3.3, as follows:

$$R_{eq} = \frac{R_1 R_2}{R_1 + R_2}$$

   b. Compute the current in amps using the user-supplied value for the voltage and Equation 3.2, as follows:

$$I = \frac{V}{R_{eq}}$$

To complete this step you must be able to write assignment statements using the basic arithmetic functions of additions, multiplication, and division. The arithmetic operations will be covered in Section 3.2.

3.1 Input statements—read and readln

Program **Circle** in Chapter 2 used assignment statements to assign values to the x and y coordinates. To compute the radius of a circle if a different point were given, you would have to edit the program and change the assignment statements. To assign values to the variables in the case study for this chapter, you cannot use the assignment statement because the values will change each time the program is executed.

Using an input statement makes a program more flexible because an input statement allows the value of each variable to be set during execution of the program. Thus, the same program can be used to solve many similar problems. For example, a program that computes the value of 2^6 is of limited use. A program that computes 2^n, where n is any input value, is more useful.

The **read** and **readln** statements are input statements; they instruct the computer to assign a value to the variable named. The computer can find the value to assign to the variable from an input list, which the user provides either at the terminal, at the end of the program, or in a file stored in long-term memory. Because the terminal is the default input device for most Pascal compilers, this section will describe interactive input—that is, input supplied by the program user at the terminal. On most systems, if you do not specify where the data is to be found, the compiler will assume interactive input. Although you may need to put special system commands at the top of your program in order to use interactive input, the syntax given here should work on most systems. If you need special commands, the documentation for your Pascal compiler, your instructor, or someone from your computer center can tell you what extra commands you need to use.

The following is a simplified syntax diagram for the **readln** statement:

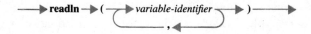

The identifiers that appear between the parentheses are often referred to as the *read-argument list*. The following is an example of a **readln** statement:

```
readln (Voltage, Resist1, Resist2)
```

Each variable identifier in a **readln** statement must be separated from the next one by a comma.

Read arguments are restricted to variable names because the purpose of the input statements is to assign values to variables. Only variable identifiers can be assigned new values; thus, if you try to read a value into an identifier that has been defined as a constant, you will get a compile-time message stating, in effect, that the value of a constant cannot be changed. The function of a **readln** statement is similar to that of an assignment statement except that in a **readln** statement a different value can be assigned to the variable each time the program is run.

When a **readln** statement is encountered during execution of the program, the program stops executing until the user enters a value at the terminal. Some systems issue a prompt such as □, ?, >, or . to indicate that the user must enter a response. When a value is entered and followed by a carriage return, that value is assigned to the variable named in the **readln** statement, and execution of the program continues. For example,

3.1 Input statements—read and readln

```
writeln (' Enter the voltage in volts.');
readln (Voltage)
```

=>
 Enter the voltage in volts.
 □5.5

After execution of this fragment, the value 5.5 is assigned to the variable **Voltage.**

Normally, for interactive input only one value is entered at a time. You can, however, read more than one value at a time by putting more than one variable in the read list. For example,

```
writeln (' Enter integers for the time in hours and minutes.');
readln (Hours, Minutes)
```

=>
 Enter integers for the time in hours and minutes.
 □

At this point, execution of the program stops until two numbers are entered. The first number is assigned to **Hours,** and the second number is assigned to **Minutes.** Each numerical value that is entered from the terminal must be separated by at least one blank from the previous value. That is, the **readln** statement will search for the next nonblank character when reading numerical data. If the following numbers were entered from the terminal

□12 2

they would be interpreted as the integer 12 and the integer 2. The 12 would be put in the storage location for **Hours,** and the 2 in the location for **Minutes.**

In the example above, both numbers could be typed on a single line or a carriage return could be entered between the numbers. Some systems require a carriage return after each value is entered. If a carriage return is entered between the numbers, another prompt will be issued indicating that more data must be entered.

The syntax for the **read** statement is identical to the syntax for the **readln** statement, except that the word **read** is substituted for the word **readln:**

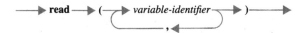

The differences between **read** and **readln** will be discussed in Chapter 10 when input files are covered. The **read** command is mentioned here because, depending on how the input and output files are implemented on a system, using the **readln** statement may cause the computer to issue a second prompt after the data has been entered. If this occurs when you use the **readln** statement, use the **read** statement instead.

REMINDERS

1. Do not use commas to separate your data.

2. Remember to enter the data after typing it. The data that you type is not transmitted to the computer until you depress either the carriage return key or the enter key, depending on the terminal.

The format of each input value must be assignment compatible with the variable to which it will be assigned. That is, if you declare a variable to be an integer, you cannot read a value of "a" into it. For example,

b 1.0

must be assigned to a character and a real variable, respectively. An integer number, such as 2, could be assigned to a variable of data type integer, real, or character. If it is assigned to an integer variable, it will be stored as the integer 2; if it is assigned to a real number, it will be stored as the real number 2.0; and if it is assigned to a character variable, it will be stored as the character '2'. Just as for the assignment statement, values must be assignment compatible with the variables into which they are read.

Care must be taken when assigning values to character variables through use of the **readln** statement. Because blanks and quotation marks are valid characters, blanks or quotation marks in the data can be read into character variables. Therefore, when characters are entered from the terminal, they must not be surrounded by quotation marks as they are when they are used in assignment statements and constant definitions. They cannot be separated by blanks either. For example, if **Char1, Char2,** and **Char3** are declared to be character variables, their values can be read interactively and then printed through use of the following statements:

```
writeln (' Enter a three letter word.');
readln (Char1, Char2, Char3);
writeln (' The word is ', Char1, Char2, Char3)
```

```
=>
Enter a three letter word.
```

3.1 Input statements—read and readln

If the characters

□dog

are entered after the prompt, the letter "d" is stored in **Char1**, the letter "o" is stored in **Char2**, the letter "g" is stored in **Char3**, and the output is

The word is dog

If the characters

□d o g

are entered after the prompt, then the letter "d" is stored in **Char1**, the character blank " " is stored in **Char2**, the letter "o" is stored in **Char3**, and the output is

The word is d o

If the characters

□'dog'

are entered, then the character "'" is stored in **Char1**, the letter "d" is stored in **Char2**, the letter "o" is stored in **Char3**, and the output is

The word is 'do

Program 3.1, **TestInput**, demonstrates interactive reading of character, integer, and real variables.

PROGRAM 3.1   TestInput

```
program TestInput (input, output);
(*      TestInput reads 3 character variables, 1 integer variable, and    *)
(*      1 real variable from the terminal and echo prints their values.   *)
var    IntVar : integer;          (* integer to be read and printed. *)
       RealVar : real;            (* real number to be read and printed. *)
       Char1, Char2, Char3 : char;  (* letters of a three letter word to be *)
                                    (* read and printed. *)

begin
  writeln (' You will be asked to enter three characters, one integer,');
  writeln (' and one real number.  The values you enter will be printed.');
```

```
        writeln;
        writeln (' Enter a three letter word');
        readln (Char1, Char2, Char3);

        writeln;
        writeln (' Enter an integer number');
        readln (IntVar);

        writeln;
        writeln (' Enter a real number');
        readln (RealVar);

        writeln;
        writeln (' The word is ', Char1, Char2, Char3);
        writeln (' The integer number is ', IntVar:10);
        writeln (' The real number is ', RealVar:10:2)
end. (* program TestInput *)


=>
You will be asked to enter three characters, one integer,
and one real number.  The values you enter will be printed.

Enter a three letter word
□cat

Enter an integer number
□1009

Enter a real number
□-5.23

The word is cat
The integer number is         1009
The real number is        -5.23
```

When you write a program in which data will be read interactively from the terminal, the messages that you include for the user of the program are important. Although you have comments within your program that explain its purpose and what the variables are to be used for, the user of the program will not see the comments. A short statement of the purpose of the program is necessary so that the user understands what the program does. The message telling the user to supply a value for a variable should include the data type of the variable to be entered (real, integer, or character) and, if appropriate, the units (such as watts, volts, feet, etc.) of the input data. Confronted with the statement "Enter a real number for the voltage in volts", most people would want to know why they are being asked for this value. The statements asking the user to supply values for the case study can be expanded to give the user more information about the program and the values that are to be supplied. These statements will be rewritten at the end of the chapter after you have learned how to write and print longer messages within your programs.

> *STYLE HINT*
>
> When writing an interactive program, always provide the user with a short statement of the purpose of the program. Whenever you request a value from the user, state how many values are expected and what the data type of each value should be.

You should also *echo,* or repeat, the values that the user has entered. The process of reporting all the input values is often referred to as *echo printing*. Without an echo print of the input, the user cannot be certain what values were used in making the computations. In the final version of the solution to the case study, the values that the user has entered for the resistance and voltage will be echoed in the statement used to report the output of the program.

3.2 Arithmetic operators and expressions

The following symbols are called arithmetic operators. They represent the basic arithmetic operations in Pascal:

+ addition
− subtration
* multiplication
/ division

To see how the arithmetic operators are used, suppose that **X, Y, Sum, Difference, Product,** and **Quotient** have all been declared to be real variables.

```
X := 4.8;
Y := 3.2;

Sum := X + Y;
Difference := X - Y;
Product := X * Y;
Quotient := X/Y
```

After the preceding code is executed, **Sum** will have a value of 8.0, **Difference** will have a value of 1.6, **Product** will have a value of 15.36, and **Quotient** will have a value of 1.5.

The spaces in the assignment statements between the variables and the arithmetic operators are not necessary. They are used to make the statements easier to read, but their use is optional.

The arithmetic operators can be used with constants and numbers as well as with variables. Suppose that you have a circle with a radius of 2.5 and that you want to write a program to compute the circumference, which is $2\pi r$; you can construct a program by writing a program heading, defining a constant for **Pi,** and declaring variables for the **Radius** and **Circumference.** Then, in the main body of the program, you could write the statements to assign a value to **Radius** and to compute the value for **Circumference,** as is done in Program 3.2, **AnotherCircle.**

PROGRAM 3.2 **AnotherCircle**

```
program AnotherCircle (input, output);
(*      This program computes a circle's circumference given its radius,   *)
(*      using the formula: circumference = 2*pi*radius.                    *)

const Pi = 3.14159;             (* used in computation of circumference. *)

var   Radius : real;            (* given radius in inches. *)
      Circumference : real;     (* computed circumference in inches. *)

begin (* main program *)
  Radius := 2.5;
  Circumference := 2 * Pi * Radius;

  writeln (' For a circle with radius', Radius:6:1, ' inches,');
  writeln (' the circumference is', Circumference:6:1, ' inches.')
end. (* program AnotherCircle *)
```

```
=>
 For a circle with radius   2.5 inches,
 the circumference is  15.7 inches.
```

Notice the use of the multiplication symbol, *, between the number 2 and the identifier for the constant, π, and between the **Pi** and the variable **Radius** in the second assignment statement. Because of the way that equations are read, it is easy to forget that the Pascal compiler does not interpret **2Pi** as **2*Pi.**

Also notice that the assignment statement for **Circumference** uses the integer number 2. This statement uses *mixed-mode arithmetic.* Mixed-mode arithmetic occurs when variables, constants, or numbers of both real and integer data types are used in the same assignment statement. Mixed-mode arithmetic is acceptable in standard Pascal as long as the variable on the

3.2 Arithmetic operators and expressions

DEBUGGING HINT

In an equation, writing two variables next to each other indicates that the multiplication operation is to be performed. In Pascal, to multiply two variables, you must include the multiplication operator between the variables. The equation

$$x = yz$$

is translated into Pascal as

$$X := Y * Z$$

---

left-hand side of the assignment statement is a real variable. Some versions of Pascal written for microcomputers do not allow any mixed-mode arithmetic. In any version of Pascal, the following statements when compiled will produce an error message:

```
RealVar := 3.0;
IntVar  := RealVar/2;

=>
     IntVar := RealVar/2;
****error****            $129


error messages :
***************

129: type conflict of operands
```

Dividing the real number 3.0 by the integer 2 results in the real number 1.5, but a real expression is not assignment compatible with an integer variable. Even if **RealVar** had been assigned a value of 2.0, the statements still would not compile. The compiler does not execute the statements; it checks whether the statements follow the rules of Pascal syntax, and the rule is that a real value cannot be assigned to an integer variable. You could use **RealVar/2** as the argument of either the function **trunc** or the function **round** and assign the resulting integer value to **IntVar.**

Div *and* Mod

Pascal has two arithmetic operators called **div** and **mod** that are valid only for values of the data type **integer. Div** and **Mod** are used for integer division.

The result of **div** or **mod** must be assigned to an integer variable. **A div B** is evaluated as the integer part of the result of **A / B**. For example,

    4 div 3   is evaluated as   1

    8 div 2   is evaluated as   4

    11 div 4   is evaluated as   2

Mod is evaluated as the remainder or the *modulus* of **A / B**. For example,

    4 mod 3   is evaluated as   1

    8 mod 2   is evaluated as   0

    11 mod 4   is evaluated as   3

There must be at least one space before and one space after the reserved words **div** and **mod**. If you leave out the space between **mod** and **4**, the Pascal compiler will interpret **mod4** as a valid identifier that has not been declared. If you leave out the space between **11** and **mod**, the compiler will interpret **11mod** as an invalid identifier.

Given the definitions and declarations

```
const Number = 5;

var   Counter, Remainder, Next, Last : integer;
```

the following assignment statements are all valid uses of **div** and **mod**:

```
Counter := Number div 2;
Remainder := Number mod 2;
Next := Counter div Remainder;
Last := 5 div 6
```

Precedence order

In Program 3.2, **AnotherCircle,** the assignment statement that was used to compute the circumference performed two multiplication operations. In multiplication, the order in which the operations are performed does not affect the final answer. But in computing the equivalent resistance for the case study, the order in which the operations are performed does affect the answer. The statement

```
EquivResist := Resist1 * Resist2 / Resist1 + Resist2
```

could be interpreted as any one of these statements:

3.2 Arithmetic operators and expressions

```
EquivResist := (Resist1 * Resist2) / (Resist1 + Resist2);

EquivResist := (Resist1 * (Resist2 / Resist1)) + Resist2;

EquivResist := ((Resist1 * Resist2) / Resist1) + Resist2;

EquivResist := Resist1 * (Resist2 / (Resist1 + Resist2))
```

Each statement may result in a different numerical value for **EquivResist.** To avoid ambiguity, Pascal has a *precedence order.* The precedence determines the order in which arithmetic operations are performed. A complete table of precedence order is given in Appendix A. The precedence order for the operators that you have learned so far is

( )	Parentheses, as used above, can be used to override the standard precedence order. Operations within parentheses are always performed first.
−	As the unary negator (that is, the negative sign on a number), the minus sign has second precedence. Thus

$$-3 + 4 = 1$$
$$-(3 + 4) = -7$$

* / **div mod**	Multiplication and division, including **div** and **mod,** have third precedence.
+ −	Addition and subtraction have last precedence.

Using the rules for precedence order, you can combine several arithmetic operations in a single expression. If two operations of the same precedence order appear in the same expression, they are evaluated from left to right. The following expressions can be evaluated using the precedence rules:

−3.0 − 6.9 * Radius	is evaluated as	(−3.0) − (6.9 * Radius)
A / B * C	is evaluated as	(A / B) * C
A * B / C	is evaluated as	(A * B) / C

If the expression for the equivalent resistance is written without parentheses, it will be evaluated according to the precedence rules. Since the multiplication and division operations have the same order of precedence, the multiplication is evaluated first because it comes first. The resulting values of **(Resist1 * Resist2)** is divided by **Resist1.** Then, the value of **Resist2** is added to **(Resist1 * Resist2) / Resist1,** and the result is stored in **EquivResist.** The order of precedence is equivalent to the statement

```
EquivResist := ((Resist1 * Resist2) / Resist1) + Resist2
```

The order of evaluation does not correspond to $(R_1 R_2)/(R_1 + R_2)$. For the expressions to be evaluated in the proper order, parentheses are needed, as follows:

```
EquivResist := Resist1 * Resist2 / (Resist1 + Resist2)
```

If you saw the assignment statement above in a program, you might have to stop to figure out the order in which the operations were evaluated. Therefore, you may include parentheses for clarity even if they are not necessary, as follows:

```
EquivResist := (Resist1 * Resist2) / (Resist1 + Resist2)
```

STYLE HINT

Use parentheses to indicate the order in which operations are performed.

$$A := (B / C) * D$$

is clearer than

$$A := B / C * D$$

which might be misinterpreted by someone reading the program as

$$A := B / (C * D)$$

3.3 Expressions

You have seen syntax diagrams in which an expression is one of the valid components of the statement. Although many examples of expressions have been given, expressions have not been defined. An expression is any valid combination of operators, characters, numbers, and identifiers. Anything that can appear on the right-hand side of an assignment statement is an expression. This includes numbers, characters, constants, and variable identifiers. Numbers and identifiers that represent numeric values can be combined with arithmetic operators to create expressions. The standard functions can also be used within expressions. The character strings that can be used in output statements are expressions. The simplified syntax diagram for an expression is

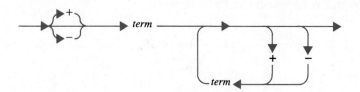

where a term is defined by the simplified syntax diagram

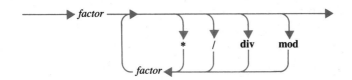

and a factor is defined by the simplified syntax diagram

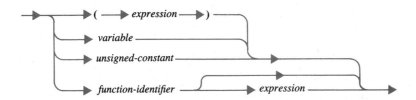

The most elementary part of an expression, a *factor* can be a variable, a constant, a function, or an expression in parentheses. Notice that the definition of a factor is another recursive definition because a factor is defined in terms of an expression, and an expression is defined in terms of a factor. A *term* can be either a single factor or a sequence of factors combined by means of the multiplication and division operators. Finally, an *expression* can be either a single term or a sequence of terms combined by means of the addition and subtraction operators.

The precedence order of the arithmetic operators can be deduced from the syntax diagrams: parentheses, which have the highest precedence, are elements of factors; multiplication and division, which have second precedence, are elements of terms; and addition and subtraction, which have the lowest precedence, are elements of expressions.

The following are all expressions, assuming that the identifiers have been properly declared:

```
Pi / 2.0                                   'character string'
Resist1 * Resist2 / Resist1 + Resist2      1.0e-03
2.0 * 4.5                                  trunc (Resist1)
sqrt(2.0)                                  4 mod IntVar
7 - 3                                      '#'
```

You should be aware that any valid expression can be created from the syntax diagrams; however, even though an expression can be created from a syntax diagram, the expression is not necessarily valid. For example, the expression

```
                'cat' * 3.2 div 5.0
```

can be created from the syntax diagram: 'cat', 3.2, and 5.0 are all factors because they are constants. These factors are combined into a term by means of * and **div,** and a term is an expression. However, the expression is nonsense and will not compile.

Printing expressions

In Chapter 2 you learned the **writeln** statement, which can be used to print the value of an expression. Now you know how to create more complicated expressions. For example,

```
writeln (' The product of 2 x 3 = ', 2 * 3);
writeln (' The square of the square root of 2 is ', sqr (sqrt(2)))
```

In the first **writeln** statement, the first expression is "The product of $2 \times 3 =$" and the second expression is "$2 * 3$". And in the second **writeln** statement, the first expression is "The square of the square root of 2 is" and the second expression is "sqr(sqrt(2))". At execution, these **writeln** statements will produce the following output:

```
=>
  The product of 2 x 3 =              6
  The square of the square root of 2 is   2.0000000000000e+00
```

Character strings that are enclosed in quotes are printed exactly as they are written except that the quotation marks are omitted. The arithmetic expression "$2 * 3$" is evaluated, and the result is printed. The arithmetic expression "sqr(sqrt(2))" is evaluated, and its result printed. Think about why the result is a real number.

Any expression can be formatted; thus, character strings can be formatted by means of a field width. If you want to line up column headings, the ability to format a character expression is useful. Program 3.3, **FormatWrite,** demonstrates the syntax for formatting real, integer, and character variables and character strings.

PROGRAM 3.3 FormatWrite

```
program FormatWrite (input, output);
   (*      FormatWrite demonstrates formatting real and character       *)
   (*      expressions in Pascal.                                       *)

   var Number : real;          (* number is the real number to be formatted. *)
```

3.3 Expressions

```
begin (* main program *)
  Number := 123.456789;      (* Set the value of the number to be formatted. *)

  writeln;
  writeln (' In default format, the number is ':50, Number);
  writeln;
  writeln (' In a field of 10, the number is ':50, Number:10);
  writeln;
  writeln (' In a field of 10:3, the number is ':50, Number:10:3)
end. (* program FormatWrite *)
```

=>
```
                In default format, the number is   1.2345678900000e+02

                In a field of 10, the number is   1.234e+02

                In a field of 10:3, the number is      123.457
```

If the field width specified for a character string is too small, most compilers truncate the string on the right, as in

```
  writeln (' In a field of 10:3, the number is':30, Number:10:3);
```

=>
```
 In a field of 10:3, the numbe      123.457
```

STYLE HINT

It is possible in Pascal to construct an entire program using only output statements without using any variables or assignment statements. Do not be tempted. Only minor arithmetic operations should be performed inside output statements. The results of arithmetic operations and the values returned by functions should be assigned to variables with descriptive names.

Long expressions in writeln *statements*

You must keep several points in mind if you want to use a long expression within a **writeln** statement. In Pascal a string cannot be extended beyond the end of the line in a program statement. The length of a program line ranges from 40 to 225 characters or more, depending on the editor you are using. Suppose you are using an editor that has an 80-character line, and you want to write as part of your output a sentence that is 80 characters long. The syntax of the **writeln** statement requires that you use 12 spaces other than the character string for the word **writeln,** the parentheses, the quotation marks, and the semicolon. Therefore you cannot write a character string longer than 68 characters on one line within a **writeln** statement. But you can put a character string on several lines, as long as you end each line with a single quote and a comma. Each part of the sentence that is enclosed within single quotes and separated from the next part by a comma is a separate expression. In the following statements, the character string is too long to be entered on one line in the program, but can be printed on one line in the output if it is broken up into several shorter expressions in the **writeln** statement.

```
writeln (' This program computes the current',
         ' in amps in a circuit',
         ' with two resistors.')
=>
 This program computes the current in amps in a circuit with two resistors.
```

If you make the series of expressions longer than a line on the output device, the output will wrap around at odd places. For example, on an 80-column terminal, the following **writeln** statement produces the output shown:

```
writeln (' This program computes the current',
         ' that flows when a voltage is applied across',
         ' two resistors connected in parallel.')
=>
 This program computes the current that flows when a voltage is applied across t
wo resistors connected in parallel.
```

Notice that there is no break between the phrases "the current" and "that flows" except for the blank that begins the second phrase.

Syntax errors are common in **writeln** statements because it is easy to forget the quotation marks, particularly when there are several strings in the same **writeln** statement. In the output statement that follows, the character expression extends over more than one line. When this statement is compiled, it will produce error messages. The exact error messages will depend on the compiler being used. The error messages from two different compilers are shown in the example to give you an idea of what you can expect from your compiler.

Compiler 1

```
  writeln (' This program prints the product of 2 X 3, the mass
***error***
$202
             in grams, and the distance in inches. ');
***error***              $104 $58 $104      $4,6
$202

error messages :
***************
    4: ")" expected.
    6: illegal symbol (possibly missing ";" in line above).
   58: error in factor (bad expression).
  104: identifier not declared.
  202: string constant may not continue past end of source line.
```

Compiler 2

```
Compiling
   6 lines

Error 55: String constant exceeds line. Press <ESC>
```

All of these error messages occur because the character expression is continued on the next line! The only difference between the statement that compiled in the previous example and the statement that produced five errors is that in the statement that compiled the character expression is broken into three expressions separated by commas.

This error demonstrates an important debugging principle: Fix the most obvious errors first. Most error messages occur because the compiler cannot interpret the code once the initial error is made. In fact, the compiler cannot even tell that the problem with the second line is a missing initial apostrophe. The compiler is trying to interpret the line as an assignment statement, hence all the error messages about undefined identifiers. The addition of an apostrophe and a comma to the first and second lines and an apostrophe to the beginning of the third line will allow the compiler to correctly interpret the rest of the program.

DEBUGGING HINT

Fix obvious syntax errors first. If you change a line that appears correct just because the compiler flagged it, you are likely to create an error where none existed.

The write *statement*

The **write** statement can also be used to create long output lines. It is similar to the **writeln** statement. A simplified syntax diagram for **write** is

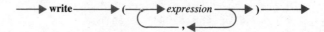

When an output statement is used to print expressions, you can envision an arrow or pointer that moves across the terminal as the characters are printed. The pointer is always positioned just after the last character that has been written and is used to indicate where the next character will be printed. The difference between a **write** and a **writeln** statement is in the position of the pointer after the statement has been executed. After the **writeln** statement has been completely evaluated, the pointer moves to the beginning of the next line. Ater a **write** statement has been executed, the pointer is still positioned after the last character that was printed. In more technical terms, a marker called the *eoln* marker is written after a **writeln** statement is evaluated, but is not written after a **write** statement is evaluated. Eoln stands for *End-of-Line*. You cannot see the eoln marker, but whenever the computer detects an eoln marker while your output is being printed, a new line is started.

The difference between **write** and **writeln** is illustrated by the following two examples:

Example 1

```
write (' This is');
write (' the end.');
writeln
```

=>
This is the end.

Example 2

```
writeln (' This is');
writeln (' the end.')
```

=>
This is
the end.

In the first example, the first two statements are **write** statements, so the output is printed on one line. After the second **write** statement is executed, the write pointer remains positioned after the letter "d". Only when the last **writeln** statement is executed does the pointer move to the next line. In the second example, after the first **writeln** statement is executed, the write pointer goes to the beginning of the next line. When the second **writeln** statement is executed, the output starts at the beginning of that line. After execution of the second **writeln** statement, the pointer will be positioned at the beginning of the next line.

If an assignment statement is written between a **write** statement and a **writeln** statement, the pointer remains positioned after the end of the output from the **write** statement until the **writeln** statement is executed. The fragment below is another illustration of the use of **write** and **writeln.**

```
write (' For a circle with radius');
Radius := 2.5;
writeln (Radius:6:1, ' inches,')

=>
For a circle with radius   2.5 inches,
```

Solution to Case Study 2

At the beginning of the chapter, the statements for Steps 1 and 4 of the algorithm were written in Pascal. Having learned the read statements, you can now complete Step 2 and write the statements to read the values, which the user has been asked to supply, for the two resistors and for the voltage.

```
writeln (' Enter a real number for the voltage in volts');
readln (Voltage);

writeln (' Enter a real number for the resistance in ohms',
         ' for the first resistor');
readln (Resist1);

writeln (' Enter a real number for the resistance in ohms',
         ' for the second resistor');
readln (Resist2)
```

Step 3 of the algorithm is the step that computes the current using the voltage and the resistances supplied by the user. Some of the statements for this step were completed in the section on arithmetic operators. The final statements for computing the current are

```
EquivResist := (Resist1 * Resist2) / (Resist1 + Resist2);
Current := Voltage / EquivResist
```

Putting the statements for the program together results in Program 3.4, **ComputeCurrent.** Notice that the output statements have been rewritten to give the program user more information about the program.

PROGRAM 3.4 ComputeCurrent

```
program ComputeCurrent (input, output);
(*      ComputeCurrent computes the current that flows when a voltage    *)
(*      is applied across two resistors connected in parallel.  The      *)
(*      values for the voltage and resistance are read interactively.    *)

var   Resist1 : real;        (* user supplied resistance 1 in ohms. *)
      Resist2 : real;        (* user supplied resistance 2 in ohms. *)
      Voltage : real;        (* user supplied voltage in volts. *)
      EquivResist : real;    (* equivalent resistance in ohms -- variable *)
                             (* intermediate to computation of current. *)
      Current : real;        (* computed current in amps. *)

begin (* main program *)
  writeln (' This program computes the current that flows when a voltage');
  writeln (' is applied across two resistors connected in parallel.');
  writeln;

  (* The next six statements ask the user to enter values for the *)
  (* voltage  and resistances. Each value is read by the computer *)
  (* as it is entered. *)

  writeln (' Enter a real number for the voltage in volts');
  readln (Voltage);

  writeln (' Enter a real number for the resistance in ohms',
           ' for the first resistor');
  readln (Resist1);

  writeln (' Enter a real number for the resistance in ohms',
           ' for the second resistor');
  readln (Resist2);

  (* The equivalent resistance and current are computed next. *)

  EquivResist := (Resist1 * Resist2) / (Resist1 + Resist2);   (* ohms *)
  Current := Voltage / EquivResist;                           (* amps *)

  (* Finally, the input values are echoed to the user and the computed *)
  (* value of the current is reported. *)

  writeln;
  writeln (' In a circuit with a voltage of', Voltage:10:2, ' volts');
  writeln (' and resistors of', Resist1:10:2, ' and', Resist2:10:2, ' ohms,');
  writeln (' the current is', Current:10:2, ' amps.')
end. (* program ComputeCurrent *)
```

Solution to Case Study 2

```
=>
 This program computes the current that flows when a voltage
 is applied across two resistors connected in parallel.

 Enter a real number for the voltage in volts
□4.2
 Enter a real number for the resistance in ohms for the first resistor
□0.5
 Enter a real number for the resistance in ohms for the second resistor
□0.8

 In a circuit with a voltage of        4.20 volts
 and resistors of         0.50 and     0.80 ohms,
 the current is       13.65 amps.
```

APPLICATION
Dimensional Analysis

CASE STUDY 3
Speed of sound in furlongs per fortnight

Speed is the distance traveled per unit of time. Speeds are usually measured in meters per second, kilometers per hour, feet per second, or miles per hour. The same speed can be expressed in any of these units or in any set of units that have the dimensions length divided by time. For example, swimmers may think of their speed in terms of laps per hour, and people walking in a city may think of their speed in terms of blocks per hour.

For many problems it is necessary to convert the speed from one set of units to another set of units. As a challenge, write a program that requests a speed in feet per second from the user and converts it into furlongs per fortnight. Use the speed of sound to test your program.

Algorithm development

The problem of converting a speed to furlongs per fortnight can be broken down into subproblems. Although the tasks of reading the speed from the terminal and reporting the result were not stated, obviously they must be included in the algorithm statement.

ALGORITHM 3.3 First iteration of the algorithm for FurlongsPerFortnight

1. Write a message to the program user explaining the purpose of the program and requesting a speed in feet per second.
2. Read **OldSpeed** in feet per second.
3. Convert **OldSpeed** from feet per second to **NewSpeed** in furlongs per fortnight.
4. Report **OldSpeed** and **NewSpeed.**

You already know how to construct all of the steps in the algorithm except for Step 3. The rest of this chapter will cover conversion of units. A complete solution to the case study is given at the end of the chapter in Program 3.7.

Systems of measurement

Many different systems of measurement have been devised in order to describe the physical world. Within any culture, a standard system of measurement is necessary so that people can describe distances, the passage of time, and the properties of objects. Each measurement system uses different units to describe three fundamental properties of the physical environment: length, mass, and time. Many other properties, such as velocity and density, can be derived from these fundamental properties. This section will introduce you to measurement systems and teach you how to convert measurements both within a system and from one system to another.

Early measurement systems were not standardized, and therefore units of measurement did not always mean the same thing to different people. Although the inch has always been defined as 1/12 of a foot, the foot was originally the length of whoever's foot was being used. Thus, the length of a ten-foot pole could vary a great deal. Modern systems use standards to define units of measure.

Three systems of measurement are commonly used by scientists and engineers: the English system, the système international (SI), and the CGS system. The basic units for the English system are the foot, the pound, and the second. SI and CGS are both metric systems. The basic units in SI are the meter, the kilogram, and the second. The basic units of the CGS system are the centimeter, the gram, and the second. The name CGS is from the first letters of the system's basic units.

Although the Metric Conversion Act was signed in 1975, most Americans are still more familiar with the English system of measurements than with the metric system. Americans tend to think of their weight in pounds and their height in feet and inches. Given someone's height in centimeters, most Americans have to convert to feet and inches before being able to tell if the person is tall or short.

The fundamental units of the English and metric systems are shown in Table 3.1. A major difference between the English and metric systems is that the metric system has mass as a basic unit whereas the English system has force, or weight. Distinguishing between the concepts of weight and mass is important. Intuitively they seem to be the same, but Isaac Newton (1642–1727) showed that they are different properties. Weight is a measure of force, and mass is a measure of the amount of matter that an object contains. Because weight is a function of the strength of the gravitational force, a person weighs less on the moon than on earth. If your weight is 150.0 pounds on earth, your weight on the moon would be 24.7 pounds, but on the earth or the moon, your mass is constant at 68.0 kilograms. Because most masses are weighed on earth, the conversion factor of 0.4536 kilogram per pound (or 2.2 pounds per kilogram) is usually the appropriate factor:

$$1 \text{ pound} = 0.4536 \text{ kilogram} \tag{3.4}$$

This statement is properly read as "The force on a 0.4536-kilogram mass is one pound."

The English unit slug is a measure of mass. If you weigh 150.0 pounds on earth, your mass is approximately 4.7 slugs. A slug under the acceleration of the earth's gravity weighs 32.174 pounds.

There is an alternative English system in which the pound is the unit of mass and the poundal is the unit of force, but this system is rarely used. You can almost always assume that "pound" means "pound force." For example, if pressure is given in pounds per square inch (psi), the dimensions express force per unit area, not mass per unit area.

The units used to measure length, time, and mass vary with the magnitude of the meaasurement. You would not measure the length of your finger in miles or the distance from Boston to New York City in inches. In the English system, to convert from inches to miles, you must know that there are 63,360 inches in a mile—that is, 12 inches per foot times 5280 feet per mile. In the metric system, conversion between units is done by multiplying or

TABLE 3.1
Fundamental Units of the Metric and English Systems

Property	Metric		English	
	Basic unit	Other units	Basic unit	Other units
Length	meter	millimeter centimeter kilometer	foot	inch yard mile
Mass	kilogram	gram	slug[a]	—
Force	newton[a]	dyne	pound	ounce ton
Time	second	minute hour	second	minute hour

[a]The newton and the slug are not fundamental units in their respective systems.

dividing by powers of ten, which is the primary advantage of the metric system.

Table 3.2 gives the standard prefixes used in the metric system. Using this table, you can conclude that a gigawatt is 10^9 watts, a kilowatt is 10^3 watts, and a milliwatt is 10^{-3} watts. Commonly used prefixes are mega-, kilo-, milli-, micro-, nano-, and pico-. Hecto-, deca-, and deci- are rarely used.

Modern measurements systems have sophisticated standards for each of the basic units. Until the early 1960s, the length of a meter was defined as the distance between two lines on a bar kept at the International Bureau of Weights and Measures in France. This required that copies of the meter be made for the standardization of other rulers. In the 1960s, the meter was redefined in terms of the wavelength of light emitted by particular atoms so that the standard could be reproduced in any laboratory. In 1983 the meter was redefined as the distance traveled by light through a vacuum in 1/299792458 of a second.

Time units are based on the divisions of the year and the day. Because there are slight variations in the length of years, an atomic standard has been devised. The second is defined as 9,192,631,770 vibrations of the radiation of a cesium atom.

The international standard for mass is a platinum-iridium cylinder housed in France. The kilogram is defined as the mass of that cylinder. The U.S. standard for the kilogram is a copy of the cylinder, kept at the National Bureau of Standards in Maryland.

TABLE 3.2
Metric Prefixes

tera-	T	10^{12}
giga-	G	10^{9}
mega-	M	10^{6}
kilo-	k	10^{3}
hecto-	h	10^{2}
deca-	da	10^{1}
deci-	d	10^{-1}
centi-	c	10^{-2}
milli-	m	10^{-3}
micro-	μ	10^{-6}
nano-	n	10^{-9}
pico-	p	10^{-12}
femto-	f	10^{-15}
atto-	a	10^{-18}

Although the properties of mass, length, and time are considered to be the three fundamental properties, some properties cannot be described in the terms of these three. Table 3.3 gives the basic properties and their units for the English system, and Table 3.4 gives the basic properties and their units for the metric system.

All other physical properties can be expressed as combinations of the basic properties. For example, the derived property of force in the metric system is defined to be mass times acceleration,

$$F = ma \qquad (3.5)$$

where acceleration is a derived unit. Acceleration has units of velocity divided by time,

$$a \propto \text{velocity/time} \qquad (3.6)$$

TABLE 3.3
Basic English Properties and Their Units

Property	Unit	Abbreviation
Length	foot	ft
Force	pound	lb
Time	second	s
Angle	degree	°
Thermodynamic temperature	degree Rankine	°R
Quantity of heat	British thermal unit	Btu
Work	horsepower	hp

TABLE 3.4
Basic Metric Properties and Their Units

Property	Unit	Abbreviation
Length	meter	m
Mass	kilogram	kg
Time	second	s
Electric current	ampere	A
Plane angle	radian	rad
Thermodynamic temperature	Kelvin	K
Amount of substance	mole	mol
Luminous intensity	candela	cd

where the symbol ∝ means proportional to. Velocity has units of length, or distance, divided by time,

$$v \propto \text{length/time} \tag{3.7}$$

Combining Equations 3.7 and 3.6 gives

$$a \propto (\text{length/time})/\text{time} \tag{3.8}$$

So, the derived property force can be defined in terms of the basic properties of mass, length, and time, or

$$F \propto \text{mass·length/time}^2 \tag{3.9}$$

Force is measured in units called newtons. A newton is one kilogram-meter per second squared, or 1 kg-m/s².

All of the properties that have been referred to have dimension or units associated with them. When using these measurements, it is important to include the dimension of the measurement. The statement "time = 4" has no meaning unless you specify that the unit of time is seconds, minutes, hours, years, centuries, or whatever.

Table 3.5 gives common derived metric properties. Each derived unit is given in terms of its basic units. For example, a pascal is the unit of pressure,

TABLE 3.5
Derived Metric Units

Quantity	Unit	Abbreviation	Basic units
Frequency	hertz	Hz	s^{-1}
Force	newton	N	$kg\ m\ s^{-2}$
Pressure	pascal	Pa	$kg\ m^{-1}\ s^{-2}$
Energy, work	joule	J	$kg\ m^2\ s^{-2}$
Power, radiant flux	watt	W	$kg\ m^2\ s^{-3}$
Electric charge	coulomb	C	$A\ s$
Electric potential, potential difference, electromotive force	volt	V	$kg\ m^2\ s^{-3}\ A^{-1}$
Capacitance	farad	F	$A^2\ s^4\ kg^{-1}\ m^{-2}$
Electric resistance	ohm	Ω	$kg\ m^2\ s^{-3}\ A^{-2}$
Conductance	siemens	S	$kg^{-1}\ m^{-2}\ s^3\ A^2$
Magnetic flux	weber	Wb	$kg\ m^2\ s^{-2}\ A^{-1}$
Magnetic flux density	tesla	T	$kg\ s^{-2}\ A^{-1}$
Inductance	henry	H	$kg\ m^2\ s^{-2}\ A^{-2}$
Volume	liter	L	m^3

and its units are newtons per meter squared or N/m². A pascal is named after Blaise Pascal (1623–1662), the same man after whom the language Pascal is named. Most of the metric units are named after people who made major contributions to their field.

Some numerical constants have no dimension. The constant π, the ratio of the circumference of a circle to its diameter, does not have a dimension, nor does the transcendental number, e. A transcendental number is a number that cannot be expressed using algebraic functions such as sum, product, or power. Table 3.6 gives approximate values for the numerical constants π and e. The number e is covered in the application section of Chapter 8.

TABLE 3.6
Numerical Constants

$\pi = 3.14159$
$e = 2.71828$

TABLE 3.7
Physical Constants

Acceleration due to gravity	g	$= 9.80665$ m/s² $= 32.174$ ft/s²
Gravitational constant	G	$= 6.67 \times 10^{-11}$ N-m²/kg²
Avogadro's number	N_A	$= 6.02252 \times 10^{23}$ molecules/mole
Boltzmann's constant	k	$= 1.38 \times 10^{-23}$ J/K
Coulomb force constant	k	$= 8.98755 \times 10^9$ N-m²/C²
Permittivity of free space	E_0	$= 8.85415 \times 10^{-12}$ C²/N-m
Permeability of free space	μ_0	$= 4\pi \times 10^{-7}$ Wb/A-m
Faraday constant	F	$= 9.6487 \times 10^4$ C/mole
Electron charge	e	$= 1.60210 \times 10^{-19}$ C
Electron mass	e_m	$= 9.11 \times 10^{-31}$ kg
Electron volt	eV	$= 1.60210 \times 10^{-19}$ J
Gas constant	R	$= 8.31434$ J/mole-K
Planck's constant	h	$= 6.62559 \times 10^{-34}$ J-s
	$\hbar$	$= 1.054494 \times 10^{-34}$ J-s
Speed of light	c	$= 2.997925 \times 10^8$ m/s
Speed of sound in air	s	$= 344$ m/s
Mass of the earth	m_e	$= 5.98 \times 10^{24}$ kg
Radius of the earth	r_e	$= 6.37 \times 10^6$ m
Earth-sun distance	d_e	$= 1.49 \times 10^{11}$ m
		$= 1$ astronomical unit (AU)
Density of water at 20°C	ρ	$= 1.00 \times 10^3$ kg/m³
Standard atmospheric pressure	p	$= 1.013 \times 10^5$ N/m²
		$= 76$ cm of mercury $= 1$ atm

Application: Dimensional Analysis

Table 3.7 gives physical constants that are commonly used in the sciences and engineering. Many are used only in specific branches of science, but some will probably become very familiar to you. These constants are included here to show you some of the constants that are often used; some of them are used in examples given later in this book.

Conversion factors

It is often necessary to convert measurements from one system of measurement to another or to convert within a system from units of one magnitude to another. Scientists use the metric systems of measurement, but engineers in the United States and Great Britain often must use the English system because they design and build systems that are used in everyday life in cultures in which the English system predominates.

Table 3.8 gives conversion factors between commonly used English and metric units. Table 3.9 gives conversion factors between basic metric units and commonly used supplementary metric units.

It is easy to convert either within a system or between systems as long as you remember that you can always multiply by 1. Suppose you want to convert a length from inches into centimeters. Looking at Table 3.8, you can find a conversion factor to convert feet into meters, and you know that there are 12 inches in a foot. If 12 inches (in) = 1 foot, then

$$1 = \frac{1 \text{ ft}}{12 \text{ in}} \quad \text{or} \quad \frac{12 \text{ in}}{1 \text{ ft}} = 1 \tag{3.10}$$

TABLE 3.8
Conversion Factors

$$
\begin{aligned}
1 \text{ foot} &= 0.3048 \text{ meter} \\
1 \text{ mile} &= 1.609 \text{ kilometers} \\
1 \text{ slug} &= 14.59 \text{ kilograms} \\
360 \text{ degrees} &= 2\pi \text{ radians} \\
°F &= °R - 459.67 \\
°F &= \tfrac{9}{5}°C + 32 \\
1 \text{ pound} &= 4.448 \text{ newtons} \\
1 \text{ Btu/hour} &= 0.293071 \text{ watt} \\
1 \text{ horsepower} &= 745.700 \text{ watts} \\
1 \text{ foot-pound} &= 1.356 \text{ joules} \\
1 \text{ Btu} &= 1.05506 \times 10^3 \text{ joules} \\
1 \text{ atmosphere} &= 1.013 \times 10^5 \text{ newtons per meter}^2
\end{aligned}
$$

TABLE 3.9
Supplementary Metric Units

$$1 \text{ dyne} = 10^{-5} \text{ newton}$$
$$1 \text{ stere} = 1.0 \text{ meter}^3$$
$$1 \text{ parsec} = 3.084 \times 10^{16} \text{ meters}$$
$$1 \text{ lightyear} = 9.46 \times 10^{15} \text{ meters}$$
$$1 \text{ erg} = 10^{-7} \text{ joule}$$
$$1 \text{ calorie} = 4.186 \text{ joules}$$
$$1 \text{ kilowatt-hour} = 3.600 \times 10^6 \text{ joules}$$
$$°C = °K - 273.15$$

To convert inches into feet, use the relationship with inches in the denominator. For example,

$$3 \text{ in} \times \frac{1 \text{ ft}}{12 \text{ in}} = ?$$

The units cancel out, and you end up with an answer in feet.

$$3 \text{ in} \times \frac{1 \text{ ft}}{12 \text{ in}} = \frac{3 \times 1 \text{ ft}}{12} = \frac{1}{4} \text{ ft}$$

Then you can convert feet into meters using the conversion factor

$$1 \text{ ft} = 0.3048 \text{ m}$$

or

$$1 = \frac{0.3048 \text{ m}}{1 \text{ ft}} \quad (3.11)$$

There are 10^2 centimeters per meter, so to convert from meters to centimeters, use the relationship

$$\frac{10^2 \text{ cm}}{1 \text{ m}} = 1 \quad (3.12)$$

One inch is converted to centimeters as follows:

$$1 \text{ in} \times \frac{1 \text{ ft}}{12 \text{ in}} \times \frac{0.3048 \text{ m}}{1 \text{ ft}} \times \frac{10^2 \text{ cm}}{1 \text{ m}} = 0.0254 \times 10^2 \text{ cm} \quad (3.13)$$

or

$$1 \text{ in} = 2.54 \text{ cm} \quad (3.14)$$

Program 3.5, **InchesToCms,** is a simple program that uses the conversion factor from inches to centimeters.

PROGRAM 3.5 InchesToCms
```
program InchesToCms (input, output);
(*       This program converts a length in inches to the equivalent    *)
(*       length in centimeters.                                        *)

const CmsPerInch = 2.54;     (* conversion factor for inches to centimeters. *)

var    Inches : real;        (* number of inches entered by user. *)
       Centimeters : real;   (* number of centimeters -- computed. *)

begin (* main program *)
   (* Write a message to the user explaining program and requesting *)
   (* data.  Read the data that is entered. *)

   writeln (' This program converts inches to centimeters.');
   writeln;
   writeln (' Enter a real value in inches to be converted to centimeters.');
   readln (Inches);

   (* Convert inches to centimeters. *)

   Centimeters := Inches * CmsPerInch;

   (* Report the converted value and the original value to the user. *)

   writeln;
   writeln(' ', Inches:6:2, ' inches = ', Centimeters:6:2, ' centimeters')
end. (* program InchesToCms *)


=>
 This program converts inches to centimeters.

 Enter a real value in inches to be converted to centimeters.
▫1.5

   1.50 inches =   3.81 centimeters
```

Consider another example. If you were traveling in Europe, you might want to convert the automobile speed limit of 55 miles (mi)/hour (h) to km/h. You could state the problem as

$$55 \frac{mi}{h} = y \frac{km}{h} \qquad (3.15)$$

and solve for y. In order to find y, you must know the conversion factor between kilometers and miles,

$$1 \text{ mi} = 1.609 \text{ km}$$

or

$$1 = \frac{1.609 \text{ km}}{1 \text{ mi}} \qquad (3.16)$$

Multiplying the left-hand side of Equation 3.15 by 1—that is, multiplying it by the ratio of Equation 3.16—gives

$$55 \frac{mi}{h} \times \frac{1.609 \text{ km}}{1 \text{ mi}} = y \frac{km}{h} \qquad (3.17)$$

Canceling the unit "mile" gives

$$55 \times \frac{1}{h} \times \frac{1.609 \text{ km}}{1} = y \frac{km}{h} \qquad (3.18)$$

or

$$88.495 \frac{km}{h} = y \frac{km}{h} \qquad (3.19)$$

Therefore, a speed limit of 55 mi/h is equivalent to a speed limit of 88.5 km/h. Program 3.6, **SpeedLimit,** converts a speed from miles per hour to kilometers per hour.

PROGRAM 3.6 SpeedLimit

```
program SpeedLimit (input, output);
   (*    SpeedLimit converts an input speed in miles/hour to kilometers/hour. *)

const KmsPerMile = 1.609;      (* conversion factor for miles to kilometers. *)

var    MilesPerHour : real;    (* speed input by the user. *)
       KmsPerHour : real;      (* speed computed by the program. *)
```

Application: Dimensional Analysis

```
begin (* main program *)
   writeln (' This program converts a speed in miles per hour to',
            ' kilometers per hour.');
   writeln (' Enter a speed in miles per hour');
   readln (MilesPerHour);

   writeln;
   write (' A speed of', MilesPerHour:10:2, ' miles/hour equals ');

   KmsPerHour := MilesPerHour * KmsPerMile;
   writeln (KmsPerHour:10:2, ' kilometers/hour')
end. (* program SpeedLimit *)
```

=>
This program converts a speed in miles per hour to kilometers per hour.
Enter a speed in miles per hour
▫55.0

A speed of   55.00 miles/hour equals   88.50 kilometers/hour

When you solve a problem involving properties that have different dimensions, you should always check that your final answer has consistent units. If you had multiplied by miles per kilometer instead of kilometers per mile, the dimensions of your answer would have been miles^2 per kilometer-hour. This dimension is clearly not a speed. But just because the units are correct, it does not necessarily follow that your answer is correct.

Solution to Case Study 3

With your knowledge of how to convert from one set of units to another, Step 3 of the algorithm for the case study can be refined into subproblems. The speed in furlongs per fortnight can be determined by multiplying the speed in feet per second by the number of seconds in a fortnight and by the number of furlongs in a foot:

$$\frac{\text{furlongs}}{\text{fortnight}} = \frac{\text{feet}}{\text{second}} \times \frac{x \text{ seconds}}{\text{fortnight}} \times \frac{y \text{ furlongs}}{\text{foot}} \qquad (3.20)$$

So Step 3 of the algorithm can be written as:

ALGORITHM 3.4   Refinement of Step 3 of Algorithm 3.3

3. Convert **OldSpeed** from feet per second to **NewSpeed** in furlongs per fortnight.
   a. Compute **SecondsPerFortnight,** the number of seconds per fortnight.
   b. Compute **FurlongsPerFoot,** the number of furlongs per foot.
   c. **NewSpeed** ← **OldSpeed** * **SecondsPerFortnight** * **FurlongsPerFoot**

Both Step 3a and Step 3b can be refined further because neither feet per furlong nor seconds per fortnight is a standard conversion factor. For Step 3a, seconds per fortnight can be written as

$$\frac{\text{seconds}}{\text{fortnight}} = \frac{\text{days}}{\text{fortnight}} \times \frac{\text{hours}}{\text{day}} \times \frac{\text{seconds}}{\text{hour}} \qquad (3.21)$$

For Step 3b, furlongs per foot can be expanded to

$$\frac{\text{furlongs}}{\text{foot}} = \frac{\text{miles}}{\text{foot}} \times \frac{\text{furlongs}}{\text{mile}} \qquad (3.22)$$

Using Equations 3.21 and 3.22, Step 3 of the algorithm is

ALGORITHM 3.5   Refinement of Steps 3a and 3b of Algorithm 3.3

3. Convert **OldSpeed** from feet per second to **NewSpeed** in furlongs per fortnight.
   a. Compute **SecondsPerFortnight,** the number of seconds per fortnight.
      SecondsPerFortnight ← days per fortnight * hours per day * seconds per hour
   b. Compute **FurlongsPerFoot,** the number of furlongs per foot.
      FurlongsPerFoot ← miles per foot * furlongs per mile

To implement Steps 3a and 3b, you will need the following conversion factors:

3600 seconds = 1 hour
   24 hours = 1 day
   14 days = 1 fortnight
  5280 feet = 1 mile
    8 furlongs = 1 mile

Application: Dimensional Analysis

These factors can be declared as the constants **SecsPerHour, HoursPerDay, DaysPerFortnight, FeetPerMile,** and **FurlongsPerMile.** The values for the number of furlongs in a mile and the number of days in a fortnight were found in a dictionary. A dictionary can be quite useful for finding both unusual and usual conversion factors. In translating Steps 3a and 3b of the algorithm into Pascal, **SecondsPerFortnight** and **FurlongsPerFoot** can be used as intermediate variables to the solution of the problem. In Pascal the conversion statements are

```
SecondsPerFortnight := DaysPerFortnight * HoursPerDay * SecsPerHour;

FurlongsPerFoot := (1.0 / FeetPerMile) * FurlongsPerMile
```

To complete Step 3, **NewSpeed,** the speed in furlongs per fortnight, can be found by multiplying **OldSpeed** by the two conversion factors. Program 3.7, **FurlongsPerFortnight,** uses constants, intermediate variables, and the arithmetic operators to convert a speed from feet per second to furlongs per fortnight.

PROGRAM 3.7   FurlongsPerFortnight

```
program FurlongsPerFortnight (input, output);
(*      Program FurlongsPerFortnight converts a speed from feet/sec to    *)
(*      the equivalent speed in furlongs/fortnight.                       *)

const SecsPerHour = 3600.0;          (* seconds per hour *)
      HoursPerDay = 24.0;            (* hours per day *)
      DaysPerFortnight = 14.0;       (* days per fortnight *)
      FeetPerMile = 5280.0;          (* feet per mile *)
      FurlongsPerMile = 8.0;         (* furlongs per mile *)

var   SecondsPerFortnight : real;    (* intermediate variable. *)
      FurlongsPerFoot : real;        (* intermediate variable. *)
      OldSpeed : real;               (* user supplied speed in feet per *)
                                     (* second. *)
      NewSpeed : real;               (* computed speed in furlongs per *)
                                     (* fortnight. *)
begin (* main program *)
  write (' This program converts a speed given in feet/sec');
  writeln (' to furlongs/fortnight.');

  writeln (' Enter a speed in feet/sec');
  readln (OldSpeed);
```

```
    writeln;
    writeln (' The speed you entered is ', OldSpeed:10:2, ' feet/sec');

    SecondsPerFortnight := DaysPerFortnight * HoursPerDay * SecsPerHour;
    FurlongsPerFoot := (1.0 / FeetPerMile) * FurlongsPerMile;
    NewSpeed := OldSpeed * SecondsPerFortnight * FurlongsPerFoot;

    writeln;
    write (' That speed is equivalent to ', NewSpeed:12:2);
    writeln (' furlongs/fortnight')
end. (* program FurlongsPerFortnight *)



=>
 This program converts a speed given in feet/sec to furlongs/fortnight.
 Enter a speed in feet/sec
□1100.0

 The speed you entered is     1100.00 feet/sec

 That speed is equivalent to     49920.00 furlongs/fortnight
```

The speed in furlongs per fortnight could have been computed in a single assignment statement combining all five constants, but the intermediate variables make the program clearer by showing the two basic conversion factors that are required. It would also have been possible to declare **SecondsPerFortnight** as a constant by computing the value beforehand. If you made a mistake in your calculations, the program would compile and execute but your answer would be wrong. You would never find the bug unless you went back and did the calculation again. By using recognizable constants, you can avoid this type of problem. For example, if you declared **SecsPerHour** to be 60.0 by mistake, you would at least have a chance of noticing and correcting your error.

Summary

The **readln** statement is used to assign values to variables during program execution. The default input device is usually the user's terminal, so a message is written to the user requesting data, and then the **readln** statement is used to retrieve the data and assign it to the variables. The input data must be assignment compatible with the data type of the variables in the read-argument list. If more than one numerical value is read by a **readln** statement, the numbers entered from the terminal must be separated by at least one blank.

The arithmetic operators enable you to translate algebraic expressions into Pascal expressions. The arithmetic operators are +, −, *, /, **div**, and **mod**. The arguments of **div** and **mod** must be integers, and the result of **div** or **mod** is always an integer value. The result when the division operator (/) is used is always a real value. The arithmetic operators are evaluated in the following order of precedence:

( )	parentheses
−	unary negator (negative sign of a number)
* / **div mod**	multiplication and division
+ −	addition and subtraction

When two operators have the same precedence, they are evaluated from left to right.

An expression in Pascal is any valid combination of factors and terms. Examples of expressions are the right-hand side of an assignment statement and the arguments in a **write** statement.

The **write** statement is similar to the **writeln** statement except that the **write** statement does not advance the write pointer to the beginning of the next line. It leaves the write pointer positioned after the last character that was written. The **write** statement can be used to print long messages and to allow the computer to evaluate statements in the middle of printing a line.

Exercises

PASCAL

1. What is the value of each of the expressions below? Be sure to indicate whether the value is real or integer.
   - (a) (3.0 + 4.2) / 6.1
   - (b) sqr(2.2) + 5.0
   - (c) abs(3.2) * abs(−1.0)
   - (d) 10 div 3
   - (e) 10 mod 3
   - (f) round(5.4 * 0.5)
   - (g) trunc(5.4 / 1.5)
   - (h) sin(3.1415)
   - (i) sqr(sqrt(5 * 5))
   - (j) 3 * 3 * 3 * 3
   - (k) ln(2.71828)
   - (l) 4 + −3

2. Correct the syntax mistakes in the following statements. **Amps, EquivRest,** and **Voltage** have been declared as real variables, **Number** has been declared as an integer variable, and **DaysPerWeek** has been declared as a constant.

   (a) `readln (5.3 + Amps);`

   (b) `writeln (The answer is ',Amps:6:2, 'amps');`

   (c) `readln (Voltage Amps);`

   (d) `readln (Amps, Number, DaysPerWeek, Voltage);`

   (e) `writeln (' The answer for circuit number',Number, 'is' Amps, ' when the voltage is', Voltage, ' volts, and the' equivalent resistance is,' EquivResist, 'ohms');`

3. Rewrite each expression below using parentheses to show how it would be evaluated using the precedence order. What is the value of each expression?

   (a) 3.0 + 5.5 * 6.0 / 7.2

   (b) −3 * −4

   (c) 3.0 / 3.0 / 3.0 / 3.0

4. Translate each of the equations below into a Pascal assignment statement. To show your understanding of the precedence rules, use as few parentheses as possible.

   (a) $F = ma$

   (b) $G = \dfrac{X_1 + X_2}{X_1 X_2}$

   (c) $\theta = \tan^{-1}(Z/B)$

   (d) $y = ax^2 + bx + c$

   (e) $w = \dfrac{\cos(2\theta)}{x_1 y_2}$

   (f) $z = |x^2 - y^2|^3$

5. What is the output from each of the following program fragments?

   (a) ```
   Temp := 4.23;
   Result := Temp / 3.2 * 5.0;
   writeln (' The result is ', Result:5:2)
   ```

   (b) ```
   Number := 6 mod sqr(2);
   Next := 4 div Number;
   writeln (' The number is ', Number:3, ' and Next is ', Next:4)
   ```

   (c) ```
   Num1 := 2.0 * (ln (3.2));
   Num2 := ln (sqr (3.2));
   Num1and2 := Num1 + Num2;
   writeln (' Num1 = ', Num1:6:2, ' Num2 = ', Num2:6:2, ' and Num1and2 = ',
       Num1and2:6:2)
   ```

6. Compute the values that would be stored in **Counter, Remainder, Next,** and **Last** after execution of the statements using **div** and **mod** on page 62 if **Number** is assigned the value 8.
   ```

APPLICATION

7. Convert the following units:
   (a) 1 centimeter to miles
   (b) 1 pound to dynes
   (c) 1 gigawatt to picowatts
   (d) 1 gram per centimeter^2 to kilograms per meter^2
   (e) 1 horsepower (hp) to Btus per hour
   (f) 1 liter to cubic inches
   (g) 1 foot-pound to joules

8. Convert each of the units below into standard English and CGS units:
   (a) hectare   (e) link   (i) minim
   (b) dram   (f) carat   (j) rod
   (c) gill   (g) barrel (k) therm
   (d) knot   (h) league (l) langley

Problems

PASCAL

1. Write a program that reads one character, three integer numbers, and three real numbers interactively. After the values have been read,
   (a) print the character in default format.
   (b) print the first integer in default format, the second in a field width of 20, and the third in a field width of 8.
   (c) print the first real number in default format, the second in exponential format in a field width of 10, and the third in a field width of 10 with 2 digits after the decimal point.

   Each time you print the value of a variable, include an explanation of the format for the variable. Right-justify the character strings that you use so that they all end in the same column.

2. Write a program that asks for the year that a person was born and then computes and reports the person's age in years.

3. Compute the volumes of two spheres, one with a radius of 0.5 cm and one with a radius of 1.5 cm. Read the values for the radii from the terminal. Declare a constant for π. Report the volumes for both spheres. Remember to label your output.

4. A computer company rents access to an information service to owners of small computers. The company charges each user a hookup fee of $10.00 per month

plus a fee of $0.50 for each minute that the service is used. Write a program that accepts as input the number of minutes that a customer has used in one month and computes the customer's bill for the month.

5. A ball is tossed vertically into the air. The height of the ball at time t is given by the quadratic equation

$$y = y_0 + v_0 t + \frac{1}{2} g t^2$$

where t is the time in seconds that the ball has been in the air, y_0 is the initial height in meters, v_0 is the initial velocity in the vertical direction in meters per second, and g is the acceleration due to gravity, which is a constant equal to -9.81 m/s².

Write a program that requests the initial height, the initial velocity, and the time, and then computes and reports the height of the ball. Test your program for several different time values.

6. A baseball is tossed straight up with an initial velocity of 15.0 m/s. The position and speed of the baseball after t seconds have elapsed are given by the following equations:

$$v = v_0 + at$$

$$d = v_0 t + \frac{1}{2} a t^2$$

where d = distance in meters
v_0 = initial velocity in meters per second
v = velocity in meters per second
a = acceleration of gravity = -9.80665 m/s²
t = time in seconds

Compute the speed and position of the baseball at two different time intervals. Read the values for the elapsed time from the terminal.

APPLICATION

7. The thermometer outside the local bank now reports the temperature only in degrees Celsius. Write a program to convert degrees Celsius into degrees Fahrenheit. The conversion formula is

$$F = 1.8C + 32.0$$

where F is the temperature in Fahrenheit and C is the temperature in Celsius. Declare the conversion factors as constants. Declare identifiers for the degrees in Celsius and the degrees in Fahrenheit as variables. Calculate and print out the Fahrenheit temperature in the following format:

   xxx.xx degrees Celsius equals yyy.yy degrees Fahrenheit

8. Write a program that computes the fuel consumption in gallons per mile for a full aircraft. You are given the seating capacity and the fuel consumption in passenger-miles per gallon. A passenger-mile per gallon is a unit used by the

airlines. Five passenger-miles per gallon would mean that five passengers are flown one mile on a gallon of fuel, or one passenger is flown five miles on a gallon of fuel, or any equivalent combination.

	passenger-miles/gallon	seating capacity
747-jet	22.0	360

9. In a Pascal program, convert the speed of light from meters per second to miles per hour. Declare the following constants:
   (a) the speed of light in meters per second
   (b) the number of miles in a meter
   (c) the number of seconds in an hour

   Declare a variable for the speed of light in miles per hour. In the main body of your program, use an assignment statement to compute the variable's value from the constants, and then write the value at the terminal.

10. A 50-g lead weight is suspended by a string. The tension in a string is given by
    $$F = mg$$
    where m is the mass in kilograms and g is the acceleration of gravity. Write a program that computes and reports the force in newtons.

11. The standard metric unit for pressure, a pascal, is equal to 1 kg/m². Standard atmospheric pressure is 14.696 lb/in². Compute and report the standard atmospheric pressure in pascals.

12. A car is driven at a constant speed for 30 minutes. Write a program that reads the speed interactively and computes and reports the distance traveled in miles and in kilometers. Use constants to define the conversion factor between miles and meters. Test your program using a speed of 30 mi/h. Test it again for a speed of 55 mi/h.

13. Compare the distance in kilometers traveled in one hour by two different rockets. One has a speed of 1234.0 mi/s, and the other has a speed of 2345.0 mi/s.

14. Convert the density of two metals from English units to SI units and to CGS units. The English units are pounds per foot³, the SI units are kilograms per meter³, and the CGS units are grams per centimeter³. Declare constants for the conversion factors from kilograms to pounds and feet to meters. Read the densities of gold (687.0 lb/ft³) and lead (1204.8 lb/ft³) from the terminal. Report the density for each metal in English, CGS, and SI units.

15. A clogged rectangular tank with a leaky faucet is 6 ft long, 22 in wide, and 11 in deep. The rate that the faucet drips is read in gallons per minute from the

terminal. Compute how many hours it will take for the tank to fill to overflowing. You need to find out the number of gallons in a cubic inch in order to solve the problem.

16. For Problem 15, also compute how much the water in the tank will weigh when the tank is full. Give the weight of the water in pounds and its mass in kilograms.

17. The cost of producing a megawatt-hour of electricity from an oil-fired power plant is the product of the fuel cost per unit heat content ($/megaBtu) and the amount of heat necessary to produce a megawatt-hour of electricity (megaBtu/MWh). Compute the cost of generating electricity in $/MWh. Test your program using a cost of $2.75/megaBtu and 11.543 megaBtu/MWh. Convert the cost to cents per kilowatt-hour.

18. The time-varying voltage, V, in a circuit with an impedance in ohms, Z, is a function of the current, I, in amperes, the time, t, in seconds, and the frequency, f, in hertz (cycles per second). The formula is

$$V = ZI \sin(2\pi ft)$$

Read the values for I, Z, and f from the terminal. Compute the voltage for $t = 0.0$ ms, $t = 0.4$ ms, and $t = 0.8$ ms. Test your program for $I = 20.0$ A, $Z = 0.25$ Ω, and $f = 60$ Hz.

19. In an assembly process a shaft must be force-fitted into a bore. The force required is given by

$$F = 2\pi r L f P$$

where r = radius of the shaft in inches

L = length of the bore in inches

f = coefficient of friction (dimensionless)

P = radial pressure in pounds per inch^2 (psi)

Compute the force in tons and in newtons for $r = 4.2$ in, $L = 5.0$, $f = 0.10$, and $P = 9000$ psi.

20. The arrival time of a rocket can be computed from its blast-off time, its speed, and the distance to its destination. The blast-off time is given in hours, minutes (min), and seconds based on a twenty-four-hour clock, the speed in kilometers per second, and the distance in kilometers. Test your program using the following data:

Blast-off time: 05 h 34 min 02 s

Rocket speed: 1.0 km/s

Distance: 12654.0 km

21. The force of attraction between two bodies is given by

$$F = \frac{G m_1 m_2}{d^2}$$

where $G = 6.67 \times 10^{-8}$ cm^3/g-s^2, the gravitational constant

m_1 = mass of the first body in grams
m_2 = mass of the second body in grams
d = distance between the bodies in centimeters

Compute the force between the sun and the earth in newtons. The earth has a mass of 6×10^{24} kg, and the sun has a mass 332,830 times larger than the earth's mass. The sun and the earth are separated by a maximum distance of 94,555,000 mi.

22. You have measured the length of a rocket traveling at 1/2 the speed of light. According to the laws of relatively, the length of the rocket measured in your frame of reference is given by

$$L = L_0 \left[1.0 - \left(\frac{v}{c}\right)^2 \right]^{1/2}$$

where L = length measured in your reference frame
L_0 = length measured in the rocket's reference frame
v = velocity of the rocket in meters per second^2
c = speed of light in meters per second^2

Write a program that computes the length of the rocket in the rocket's reference frame.

23. When a train travels over a straight section of track, it exerts a downward force on the rails. When it rounds a level curve, it also exerts a horizontal force outward on the rails. Both forces must be considered in the design of the track. The downward force is equivalent to the weight of the train. The horizontal force is called the centrifugal force and is a function of three parameters: the weight of the train, the speed of the train as it rounds the curve, and the radius of the curve. The equation to compute the horizontal force in pounds is

$$F = \frac{mv^2}{r}$$

where m = mass of the train in slugs
v = speed of the train in feet per second
r = radius of the curve in feet

Write a Pascal program to compute and report the centrifugal force if the weight of the train is 405.7 tons (t), the speed of the train is 30.5 mi/h, and the radius of the curve is 2005.33 ft.

CHAPTER 4

Procedures and Program Development

In this chapter the entire process of developing a program will be covered, from defining the problem to producing a program that runs correctly and that can be used by other people. A type of subprogram called a *procedure* will be introduced. Procedures, when used in conjunction with top-down design, make writing correct programs easier. Also covered are debugging strategies to help you find the mistakes when your program does not work. Finally, program documentation, which is important both to the users of a program and to programmers, will be discussed.

CASE STUDY 4
Computing the current in a circuit connected in series

Resistors in a circuit can be connected either in parallel, as they were in Chapter 3, or in series. By comparing Figure 4.1 with Figure 3.2, you can see the difference between circuits with resistors connected in parallel and those with resistors connected in series. When the resistors are connected in series, the equivalent resistance is determined in a different way from when they are connected in parallel. The equation for determining the equivalent resistance for a series connection is

$$R_{eq} = R_1 + R_2 \tag{4.1}$$

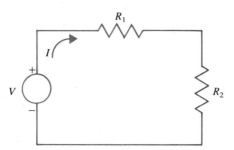

FIGURE 4.1
Resistors in a circuit connected in series.

And the current that flows when a voltage V is applied to a circuit with resistors connected in series is given by

$$I = \frac{V}{R_{eq}} \qquad (4.2)$$

You are given three circuits, each with different resistors and voltages. The resistors in these circuits may be connected either in series or in parallel. Write a Pascal program to compute the current for each circuit when the resistors are connected in series and when the resistors are connected in parallel. The data is to be supplied interactively.

4.1 Algorithm development

Algorithm development is necessary to write a program in any computer language, but the language used in an algorithm often reflects the language in which the program will be written. You are now familiar with how to use top-down design to write algorithms for simple problems and how to develop a Pascal program from an algorithm. For most of the examples used so far, it has not been difficult to state what the problem is and to outline the steps for its solution. Often, stating the problem clearly and stating the sequence of steps involved in solving it are the most difficult parts of programming. If an algorithm has been carefully developed, the translation into Pascal or another computer language is usually easy.

Before writing a computer program in any language, you need to have a good sense of what the problem is and how you are going to solve it. If you just start writing computer code without any strategy, your program will take much longer to write and will have more bugs. The bugs will be harder to find, and the code will not make any sense to you a week after you have written it.

When you have a problem to solve on a computer, restate the problem in your own words and be sure that you understand what the purpose of the program is. Begin by stating what must be done before you decide how to do it. You should note specific requirements of the problem. If the program is supposed to calculate values for positive and negative input values, the algorithm you develop must work for both positive and negative numbers. Discuss the problem with at least one other person to be sure that you have not misunderstood the question or overlooked an essential part of it.

DEBUGGING HINT

Always write your algorithms down on paper first. It is much easier to revise an algorithm written in English on paper than to revise a program written in Pascal and stored in a computer file.

For the case study you can restate the problem as follows: Write a program that computes and reports the current when a voltage is applied to each of three circuits containing two resistors. Compute the current for both the series and the parallel connection in each circuit.

Once you have a clear understanding of the statement of the problem, you can begin to list the tasks that must be accomplished in order to reach a solution to the problem. The tasks must be in an ordered sequence; that is, the order in which the tasks are to be performed must be clear. There must be a single beginning point for the algorithm. As you will see later, some Pascal constructions allow alternative series of code to be carried out under differing conditions, but each pathway must lead to the end of the program.

Most people use an informal approach to outlining a problem before beginning to write a computer program. You will develop your own style that will be a mixture of English, math, and whatever programming language you are using. Development languages that combine these elements are called *pseudocodes*. Pseudocodes incorporate features such as the assignment operator and operations such as read, write, square, and square root that are common to most computer languages. Values are referred to by the identifiers that they will have within the program. When an algorithm is written in pseudocode, the structure of the algorithm is similar to the structure that the program will have, but the syntax is not as exacting.

Whereas informal pseudocodes are normally used, formal pseudocodes with prescribed syntax have been developed for use when a group of people work together on the development of an algorithm. When a large problem is being worked on, often one person writes an algorithm that someone else translates into a computer language. Even if you do not use formal pseudocode when you write a program outline, the outline should be clear enough that someone else could take your algorithm and turn it into a computer program.

Looking at the problem for the case study, you can write an outline of what must be done to solve it.

4.1 Algorithm development

ALGORITHM 4.1   First iteration of algorithm for Case Study 4

begin
1. Write a message to the user asking for the values of **Voltage, Resist1, Resist2** for each circuit
2. Read **Voltage, Resist1, Resist2** for each circuit
3. Compute **ParallelCurrent,** the current for each circuit with the resistors in parallel
4. Compute **SeriesCurrent,** the current for each circuit with the resistors in series
5. Report **ParallelCurrent, SeriesCurrent** for each circuit

end

At this point you must make a decision: Will you create a separate variable for the voltages and resistances in each circuit (**Voltage1, Voltage2, Voltage3, Resist1Circ1, Resist2Circ1,** etc.), or will you deal with the circuits one at a time and use the same variable names for each circuit (**Voltage, Resist1, Resist2**)? The second solution is simpler, and foresight shows it to be the better solution. To implement this solution, you must rearrange the steps of Algorithm 4.1 so that the steps to solve each circuit are grouped together. In this strategy, the variables initially represent the values for the first circuit. The computations using these values are performed, the answer is reported, and then the variables can be used to represent the values for the next circuit.

ALGORITHM 4.2   Second iteration of algorithm for AnalyzeThreeCircuits

begin algorithm
1. Solve Circuit 1
    begin
    a. Write a message asking the user to enter **Voltage, Resist1, Resist2**
    b. Read **Voltage, Resist1, Resist2**
    c. Compute **ParallelCurrent** for the resistors in parallel
    d. Compute **SeriesCurrent** for the resistors in series
    e. Report **ParallelCurrent, SeriesCurrent**
    end of Circuit 1
2. Solve Circuit 2
    begin
    a. Write a message asking the user to enter **Voltage, Resist1, Resist2**

b. Read **Voltage, Resist1, Resist2**
   c. Compute **ParallelCurrent** for the resistors in parallel
   d. Compute **SeriesCurrent** for the resistors in series
   e. Report **ParallelCurrent, SeriesCurrent**
   end of Circuit 2
3. Solve Circuit 3
   begin
   a. Write a message asking the user to enter **Voltage, Resist1, Resist2**
   b. Read **Voltage, Resist1, Resist2**
   c. Compute **ParallelCurrent** for the resistors in parallel
   d. Compute **SeriesCurrent** for the resistors in series
   e. Report **ParallelCurrent, SeriesCurrent**
   end of Circuit 3
end of algorithm

The problem has been refined into three sets of tasks that each have the same five steps. Step c, computing the current for a parallel connection, was solved in Chapter 3. For Step d, computing the current for a series connection, Equations 4.1 and 4.2 can be combined into the algorithmic statement

EquivResist ← **Resist1** + **Resist2**
SeriesCurrent ← **Voltage** / **EquivResist**

The complete algorithm to solve the case study is now

ALGORITHM 4.3   Completed algorithm for AnalyzeThreeCircuits

begin algorithm
1. Solve Circuit 1
   begin
   a. Write a message asking the user to enter **Voltage, Resist1, Resist2**
   b. Read **Voltage, Resist1, Resist2**
   c. Compute **ParallelCurrent** for the resistors in parallel
      i. **EquivResist** ← (**Resist1** ∗ **Resist2**) / (**Resist1** + **Resist2**)
      ii. **ParallelCurrent** ← **Voltage** / **EquivResist**
   d. Compute **SeriesCurrent** for the resistors in series
      i. **EquivResist** ← **Resist1** + **Resist2**
      ii. **SeriesCurrent** ← **Voltage** / **EquivResist**
   e. Report **ParallelCurrent, SeriesCurrent**
   end of Circuit 1
2. Repeat Steps a through e for Circuits 2 and 3
end of algorithm

All of the steps in the algorithm could now be written in Pascal, but the solution to the case study will not be completed until the end of the chapter, after procedures have been covered.

4.2 Walkthroughs

Once you have developed an algorithm that you think will solve a problem, you can test the algorithm by making a chart of the variables in your solution and executing the steps of the algorithm as if you were the computer. You can keep track of the values of the variables as they change to see if the program will do what you intended it to do. This testing process is called a *walkthrough*. When you do a walkthrough, you must be extremely careful to do what the computer *will* do, not what you think it *should* do.

You could perform a walkthrough of the algorithm for the case study to assure yourself that the values for each circuit are never required after they have been reported. However, instead of using the case study for the demonstration of a walkthrough, a walkthrough will be performed on the common but tricky task of exchanging variables. The walkthrough below demonstrates how walkthroughs can help you write correct algorithms. The exchange algorithm developed in this section will be used in the case study of Chapter 5.

In an exchange, if the memory locations originally are

X	Y
4	8

then after the exchange the memory locations should be

X	Y
8	4

To make this exchange, you might write the algorithm

ALGORITHM 4.4 Initial algorithm for exchanging two variables

begin
    1. **X** ← **Y**
    2. **Y** ← **X**
end

If the initial value of **X** is 4 and the initial value of **Y** is 8, then the first entry in the walkthrough chart is

	X	Y
Initial values	4	8

When Step 1 is executed, the value stored in **Y** is assigned to the variable **X**. Now the chart looks like

	X	Y
Initial values	4	8
X←Y	8	8

When Step 2 is executed, the value stored in **X** is assigned to the variable **Y**:

	X	Y
Initial values	4	8
X←Y	8	8
Y←X	8	8

When both steps have completed execution, the memory locations are

```
   X          Y
 [ 8 ]      [ 8 ]
```

The values have not been exchanged! The bug occurs in the first assignment statement when the value 4 stored in **X** is replaced by the value 8 stored in **Y**. Once the value 4 has been replaced by the value 8, the value 4 cannot be recovered. To save the value 4 requires that an intermediate variable be used. With an intermediate variable to save the value in **X**, the new algorithm is

ALGORITHM 4.5 Second iteration for the exchange algorithm

begin
   1. **Save ← X**
   2. **X ← Y**
   3. **Y ← Save**
end

If the storage locations at the beginning of the program are

X	Y	Save
4	8	

then the first entry in the walkthrough chart is

	X	Y	Save
Initial values	4	8	—

When Step 1 is executed, the chart becomes

	X	Y	Save
Initial values	4	8	—
Save←X	4	8	4

After Step 2 is executed, the chart is

	X	Y	Save
Initial values	4	8	—
Save←X	4	8	4
X←Y	8	8	4

Finally, after Step 3 is executed, the chart is

	X	Y	Save
Initial values	4	8	—
Save←X	4	8	4
X←Y	8	8	4
Y←Save	8	4	4

The final values for the memory locations are

X	Y	Save
4	8	4

The original values stored in **X** and **Y** have been exchanged.

Performing a walkthrough can help prevent you from writing an algorithm that is logically incorrect. Walkthroughs can also give you insights into solving a problem.

4.3 Procedures

Looking at the algorithm for the case study, you can see that translating the algorithm would involve writing the same set of instructions three times, once for each circuit. To avoid having to repeat the code for each circuit, you can create a Pascal *procedure*. Procedures allow you to write a set of instructions once and use it many times in the same program.

A procedure defines a sequence of instructions that performs a well-defined subtask (for example, reading in data, finding the mean, or computing the electric bill). The series of statements that performs the task is given an identifying name. Instead of retyping the same instructions each time you want to perform the task, all you have to do is write the identifying name. Writing the procedure name is the same as writing all of the statements in the body of the procedure. The same sequence of statements is executed every time you use the procedure name.

You already know four standard Pascal procedures: **read, readln, write,** and **writeln.** When these procedures are used, they each perform a task—reading or writing data. If these procedures did not exist, you would have to learn the details of how data is read and written in order to instruct the computer to read your input or write your output. Remember that even though a standard procedure looks like a single statement, it is actually a sequence of statements. Each time you use the procedure **read,** the same set of instructions is executed, causing the data to be read from the terminal and to be assigned to the appropriate variable names.

Procedures are used to organize, clarify, and simplify programs. If you think of your program as analogous to an English essay, procedures are the paragraphs that organize your ideas. An essay written in several well-thought-out paragraphs is much easier to read than one written in a single paragraph. In Pascal each separate subtask, or paragraph, can be defined and given a name.

Declaring Procedures

The procedure declaration is a definition; it equates the identifier of the procedure with a sequence of statements. Because it is a definition, it is placed in the declarative part of the program prior to the main program. The structure of a Pascal procedure is the same as the structure of a Pascal program:

heading

declarative part

body

A simplified syntax diagram for a procedure heading is

Procedure is a reserved word in Pascal. The *identifier* is the name that the instructions in the body of the procedure will be called. The rules for creating identifier names for procedures are the same as the rules for creating other identifier names. An example of a procedure heading is

```
procedure ReadVoltageAndResist;
```

Like identifier names for variables and programs, the identifier name for a procedure should be self-descriptive. For example, **ReadData, ComputeMean,** and **CheckForYes** are all self-descriptive; **House, Ronald,** and **Joe** are not. Suppose the standard procedures for **read, readln, write,** and **writeln** had been called **anne, jane, john,** and **louise.** Every time you wanted to read or write, you would have to remember which procedure performed which task.

The *declarative part* of a procedure is the same as the declarative part of a program. It may contain constants and/or variables. The constants and variables that are defined in the declarative part of the procedure are available for use only within the body of the procedure. If a variable is declared in the main program, it can be used within a procedure, but should not be declared again in the procedure. This topic will be covered in more detail in Chapter 6.

The *body* of a procedure contains the statements that are executed when the procedure is invoked. The body is set off by a **begin-end** pair, just as the main body of a program is. A procedure ends in a semicolon instead of a period because the period is used only to indicate the end of the entire program.

The procedure definition consists of the entire procedure, including the heading, the declarative part, the **begin,** the sequence of statements, and the **end;**:

```
procedure ReadVoltageAndResist;
   (*    This procedure reads values for the voltage and two resistances    *)
   (*    for each circuit.                                                  *)

begin  (* procedure ReadVoltageAndResist *)
  writeln;
  writeln (' Enter a real number for the voltage in volts');
  readln (Voltage);
```

```
    writeln (' Enter a real number for the resistance in ohms of the first',
             ' resistor');
    readln (Resist1);

    writeln (' Enter a real number for the resistance in ohms of the second',
             ' resistor');
    readln (Resist2)
end; (* procedure ReadVoltageAndResist *)
```

Because a procedure is a definition, it must be included in the declarative part of the program. Procedure definitions appear after the **const** and **var** declarations in the main program. The declarative part of a program or a procedure may also contain other kinds of Pascal declarations that will be covered later in the text.

Using procedures

Once a procedure has been declared, its name can be used as an executable statement. During program execution, if a procedure identifier is encountered as an executable statement, the procedure is *invoked* or *called*, causing the statements in the body of the procedure to be executed. Just as using the standard procedure **writeln** causes a particular sequence of instructions to be executed, using the identifier for your own procedure causes your sequence of instructions to be executed. When the statement

```
ReadVoltageAndResist
```

is encountered during program execution, the statements in the main body of **ReadVoltageAndResist** are executed. Program 4.1, **ShowProcedure,** shows the placement of the declaration of **ReadVoltageAndResist** and its invocation.

DEBUGGING HINT

Do not call your procedure **read.** If you use the identifier **read** as the name of your procedure, you will not be able to use the standard Pascal procedure **read.** Because **read** is a standard Pascal word, not a reserved word, the compiler will allow you to redefine **read.** Since Pascal will allow you to redefine any standard Pascal identifier, be careful when choosing names for procedures.

PROGRAM 4.1 ShowProcedure

```
program ShowProcedure (input, output);    (* PROGRAM HEADING *)
(*      This program shows the definition and placement of a procedure    *)
(*      within a program and its invocation.                              *)

(* BEGIN DECLARATIVE PART OF PROGRAM *)

var    Voltage : real;      (* user supplied voltage in volts. *)
       Resist1 : real;      (* user supplied resistance 1 in ohms. *)
       Resist2 : real;      (* user supplied resistance 2 in ohms. *)


   procedure ReadVoltageAndResist;
   (*      ReadVoltageAndResist reads values for the voltage and the two   *)
   (*      resistances for each circuit.                                   *)

   begin (* procedure ReadVoltageAndResist *)
      writeln;
      writeln (' Enter a real number for the voltage in volts');
      readln (Voltage);

      writeln (' Enter a real number for the resistance in ohms',
               ' of the first resistor');
      readln (Resist1);

      writeln (' Enter a real number for the resistance in ohms',
               ' of the second resistor');
      readln (Resist2)
   end; (* procedure ReadVoltageAndResist *)

(* END DECLARATIVE PART OF PROGRAM *)


(*-------------------------BODY OF MAIN PROGRAM ------------------------*)
begin (* main program *)
   ReadVoltageAndResist            (* executable statement to invoke procedure *)
end. (* program ShowProcedure *)

=>

 Enter a real number for the voltage in volts
□10.0
 Enter a real number for the resistance in ohms of the first resistor
□3.0
 Enter a real number for the resistance in ohms of the second resistor
□3.0
```

To determine what a program does, begin by reading the main body of the program. This is equivalent to reading the main steps in an outline before reading the refinements of each step. A common mistake is to read a program from the top, assuming that the procedures are invoked in the order they are declared. Declaring procedures is no different from declaring variables. Procedures do not have to be invoked in the order in which they are declared. A procedure can be invoked once, many times, or not at all.

Program 4.2, **OnceTwice,** illustrates a procedure that writes one line. In the main program, the procedure is invoked twice, so the line is written twice.

PROGRAM 4.2   OnceTwice

```
program OnceTwice (input, output);
(*     Program OnceTwice defines a procedure and then invokes it twice.      *)

  procedure WriteOne;
  begin
    writeln (' One = 1')
  end; (* procedure WriteOne *)


(*------------------------------------------------------------------------*)
begin (* main program *)
  WriteOne;
  WriteOne
end. (* program OnceTwice *)



=>
 One = 1
 One = 1
```

> *DEBUGGING HINT*
>
> Always look at the main body of the program to find out what a program does. If a procedure is declared but not invoked, it will not be used. If you are debugging a program and cannot figure out why a procedure was not executed, be sure to check that the procedure name appears as an executable statement.

Program 4.3, **ParallelCircuit,** is the same as Program 3.4, **ComputeCurrent,** except that it uses procedures. Comparing the two programs, you can see that the program using procedures is longer than the original program. However, by going to the main body of Program **ParallelCircuit,** you can immediately determine the purpose of the program.

PROGRAM 4.3 ParallelCircuit

```pascal
program ParallelCircuit (input, output);
(*     ParallelCircuit computes the current that flows when a voltage   *)
(*     is applied across two resistors connected in parallel.  The      *)
(*     values for the voltage and resistance are read interactively.    *)


var    Resist1 : real;         (* user supplied resistance 1 in ohms. *)
       Resist2 : real;         (* user supplied resistance 2 in ohms. *)
       Voltage : real;         (* user supplied voltage in volts. *)
       EquivResist : real;     (* equivalent resistance in ohms -- *)
                               (* intermediate variable *)
       Current : real;         (* computed current in amps. *)



   procedure ReadVoltageAndResist;
   (*     ReadVoltageAndResist reads values for the voltage and the two   *)
   (*     resistances for each circuit.                                   *)



   begin
     writeln;
     writeln (' Enter a real number for the voltage in volts');
     readln (Voltage);

     writeln (' Enter a real number for the resistance in ohms of the first',
              ' resistor');
     readln (Resist1);

     writeln(' Enter a real number for the resistance in ohms of the second',
             ' resistor');
     readln (Resist2)
   end; (* procedure ReadVoltageAndResist *)



   procedure ComputeParallelCurrent;
   (*     This procedure computes the equivalent resistance and the       *)
   (*     current that flows in a circuit connected in parallel.          *)



   begin
     EquivResist := (Resist1 * Resist2) / (Resist1 + Resist2);   (* ohms *)
     Current := Voltage / EquivResist                            (* amps *)
   end; (* procedure ComputeParallelCurrent *)
```

```
procedure ReportCurrent;
(*      Procedure ReportCurrent reports the current.                        *)

begin
  writeln;
  writeln (' In a circuit with a voltage of', Voltage:10:2, ' volts');
  writeln (' and resistors of', Resist1:10:2, ' and', Resist2:10:2,
           ' ohms,');
  writeln (' the current is', Current:10:2, ' amps.')
end; (* procedure ReportCurrent *)



(*--------------------------------------------------------------------------*)
begin (* main program *)
  writeln (' This program computes the current that flows when a voltage');
  writeln (' is applied across two resistors connected in parallel.');

  ReadVoltageAndResist;
  ComputeParallelCurrent;
  ReportCurrent
end. (* program ParallelCircuit *)



=>
This program computes the current that flows when a voltage
is applied across two resistors connected in parallel.

Enter a real number for the voltage in volts
□10.0
Enter a real number for the resistance in ohms of the first resistor
□3.0
Enter a real number for the resistance in ohms of the second resistor
□3.0

In a circuit with a voltage of        10.00 volts
and resistors of         3.00 and      3.00 ohms,
the current is           6.67 amps.
```

4.4 Block structure

Pascal programs are organized in a *block* structure. Programs and procedures are both examples of blocks. A block has the following elements:

a. heading (a program or procedure statement)

b. a sequence of declarative statements

c. a **begin**

d. a sequence of executable statements

e. an **end**

The declarative part of a block may include the definitions for one or more other blocks, so a procedure block appears within the declarative part of the program block. The main program is called the *outer block,* and the procedures are called *inner blocks.* For beginning programmers, the most common block structure is a main block that contains several procedure blocks all defined in the declarative part of the main block. This structure can be visualized as illustrated in Figure 4.2.

To follow the statements that will be executed, go to the main program first. Do not be distracted by the variable and procedure declarations. If the main program invokes a procedure, go to the body of that procedure. If that procedure invokes another procedure, go to the body of that procedure. When you reach the **end;** of a procedure, return to the body of the program or procedure from which the procedure you are reading was invoked and proceed to the next statement. Eventually, you will return to the outermost block and reach the **end.** of the program.

When procedures are declared as separate blocks within a main block, as illustrated in Figure 4.2, a procedure can be invoked in any procedure block that comes after it. That is, a procedure can be invoked any time after it has been declared. Procedure **AnalyzeACircuit** in the solution to the case study at the end of the chapter is an example of a procedure that invokes other procedures at the same level.

```
program Main (input, output);
const MainConst = 3.45;
var MainVar : integer;
    procedure A;
    var AVar : integer;
    begin
       (* A body *)
    end; (* procedure A *)

    procedure B;
    var BVar : integer;
    begin
       (* B body *)
    end; (* procedure B *)
begin (* main program *)
    A;  (* main body *)
    B
end. (* program Main *)
```

FIGURE 4.2
Main block with procedure blocks at the same level.

The variables that are declared immediately after your program heading are called *global variables*. Variables that are declared within a procedure are called *local variables*. The value stored in a global variable can be accessed by the main program and by any procedure within the program. A variable can never be accessed in a block that is at a higher level than the level at which it is declared, so the variables that are local to a procedure cannot be accessed by the main program. An easier way to think of this is that you can look out of a block, but you cannot look in.

Because both programs and procedures have declarative parts, the same variable name may be declared more than once in a program. If the same variable name is declared both in the main program and in a procedure, the local variable takes precedence over the global variable inside that procedure. To avoid conflicts, Pascal uses the following rules:

1. A variable declared at the beginning of a block is accessible to all executable statements that are part of the block, including statements belonging to inner blocks, but to no others.

2. The innermost or most local declaration of a variable is always used.

The rules above are said to define the *scope* of a variable—that is, the block or blocks that can access the variable. A procedure declaration also has a scope. The rules for the scope of a procedure are essentially the same as for variables:

1. A procedure declared within a block is accessible to all executable statements that are part of the block and that follow the procedure declaration.

2. The innermost or most local declaration of a procedure is always used.

Because the declarative part of a block can include the declaration of another procedure, one block can be *nested* inside another. A procedure may be an inner block relative to the main program and an outer block relative to another procedure. A nested block structure is illustrated in Figure 4.3. When procedures are nested, the rules for the scope of the identifiers must be followed closely. Procedure **SubA** is local to **ProcedureA,** so the main program cannot call **SubA** directly.

The rules for name precedence for procedures and variables were developed so that several programmers could work on different parts of a large program without having to worry about using conflicting identifier names. If you are writing short programs, it is best not to nest your procedures. With nested procedures, it is easy to become confused about which procedures can call which other procedures. Defining all of your procedures at the level of the main program makes the program less confusing.

```
program Main (input, output);
const MainConst = 3.45;
var MainVar : integer;
   procedure A;
   var AVar : integer;
      procedure SubA;
      var SubAVar : integer;
      begin
         (* SubA body *)
      end; (* procedure SubA *)
   begin
      SubA   (* body of A *)
   end; (* procedure A *)
begin (* main program *)
   A    (* main body *)
end. (* program Main *)
```

FIGURE 4.3
Main block with nested procedure blocks.

4.5 Program development

Once you have written an algorithm and have performed a walkthrough of the algorithm to check its logic, you can start translating the program into Pascal. The main program should be a restatement of the main steps of the algorithm. You can write the main program first by naming the major procedures that you will need in your program. Once you have written the main program, you can attack the subproblems one at a time as you write each procedure.

Because the algorithm for the case study involves repeating the same five steps for each circuit, you can write a procedure for each step and then invoke each procedure three times. Step a, writing a message to the user, and Step b, reading the values that are entered, can be combined in one procedure that you could call **ReadVoltageAndResist.** Step c can be implemented in a procedure called **ComputeParallelCurrent.** Step d can be implemented in a procedure called **ComputeSeriesCurrent,** and Step e can be implemented in a procedure called **ReportCurrents.**

DEBUGGING HINT

As a way of getting organized, you may find it helpful to write the program heading and variable declarations on one sheet of paper, the main program on another sheet, and each procedure on a separate sheet. Organizing your program in this way makes it easy to add or delete variables and procedures, to refer to any part of the program at any time, and to arrange the procedures in any desired order for entry into the computer.

PROGRAM 4.4 AnalyzeThreeCircuits1

```
program AnalyzeThreeCircuits1 (input, output);
(*      AnalyzeThreeCircuits1 is a program to test the design of an      *)
(*      algorithm to find the current that flows when a voltage is       *)
(*      applied across each of three circuits in which two resistors     *)
(*      may be connected either in series or in parallel.                *)

var     Voltage : real;                 (* user supplied voltage in volts. *)
        Resist1 : real;                 (* user supplied resistance 1 in ohms. *)
        Resist2 : real;                 (* user supplied resistance 2 in ohms. *)
        ParallelCurrent : real;         (* computed current in amps for parallel *)
                                        (* connection of resistors. *)
        SeriesCurrent : real;           (* computed current in amps for series *)
                                        (* connection of resistors. *)
        EquivResist : real;             (* intermediate variable for computing *)
                                        (* the equivalent resistance. *)


(* dummy procedures *)

   procedure ReadVoltageAndResist;
     begin end;

   procedure ComputeParallelCurrent;
     begin end;

   procedure ComputeSeriesCurrent;
     begin end;

   procedure ReportCurrents;
     begin end;


(*---------------------------------------------------------------------*)

begin (* main program *)
   ReadVoltageAndResist;            (* Circuit 1 *)
   ComputeParallelCurrent;
   ComputeSeriesCurrent;
   ReportCurrents;

   ReadVoltageAndResist;            (* Circuit 2 *)
   ComputeParallelCurrent;
   ComputeSeriesCurrent;
   ReportCurrents;

   ReadVoltageAndResist;            (* Circuit 3 *)
   ComputeParallelCurrent;
   ComputeSeriesCurrent;
   ReportCurrents
end.  (* program AnalyzeThreeCircuits1 *)
```

4.5 Program development

As you write a program using top-down design, you should always have a program that almost works. Because code is most easily debugged in small sections, it is a good idea to start by entering just the main program into the computer, rather than typing in the entire program all at once. But if you enter only the main program, the procedure names used in the main program will be unknown to the compiler. To compile the main program, you can include *dummy* procedures for each of the procedures named in the main program. A dummy procedure does not do anything. It consists of just a procedure heading and a **begin-end;** pair.

```
procedure ComputeSeriesCurrent;
begin
end;
```

Using dummy procedures, you can create a complete Pascal program that can be compiled and tested. Rarely does a program work the first time it is entered into the computer. Even when the logic is correct and the algorithm has been translated correctly, syntax errors, such as missing semicolons or parentheses, often prevent the program from being translated into machine language. Program 4.4, **AnalyzeThreeCircuits1,** does not perform any calculations, but it can be used to test the original design of the algorithm.

An alternative to including dummy procedures is to put comment brackets around the procedures that have not been entered so that they are ignored by the compiler. For example,

```
program AnalyzeThreeCircuits1 (input, output);
(*      AnalyzeThreeCircuits1 is a program to test the design of an    *)
(*      algorithm to find the current that flows when a voltage is     *)
(*      applied across each of three circuits in which two resistors   *)
(*      may be connected either in series or in parallel.              *)

var     Voltage : real;          (* user supplied voltage in volts. *)
        Resist1 : real;          (* user supplied resistance 1 in ohms. *)
        Resist2 : real;          (* user supplied resistance 2 in ohms. *)
        ParallelCurrent : real;  (* computed current in amps for parallel *)
                                 (* connection of resistors. *)
        SeriesCurrent : real;    (* computed current in amps for series *)
                                 (* connection of resistors. *)
        EquivResist : real;      (* intermediate variable for storing *)
                                 (* the equivalent resistance. *)

(*-----------------------------------------------------------------*)
begin (* main program *)
   (* ReadVoltageAndResist; *)       (* Circuit 1 *)
   (* ComputeParallelCurrent; *)
   (* ComputeSeriesCurrent; *)
   (* ReportCurrents; *)
```

```
      (* ReadVoltageAndResist; *)        (* Circuit 2 *)
      (* ComputeParallelCurrent; *)
      (* ComputeSeriesCurrent; *)
      (* ReportCurrents; *)

      (* ReadVoltageAndResist; *)        (* Circuit 3 *)
      (* ComputeParallelCurrent; *)
      (* ComputeSeriesCurrent; *)
      (* ReportCurrents *)
end.    (* program AnalyzeThreeCircuits1 *)
```

Either by *commenting out* the procedures or statements that you know will not compile or by using dummy procedures, you can compile and test the rest of your program. The outline of the algorithm will remain visible so that you know what you have to do next.

The main program as it is written so far does not seem to do very much. However, in creating the main program above, many decisions have been made. For example, think about how the data will be stored. Because all of the procedures need to access the values stored in **Voltage, Resist1,** and **Resist2,** they must be declared as global variables. In Chapter 6 you will learn a better way to access values within procedures.

After the main program has been entered and compiled, you can begin to add the procedures one at a time. If the main program has compiled without error messages, you know that any error messages encountered are generated by the new code that you have entered. The main program produces no output, so you could type in the procedures for input and output first to give yourself proof that the program is actually doing something. For the output procedure to work before the rest of the program has been written, arbitrary values must be assigned to the output variables.

4.6 Debugging strategies

As your program develops from an algorithm to a completed program, it will go through several stages. Initially, bugs may exist in the algorithm itself. Using walkthroughs or other logical procedures, you must test the algorithm. Never translate an algorithm into code unless you are certain that the algorithm is correct; otherwise you can spend large amounts of time on a program that will never work. Once you have a correct algorithm, you can start translating it into Pascal. Usually, the first time you compile a program, there are typographical errors that must be corrected. If you use top-down design and add procedures one at a time, your debugging may be limited to correcting typing mistakes. However, most programmers eventually create bugs that are more complicated than missing semicolons. This section gives hints for preventing and finding such bugs.

The first rule of debugging is: If it isn't broken, don't fix it! Many programmers find it difficult to leave a program alone once it works. Improvements can always be made to make a program better in some way or another; you must know when to stop. If you cannot resist the temptation to improve a program, make a copy of the working program before you begin making changes. Frequently improvements result in programs that will not run and that are difficult to debug. If you have saved a copy, you can always go back to the original version.

A working program that is understandable, even if it is inefficient, is better than an efficient program that does not make sense, even if it works. When you are learning to program, you should concentrate on writing programs that solve problems correctly. For small programs, the gains from increasing efficiency are negligible, and the bugs from increasing efficiency can be horrendous.

Preventing bugs

Write procedures that do only one thing and do it clearly. A procedure that is more than a page long should be broken down into several procedures. You can tell if a procedure is doing one thing correctly more easily than you can tell if it is doing several things correctly. The topic of testing procedures for correctness is covered in Chapter 12.

Use constants where appropriate so that changes are easy to make. For example, **NumberOfStudents** could be a constant set to 10 for debugging and set to 1000 for runs. If you are using complicated algebraic expressions, break them down into manageable parts. It is better to use intermediate variables than to have an expression that goes on for three lines and is impossible to figure out. Know the operator precedence rules, but use parentheses, even if they are not necessary, if they make the expression clearer.

Indent your programs so that the structure of the program can be seen. The Pascal compiler will accept your program no matter where the blank spaces are, but the program is easier for you to read if the format is easy to follow. Programs 4.5 and 4.6 are two versions of the same program. In Program 4.5, none of the statements are indented, and there are no blank lines. Program 4.6 uses indentation and spacing to make the program easier to read. Both programs are interpreted the same way by the compiler.*

---

*Program 4.6 was created by running a program called **Prettify** using Program 4.5 as its input. Many mainframe computers have programs like **Prettify** that will reformat a Pascal program. These programs use the reserved words in the program as signals for indentation, so if your program has syntax errors, programs like **Prettify** will not run correctly.

PROGRAM 4.5 Demonstrate

```
program Demonstrate(input,output);
(* This program demonstrates printing integer and real variables. *)
const Pi=3.14; (* constant of data type real. *)
Two =2;(* constant of data type integer. *)
var RealVar:real;(* will take the value of the real constant. *)
IntVar:integer;(* will take the value of the integer constant. *)
begin (* main program *)
RealVar:=Pi;
IntVar:=Two;
(* write out variables *)
writeln(' The real variable is ',RealVar:5:2);
writeln(' The integer variable is ',IntVar:2)
end. (*program Demonstrate*)
```

```
=>
 The real variable is  3.14
 The integer variable is  2
```

PROGRAM 4.6 Demonstrate

```
program Demonstrate(input, output);

(* This program demonstrates printing integer and real variables. *)

  const  Pi = 3.14;  (* constant of data type real. *)
         Two = 2;    (* constant of data type integer. *)

  var  RealVar : real;  (* will take the value of the real constant. *)
       IntVar : integer;  (* will take the value of the integer constant. *)

  begin (* main program *)
    RealVar := Pi;
    IntVar := Two;

(* write out variables *)

    writeln(' The real variable is ', RealVar : 5 : 2);
    writeln(' The integer variable is ', IntVar : 2)
  end. (* program Demonstrate *)
```

```
=>
 The real variable is  3.14
 The integer variable is  2
```

Finding bugs

If the compiler generates many error messages, fix only the statements that you are certain are incorrect. As the error messages from the missing apostrophe in Chapter 2 illustrate, the compiler will frequently generate several error messages that can be traced back to a single typing mistake. Do not start adding parentheses or semicolons just because the error message said you might be missing one.

Be aware of the largest and smallest numbers that the computer you are using can store. On some systems, you are given no warning that you have exceeded these limits. Your only clue may be that your output makes no sense. This problem is demonstrated in Program 4.7, in which dividing by a very small number creates a number that is too large to store.

PROGRAM 4.7 SmallDivide

```
program SmallDivide (input, output);
(*     This program illustrates what happens if you divide by a small   *)
(*     number and create a number too large to store.                   *)

var    Distance : real;      (* to be assigned a relatively large number. *)
       Time : real;          (* to be assigned a small number. *)
       Velocity : real;      (* equal to distance divided by time. *)

begin
  Distance := 200000.0;
  Time := 1.0e-72;
  Velocity := Distance / Time;
  writeln (' Velocity = ', Velocity:10:2)
end. (* program SmallDivide *)
```

Compiler 1

=>

```
                                       pascal termination log
                                       ------ ----------- ---
*** error ***          program interrupt - exponent overflow exception
old psw:    03f5000c 6e00b778
r0-r7    e3404040 000107e8 fefefefe fefefefe  fefefefe fefefefe fefefefe 0003ddd0
r8-r15   00010f98 6e00b76e fefefefe fefefefe  fefefefe fefefefe 0000b6e8 00010e68

storage near interrupt:
00b740:  41f01680 459010f0 00206800 e1086000  16b86800 e1006000 16a86800 16b80590
00b760:  690011c8 4780112c 682016a8 05906920  11c84780 112c2d02 600016b0 41a01694
00b780:  18fa41a0 e0ec41b0 000e4100 0eab4590  10f00070 680016b0 05906900 11c84780
```

```
stack base = 0107e8, heap top = 03df58, runtime support data = 00b8b8, entry pt
= 00b6d0 (p0start)

interrupt from compiled code
interrupt at offset 000090 in main program

local vars: distance=  2.000000000000e+05 time    = 1.000000000000e-72 velocit
y= <undefined>
```

Compiler 2

```
=>
Run-time error 01, PC=004B
Program Aborted

Searching
10 lines

Run-time error position found. Press <ESC>
              Gothic 4.12 Program 4.7 SmallDivide
```

A similar error is created in Program 4.8, in which 1 is added to **maxint,** the largest integer that can be stored. In this case, the second compiler does not flag the number as being too large to store, so the program prints an answer even though it is nonsense.

PROGRAM 4.8 PlusOne

```
program PlusOne (input, output);
(*     This program shows what happens if you create a number that is    *)
(*     bigger than maxint, the largest integer the computer can store.   *)

var    Number : integer;


begin
  writeln (' maxint = ', maxint);
  Number := maxint + 1;
  writeln (' maxint + 1 = ', Number)
end. (* program PlusOne *)
```

Compiler 1

```
=>
 maxint =   2147483647
```

```
                                          pascal termination log
                                          ------ ----------- ---
*** error ***         program interrupt -  fixed point overflow exception
old psw:    03f50008 be008776

r0-r7    7fffffff 0000d7c0 fefefefe fefefefe  fefefefe fefefefe fefefefe 0003d
r8-r15   0000df60 9e008772 80000000 0000000c  fefefefe fefefefe 000086c8 0000d

storage near interrupt:

08740:  410009ab 459010f0 007058a0 16a8180a   459011a0 41b0000c 410000ab 459010f0
08760:  00584590 10f0005c 58a016a8 180a4590   11a05aa0 e10c50a0 16a841a0 169418fa
08780:  41a0e104 41b00005 410005ab 459010f0   007058a0 16a8180a 459011a0 41b0000c
```

Compiler 2

```
=>
 maxint = 32767
 maxint + 1 = -32768
```

Program 4.9 illustrates a run-time error that occurs not because there is a problem with the program, but because the data entered is of the wrong data type. Given what you have learned so far, the only solution to this problem is to run the program again, entering the right type of data. In Chapters 8 and 12, we will write an input routine to prevent this run-time error.

PROGRAM 4.9 BadData

```
program BadData (input, output);
(*     Program BadData shows what happens when the input data is of a    *)
(*     data type that does not agree with the declared data type of      *)
(*     the variable.                                                     *)

var    Number : integer;       (* integer variable into which the user will *)
                               (* attempt to enter a character. *)
```

```
begin
  writeln (' Enter a number');
  readln (Number);
  writeln (' The number is ', Number:10)
end. (* program BadData *)
```

Compiler 1

```
=>
 Enter an integer
□a
```

```
                                              pascal termination log
                                              ------ ----------- ---
*** error ***            file 'input   ', read format error - digit expected

program terminated at offset 0000b8 in main program

local vars: number  = <undefined>
```

Compiler 2

```
=>
 Enter an integer
□a

I/O error 10, PC=003D
Program Aborted

Searching
9 lines

Run-time error position found.  Press <ESC>
```

One of the most common causes of run-time errors is an undefined variable. If your compiler does not flag undefined variables, you must first recognize that your output is incorrect and deduce that the cause is an undefined variable. If your compiler does flag undefined variables, you will at least know which variable is undefined and where it is undefined. Once you know which variable is undefined, check to be sure that it was assigned a value. If you have assigned the variable a value, check to be sure that you have not used the same identifier name for both a global variable and a local variable. Program 4.10, **WrongScope,** has an undefined variable because

4.6 Debugging strategies

Number is declared as both a local and a global variable. The local variable **Number** is assigned a value in procedure **ReadANumber,** and the global variable **Number** is printed in the main program. Again, the second compiler does not flag the undefined variable and prints a nonsense answer.

PROGRAM 4.10 WrongScope

```
program WrongScope (input, output);
(*      Program WrongScope has an undefined variable in the main program  *)
(*      because Number is declared as both a local and a global variable. *)
(*      The local variable Number is assigned a value in ReadANumber, but *)
(*      it is the global variable Number that is printed.                 *)
var    Number : integer;        (* number declared as a global variable. *)


  procedure ReadANumber;
  var    Number : integer;        (* number declared as a variable local to *)
                                  (* procedure ReadANumber. *)

  begin
    writeln (' Enter an integer');
    readln (Number)
  end; (* procedure ReadANumber *)



(*----------------------------------------------------------------------*)

begin (* main program *)
  ReadANumber;
  writeln (' The number is', Number:5)
end. (* program WrongScope *)
```


Compiler 1

```
=>
 Enter an integer
□5
 The number is

                                            pascal termination log
                                            ------ ----------- ---
*** error ***         undefined variable used in expression

program terminated at offset 000088 in main program

local vars: number  = <undefined>
```

Compiler 2

```
=>
 Enter an integer
□5
 The  number is   67
```

Too often, students assume that once their program has no compile-time errors and no run-time errors, it is correct. This assumption is not valid. Although the program may be syntactically correct and may even be a solution to some problem, the program is incorrect if it does not solve the problem as stated. Once your program runs without errors, you must look at your output critically. You should ask yourself whether the output makes sense. If the output does not make sense, you must go back and look at your logic to see where you went wrong. Even if your output passes the first test—that is, it sounds reasonable—you must still do further testing to be sure that the answer is right. If possible, you should compute the answer in a different way. You could use a calculator to see whether both methods give the same answer for at least a limited number of cases.

The worst type of computer bugs are those that give reasonable, and even correct, answers most of the time. When a program seems to give reasonable answers and is correct for every case you test, you begin to trust it. Then, when it gives you total nonsense, you believe it. Even if you have been running a program for years, you should never stop questioning its results.

If you know your program is not working, *think* about what the problem could be; do not make random changes. If you are debugging your program using an interactive editor, doing things like changing multiplication to division just to see if the answer comes out right is easy. But do not give in to the temptation. If you make changes, have a good reason.

If you do not know precisely what is wrong, insert output statements into the program to report the values of the relevant variables. Be sure to label the variables that you are printing. From the printed values of the variables, you should be able to find the section of code where the problem first occurs. If you still cannot figure out what the problem is, add more output statements in the area where you know the problem lies. In this way, you can zero in on the code that contains the bug.

If you still do not understand why your program does not work, write a separate short program containing only the statements that you suspect are causing the bug. If this program does not exhibit the bug, add a few statements at a time until the program again exhibits the bug. In this way, you can isolate the bug. If the bug has been isolated but you still cannot precisely locate it, show your test program to someone else. If you have followed the steps above, you will be able to present someone with an organized discussion of the problem and your approach to solving it, which will help the other person to find the bug.

4.7 Documentation

Program documentation is of two different types. One type is information that explains the program to the user of the program. Messages that appear on the terminal during execution of a program are an example of this type of documentation. The other type of documentation is intended for programmers who need to know how the program works in order to correct or expand it. Comments that you include within your programs are an example of documentation for other programmers. For large programs both forms of documentation are often published in manuals.

A user's guide should tell the user what the program can be expected to do and provide instruction in how to use the program. Many large programs are designed for use by people who are not programmers and who may have little knowledge of how computers work. Software for word processing, data base management, videotex, spread sheets, and games falls into this category. The user's guide must tell the user exactly what to do and must not assume that the user knows the steps involved in getting the program to produce the desired output. The guide should be task oriented, telling the user the steps that need to be performed and the order in which they must be performed to accomplish a particular task. For example, when using a word-processing program, the user might want to delete an incorrectly entered word. The user's guide should describe what keys on the keyboard must be pressed in order to accomplish this task. The user's guide should also include a list of error messages that the user may encounter while using the program and tell the user how to correct the error that caused the problem.

The programmer's guide describes how the program works. It includes an algorithm or a chart describing the overall logic of the program. Names and descriptions of all global variables must be included, as well as descriptions of procedures and the local variables used by the procedures. Some programmer's guides also include a technical discussion of the algorithm. Suppose that the users of the word-processing program find that they are unable to retrieve words that have been unintentionally deleted. The programmer's guide should give the programmer enough information about how the program works so that the program can be modified to allow the user to retrieve deleted words. A complete description of error messages generated by the software and their probable causes should be included in the programmer's guide.

Solution to Case Study 4

At the beginning of the chapter, Algorithm 4.3 was developed for the solution to the case study. The main program and some of the procedures for the algorithm have been developed throughout the chapter.

Comparing Program 4.3, **ParallelCircuit,** to the main program of **AnalyzeThreeCircuits1,** you can see that with minor changes all the procedures in **ParallelCircuit** can be used in the program to solve the case study. Program 4.11, **AnalyzeThreeCircuits2,** is the final program that results from modifying the procedures from **ParallelCircuit** and adding the procedure to compute **SeriesCurrent.**

PROGRAM 4.11   AnalyzeThreeCircuits2

```
program AnalyzeThreeCircuits2 (input, output);
(*      Program AnalyzeThreeCircuits2 computes the current in each of     *)
(*      three different circuits when a voltage is applied across two     *)
(*      resistors connected in series and in parallel.                    *)

var   Voltage : real;              (* user supplied voltage in volts. *)
      Resist1 : real;              (* user supplied resistance 1 in ohms. *)
      Resist2 : real;              (* user supplied resistance 2 in ohms. *)
      ParallelCurrent : real;      (* computed current in amps for parallel *)
                                   (* connection of resistors. *)
      SeriesCurrent : real;        (* computed current in amps for series *)
                                   (* connection of resistors. *)


  procedure FirstMessage;
  (*      Procedure FirstMessage tells the user the purpose of program,   *)
  (*      AnalyzeThreeCircuits2.                                          *)

  begin
    writeln (' This program computes the current that flows when a');
    writeln (' voltage is applied across each of three circuits with');
    writeln (' two resistors.  The current is computed for both the');
    writeln (' parallel and series connection of the resistors.')
  end; (* procedure FirstMessage *)



  procedure ReadVoltageAndResist;
  (*      This procedure reads values for the voltage and two resistances  *)
  (*      for each circuit.                                                *)

  begin  (* procedure ReadVoltageAndResist *)
    writeln;
    writeln (' Enter a real number for the voltage in volts');
    readln (Voltage);
```

```
    writeln (' Enter a real number for the resistance in ohms of the first',
             ' resistor');
    readln (Resist1);

    writeln (' Enter a real number for the resistance in ohms of the second',
             ' resistor');
    readln (Resist2)
  end; (* procedure ReadVoltageAndResist *)

  procedure ComputeParallelCurrent;
  (*      This procedure computes the current when the resistors are     *)
  (*      connected in parallel.  The values for Voltage, Resist1, and   *)
  (*      Resist2 must be assigned prior to invoking this procedure.     *)

    var    EquivResist : real;      (* Computed equivalent resistance in ohms *)
                                    (* -- an intermediate variable that is *)
                                    (* local to this procedure. *)

  begin
    EquivResist := (Resist1 * Resist2) / (Resist1 + Resist2);
    ParallelCurrent := Voltage / EquivResist
  end; (* procedure ComputeParallelCurrent *)

  procedure ComputeSeriesCurrent;
  (*      This procedure computes the current when the resistors are     *)
  (*      connected in series.  The values for Voltage, Resist1, and     *)
  (*      Resist2 must be assigned prior to invoking this procedure.     *)

    var    EquivResist : real;      (* Computed equivalent resistance in ohms *)
                                    (* -- an intermediate variable that is *)
                                    (* local to this procedure. *)

  begin
    EquivResist := Resist1 + Resist2;
    SeriesCurrent := Voltage / EquivResist
  end; (* procedure ComputeSeriesCurrent *)

  procedure ReportCurrents;
  (*      This procedure reports the current in amps when a voltage is   *)
  (*      applied across a circuit with two resistors connected in       *)
  (*      parallel and in series.                                        *)

  begin
    writeln;
    writeln (' In a circuit with a voltage of', Voltage:10:2, ' volts');
    writeln (' and resistors of', Resist1:10:2, ' and', Resist2:10:2,
             ' ohms,');
    writeln (' the current for the series connection is', SeriesCurrent:10:2,
             ' amps,');
    writeln (' and the current for the parallel connection is',
             ParallelCurrent:10:2, ' amps.')
  end; (* procedure ReportCurrents *)
```

```
(*---------------------------------------------------------------------*)
begin (* main program *)
  FirstMessage;

  ReadVoltageAndResist;              (* Circuit 1 *)
  ComputeParallelCurrent;
  ComputeSeriesCurrent;
  ReportCurrents;

  ReadVoltageAndResist;              (* Circuit 2 *)
  ComputeParallelCurrent;
  ComputeSeriesCurrent;
  ReportCurrents;

  ReadVoltageAndResist;              (* Circuit 3 *)
  ComputeParallelCurrent;
  ComputeSeriesCurrent;
  ReportCurrents
end. (* program AnalyzeThreeCircuits2 *)
```

=>
This program computes the current that flows when a
voltage is applied across each of three circuits with
two resistors. The current is computed for both the
parallel and series connection of the resistors.

Enter a real number for the voltage in volts
☐10.0
Enter a real number for the resistance in ohms for the first resistor
☐3.0
Enter a real number for the resistance in ohms for the second resistor
☐3.0

In a circuit with a voltage of   10.00 volts
and resistors of   3.00 and   3.00 ohms,
the current for the series connection is   1.67 amps,
and the current for the parallel connection is   6.67 amps.

Enter a real number for the voltage in volts
☐10.0
Enter a real number for the resistance in ohms for the first resistor
☐4.0
Enter a real number for the resistance in ohms for the second resistor
☐2.0

In a circuit with a voltage of   10.00 volts
and resistors of   4.00 and   2.00 ohms,
the current for the series connection is   1.67 amps,
and the current for the parallel connection is   7.50 amps.

```
Enter a real number for the voltage in volts
□10.0
Enter a real number for the resistance in ohms for the first resistor
□5.0
Enter a real number for the resistance in ohms for the second resistor
□1.0

In a circuit with a voltage of      10.00 volts
and resistors of        5.00 and      1.00 ohms,
the current for the series connection is       1.67 amps,
and the current for the parallel connection is      12.00 amps.
```

In the procedures **ComputeParallelCurrent** and **ComputeSeriesCurrent**, the decision was made to make **EquivResist** a local variable. This decision was made because the equivalent resistance is used as an intermediate variable in computing the current for resistors connected in parallel and in series, but the formula for the equivalent resistance is different for each type of circuit.

In programs **ParallelCircuit** and **AnalyzeThreeCircuits2,** the procedure **ReadVoltageAndResist** is identical. If you have written a procedure for one program that can be used in another program, you do not have to rewrite the entire procedure. Compiler options, which are system dependent, can be used to change the defaults for the compiler. When a procedure has been developed previously and is stored in a separate computer file, the compiler option called *include* can be used. The syntax that will be used throughout this text for the include compiler option is

```
(*$i filename *)
```

where **filename** is the name of a computer file that is accessible to the compiler. The effect of the include compiler option is the same as if the contents of **filename** had been entered in the program. To avoid confusion, for **filename** we will always use the name of the procedure.

Compiler options are always enclosed between (*$ and *) or, equivalently, between {$ and }. When reading a program, do not confuse compiler options with comment statements, and when writing a program, do not begin a comment statement with a $ immediately after the * or it will be interpreted as a compiler option. The compiler options available for each compiler are given in Appendices F through H.

In **AnalyzeThreeCircuits2** the compiler option to include files is used. You should interpret the include statement to mean that the procedure **ReadVoltageAndResist** should be inserted in the program where the include statement appears.

```
(*$i readvoltageandresist *)
```

The program **AnalyzeThreeCircuits2** works, but it could be modified and made more efficient. You may notice that repeating the same four statements three times is not very efficient. After saving the version of the program that works, you can try to modify the approach. You might write a procedure called **AnalyzeACircuit** that invokes the other procedures and call it three times in the main program. If this approach works, you will have two versions of the program to choose from. If it does not, you will still have the original version.

PROGRAM 4.12   AnalyzeThreeCircuits3

```
program AnalyzeThreeCircuits3 (input, output);
(*      This program computes the current in three different circuits    *)
(*      when a voltage is applied  across two resistors connected in     *)
(*      series and in parallel.                                          *)

var    Voltage : real;            (* user supplied voltage in volts. *)
       Resist1 : real;            (* user supplied resistance 1 in ohms. *)
       Resist2 : real;            (* user supplied resistance 2 in ohms. *)
       ParallelCurrent : real;    (* computed current in amps for parallel *)
                                  (* connection of resistors. *)
       SeriesCurrent : real;      (* computed current in amps for series *)
                                  (* connection of resistors. *)

(*$i firstmessage *)
(*$i readvoltageandresist *)
(*$i computeparallelcurrent *)
(*$i computeseriescurrent *)
(*$i reportcurrents *)

  procedure AnalyzeACircuit;
     (*    This procedure calls four other procedures that perform the  *)
     (*    tasks involved in analyzing a circuit.                       *)
  begin
    ReadVoltageAndResist;
    ComputeParallelCurrent;
    ComputeSeriesCurrent;
    ReportCurrents
  end; (* procedure AnalyzeACircuit *)

(*-------------------------------------------------------------------*)

begin (* main program *)
  FirstMessage;

  AnalyzeACircuit;              (* Circuit 1 *)
  AnalyzeACircuit;              (* Circuit 2 *)
  AnalyzeACircuit               (* Circuit 3 *)
end.  (* program AnalyzeThreeCircuits3 *)
```

```
=>
This program computes the current that flows when a
voltage is applied across each of three circuits with
two resistors.  The current is computed for both the
parallel and series connection of the resistors.

Enter a real number for the voltage in volts
□10.0
Enter a real number for the resistance in ohms for the first resistor
□3.0
Enter a real number for the resistance in ohms for the second resistor
□3.0

In a circuit with a voltage of     10.00 volts
and resistors of      3.00 and     3.00 ohms,
the current for the series connection is    1.67 amps,
and the current for the parallel connection is    6.67 amps.

Enter a real number for the voltage in volts
□10.0
Enter a real number for the resistance in ohms for the first resistor
□4.0
Enter a real number for the resistance in ohms for the second resistor
□2.0

In a circuit with a voltage of     10.00 volts
and resistors of      4.00 and     2.00 ohms,
the current for the series connection is    1.67 amps,
and the current for the parallel connection is    7.50 amps.

Enter a real number for the voltage in volts
□10.0
Enter a real number for the resistance in ohms for the first resistor
□5.0
Enter a real number for the resistance in ohms for the second resistor
□1.0

In a circuit with a voltage of     10.00 volts
and resistors of      5.00 and     1.00 ohms,
the current for the series connection is    1.67 amps,
and the current for the parallel connection is   12.00 amps.
```

The main program for **AnalyzeThreeCircuits3** could be refined further, because the same statement is repeated three times. If there were ten circuits instead of three circuits, the main program would be rather awkward. In Chapter 7 you will learn a construction that will make repetition easier.

Summary

Before a computer program can be written, an algorithm must be developed. Algorithms can be written in pseudocodes that combine English and computer languages. Through use of top-down design, each step in the algorithm can be refined in a step-wise manner until the problem is solved.

Used in conjunction with top-down design, procedures are a powerful tool for creating well-organized, easy-to-read, correct programs. The algorithm can be translated into Pascal, with each of the major steps implemented as a procedure. Procedures allow a sequence of statements to be identified by a name. The structure of a procedure is similar to the structure of a program; that is, a procedure has a heading, a declarative part, and a body. The definition of a procedure appears in the declarative part of the program. Once a procedure has been declared, its name becomes an executable statement. Every time a procedure is invoked, the statements in its body are executed.

Pascal programs have a block structure. The overall program is a block, and the procedures are blocks within the program block or within other procedure blocks. The rules for identifier precedence within and among blocks allow many programmers to work independently on blocks within a large program.

A vital, but often neglected, element of any well-designed program is its documentation. Documentation that explains how to use a program must be written for the users of a program. Documentation that explains how a program works must be written for other programmers who might have to extend or modify the program. A program without documentation is useless.

Exercises

PASCAL

1. What is the output from the following program?

```
program SimpleProcedures (input, output);
(*      This program computes the volume and surface area of a right   *)
(*      circular cylinder (a tube) with a length of 10.3 inches and    *)
(*      a radius of 5.4 inches.                                        *)

const Pi = 3.14159;           (* used in computation of volume and surface *)
                              (* area of right circular cylinder. *)

var   Radius : real;          (* radius of right circular cylinder. *)
      Length : real;          (* length of right circular cylinder. *)
      Volume : real;          (* computed volume of right circular *)
                              (* cylinder. *)
      SurfaceArea : real;     (* computed surface area of right circular *)
                              (* cylinder. *)
```

```
procedure SetData;
(*    This procedure sets the values of the length and radius of the    *)
(*    right circular cylinder.                                          *)

begin
  Length := 10.3;
  Radius :=  5.4
end; (* procedure SetData *)




procedure ComputeVolume;
(*    This procedure computes the volume in cubic inches of a right     *)
(*    circular cylinder.                                                *)
begin
  Volume := Pi * sqr (Radius) * Length       (* cubic inches *)
end; (* procedure ComputeVolume *)




procedure ComputeSurfaceArea;
(*    This procedure computes the surface area in square inches of      *)
(*    a right circular cylinder.                                        *)
begin
  SurfaceArea := 2.0 * Pi * Radius * Length     (* square inches *)
end; (* procedure ComputeSurfaceArea *)




procedure ReportAnswers;
(*    This procedure reports the volume and surface area of the         *)
(*    right circular cylinder.                                          *)

begin
  writeln (' For a tube with a radius of ', Radius:10:2, ' inches');
  writeln ('             and a length of ', Length:10:2, ' inches,');
  writeln (' the surface area is ', SurfaceArea:10:2, ' inches squared');
  writeln ('   and the volume is ', Volume:10:2,' inches cubed.')
end; (* procedure ReportAnswers *)

(*--------------------------------------------------------------------*)
begin (* main program *)
  SetData;
  ComputeSurfaceArea;
  ComputeVolume;
  ReportAnswers
end. (* program SimpleProcedures *)
```

2. Rewrite the title page program from Problem 6 in Chapter 2 as a procedure. Write a program that invokes the procedure.

3. Perform a walkthrough for the following obscure program. Create a table of values. Be sure to make a new table entry any time the value of any variable changes.

```
program Obscure (input, output);
(*    Program Obscure is deliberately difficult to follow.  Create a    *)
(*    table that gives the value of each variable after each            *)
(*    assignment statement.                                             *)

const Const1 = '1';
      Const2 = 534;

var   CharVar : char;
      RealVar1, RealVar2 : real;
      IntVar1, IntVar2 : integer;

begin
  IntVar1 := Const2 div 5;
  IntVar2 := IntVar1 + 3;
  RealVar1 := sqrt (IntVar2);
  RealVar2 := RealVar1;
  IntVar1 := trunc (RealVar2);
  RealVar2 := 5.2;
  CharVar := Const1
end. (* program Obscure *)
```

4. What is the output from this program with nested procedures and duplicate identifiers?

```
program Nested (input, output);
(*    Program Nested is confusing because it contains nested procedures   *)
(*    with duplicate identifier names.  If you can determine its          *)
(*    output, you have an understanding of the Pascal rules of scope.     *)
(*    The procedure declarations are not indented or spaced in the way    *)
(*    that they have been in other programs, so you have only your        *)
(*    understanding of the Pascal syntax to guide you.                    *)

var X, Y : integer;

procedure B;

var Y : integer;

procedure C;

var X, Y : integer;
```

```
begin
  Y := 4;
  X := sqr(Y);
  writeln (' In procedure C, X =', X:5, ' Y =', Y:5)
end; (* procedure C *)


begin
  Y := 1;
  C;
  writeln (' In procedure B, X =', X:5, ' Y =', Y:5)
end; (* procedure B *)

procedure C;


begin
  X := 6;
  writeln (' In procedure C, X =', X:5, ' Y =', Y:5)
end; (* procedure C *)

begin (* main program *)
  X := 3;
  Y := -3;
  B;
  C;
  writeln (' In the main program, X =', X:5, ' Y =', Y:5)
end.
```

5. Find and correct the errors in the following program.

```
program ProcedureErrors (input, output);
(*      There are errors in the procedures in this program.  Find them     *)
(*      correct them.                                                       *)

var   Another : integer;
      Next : integer;


   procedure ReadNumbers
   var    Number : integer;

   begin;
     writeln (' Enter a number please');
     readln (Number);
     writeln (' Enter another');
     readln (Another);
   end; (* procedure ReadNumbers *)
```

```
procedure Next;
begin
  Next := Number + Another
  writeln (' Next is', Next:10:2)
end.

var Number : integer;

begin
  ReadNumbers;
  Next
end.
```

Problems

PASCAL

1. Using procedures, rewrite the program for Problem 3 in Chapter 3 to compute the volume of two spheres.

2. The number of bacteria in a culture after one generation is given by the formula

$$y = x(1.0 + r)$$

where x = the number of bacteria in the original population
   y = the number of bacteria in the total new population
   r = the growth rate (births − deaths)

Write a program that computes the new population for four different bacteria cultures. Test your program on the following data:

x	10000.0	10000.0	356.0	356.0
r	0.10	0.05	0.03	0.18

3. Rewrite the baseball program from Problem 6 in Chapter 3 so that it uses procedures.

4. You have just bought a new toaster that is rated at 1300 W. Before you plug it in, you want to know if it will blow the fuse on your 10A circuit. The relationship between watts (power) and amperes (current) is

$$P = VI$$

where P = power in watts
   V = potential force in volts
   I = current in amperes

Standard household current is 110 V. Compute the current in amperes that the toaster draws. Read the wattage from the terminal.

5. For a project you are building, you have marked off a right triangle. The hypotenuse and the height of the triangle have already been determined. Write a program that will compute the length of the base and the area enclosed by the triangle. The base can be computed from the Pythagorean theorem. The area is one half the base times the height.

APPLICATION

6. Rewrite the program from Problem 17 in Chapter 3 to compute the fuel cost of electricity for four different types of power plants. Make separate procedures for reading in the data, computing the cost, and writing out the results. Add the following three power plants to the original data:

	$/mega Btu	mega Btu/MWh
Coal	0.90	8.450
Gas turbine	2.45	14.000
Nuclear	0.67	10.400

7. The total momentum of three falling apples is the sum of the values for the momentum of each of the apples. The momentum of a falling apple is given by

$$p = mv \quad \text{and} \quad v = at$$

where p = momentum in kilogram-meters per second
  m = mass in kilograms
  v = velocity in meters per second
  t = time in seconds
  a = acceleration of gravity in meters per second squared

Read the number of seconds that the apples have been falling and the masses of each of the three apples from the terminal. Report the momentum of each apple and the total momentum.

8. The coefficient of performance (COP) for a refrigerator is given by the formula

$$\text{COP} = \frac{T_i}{T_a - T_i}$$

where T_a = ambient (room) temperature in degrees Rankine (R)
  T_i = temperature inside the refrigerator in degrees Rankine
  °R = °F + 459.7

The inside temperature for the refrigerator is always set at $T_i = 25°F$, but the ambient temperature depends on whether it is summer or winter. Read the values for the summer and winter room temperature in degrees Fahrenheit. Compute the COP for the refrigerator in both summer and winter. Include in your output the following values:

(a) the refrigerator temperature
(b) the summer and winter room temperatures
(c) the COP for the refrigerator in summer and winter

9. The force on a mass that is being accelerated by gravity is given by

$$F = ma$$

The kinetic energy of a mass being accelerated by gravity is

$$K = \frac{1}{2}mv^2 \quad \text{and} \quad v = at$$

where a = acceleration of gravity in meters per second squared
v = velocity in meters per second
t = time in seconds
m = mass in kilograms

Two masses are dropped. The first mass weighs 100 g, and the second mass weighs 15 g. Evaluate the kinetic energy for both masses at $t = 5.0$ s. Include in your output the following values:

(a) the acceleration of gravity
(b) the time
(c) the masses
(d) the force and kinetic energy for each mass

Remember to include the units for each output value.

10. The thermal energy Q, in joules, lost through the walls of a hot water tank at temperature T_1 into a room at temperature T_2 over a time period of t minutes is

$$Q = UAt(T_1 - T_2)$$

where Q = quantity of heat in joules
U = loss coefficient of the hot water tank walls in joules per minute per meter squared per degree Celsius
A = surface area of the tank in meters squared
t = time in minutes

Write a program that reports the cumulative heat lost each minute over a period of ten minutes. The formula above cannot be used for long lengths of time, because as heat is lost the temperature in the tank drops and T_1 must be adjusted. Read the values for the temperatures, the loss coefficient, and the surface area from the terminal. Test your program using the following data: The temperature in the tank is 50°C, the temperature in the room is 21°C, the loss coefficient is 0.17 J/min/m²/°C, and the surface area of the tank is 4.0 m².

CHAPTER 5

Boolean Expressions and Conditional Statements

In this chapter you will learn the Pascal constructions that allow you to test whether certain conditions are met. You can then control whether or not statements in your program are executed based on the outcome of the test. The ability to *branch*—that is, to choose which instructions will be executed—is included in all programming languages.

CASE STUDY 5
Highest and lowest temperatures to date

There are many problems in science for which you might need to record environmental factors such as temperature and rainfall. Ecology and meteorology are branches of science in which such information could be needed. For example, meteorologists often keep records of the highest and lowest temperature in a given location for a time period. Records are also kept of the total amount of rainfall over a period of time.

A large program being developed for meteorologists will allow them to keep track of weather data over extended periods of time. Among other things, the program will be able to keep track of the highest temperature, the lowest temperature, and the total rainfall for a period of days. You are asked to write a set of procedures that performs these tasks for a single day. You will also need to write a procedure that allows the meteorologist to select one of the three tasks to perform. These procedures must be integrated with the rest of the program.

The input for the program will be given to you at the end of each day. You will be given the high and low temperatures for the day and the amount of rainfall for the day. When the large program is complete, the values from the previous days will be supplied by another part of the program, but in order to test the procedure you will need to write a test program that sets the initial values, calls the procedures, and prints the results.

Algorithm development

In developing a solution to the problem, we will use a procedure to exchange the values of the temperatures if the high temperature for today is higher than the previous high temperature. This will result in the value of the previous high temperature being stored in the location for today's high temperature. This may seem to be an inappropriate method of solving the problem, since it is obvious that today's high temperature does not change. The reason for this approach is that many algorithms for problems such as the one in the case study involve sorting the values into ascending or descending order. By using an exchange, you will be learning the first step in a sorting algorithm.

The case study problem can be restated as follows: Write a program that gives the user the choice of performing one of the following tasks:

1. Determine whether today's high temperature is higher than the previous high temperature. If it is, exchange it with the previous high temperature.

2. Determine whether today's low temperature is lower than the previous low temperature. If it is, exchange it with the previous low temperature.

3. Add today's rainfall to the total rainfall.

Depending on the choice, the program must report the high temperature, low temperature, or total rainfall.

In an algorithm, it is often necessary to take one of two or more paths, depending on a particular condition. The program to solve the problem in the case study must be able to execute one of three alternatives based on input from the meteorologist. The subsequent actions taken by the computer will depend on that choice. Once an alternative has been selected, the program must be able to make comparisons between the current temperature and the highest or lowest temperature so far, and it must be able to keep a running total of the rainfall. Algorithm 5.1 is a rewriting of these requirements in algorithmic form.

ALGORITHM 5.1   First iteration of an algorithm for Case Study 5

begin algorithm
  1. Write a message to the user stating what the program does

2. Write a message to the user listing the choices of actions and read **Choice,** a variable that indicates the action to be taken
   begin
       a. Write a message asking the user to select whether to determine Highest temperature, Lowest temperature, or total Rainfall
       b. Read **Choice**
   end
3. Execute one of the following actions, depending on **Choice:**
   begin
       a. If **Choice** is Highest, then execute a procedure to compare the current highest temperature with today's high temperature. If today's temperature is higher, then **Exchange**
       b. If **Choice** is Lowest, then execute a procedure to compare the current lowest temperature with today's low temperature. If today's temperature is lower, then **Switch**
       c. If **Choice** is Rainfall, then execute a procedure to add today's rainfall to the total rainfall
   end
4. Report the highest temperature so far, the lowest temperature so far, and the total rainfall

end algorithm

You already know how to perform the exchange for Step 3a using the fragment developed in Chapter 4 in the section on walkthroughs and you can write a procedure, which can be called **Exchange,** to perform the exchange. Now you need to be able to control when the exchange is executed and when it is not. You must ensure that the higher of two temperatures is stored in a variable called **Highest.** If the values of the variables are

Highest	HighToday
78.5	65.3

where **Highest** is the highest temperature so far and **HighToday** is the temperature that is to be compared to **Highest,** the procedure to exchange the values should not be executed.

If the values of the variables are

Highest	HighToday
78.5	79.1

the procedure should be executed, and the values should be exchanged. You must be able to test whether the value of **HighToday** is greater than the value stored in **Highest.** If it is greater, you must exchange the value stored in **HighToday** with the value stored in **Highest;** if it is not greater, you do not have to do anything.

The algorithm for Step 3a can be rewritten as follows:

ALGORITHM 5.2 Refinement of Step 3a of Algorithm 5.1

3a. If **Choice** is Highest, then execute a procedure to compare the current highest temperature with today's high temperature
begin
    i. Write 'What is today's high temperature?'
    ii. Read **HighToday**
    iii. If **HighToday** > **Highest,** then execute procedure **Exchange**
    procedure **Exchange**
    begin
        Save ← **Highest**
        Highest ← **HighToday**
        HighToday ← Save
    end **Exchange**
end

In developing the algorithm, you must remember to make sure that your solution will work for all possible situations. Think about what will happen when you enter the values for the first day: **Highest** will be undefined because no value has been put into that storage location yet. You must *initialize,* or assign an initial value to, the variable. If the program is being run for the first time, you can assign an extreme value to **Highest.** On the first day of observation, the temperature will be the highest so far. If the initial value assigned to **Highest** is an impossibly low temperature, the first day's temperature will be exchanged with the value stored in **Highest** and become the current highest temperature. If you assign an initial value of −1000.0 to **Highest,** you can be sure that the value will be replaced by **HighToday.*** To test the procedure, you can also initialize **Highest** to a temperature that is a reasonable value for the highest temperature.

Because you do not want **Highest** to be initialized each time the procedure is executed, you must initialize the variable prior to the procedure. Usually, when a program uses variables that must be initialized, the initial values are assigned in the first statements of the program. So you could include the initialization of variables in Step 1 of the algorithm.

---

*When this reasoning is used for integer variables, ± **maxint** is often assigned as an extreme value during initialization.

Step 3b of the algorithm can be solved in the same way as Step 3a, except that the values will be exchanged if **LowToday** is lower than the current **Lowest** temperature. The value of **Lowest** must be initialized. Using the same reasoning as above, you can assign it an extremely high value. A separate procedure to exchange the values for the low temperatures (call it **Switch**) must also be written. In Chapter 6 you will learn to write procedures that will operate on variables with different names so that you need write a procedure only once if the same action is to be carried out with more than one set of variables.

For Step 3c of the algorithm, the total rainfall must be computed. To determine the amount of rainfall over a given time period that is longer than a day, the rainfall for the current day must be added to the total rainfall that has been previously recorded. The total rainfall so far is referred to as a *running total*. A running total is kept by adding the current value to the total so far and placing the sum in the storage location for the running total. In refining Step 3c of the algorithm, you could call the rainfall measured today **Rainfall** and the total amount of rainfall **TotalRain.**

ALGORITHM 5.3   Refinement of Step 3c of Algorithm 5.1

3c. If **Choice** is Rainfall, then execute a procedure to add today's rainfall to the total rainfall
   begin
       a. Write 'What is today's rainfall?'
       b. read **Rainfall**
       c. **TotalRain ← TotalRain + Rainfall**
   end

The first time the program is run, the value of **TotalRain** must be 0.0. When you test your program, you can initialize **TotalRain** to 0.0 in some runs and to larger values in other runs.

The complete algorithm for the case study is as follows:

ALGORITHM 5.4   Final algorithm for Case Study 5

begin algorithm
1. procedure **Initialize**
   begin
       a. Set **Highest** = −1000.0
       b. Set **Lowest** = 1000.0
       c. Set **TotalRain** = 0.0
       d. Write 'This program allows you to determine the highest or lowest temperature or the total amount of rainfall for an observation period. You will be asked which you want to determine.'

end
2. procedure **ReadChoice**
   begin
      a. Write a message asking the user to select whether to determine Highest temperature, Lowest temperature, or total Rainfall
      b. Read **Choice**
   end
3. procedure **Choose**
   begin
      a. If **Choice** is Highest, then execute procedure **HighCompare**
         procedure **HighCompare**
         begin
            i. Write 'What is today's high temperature?'
            ii. Read **HighToday**
            iii. If **HighToday** > **Highest**, then execute procedure **Exchange**
            procedure **Exchange**
            begin   Save ← **Highest**
               Highest ← **HighToday**
               HighToday ← Save
            end **Exchange**
         end
      b. If **Choice** is Lowest, then execute procedure **LowCompare**
         procedure **LowCompare**
         begin
            i. Write 'What is today's low temperature?'
            ii. Read **LowToday**
            iii. If **LowToday** < **Lowest**, then execute procedure **Switch**
            procedure **Switch**
            begin
               Save ← **Lowest**
               Lowest ← **LowToday**
               LowToday ← Save
            end **Switch**
         end
      c. If **Choice** is Rainfall, then execute procedure **Precipitation**
         procedure **Precipitation**
         begin
            i. Write 'What is today's rainfall?'
            ii. Read **Rainfall**
            iii. **TotalRain** ← **TotalRain** + **Rainfall**
         end

4. Report the highest temperature so far, the lowest temperature so far, and the total rainfall
end algorithm

5.1 Applications of boolean algebra to programming

Boolean algebra is a branch of mathematics devised by the English mathematician George Boole (1815–1864). In boolean algebra, all variables and expressions have a value of either true or false. Boolean algebra is also called *symbolic logic* because it is used to test the logical truth of expressions. Because there are only two possible values that the variable can take, boolean variables are binary variables. Boolean algebra is closely linked to the design and operation of digital computers because computers are binary machines. Although booleans may appear to have rather limited usefulness for programming, you will discover that they are quite powerful tools in programming languages.

You have learned the data types **real, integer,** and **char. Boolean** is the fourth simple data type in Pascal. In Pascal a boolean variable or a boolean expression can take only one of two values: true or false. The first part of this section will discuss boolean expressions that are made up of the relational operators and the boolean operators in conjunction with expressions of the data types that you already know. Boolean expressions are used to test whether a condition is true or false; they allow you to include statements in a program that will be executed only under certain conditions. Boolean expressions are used in many Pascal constructions, but boolean variables are not as commonly used, so boolean variables will be discussed separately after the discussion of boolean expressions.

Relational operators

The *relational operators* are used to compare the values of two variables or to compare a value with the value of a variable. The result of the comparison is always a boolean value of true or false. For two variables to be compared, the variables must be of compatible data types. **Integer** and **real** values can be compared to each other (on most, but not all, Pascal implementations), but values of data type **char** or **boolean** must be compared only to values of the same data type.

The following operators are valid relational operators:

=	equal
<	less than
>	greater than
<=	less than or equal to
>=	greater than or equal to
<>	not equal to

Note that the Pascal symbols for less than or equal to, greater than or equal to, and not equal to are different from the standard mathematical symbols ≤, ≥, ≠.

If **Num1** has been assigned a value of 5 and **Num2** has been assigned a value of 3, the following boolean expressions can be evaluated:

Boolean expression	Result of evaluation
Num1 = Num2	false
Num1 < Num2	false
Num1 > Num2	true
Num1 <= Num2	false
Num1 >= Num2	true
Num1 <> Num2	true

The following are also valid uses of the relational operators:

```
Denominator <> 0
5 >= Num1
CharVar1 > CharVar2
a <= CharVar1
```

When character values are compared, the result of the comparison depends on whether the character set is ASCII or EBCDIC. Characters are arranged sequentially in each character set, and each character has a corresponding number. This number is referred to as the *ordinal value* of the character. The relational operators compare the ordinal values of characters. Because characters are ordered differently in the two character sets, the results of comparisons of characters may differ depending on which character set is used. Appendix A lists the two character sets and their ordinal values.

For the case study, the values stored in the variables **HighToday** and **Highest** can be compared through use of the following boolean expression:

```
HighToday > Highest
```

If the value stored in **HighToday** is greater than the value stored in the variable **Highest,** the boolean expression has a value of true; otherwise, it has a value of false.

Notice the difference between the assignment operator and the relational operator. When the equal sign is used as a relational operator, as in the following expression;

```
Num1 = Num2
```

the equal sign asks, Is **Num1** equal to **Num2**?

5.1 Applications of boolean algebra to programming

> *DEBUGGING HINT*
>
> Do not confuse the comparative operator = with the assignment operator :=.

Boolean operators

The boolean operators are **and, or,** and **not.** The result of a boolean operation is always a boolean value of true or false. Boolean operators act only on expressions or variables that take values of true or false. The boolean operators can be used with simple boolean expressions like those created before to form more complicated boolean expressions or with boolean variables that you will learn later.

The following tables show the results of applying the boolean operators to boolean variables or expressions. These tables are called *truth tables*. To see why the boolean operators have the stated results, try substituting English statements for the boolean expressions represented by **A, B,** and **C**; however, do not depend on English to get the logic correct. As you will see when you work with the truth tables, an English sentence sometimes does not convey the same meaning as the corresponding symbolic logic statement. If **A** and **C** represent boolean expressions, the statement

$$C := \text{not } A$$

means that if **A** is true, then **C** is false, and if **A** is false, then **C** is true. Written in a table, the values are

A	C = not A
true	false
false	true

To see how the table works, substitute the phrase "things are OK" for **A** and "things are not OK" for **C**. If **A** is true, then **C** is false. If **A** is false, then **C** is true.

The statement

$$C := A \text{ and } B$$

means that if **A** is true and **B** is true, then **C** is true; otherwise, **C** is false. The truth table for the statement is as follows:

A	B	C = A and B
true	true	true
true	false	false
false	true	false
false	false	false

Substitute the phrase "the sky is blue" for **A**, "the grass is green" for **B**, and "things are OK" for **C**. If **A** and **B** are true (the sky is blue and the grass is green), then **C** is true (things are OK). If **A** and/or **B** is false (the sky is not blue and/or the grass is not green), then **C** is false (things are not OK).

The statement

$$C := A \text{ or } B$$

means if either **A** is true or **B** is true, then **C** is true. **C** is false only if both **A** and **B** are false, as the following table shows:

A	B	C = A or B
true	true	true
true	false	true
false	true	true
false	false	false

Substitute the phrase "Mary is home" for **A**, "Jane is home" for **B**, and "someone is home" for **C**. (Jane and Mary are roommates, and no one else lives with them.) If **A** and/or **B** is true (Mary and/or Jane is home), then **C** is true (someone is home). If **A** and **B** are false (neither Mary nor Jane is home), then **C** is false (no one is home).

> *REMINDER*
>
> In logic, A *or* B means A and/or B; A *and* B means both A and B. There are many times when the popular meaning of an English sentence does not correspond to its logical meaning. For example, the sentence "John or Mary will go" would be interpreted by most people as meaning that either John or Mary will go but not both. In logic, John or Mary means John and/or Mary.

5.1 Applications of boolean algebra to programming

The use of the boolean operators with boolean expressions that include relational operators is illustrated by the following expressions. In these examples, the expression containing the relational operator is used as a boolean subexpression in a more complex boolean expression. Remember that **Num1** equals 5 and **Num2** equals 3.

Boolean expression	Result of evaluation
not (Num1 = Num2)	true
not (Num1 <> Num2)	false
(Num1 < Num2) and (Num1 > 0)	false
(Num1 > Num2) and (Num2 > 0)	true
(Num1 <= Num2) or (Num1 > 0)	true
(Num1 >= Num2) or (Num2 > 0)	true
(Num1 < Num2) or (Num2 > 6)	false

The distributive laws and DeMorgan's laws can be used to simplify compound expressions containing the boolean operators. In the following examples, the boolean expression on the right is equivalent to the expression on the left. The first two examples show the distributive laws, and the last two illustrate DeMorgan's laws. Assume that **A, B,** and **C** represent boolean subexpressions.

Boolean expression	Equivalent simplified expression
(A or C) and (B or C)	(A and B) or C
(A and C) or (B and C)	(A or B) and C
(not A) and (not B)	not (A or B)
(not A) or (not B)	not (A and B)

The precedence of the relational operators, boolean operators, and arithmetic operators is

highest	parentheses
·	**not**
·	*****, **/**, **div, mod, and**
·	**+, −, or**
lowest	**=, <>, <, >, <=, >=**

Notice that when a boolean expression such as **Num1 = Num2** is used as a subexpression in a larger boolean expression, it must be enclosed in parentheses. For example, in the expression **not (Num1 = Num2)**, the parentheses are required because **not Num1 = Num2** would be evaluated as **(not Num1) =**

Num2. Because **Num1** does not have a data type of **boolean,** the boolean operator cannot act on it. It can operate on the boolean value resulting from the comparison of **Num1** and **Num2.**

Knowing the relational operators and the boolean operators, you can create a more complete syntax diagram for an expression. The simplified syntax diagram for an expression is

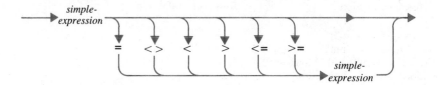

where *simple expression* has the syntax diagram

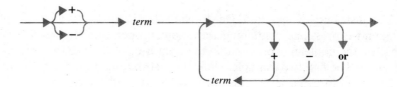

term has the syntax diagram

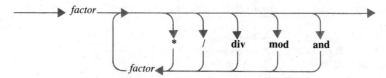

and *factor* has the simplified syntax diagram

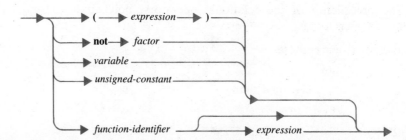

These syntax diagrams may look intimidating, but they are merely refined versions of the diagrams you saw in Chapter 3. When evaluating an expression, start with the diagram for a factor and work your way up.

The compiler evaluates boolean subexpressions that are at the same precedence level from left to right. As usual, when parentheses are used the computer starts from the innermost set of parentheses and works out. On some Pascal implementations, the entire expression is evaluated; on other implementations, only part of a compound expression may be evaluated. For example, in an expression such as

```
(Divisor <> 0.0) and (Number / Divisor > 1.0)
```

if the first subexpression is false, the entire expression will be false. Some Pascal implementations will stop evaluation of the expression at this point; others will continue to evaluate the entire expression. In the above example, evaluation of the entire expression would be impossible if **Divisor** were equal to 0.0, as the evaluation would involve division by 0.0. After you learn the **if** statement in Section 5.2, you can use the expression above to determine whether the Pascal implementation that you are using evaluates the entire expression or only enough of it to determine the boolean value of the entire expression. If you set **Divisor** equal to 0.0 and the entire expression is evaluated by your computer, you will get an error message telling you that you cannot divide by zero when the above expression is executed. If you want to be able to run your programs on all Pascal compilers, write boolean expressions as if all the subexpressions will be evaluated.

Boolean variables

A boolean variable must be declared to be of data type **boolean.** Variables of the data type **boolean** have a value of either true or false. The values **true** and **false** are standard constants in Pascal. An example of the declaration of a boolean variable is

```
var    Check : boolean;
```

A boolean variable can be assigned a value of **true** or **false** in an assignment statement and its value can be printed in an output statement, as in the following fragment:

```
var    Test : boolean;

begin
  Test := false;
  writeln (' The value of Test is ', Test);
  Test := true;
  writeln (' The new value is ', Test)
end.

=>
 The value of Test is false
 The new value is true
```

You cannot, however, use a **read** statement to assign a value to a boolean variable. If **Test** is declared to be a **boolean** variable, the following statement is invalid and will produce an error message similar to the one shown:

```
readln (Test);
***error***    $153

error messages:
**************
153: illegal parameter type for read.
```

If **Answer** and **Check** have both been declared as boolean variables, and if the value of **Check** is false and the value of **Answer** is true, the following are valid boolean expressions containing boolean variables and boolean operators:

Boolean expression	Result of evaluation
not Check	true
not Answer	false
Check and Answer	false
Answer or Check	true

The relational operators are rarely used with boolean variables. If you do use the relational operators, you should know that false has an ordinal value of 0 and true has an ordinal value of 1, so false is less than true:

Boolean expression	Result of evaluation
Answer = Check	false
Check < Answer	true
Answer >= Check	true

You can also write an assignment statement such as

```
Check := Num1 > Num2
```

where the boolean variable **Check** is assigned the value resulting from the comparison of **Num1** and **Num2**. You should recognize such statements as valid statements when you encounter them, but they can be confusing and should usually be avoided. Program 5.1, **LogicalTest,** uses boolean expressions to assign values to the **boolean** variable **Test.**

PROGRAM 5.1 LogicalTest

```
program LogicalTest (input, output);
(*      This program uses the boolean and relational operators to set    *)
(*      the value of a boolean variable.                                 *)

var   X : integer;          (* variable to be compared to Y. *)
      Y : integer;          (* variable to be compared to X. *)
      Test : boolean;       (* stores the result of comparison of X and Y. *)

begin (* main program *)
  X := 1;
  Y := 2;

  Test := X < Y;
  writeln;
  writeln (X:3, ' < ', Y:2, ' is ', Test);

  Test := (X > Y) or (X <= Y);
  writeln (' (', X:2, ' > ', Y:2, ') or (', X:2, ' <= ', Y:2, ') is ',
           Test);

  Test := (X > Y) and (X <= Y);
  writeln (' (', X:2, ' > ', Y:2, ') and (', X:2, ' <= ', Y:2, ') is ',
           Test);

  Test := not ((X > Y) or (X <= Y));
  writeln (' not ((', X:2, ' > ', Y:2, ') or (', X:2, ' <= ', Y:2, ')) is ',
           Test);

  Test := not (X > Y) or (X <= Y);
  writeln (' not (', X:2, ' > ', Y:2, ') or (', X:2, ' <= ', Y:2, ') is ',
           Test)
end. (* program LogicalTest *)

=>
 1 <  2 is true
 ( 1 >  2)  or  ( 1 <=  2) is true
 ( 1 >  2) and  ( 1 <=  2) is false
 not (( 1 >  2) or ( 1 <=  2)) is false
 not ( 1 >  2)  or  ( 1 <=  2)  is true
```

Standard boolean functions

There are three standard boolean functions that return a value of true or false: **odd** (X), **eoln,** and **eof.** The first is used to determine whether a variable has an odd or an even value. The argument of the function **odd** must be an integer or an integer variable. If the argument is odd, the value returned by

odd is true. If the argument is even, **odd** will return a value of false. For example,

```
BoolVar := odd(3);
writeln (' It is ', BoolVar, ' that the integer 3 is odd.');

X := 16;
BoolVar := odd(X);
writeln (' It is ', BoolVar, ' that the integer ', X:2, ' is odd.')
```

```
=>
It is  true that the integer 3 is odd.
It is false that the integer 16 is odd.
```

The other two boolean functions, **eoln** and **eof,** are used with data files. **Eoln** stands for end-of-line and **eof** stands for end-of-file. These functions will be covered in Chapter 10.

5.2 The if statement

If statements are used to control the operation of a program based on the value of a boolean expression. With an **if** statement, you can evaluate a boolean expression and then, based on the results, execute one section of your program or another.

The syntax diagram for a simple **if** statement is

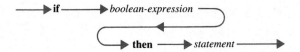

An **if** statement tests the value of a boolean expression. If the boolean expression is true, the statement following the **then** is executed. If the boolean expression is false, the program execution goes on to the next statement. The boolean expression in an **if** statement can be a simple boolean expression, as in

```
if X > 0 then
   writeln (' X is greater than zero.')
```

or a boolean variable, as in

```
Test := X > 0;
if Test then
   writeln (' X is greater than zero.')
```

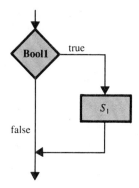

FIGURE 5.1 Flowchart of an **if** statement.

or a compound boolean expression, as in

```
if ((X > 0) and (Y > 0)) then
  writeln (' Both X and Y are greater than 0.')
```

Figure 5.1 is a *flowchart* that depicts the action taken by the computer when an **if** statement is encountered. Flowcharts are sometimes used either in addition to pseudocode or instead of pseudocode to describe the actions to be taken by the computer. When the execution of alternative sections of code depends on the outcome of a test, it is sometimes easier to look at a flowchart rather than an algorithm to see which statements will be executed next. In a flowchart, boxes of different shapes are used to indicate different types of operations. There are standards that specify which shapes are to be used.

In a flowchart, the diamond shape indicates a statement in which a decision is made. In this case the boolean expression **Bool1** is evaluated, and the outcome of the evaluation is a value of either true or false. The path taken in the execution of the program depends on this value. If it is true, the path leading to the rectangle is followed; thus, if **Bool1** is true, the statement or statements indicated by S_1 are executed, then the next statement in the program is executed. If **Bool1** is false, then statement S_1 is not executed, and the computer goes directly to the next statement in the program.

Compound statements

In the **if** statements above, only one statement is executed when the boolean variable is true. For many algorithms, and for the case study in particular, it is necessary to execute more than one statement when a condition is met. When the value stored in **HighToday** is greater than the value stored in **Highest,** you need to execute several instructions in order to exchange the values. The execution of more than one statement as part of an **if** statement requires a compound statement.

A *compound statement* is a sequence of statements surrounded by a **begin-end** pair. The statements within the **begin-end** pair are treated as a group. If any one of them is executed, they all are executed. The words **begin-end** act like parentheses, grouping the statements together. The syntax diagram for a compound statement is

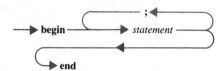

For example, the following **if** statement followed by a compound statement could be used in the solution to the case study:

```
if HighToday > Highest then
   begin
      Save := Highest;
      Highest := HighToday;
      HighToday := Save
   end
```

Notice the punctuation for the compound statement. The three executable statements within the **begin-end** pair are separated by semicolons. The last executable statement does not require a semicolon because it is followed by **end**, which is not an executable statement. Putting a semicolon after the third executable statement, however, will not result in an error. The semicolon will be interpreted as separating the preceding statement from the *null statement*. The null or empty statement tells the computer to do nothing. The computer does nothing and then goes on to the next statement. Thus the code will execute in the same way whether or not there is a semicolon before a nonexecutable statement.

An alternative to using a compound statement is creating a procedure that contains the statements in the compound statement. If the boolean expression in the **if** statement is true, the procedure is invoked, as in

```
if HighToday > Highest then Exchange
```

where **Exchange** has been defined as

```
procedure Exchange;
   (*    This procedure exchanges the values that are stored in Highest    *)
   (*    and HighToday using the intermediate variable Save.                *)

var Save : real;
```

```
begin
  Save := Highest;
  Highest := HighToday;
  HighToday := Save
end;   (* procedure Exchange *)
```

For the case study, the latter solution will be used.

Precautions must be taken when real variables are to be compared in boolean expressions. Because of rounding errors, real numbers are seldom identically equal. To test whether the value of the real variable **Energy** is equal to the value of the real variable **Work,** you could write the following statement:

```
if Work = Energy then ComputeEntropy
```

However, the boolean expression **Work = Energy** will almost always be false when evaluated by the computer. If the difference between the two variables is only due to rounding errors and the variables are actually equal, your program will not execute the way you expect. Using the following sequence of instructions, you can instruct the computer to consider the variables to be equal if the difference between them is very small:

```
const Small = 0.000006;

var    Energy, Work, Diff : real;
          .
          .
          .
    Diff := Work - Energy;
    if abs (Diff) <= Small then ComputeEntropy
```

where **abs** is the standard Pascal function that takes the absolute value of the argument in parentheses. These statements are equivalent to stating that if the value of **Work** is equal to the value of **Energy ± Small,** the procedure **ComputeEntropy** should be executed.

5.3 The if-then-else statement

Frequently, you will want to carry out one action if the boolean expression in an **if** statement is true and another action if it is false. Rather than performing two tests, you can use an **if-then-else** statement. In an **if-then-else** statement, if the boolean is true, the first statement is executed; otherwise, the second statement is executed.

The syntax diagram for an **if-then-else** statement is

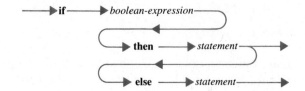

For example,

```
if Acceleration > 0.0 then
   writeln (' The speed is increasing.')
else
   writeln (' The speed is steady or decreasing.')
```

The **if-then-else** statement above is equivalent to the two statements

```
if Acceleration > 0.0 then
    writeln (' The speed is increasing.');
if not (Acceleration > 0.0) then
    writeln (' The speed is steady or decreasing.')
```

Because the boolean expression in the **if-then-else** statement is either true or false, either the first statement or the second statement, but never both, will be executed. A flowchart for the **if-then-else** statement is given in Figure 5.2.

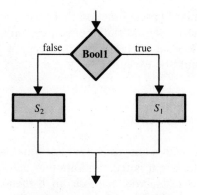

FIGURE 5.2 Flowchart for the **if-then-else** statement.

> *DEBUGGING HINT*
>
> There is no punctuation before **else. If-then-else** is a single executable statement. Because a semicolon is used to separate statements, a semicolon after the end of the first statement signals the end of the **then** statement. If an **else** follows, it is called *disconnected* or *dangling* because it has been separated from its **if.**

Because a boolean variable cannot be assigned a value in a **read** statement, a *proxy* or *dummy* variable can take the place of the boolean variable in the **read** statement. The value of the proxy variable can then be used to assign a value to the boolean variable. The following fragment uses an **if-then-else** statement to ask the user of a program if more information is to be entered:

```
var   Response : char;        (* used as proxy for answer to be read. *)
      Answer : boolean;       (* to take value of true if Response is 'y' *)
                              (* and value of false otherwise. *)
          .
          .
          .
      writeln (' Do you want to enter more data?  Enter y or n');
      readln (Response);

      if Response = 'y' then
         Answer := true
      else
         Answer := false;
```

5.4 Compound if statements

An **if** or an **if-then-else** can have another **if** or **if-then-else** statement as its executable statement. If **Temperature** has been declared to be a **real** variable representing the temperature of H_2O in degrees Celsius, the following statements can be written:

```
            if Temperature < 100.0 then
               if Temperature > 0.0 then
                  Water;
            NextProcedure
```

The second comparison is made only if the first boolean expression has a value of true. The fragment executes as follows: If **Temperature** is less than

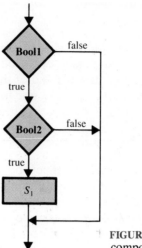

FIGURE 5.3 Flowchart for a compound **if** statement.

100.0, the second comparison is made. If **Temperature** is greater than 0.0, the procedure **Water** is executed, and then **NextProcedure** is executed. If the first **if** statement is false—that is, if **Temperature** is greater than or equal to 100.0—the second **if** statement is not evaluated, and **NextProcedure** is executed. A diagram for the flow of a compound **if** statement is shown in Figure 5.3.

An easier way to accomplish the same result as is accomplished with a compound **if** statement is to use the syntax

```
if (Temperature < 100.0) and (Temperature > 0.0) then
   Water;
NextProcedure
```

However, as mentioned in the previous section, some compilers evaluate all the expressions in a compound boolean expression. If any of the expressions cannot be evaluated, this construction may result in a run-time error. For example, if **Bottom** is equal to or very close to 0.0, then **Top/Bottom** in the following fragment cannot be evaluated:

```
if (abs (Bottom) > Small) and (Top / Bottom > 1.0) then
   ComputeRatio
```

Testing the value of **Bottom** in the first expression does not prevent some compilers from attempting to evaluate the second expression. A compound **if** statement should be used in this situation:

5.4 Compound if statement

```
if abs (Bottom) > Small then
  if Top / Bottom > 1.0 then
    ComputeRatio
```

In a compound **if** statement, either **if** can have an **else** associated with it. An **else** is always associated with the nearest **if** preceding it that does not already have an **else**. For example,

```
if Temperature < 100.0 then
  if Temperature > 0.0 then
    Water
  else
    Ice
else
  Steam;
NextProcedure
```

executes as follows: If **Temperature** is less than 100.0, the next **if** statement is evaluated. If **Temperature** is greater than 0.0, the procedure **Water** is executed, and then **NextProcedure** is executed. If **Temperature** is less than or equal to 0.0, procedure **Ice** is executed, and then **NextProcedure** is executed. If the first **if** statement is false, the **else** statement that invokes procedure **Steam** is executed, and then **NextProcedure** is executed.

The compound **if** statement invites *dangling* and *disconnected elses*. These are **else**s that are associated either with no **if** or with a different **if** than intended. The following program fragment is *syntactically correct,* but it is *logically incorrect* because it has a disconnected else. The **else** is not associated with the statement that it logically should be associated with. Assume that **Winged, Warmblooded,** and **Bird** have been declared as boolean variables.

```
if Winged then
  if Warmblooded then
    if not Bird then
      writeln (' This animal is a bat.')
else
  writeln (' The organism is an insect or an extinct flying dinosaur.')
```

If **Winged, Warmblooded,** and **Bird** are all true, this fragment will produce the following output:

```
=>
  The organism is an insect or an extinct flying dinosaur.
```

Clearly, insects and flying dinosaurs are not warmblooded birds. The output statement is not consistent with the logical intent. Indentation of the

if-then-else statement can be used to make the statement easier to read, but it does not affect the way the statement executes. A **begin** and **end** can be used to mark off the statements that are to be associated with the **if** or the **else.** The next program fragment is a corrected version of the preceding fragment:

```
if Winged then
  if Warmblooded then
    begin
      if not Bird then
        writeln (' This animal is a bat.')
    end
  else
    writeln (' The organism is an insect or an extinct flying dinosaur.')
```

Now, if **Winged** is true and **Warmblooded** is false, this fragment prints the output statement about insects and flying dinosaurs.

Notice that there is never any punctuation before an **else.** Putting a semicolon before an **else** will result in a compile-time error:

```
            if Temperature < 100.0 then
              if Temperature > 0.0 then
                Water
              else
                Ice;
***error***      $6

            else
              Steam;
            NextProcedure

error messages :
***************
    6: illegal symbol (possibly missing ";" in line above).
```

Remember that a semicolon separates the next executable statement from the current executable statement. **If-then-else** is a single executable statement. Any semicolons within an **if-then-else** statement must be inside compound statements that are marked off by **begin-end** pairs.

> REMINDER
>
> Indentation and spacing are used to make a Pascal program easier to read. Indentation within an **if-then-else** statement can be used to make it clear to the reader which **if** is associated with which **else**. The indentation and spacing do not, however, affect the way a program executes. Indentation can be used to clarify (for human readers) a statement that is syntactically and logically correct, but the computer reads only the syntax.

5.5 The case statement

A **case** statement can be used when the value of a single expression determines which of several statements will be executed and when all the possible values that the expression can assume in a particular situation can be explicitly stated. For example, an automatic money machine at a bank executes the dispense-money, deposit-money, or check-balance routine depending on your response to the question, "What do you want to do?" The money machine could be programmed with a **case** statement.

The syntax diagram for the **case** statement is

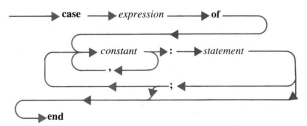

If the variable **Answer** has been declared to be a character variable and the procedures **Balance, SavingDeposit, CheckingDeposit,** and **DispenseMoney** have been defined, the following **case** statement can be formed from the syntax diagram:

```
writeln (' What do you want to do?');
writeln (' Enter');
writeln ('        b  to get your balance');
writeln ('        s  to deposit money to your savings account');
writeln ('        c  to deposit money to your checking account');
writeln ('        r  to receive money');

readln (Answer);
```

```
case Answer of
  'b' : Balance;
  's' : SavingDeposit;
  'c' : CheckingDeposit;
  'r' : DispenseMoney
end
```

If the value of the **Answer** is 'b', the procedure **Balance** is executed; if the value is 's', the procedure **SavingDeposit** is executed; and so on. The expression that determines the execution of the **case** statement (the variable **Answer** in the example above) is called the *control expression*. The constants ('b', 's', 'c', and 'r') are called the *case-label list*. All values that it is possible for the control expression to assume at the time of execution of the **case** statement must be in the case-label list. Note that the case-label list contains only constants. You cannot put variables in a case-label list.

The **case** statement is executed as a sequence of **if-then-else** statements. The preceding example is equivalent to

```
if Answer = 'b' then
  Balance
else
  if Answer = 's' then
    SavingDeposit
  else
    if Answer = 'c' then
      CheckingDeposit
    else
      if Answer = 'r' then
        DispenseMoney
```

Notice that each time a **case** statement is executed only one of the statements in the list is executed, because the control expression can be equal to only one of the constants in the list. The following procedure illustrates a **case** statement that, when executed, executes one of four statements, depending on the value that is entered at the terminal:

```
procedure NumberNames;
(*    This procedure writes the word for the number entered using     *)
(*    a case statement to control which message is printed.           *)

var   Control : integer;
```

```
begin
  writeln;
  writeln (' Enter a number between 0 and 3');
  readln (Control);

  case Control of
     0 : writeln (' zero');
     1 : writeln (' one');
     2 : writeln (' two');
     3 : writeln (' three')
  end (* case *)
end; (* procedure NumberNames *)

=>
 Enter a number between 0 and 3
 ▫2
 two
```

The statements after the constants in the case-label list can be simple or compound statements. They can be procedures, **if** statements, or even other **case** statements. If the same statement is to be executed for more than one value of the control expression, the constants can be listed with commas between them. For example,

```
case Index of
   1, 2, 3, 4 : writeln (Index:3, ' is less than 5.');
           5 : begin
                 writeln (' That''s a five.');
                 writeln (' And that''s the end.')
               end    (* case 5 *)
end (* case *)
```

A **case** statement must end with the word **end**, but it does not start with **begin**. Therefore, be careful if your statements are compound statements with **begin**s and **end**s. It may appear that you have an extra **end** when in fact the **end** is not extra—it is necessary.

The control expression must take values of an ordinal data type (**integer**, **boolean**, or **char**), and the constants in the case-label list must be of the same data type as the value of the control expression. It is impossible to list all possible values of a real variable; therefore the control expression cannot be of the data type **real**.

If the control expression equals a value that does not appear in the case-label list, the result of the **case** statement is undefined. Some compilers execute the next statement. Others issue a run-time error message. For example,

```
=>
□6
                                                          pascal termination log
                                                          ------ ----------- ---
*** error ***              undefined case label

program terminated at offset 000116 in segment caseind

local vars: index = 6
```

It is a good idea to put an **if** statement in front of a case statement to prevent the case statement from being executed when the control variable is not within the correct range. An **if** statement precedes the following **case** statement:

```
if (Index < 1) or (Index > 5) then
   writeln (' Index must be between 1 and 5')
else
   case Index of
      1, 2, 3, 4 : writeln (Index:3, ' is less than 5.');
               5 : begin
                      writeln (' That''s a five.');
                      writeln (' And that''s the end.')
                   end   (* case 5 *)
   end (* case *)
```

As an extension to standard Pascal, many Pascal compilers support the use of **otherwise** in the case statement. The **otherwise** allows you to specify the action to be taken if the control variable does not match any of the values in the case-label list. You will have to check the documentation for your compiler to find out if it supports **otherwise.** When using any extensions to standard Pascal, be aware that your program will not work on all Pascal compilers. If you want your program to be portable to other compilers, use only standard Pascal constructions.

▌Solution to Case Study 5

The algorithm for the solution to the case study involves the use of procedures that invoke other procedures. The main program invokes the procedures **Initialize** and **Choose. HighCompare, LowCompare,** and **Precipitation** are invoked from procedure **Choose. Switch** is invoked from **LowCompare,** and **Exchange** from **HighCompare.** Because **Highest, Low-**

est, and **TotalRain** are used both within the procedure **Initialize** and within the block of procedure **Choose,** they must be declared as global variables.

In reading the program below, remember to go first to the main program. Pretend that the main program and each of the procedures are written on separate pieces of paper so that you can see the development of each part of the program. The main program is written first, then the declaration of global constants and variables. The individual procedures are then written one at a time. The pieces that have been written are put together, and then the program is typed into the computer.

As you look at the program, compare it with Algorithm 5.4. Except for the addition of variable declarations, the refinement of Step 3 which uses the **case** statement, and the decision to use a **case** statement in the last step, the program statements are almost direct translations of the algorithmic statements.

PROGRAM 5.2   Meteorology

```
program Meteorology (input, output);
(*      This program tests a set of procedures that are to be included   *)
(*      in a large program designed for meteorologists who must keep     *)
(*      track of large amounts of data.  The procedures included in      *)
(*      this test program keep track of the highest temperature,         *)
(*      the lowest temperature, and the total rainfall over a period of  *)
(*      days.  In the final version, the values for the previous days    *)
(*      will be supplied by another part of the program, but for the     *)
(*      test program, they must be initialized.                          *)


   var  Highest : real;      (* highest temperature in the observation *)
                             (* period so far - in degrees C. *)
        Lowest : real;       (* lowest temperature in the observation *)
                             (* period so far - in degrees C. *)
        TotalRain : real;    (* total rain that has fallen during the *)
                             (* observation period - in cm. *)
        HighToday : real;    (* highest temperature of the current day - in *)
                             (* degrees C. *)
        LowToday : real;     (* lowest temperature of the current day - in *)
                             (* degrees C. *)
        RainFall : real;     (* rain fall today - in cm. *)
        Choice : char;       (* character entered by the user indicating *)
                             (* which procedure should be executed. *)
```

```pascal
procedure Initialize;
(*      Initialize writes a message to the user and initializes the     *)
(*      variables Highest, Lowest, and TotalRain to their initial       *)
(*      values.  In the final program, these values would be passed     *)
(*      from  another procedure.                                        *)

begin
  writeln;
  writeln (' This program allows you to determine the highest',
           ' temperature, the');
  writeln (' lowest temperature, or the total amount of rainfall for');
           ' a given');
  writeln (' observation period.  You will be asked which you want to',
           ' determine.');
  writeln;

  Highest := -1000.0;
  Lowest := 1000.0;
  TotalRain := 0.0
end; (* procedure Initialize *)

(*$i exchange *)

procedure Switch;
(*      Procedure Switch exchanges the values that are stored in Lowest *)
(*      and LowToday using the intermediate variable Save.              *)

var  Save : real;

begin
  Save := Lowest;
  Lowest := LowToday;
  LowToday := Save
end; (* procedure Switch *)


procedure HighCompare;
(*      This procedure reads today's high temperature and compares it   *)
(*      to the highest temperature so far.  If today's temperature is   *)
(*      higher, then the values are exchanged by the procedure Exchange.*)

begin
  writeln (' What was today''s highest temperature in degrees Celsius?');
  readln (HighToday);

  if HighToday > Highest then
    Exchange
end; (* procedure HighCompare *)
```

```pascal
procedure LowCompare;
(*      This procedure reads today's low temperature and compares it      *)
(*      to the lowest temperature so far.  If today's temperature is      *)
(*      lower, then the values are exchanged using the procedure Switch.  *)
begin
  writeln (' What was today''s lowest temperature in degrees Celsius?');
  readln (LowToday);

  if LowToday < Lowest then
    Switch
end; (* procedure LowCompare *)

procedure Precipitation;
(*      This procedure reads today's rainfall and adds it to the          *)
(*      total rain that has fallen during the observation period.         *)

begin
  writeln (' What was today''s rainfall in cm? ');
  readln (RainFall);

  TotalRain := TotalRain + RainFall
end; (* procedure Precipitation *)

procedure ReadChoice;
(*      ReadChoice presents the menu of choices to the user, asks which   *)
(*      function is to be performed, and reads the response from the      *)
(*      user.                                                             *)

begin
  writeln (' What do you want to do?');
  writeln (' Enter');
  writeln ('         h  to determine the highest temperature');
  writeln ('         l  to determine the lowest temperature');
  writeln ('         r  to determine the rainfall');
  writeln;

  readln (Choice)
end; (* procedure ReadChoice *)

procedure Choose;
(*      Choose uses a case statement to make the choice among the         *)
(*      alternatives presented to the user. One of three procedures       *)
(*      will be executed depending on the value of the character          *)
(*      variable Choice.                                                  *)
begin
  case Choice of
    'H', 'h' : HighCompare;
    'L', 'l' : LowCompare;
    'R', 'r' : Precipitation
  end (* case *)
end; (* procedure Choose *)
```

```
    procedure Report;
    (*     This procedure reports the current values of Highest, Lowest,     *)
    (*     or TotalRain depending on which option was chosen by the user.    *)

    begin
      writeln;
      writeln (' For the current observation period, ');

      if (Choice = 'H') or (Choice = 'h') then
        writeln (' the highest temperature so far is', Highest:7:2,
                 ' degrees C');

      if (Choice = 'L') or (Choice = 'l') then
        writeln (' the lowest temperature so far is', Lowest:7:2,
                 ' degrees C');

      if (Choice = 'R') or (Choice = 'r') then
        writeln (' the total rain fall has been',TotalRain:7:2,' cm');
      writeln
    end; (* procedure Report *)
(*---------------------------------------------------------------------*)

begin (* main program *)
  Initialize;
  ReadChoice;
  Choose;
  Report
end. (* program Meteorology *)

=>

This program allows you to determine the highest and lowest
temperature, or the total amount of rainfall for a given
observation period.  You will be asked which you want to determine.

What do you want to do?
Enter
        h  to determine the highest temperature
        l  to determine the lowest temperature
        r  to determine the rainfall
□L
What was today's lowest temperature in degrees Celsius?
□12.0

For the current observation period,
the lowest temperature so far is  12.00 degrees C
```

APPLICATION
Computational Errors

There are many sources of error in computer calculations. Some types of errors are inherent to the mathematical computations; others arise more specifically in computer calculations. For example, in calculations done with data obtained from experimental measurements, errors due to uncertainty of the measurements will occur whether you perform your calculations on a computer, on paper, or in your head. On the other hand, errors due to limitations on the size of a number that can be stored in a computer's memory are a problem exclusive to computers.

CASE STUDY 6
Errors in computing the travel time of a ball

A ball is tossed straight up into the air. The equation that describes the motion of the body is

$$y = y_0 + v_0 t + gt^2 \qquad (5.1)$$

where
y = vertical displacement in meters
y_0 = initial height in meters
v_0 = initial velocity in meters per second
g = acceleration of gravity in meters per second squared
t = time in seconds

The value entered for g must be a negative number because the force of gravity acts in a downward direction. Figure 5.4 illustrates the vertical displacement of the ball as a function of time. How long does it take for the ball to fall to the ground ($y = 0$)? What are the sources and effects of errors in the computation of the time required for the ball to reach the ground?

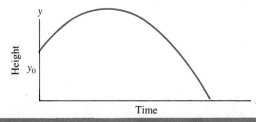

FIGURE 5.4
Vertical displacement of a ball tossed into the air as a function of time.

Algorithm development

Through use of the quadratic formula, Equation 5.1 can be rewritten so that time is given in terms of the known variables, y_0, y, v_0, and g:

$$t = \frac{-v_0 \pm \sqrt{v_0^2 - 4g(y_0 - y)}}{2g} \quad (5.2)$$

Figure 5.4 shows that the curve of the ball meets the time axis once. The value of time t at which $y = 0$ is called a *root* of the equation. If the curve of the ball were continued to the left, it would also intersect the time axis at a negative value, so the curve has two roots. The solution to Equation 5.2 gives the same result, a positive and negative root:

$$t_1 = \frac{-v_0 + \sqrt{v_0^2 - 4gy_0}}{2g}$$

and (5.3)

$$t_2 = \frac{-v_0 - \sqrt{v_0^2 - 4gy_0}}{2g}$$

It is clear that for this application only the positive root of the quadratic equation is of interest. A ball cannot be thrown into the air and land in negative time. Because of the nature of the problem, there are no complex roots. Complex roots, which will be covered in Chapter 13, occur when the argument of the square root is negative.

The first part of the case study can be restated as follows: Use Equations 5.3 to find the positive value of time t at which $y = 0$. This will be either t_1 or t_2. The initial **Height,** y_0, and the initial **Velocity,** v_0, are supplied by the user. The acceleration due to **Gravity,** g, can be either declared as a constant or supplied by the user. If the coefficients of the quadratic formula, usually referred to as a, b, and c, were all supplied by the user (in the case study these are **Gravity, Velocity,** and **Height**), the algorithm could be used as a general solution to the quadratic formula. However, because the input and output will be about heights, velocities, and times rather than a, b, and roots, and because **Gravity** is a constant, the program will solve just the case study rather than attempting to be a general solution. From the program, you should be able to write a general solution to the quadratic formula. The algorithm for the case study can be stated as follows:

ALGORITHM 5.5   First iteration of the algorithm for Case Study 6

begin algorithm
1. Initialize the program
2. Read **Height** and **Velocity**
3. Find the roots of the equation, **Time1** and **Time2,** using Equations 5.3
4. Report the positive root

end algorithm

Step 3 must be refined to account for the fact that the argument of the square root might be negative. If it is negative, a message stating that the equation has complex roots and that the computation cannot continue must be written to the user. Algorithm 5.6 is a refinement of Algorithm 5.5

ALGORITHM 5.6   Second iteration of the algorithm for Case Study 6

begin algorithm
1. Initialize the program
   Write 'This program computes the time in seconds that it takes a ball tossed into the air to reach the ground.'
2. Read **Height** and **Velocity**
   begin
     Write 'Enter the height in meters from which the ball is tossed'
     Read **Height**
     Write 'Enter the initial velocity of the ball in meters per second'
     Read **Velocity**
   end
3. Find the roots of the equation, **Time1** and **Time2,** using Equations 5.3
   begin
     Radical $\leftarrow$ sqr(**Velocity**) $-$ 4.0 $*$ **Gravity** $*$ **Height**
     if **Radical** $<$ 0.0 then
       write 'The quadratic equation has complex roots, which this program cannot compute'
     else
       begin
         Time1 $\leftarrow$ ($-$**Velocity** $-$ sqrt(**Radical**)) / (2.0 $*$ **Gravity**)

$$\textbf{Time2} \leftarrow (-\textbf{Velocity} + \text{sqrt}(\textbf{Radical})) / (2.0 * \textbf{Gravity})$$
    end else
  end
 4. Report the positive root
   begin
    if **Time1** > 0.0 then
     write 'The time for the ball to reach the ground is', **Time1**, 'seconds'
    else
     write 'The time for the ball to reach the ground is', **Time2**, 'seconds'
   end
 end algorithm

One problem with Algorithm 5.6 is that Step 4 is executed even if **Radical** < 0.0 and **Time1** and **Time2** have not been computed. A method is needed to prevent Step 4 from being executed if there are no real roots. There are several alternatives. One alternative is to include Step 4 within the **else** statement of Step 3. Another alternative is to set a boolean variable, called an *error flag*, when the error condition is encountered in Step 3. Then before Step 4 is executed, the error flag can be tested. Algorithm 5.7 includes this refinement. Notice that the error flag should be initialized at the beginning of the program.

ALGORITHM 5.7   Third iteration of the algorithm for Case Study 6

begin algorithm

1. Initialize the program
 Write 'This program computes the time in seconds that it takes a ball tossed into the air to reach the ground.'
 Initialize the error flag to false
  Error ← false

2. Read **Height** and **Velocity**

3. Find the roots of the equation, **Time1** and **Time2,** using Equations 5.3
 begin
  Radical ← sqr(**Velocity**) − 4.0 * **Gravity** * **Height**
  if **Radical** < 0.0 then
   begin
    write 'The quadratic equation has complex roots, which

                    this program cannot compute'
                Error ← true
            end if
        else
            begin
                Time1 ← (−**Velocity** − sqrt(**Radical**)) / (2.0 ∗ **Gravity**)
                Time2 ← (−**Velocity** + sqrt(**Radical**)) / (2.0 ∗ **Gravity**)
            end else
    end
4. Report the positive root
    begin
        if not **Error** then
            begin
                if **Time1** > 0.0 then
                    write 'The time for the ball to reach the ground is',
                        Time1, 'seconds'
                else
                    write 'The time for the ball to reach the ground is',
                        Time2, 'seconds'
            end if
        end
end algorithm

The remainder of the chapter will discuss the second question posed in the case study: the determination of the sources and effects of errors in the computation of the time required for the ball to reach the ground.

Human errors

Errors may arise from flaws either in the logic of an algorithm or in the coding of a program. These types of errors are due to human mistakes. Methods for dealing with such programmer errors are covered throughout the book. Errors may also arise from mistakes in the hardware or software of the computer system itself. Such system errors, also due to human error, are less common but can affect many people, as illustrated in the example below.

A compiler was released with a bug in it: it did not multiply by zero correctly. Because all other multiplication was correct and because it is very unusual to explicitly multiply by zero, the bug went undetected. The bug only exhibited itself when a variable that became zero inside a program was multiplied by another number. Unless all the intermediate values were printed out, there was no way to know when the answer was garbage and when it was not.

This type of bug is rare. Logic errors are a much more common source of computer errors. The other errors arising in programming are due mainly to truncation errors, inherent errors (measurement errors), and round-off errors, all of which are to some extent unavoidable.

Truncation error

Truncation errors arise when an infinite process is approximated by a finite process. Many functions such as the sine and square root are equal to the sum of an infinite series of numbers. To evaluate these functions, a computer must drop or truncate some of the terms of the infinite series; otherwise the computation would continue forever. The error resulting from the dropping of terms in referred to as truncation error.

Truncation error also arises from the decisions that are made about which equations to use to describe a system. Because a finite set of equations can never represent every aspect of a system, the computer program will always be an approximation to the physical system. For example, in the case study, the motion of a ball is described with a simple quadratic equation. There are, however, other factors that influence the time that it takes for the ball to reach the ground. Equation 5.1 ignores factors such as the wind resistance of the ball and the effect of wind velocity on the ball. These factors may be negligible, or they may overwhelm the factors included in Equation 5.1. To use a model intelligently, you must be aware of the factors that have been omitted so that you know the range of valid input values. Would you expect Equation 5.1 to be valid if the ball were tossed from a height of 5 kilometers?

Inherent errors

Inherent errors are errors due to uncertainty of measurement. Any measured value of a variable is subject to error. Most engineering and scientific programs are mathematical models of physical processes that require input data from experiments. Measurements obtained from experiments are subject to unknown errors due both to the design of an experiment and to miscalibration and misreading of instruments. Sometimes available data are insufficient, and the input data must be approximated by another computer model. For example, in the case study, the initial height and the initial velocity may not be precisely known.

There is nothing you can do in the writing of a computer program to repair inherent errors. You can, however, perform tests that measure the *sensitivity* of the results of your program to errors in the input data. You can determine the effect of a small change in a variable, such as the initial velocity in the case study. If a small change in the initial velocity results in a large difference in the computed time, you know that the accuracy of the measurement of the initial velocity is critical to the results. If you can show that this measurement is important to the program results, you might consider investing in a new instrument that will improve the accuracy of the measurement of the initial velocity.

Significant digits

A *significant digit* is a digit that contributes to the accuracy or precision of a number. The *most-significant digit* is the digit that contributes the most value to the number, and the *least-significant digit* is the digit that contributes the least value to the number. In the number 1.2345, the digit 1 is the most-significant digit and the digit 5 is the least-significant digit. Determining the significant digits in a number such as 5000 is more difficult. The number 5000 may be a rough estimate, with the zeros used only to indicate the placing of the decimal point, or the number 5000 may be a more precisely measured value that is between 4999 and 5001. In the first case, the number 5000 has only one significant digit. In the second case, the number 5000 has four significant digits. Scientific notation can be used to differentiate between the two values. In scientific notation, only the significant digits are recorded. So, the number 5×10^3 has one significant digit, and the number 5.000×10^3 has four significant digits.

When arithmetic operations are being performed, it is important to pay attention to the number of significant digits. If you perform several calculations based on the approximate number 5×10^3, the final answer will be no more accurate than the original estimate. It is easy to lose sight of this fact when you are using a calculator or computer that can report numbers to 8, 10, or more decimal places. Reporting nonsignificant digits implies an accuracy that does not really exist. The standard practice for reporting significant digits is to use the digit to the right of the least-significant digit: If this digit is greater than or equal to 5, the least-significant digit is rounded up to the next integer; otherwise, the least significant digit is left as is.

The following guidelines can help you to judge which digits in an answer are significant:

1. *Multiplication and Division:* The number of significant digits in the

product or quotient is the number of significant digits in the number with the fewest significant digits.

For example, the product

$$5.2378 \times 4.38 = 22.94156$$

should be reported as 2.29×10^1 because 4.38 has only three significant digits. Remember that the number of significant digits is *not* the number of decimal places. It is the number of digits that contribute to the value of a number. The quotient

$$8.95 \times 10^4 \div 3.2 = 27968.75$$

should be reported as 2.8×10^4 because 3.2 has only two significant digits. Notice that the answer is rounded up to 2.8 when it is reported.

2. *Addition and Subtraction:* In addition and subtraction, the result should not be carried out to the right beyond a column that does not contain a significant digit.

For example, the sum

$$3.78 + 654.9 = 658.68$$

should be reported as 6.587×10^2 because the tenths column is the last column in which every number is significant. Notice that the answer has four significant digits, but that the number 3.78 has only three significant digits. The difference

$$5 \times 10^3 - 1 = 4999$$

should be reported as 5×10^3 because the thousands column is the last column in which all the digits are significant. Because the number 5×10^3 has only one significant digit, it could represent any number between 4500 and 5499, so subtracting one from it does not change the reported value.

When calculations are performed by hand, it is standard practice to round intermediate values to the number of significant digits plus one extra digit. Rounding is useful in hand calculations because it eliminates unnecessary computations. When calculations are performed with computers, however, rounding of intermediate variables is cumbersome and unnecessary. It is standard practice not to round numbers until the last step, when the numbers are reported.

A *transcendental number* such as π or *e* has no exact representation in any number base, so all digits are significant. If you enter a transcendental number as a constant, you can enter as many digits as the computer will store. Because all of your other data is unlikely to be that accurate, you only need to enter as many significant digits as your least accurate data value. If you enter too many digits, the number will be truncated.

For other numbers, such as the square root of 2, you can use the built-in functions of the computer, which should give the same accuracy as if you had typed in the number. For some physical constants that have been measured accurately to many places, such as the speed of light in a vacuum and the charge on an electron, you only need to enter as many significant digits as the number of significant digits in the other variables in the program. For other physical constants that vary depending on when and where they are measured, such as the force of gravity and the speed of sound, you should only enter three or four significant digits. Entering more digits implies an accuracy that does not really exist.

Floating-point arithmetic

As you have already learned, real numbers are represented in the computer by their mantissa and their exponent. The mantissa contains the significant digits, and the exponent contains the power of 10 by which the number is multiplied. For example,

	Mantissa	Exponent
$105.2 = 0.1052 \times 10^3$	.1052	3
$0.0006 = 0.6000 \times 10^{-3}$	.6000	-3
$-0.01 = -0.1000 \times 10^{-1}$	$-.1000$	-1
$-1234.0 = -0.1234 \times 10^4$	$-.1234$	4

The mantissa is stored in *normalized form;* that is, the leading digit next to the decimal point is the first nonzero digit in the number. The normalized form ensures that as many significant digits as possible are stored.

Most computers include a *guard bit* for the mantissa. The guard bit is an extra digit that is never reported, but is used to test whether a number should be rounded up or down. In computers that do not use a guard bit, the number is chopped off after the last digit of the mantissa. The proposed IEEE standard for floating-point arithmetic requires not only a guard bit, but also a round bit and a sticky bit, both of which are necessary to ensure accuracy in rounding of numbers. (See Reference 5 at the end of the chapter.) In the examples that are given below, to emphasize the effects of the limited word size, the numbers will be chopped rather than rounded.

Arithmetic operations on real numbers are performed in *floating-point* arithmetic. To add two numbers, for instance, the decimal points must be aligned so that the ones columns are added together, the tens columns are added together, and so on. For example, to line up the decimal points of the numbers

$$.1052 \times 10^3$$
$$.6000 \times 10^{-3}$$

multiply the second, or smaller, number by 10^6. Both numbers will be raised to the same power, and the mantissas can be added:

$$.1052 \times 10^3 + .0000006 \times 10^3 = .1052006 \times 10^3$$

The power of 10 that the smaller number is multiplied by (10^6) is just the difference between the two exponents $(3 - (-3))$.

Round-off errors

The number of digits that can be used to represent a number inside a digital computer is fixed by the word size of the computer. The number of ways the bits can be arranged is finite; therefore, only a finite number of numbers can be represented. Every number must be mapped into one of the available patterns.

Suppose there were a computer operating in base 10 that could store only the sign of the mantissa, up to four digits for the mantissa, the sign of the exponent, and two digits for the exponent. On this computer, the following numbers would all be stored as 0.1234 or $+1234e-04$:

0.1234

0.12341

0.123431

0.123434

0.1234302

etc.

The fixed word size of computers affects the outcome of all the arithmetic operations. If the hypothetical computer that stores only four significant digits

TABLE 5.1
Errors in the Computation of t_1 and t_2

v_0	y_0	t_1 computed	t_1 actual	t_2 computed	t_2 actual
0.0	1.0	−0.3192	−0.319275	0.3192	0.31928
15.0	1.0	−0.06371	−0.063989	1.592	1.5930
25.0	1.0	−0.03925	−0.039391	2.587	2.5878
35.0	1.0	−0.02854	−0.028346	3.596	3.5961
45.0	1.0	−0.02192	−0.022116	4.609	4.6093
55.0	1.0	−0.01784	−0.018123	5.621	5.6246

were to add the numbers 0.1052×10^3 and 0.6000×10^{-3} the result would be 0.1052×10^3, the same as the first number. Because of the fixed word size, $A + B$ is equal to A. This effect is not the same as that of rounding for significant digits. Both numbers might have more significant digits, but because of the limited word size not all the significant digits can be reported. This effect is referred to as *round-off error*.

To see the effect of round-off error in a problem, assume that the case study is to be solved on the computer that stores four significant digits of a real number. Table 5.1 gives values of v_0 and y_0, the computed values of t_1 and t_2, and the actual values of t_1 and t_2. The round-off errors are largest when v_0 is much larger than y_0. Under these conditions, the square root is approximately equal to v_0, so t_1 is equal to the difference between two almost-equal numbers.

Representation errors

The examples above have been given in base 10 so that they would be understood, but computers operate in base 2 rather than base 10. This fact leads to another problem: conversion from base 10 to base 2. If you enter the number 0.10000 in your data, this number appears to be very precise. It has a *terminating* representation in base 10. However, when 0.10000 is converted to base 2 to be stored in the computer's memory, the resulting number is 0.000110011001100, a *nonterminating, repeating* representation in binary. The same problem also occurs in reverse. Within the computer, all the computations are performed in base 2. Before the numbers are reported to the user, they must be converted back to base 10.

Quantification of errors

The *absolute error*, e_a, in a calculation is defined to be the difference between the true value and the computed value:

$$e_a = x_t - x_c \tag{5.3}$$

where x_t is the true value and x_c is the computed value. In the example concerning floating-point arithmetic, the absolute error is $105.2006 - 105.2 = 0.0006$.

Of course, the true value is rarely known. If you knew what the true value was, you would use the true value instead of the calculated value. Errors can be quantified, however, without the true value. Often an answer is known to lie between an upper and lower bound:

$$T_{\min} < x_t < T_{\max} \tag{5.4}$$

For some mathematical problems, the upper and lower bounds can be found through numerical analysis. Usually, however, these bounds are conservative; that is, they are much larger than the actual errors in the computation.

An *error band,* which is the difference between the upper and lower bounds, can also be used to define the error. Absolute errors are frequently given in the form

$$x_t = x_c \pm 0.0006 \tag{5.5}$$

Equation 5.5 is read as "The true value of x is within plus or minus 0.0006 of the computed value."

How good or bad the error band is depends on how large x_t is. If x_t is on the order of 10^{12} and the error is 0.001, then the error is relatively small. If however, x_t is on the order of 10^{-12} and the error is 0.001, then the error is large because it is bigger than what is being measured. Therefore, the *relative error* is the more common measurement for error. The relative error is the absolute error divided by the computed value:

$$e_r = \frac{x_t - x_c}{x_c} \tag{5.6}$$

For the last row of Table 5.1, the relative error in t_1 is

$$e_r = \frac{-0.018123 - (-0.01784)}{-0.01784} = 0.0159$$

Normally, the relative error is given as a percentage so that it can be easily distinguished from the absolute error. For example,

$$x_t = x_c \pm 1.59\% \tag{5.7}$$

To convert the relative error into percentage error, multiply the relative error by 100.0.

Most error bounds cannot be stated so precisely as they are in Equations 5.6 and 5.7. Because any measured value of a variable is subject to unknown errors, the probability that a measured value of x lies between two bounds is usually given. Probability and statistics, which will be covered in Chapter 10, can be used to quantify errors in experimental measurements.

Propagation of errors

Another area of concern is the propagation of errors in a sequence of calculations. Some errors may be amplified as further calculations are carried out, whereas other errors may decrease. Whether the initial errors are due to round-off error or inherent error, the interaction of errors in calculations can be difficult to quantify or detect. A complete discussion of the propagation of errors is beyond the scope of this book. References 1 and 2 have individual chapters on errors and error propagation. References 3 and 4 contain sophisticated mathematical discussions of the effects of round-off errors on numerical computations.

Solution to Case Study 6

Program 5.3, **TimeToFall,** uses the algorithm developed at the beginning of the chapter to compute the time required for a ball to reach the ground.

The sources of error in this program are round-off errors in the computer calculations and inherent errors in the measurement of the initial height and the initial velocity. As shown in Table 5.1, the positive root of the quadratic equation is relatively insensitive to truncation errors. The two sets of output from Program 5.3 show the relative sensitivity of the positive root to a small change in the velocity and to a small change in the height.

PROGRAM 5.3 TimeToFall

```
program TimeToFall( input, output);
(*     This program computes the time that it takes for a ball, tossed    *)
(*     straight up into the air, to fall to the ground.  The height from  *)
(*     which the ball is tossed and the velocity with which it is tossed  *)
(*     are read from the user.                                            *)

const Gravity = -9.8;      (* force of gravity in meters per second. *)
                           (* the force is negative because gravity acts *)
                           (* in a downward direction. *)

var  Height : real;        (* height in meters from which the ball is *)
                           (* tossed -- given. *)
     Velocity : real;      (* velocity in meters per second with which the *)
                           (* ball is tossed -- given *)
     Time1 : real;         (* first root of the equation -- computed. *)
     Time2 : real;         (* second root of the equation -- computed. *)
     Error : boolean;      (* Error is set to true if the roots of the *)
                           (* equation are imaginary. *)

  procedure Initialize;
  (*     This procedure writes the message to the user stating the        *)
  (*     purpose of the program and initializes the error flag to false.  *)
  begin
    writeln (' This program computes the time in seconds that it takes a');
    writeln (' ball tossed into the air to reach the ground.');

    Error := false
  end; (* procedure Initialize *)


  procedure ReadInitialData;
  (*     This procedure reads the initial height and velocity from the    *)
  (*     user.                                                            *)
  begin
    writeln;
    writeln (' Enter the height in meters from which the ball is tossed.');
    readln (Height);

    writeln (' Enter the initial velocity of the ball in meters per',
             ' second.');
    readln (Velocity)
  end; (* procedure ReadInitialData *)
```

```
procedure FindRoots;
  (*    This procedure finds the roots of the equation that describes    *)
  (*    travel time of the ball.  The equation is a quadratic equation   *)
  (*    that can be solved using Equation 5.3.                           *)

  var  Radical : real;      (* intermediate variable to store the value of *)
                            (* the argument of the square root. *)
begin
  Radical := sqr (Velocity) - 4.0 * Gravity * Height;

  if Radical < 0.0 then
    begin
      writeln (' The quadratic equation has complex roots, which',
               ' this program cannot compute.');
      Error := true
    end
  else
    begin
      Time1 := (-Velocity - sqrt (Radical)) / (2.0 * Gravity);
      Time2 := (-Velocity + sqrt (Radical)) / (2.0 * Gravity)
    end
end; (* procedure FindRoots *)


procedure ReportPositiveRoot;
  (*    This procedure reports the positive root of the quadratic        *)
  (*    equation, if it exists.                                          *)
begin
  if not Error then
    begin
      writeln;
      writeln (' When a ball is tossed from an initial height of ',
               Height:10:2, ' meters');
      writeln (' at a velocity of ',Velocity:10:2, ' meters per second,');

      if Time1 > 0.0 then
        writeln (' the time for the ball to reach the ground is',
                 Time1:10:2, ' seconds')
      else
        writeln (' the time for the ball to reach the ground is',
                 Time2:10:2, ' seconds')
    end (* if not Error *)
end; (* procedure ReportPositiveRoot *)
```

```
(*--------------------------------------------------------------------*)
begin
  Initialize;
  ReadInitialData;
  FindRoots;
  ReportPositiveRoot
end. (* program TimeToFall *)
```

```
=>
 This program computes the time in seconds that it takes a
 ball tossed into the air to reach the ground.

 Enter the height in meters from which the ball is tossed.
□10.0
 Enter the initial velocity of the ball in meters per second.
□5.0

 When a ball is tossed from an initial height of      10.00 meters
 at a velocity of       5.00 meters per second,
 the time for the ball to reach the ground is      1.30 seconds


=>
 This program computes the time in seconds that it takes a
 ball tossed into the air to reach the ground.

 Enter the height in meters from which the ball is tossed.
□10.1
 Enter the initial velocity of the ball in meters per second.
□5.1

 When a ball is tossed from an initial height of      10.10 meters
 at a velocity of       5.10 meters per second,
 the time for the ball to reach the ground is      1.31 seconds
```

Summary

Boolean is the fourth simple data type in Pascal. Boolean variables take values of true or false. Boolean expressions, which also take values of true or false, are formed of variables, relational operators, and boolean operators. The relational operators are =, <, >, <=, >=, <>. The boolean operators are **and, or, not.**

The use of conditional statements allows you to write programs that branch. The value of a boolean expression can be used to determine which sections of code are executed. An **if** statement tests the value of a boolean expression, executing a section of code if the expression's value is true and not executing the section if the expression's value is false. An **if-then-else** statement is used when one section of code is to be executed if the boolean expression is true and another is to be executed if the boolean expression is false.

The **case** statement is used when all the values that an expression can assume in a particular situation can be stated. The values that the expression can take are listed in a case-label list, and the action to be taken is stated for each possible value of the expression. The expression cannot take values of the data type **real.**

References

1. W. S. Dorn and D. D. McCracken, *Numerical Methods with Fortran IV Case Studies,* New York: John Wiley and Sons, 1972.

2. J. R. Rice, *Numerical Methods, Software, and Analysis: IMSL Reference Edition,* New York: McGraw-Hill Book Company, 1983.

3. P. H. Sterbenz, *Floating-Point Computation,* Englewood Cliffs, N.J.: Prentice-Hall, 1974.

4. J. H. Wilkinson, *Rounding Errors in Algebraic Processes,* Englewood Cliffs, N.J.: Prentice-Hall, 1963.

5. "A Proposed Standard for Binary Floating-Point Arithmetic," Draft 8.0 of IEEE Task P754, *Computer,* 1981, pp. 51–62.

Exercises

PASCAL

1. Correct the syntax error or errors in each section below. Assume that all variables and procedures have been declared.

(a) if First < Second then;
    First := First + Second

(b) if Number > '3' then
    Number := sqrt (Number)

(c) case RealVar of
      1.0 : Next := RealVar * 2.0;
      2.0 : begin
              if Next < Realvar then
                begin
                  Exchange;
                  Report
                end;
      3.0 : if Next > RealVar then
              Report;
    ReportAfterCase

(d) case Number of
      begin
        X : ComputeX;
        Y : ComputeY;
      end

(e) if Test = true then
      writeln (The value of',
               ' Test is ', Test)

2. Draw a diagram of the block structure of the solution to the case study. Use the diagrams in Chapter 4 as models.

3. What is the output for each fragment below? Assume that **Num1** = 1, **Num2** = 4, and **Num3** = 0. Do not be misled by the indentation. Repeat the problem for **Num1** = 4, **Num2** = −1, and **Num3** = 2.

(a) if Num1 < Num2 then
      begin
        if Num1 > 2 then
          writeln (' Num1 = ', Num1:5)
      end
    else
      writeln (' Logically that''s:',
               Num2 + 2:7)

(b) if not (Num1 >= Num2 + 2) then
      writeln (' True')
    else
      writeln (' False')

(c) if (Num1 < 8) and (Num3 < 8) then
      writeln (' They''re less')
    else
      writeln (' They''re more')

(d) if Num1 < Num2 then
      if Num3 > Num2 then
        writeln (' One more')
      else
        writeln (' Done')

(e) if Num1 < Num2 then
      begin
        writeln (' Num1 is', Num1:6,
                 ' and Num2 is', Num2:5);
        Num3 := 2
      end
    else
      writeln (' Num1 is greater ',
               ' than Num2');
    if Num1 < Num3 then
      writeln (' Num3 =', Num3:4)
    else
      writeln (' Num1 =', Num1:5)

(f) Test := ((Num1 < 8) or (Num2 > 8))
            and (Num3 > -1);
    if Test then
      writeln (' More or less')
    else
      writeln (' Less is more')

4. What is the range for the control variable **Index** in the following **case** statement? What is the output for each value?

```
case Index of
  5 : writeln (' The number is', Index:5);
  3 : writeln (' That was a ', Index:4);
  1 : writeln (' This is the first one');
  4 : writeln;
  2 : writeln (' The next number will be', Index + 1:3)
end  (* case statement *)
```

5. What values of the variable **Number** make the following **case** statement invalid? For each valid value of **Number**, what is the output of the **case** statement?

```
Control := Number * 2;
case Control of
  2,6 : begin
           Num1 := Control * 2.0;
           writeln (' Num1 =', Num1:8:2)
        end;
    8 : writeln (' Control is now', Control:3);
    4 : begin
           Num2 := Control div 3;
           writeln (' Num2 is ', Num2:4)
        end;
   10 : writeln (' Done')
end (* case *)
```

6. The control expression for a **case** statement can be a boolean expression. Would you ever write a **case** statement using a boolean expression as the control expression? Why or why not?

APPLICATION

7. Ask six different people to measure a piece of string with the same ruler, and record their answers. What is the error band? What is the average value of the measurements? Can you determine the true value? Can you determine the percentage error in the measurements?

8. Reconstruct Table 5.1 for a computer that stores only two significant digits for each number.

9. A shaft is to be inserted into a hole. The outer diameter of the shaft is 7.43 cm $\pm$ 0.05%. The diameter of the hole is 7.5 cm. How large can the error on the hole be before the largest shaft will no longer fit into the smallest hole? Give both the percentage error and the absolute error for the diameter of the hole.

Problems

PASCAL

1. Using a **case** statement, write a program to print one of four different messages depending on whether the value of an input variable is 0, 1, 2, or 3. If the input variable is outside the valid range of 0 to 3, a message should be written to the user and the program halted.

2. Write a program that reads four values into four variables. At the end of the program, the value that was read into the first variable will be stored in the last variable, the value that was read into the second variable will be stored in the third variable, the third in the second, and the fourth in the first. Print out the values of the variables at the end of the program.

3. The quadrant of a point (x, y) can be determined by the sign of x and y:

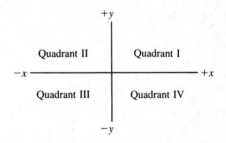

   Write a program that reads the x and y coordinates of a point and reports the quadrant in which the point is located.

4. The **arctan** function returns the principal angle, which is between $-\pi/2$ and $\pi/2$. That is, if y/x is positive, **arctan** returns an angle between 0 and $\pi/2$. If y/x is negative, **arctan** returns an angle between $-\pi/2$ and 0. Write a program that returns the angle, measured in a counterclockwise direction from the positive x axis, for a point in any quadrant. Be sure that your program returns the correct answer when x is equal to 0.

5. (a) Compute the grade for one student in a class with the following grading policy:

Quiz 1	30%
Final exam	45%
The best 9 out of 10 homework scores	25%

   Test your program on the following data:

   Quiz 1:   86
   Final exam: 78
   Homework: 90 85 95 90 80 00 85 90 70 85

(b) Rewrite the program assuming that the best 8 out of 10 homework scores are counted. That is, you must drop the two lowest grades.

6. Write a program to sort a list of three numbers into descending order. Report the values stored in the three variables after each exchange.

7. Write a program based on Program 3.4, **ComputeCurrent,** that allows the user to choose between analyzing a single circuit with a parallel connection and a single circuit with a series connection.

8. Write a program to compute the distances between three pairs of points. The formula for the distance between each pair of points (x_1, y_1) and (x_2, y_2) is given by

$$d = ((x_2 - x_1)^2 + (y_2 - y_1)^2)^{1/2}$$

Compute the distance between each pair of points, and then print the shortest distance. Test your program on the following points: [(1, 0), (0, 1)], [(2, 4), (3, 2)], [(−1, 2), (5, 7)].

9. Write a program that reads three positive integer numbers and tests whether any of the numbers are equal. Report how many numbers are equal and what they are. For example, if the input were 5 5 5, the output would be "There are 3 numbers equal to 5." If the input were 4 6 4, the output would be "There are 2 numbers equal to 4." If the input were 0 1 2, the output would be "There are no equal numbers."

10. The value of Y is defined for integer x by the function

$$Y = \begin{cases} x(x + 20) & \text{if } 2 \leq x \leq 5 \\ 2x & \text{if } -1 \leq x < 2 \\ 0 & \text{if } x > 5 \text{ or } x < -1 \end{cases}$$

Write a program that evaluates Y, given an integer value of x read from the terminal.

11. For an experiment you are running, only positive values can occur. Any negative values are data errors. Write a program that requests 20 experimental measurements from the user, tests the sign of each number, and writes out the position of any negative values (for example, "data point 4 is negative"). If there are no negative numbers in the list, at the end the program should write a message stating that all numbers are positive. Think about how you can tell, at the end of the program, whether all the values have been positive. Hint: Use a boolean variable or a counter.

12. Write a program that reads in 20 experimental data values and finds the number of *zero-crossings* in the data. A zero-crossing occurs when the algebraic sign (+ or −) of a data value is different from that of the preceding data value. For example, in the data values 10 12 −2 −3 1, a zero-crossing occurs in position 3, where the sign changes from +12 to −2, and in position 5, where the sign changes from −3 to +1. Consider 0 (zero) to be positive. Request 20 data values from the user, and then report the position (1 to 20) of each data value immediately following a zero-crossing. Also report the total number of zero-crossings.

APPLICATION

13. For your computer, determine
    (a) the smallest positive number, N, such that $1.0 + N > 1.0$
    (b) the smallest positive real number
    (c) the smallest negative real number
    (d) the largest positive real number
    (e) the largest negative real number

14. Write a program that computes the value of $x^2 - y^2$ using two different methods. Discuss the relative errors for each method (a) when the values of x and y are close, (b) when $x \gg y$, and (c) when x and y have opposite signs.

15. The resistors used in the circuit for Program 4.12, **AnalyzeThreeCircuits3**, are known to have an error band of $\pm 10\%$. Perform a sensitivity analysis for the computed current in one circuit when a 5.0-ohm resistor and a 10.0-ohm resistor are connected in parallel and when they are connected in series.

16. The force of gravity, relative to that of earth, on each of the nine planets is given below, along with an estimate of the error in the measurement. Write a program that requests the user's weight on earth in pounds and the name of a planet, and then determines the upper and lower bounds for the user's weight on the chosen planet. Assume that the absolute error in the user's weight is ± 5 lbs.

    | Mercury | $0.38 \pm 0.05\%$ |
    | Venus | $0.78 \pm 0.05\%$ |
    | Earth | 1.00 |
    | Mars | $0.39 \pm 0.05\%$ |
    | Jupiter | $2.65 \pm 0.05\%$ |
    | Saturn | $1.17 \pm 0.05\%$ |
    | Uranus | $1.05 \pm 0.10\%$ |
    | Neptune | $1.23 \pm 5.00\%$ |
    | Pluto | $0.05 \pm 25.0\%$ |

17. A computer salesperson has a choice between two ways of being paid: (1) an hourly wage of $7.00 per hour plus a 10% commission for each computer sold or (2) a 20% commission. The computers sell for $2000.00 each. The salesperson works a 40-hour week. Write a program that reads in the number of computers the salesperson expects to sell in a week and reports which payment scheme results in a higher salary. The program should also perform an error analysis for the salesperson, assuming that the expected number of computers sold could be three times actual sales or one-half actual sales.

18. Rewrite the airplane program from Problem 8 in Chapter 3 using the following additional data:

	passenger-miles/ gallon	seating capacity
747-jet	22.0	360
707-jet	21.0	136
SST	18.0	250
Helicopter	7.5	78

Ask the user what kind of an aircraft will be flown and how many passengers will be on board. If the user enters a number that is greater than the seating capacity for the chosen aircraft, write the user an error message. Report the fuel consumption in miles per gallon. Assuming that the passenger-miles per gallon can vary by 20% depending on the tailwinds and that the number of passengers can vary by 10% depending on the number of standbys and no-shows, report the upper and lower bounds on the fuel consumption. Do not forget that the number of passengers on board must be an integer number and that the plane will not hold more than the seating capacity.

CHAPTER 6

Subprograms with Parameters

You have learned to use the standard procedures, **read** and **write,** and the standard functions, **odd** and **sqrt.** You have also learned to define your own procedures. In this chapter you will learn to define your own functions and to write modular subprograms. Using modular subprograms will allow you to write subprograms that can be used in many different programs. Modular subprograms use parameters to access and change the values of global identifiers, eliminating the need to refer to the global identifier names. Because you are already familiar with procedures, writing modular procedures will be covered first.

> CASE STUDY 7
> **Minimum and maximum temperatures**
>
> In the case study in Chapter 5, a program called **Meteorology,** which found the highest and lowest temperatures and the total rainfall, was developed. In that program, two procedures, **Switch** and **Exchange,** perform the same actions with two sets of variables. Write a single procedure called **Exchange** to replace the two procedures in **Meteorology.** The new subprogram should exchange any two real variables regardless of their names.

Algorithm development

The procedures **Exchange** and **Switch** that were written for the case study in Chapter 5 can only exchange two variables if they are called either **Highest** and **HighToday** or **Lowest** and **LowToday**. Since exchanging two variables is a common task in programming, a general procedure that exchanges variables can be used in many different Pascal programs. Having written the generalized procedure once, you never have to write it again. Such procedures are often called *utility procedures*. As you write more programs that contain modular subprograms, you will develop your own library of utility subprograms that perform tasks such as finding an average, searching for a value, or sorting a list.

The algorithm for the solution to this problem is the same as the algorithm used in Chapter 5 with the following three exceptions:

1. Step 3, in which the procedure **Switch** was invoked, is changed to

    if **LowToday** < **Lowest,** then **Exchange**

2. The variable names used in procedure **Exchange** are changed to refer to any two real numbers rather than to the values of two specific variables, as follows:

    procedure **Exchange**
      begin
        Save ← **First**
        First ← **Second**
        Second ← **Save**
      end

3. Procedure **Switch** is eliminated.

The next section will discuss how the procedure **Exchange** can be modified to refer to any two real numbers.

6.1 Parameters

Using global variables can be convenient because any variable can be referenced or changed by any procedure within the program; however, using a global variable within a procedure violates the stylistic principle of *modular design* for programming. In modular design, each subprogram is a self-contained unit. A modular subprogram uses only variables that are declared within the subprogram's block. A subprogram, however, cannot exist entirely on its own; there must be a method for passing information from the main program to the subprograms, and for returning information from subprograms back to the main program. In Pascal, *parameters* are used for this purpose.

Restricting the scope of variables to the blocks in which they are declared is a matter of programming style. The restriction is not required by the Pascal syntax. Pascal syntax requires only that you follow the rules for scope that were explained in Chapter 4.

The two types of parameters that can be used in Pascal procedures are *variable parameters* and *value parameters.* Variable parameters can be used both to send values from an outer block to a procedure and to return values to the outer block. A variable parameter is used when a procedure changes the value of a parameter and the change must be returned to the outer block. For example, in the procedure to exchange two numbers, you want the values that are assigned to the parameters by means of the exchange to become the new values of the variables in the main program. On the other hand, value parameters can be used only to send values into a procedure. Information cannot be sent back to the outer block through a value parameter. For example, value parameters are used in a reporting procedure whose sole purpose is to print the current values of variables. A reporting procedure uses the values of the variables, but it should not change the values of the variables.

Constants, values, and even other subprograms can be passed into procedures or functions; thus the following discussion refers to expressions that are passed into procedures. Keep in mind that a constant, a variable, or a value is a valid expression.

Like a local variable, a parameter does not have a value, or even any storage allocation, until the subprogram is invoked. When a subprogram is invoked, expressions are passed to the subprogram through the parameters of the subprogram. So, in the program for the case study, the variable **Highest** will be passed into the procedure through the parameter **First.** The expressions that are passed in are called *arguments*. This term was used when standard functions were discussed in Chapter 2. For standard functions, the argument of the function is the constant or variable on which it acts. For example, in the expression **sin(x),** the argument is **x.** For procedures, the arguments are the expressions, such as **Highest,** on which the procedure acts.

6.2 Procedures with variable parameters

The syntax diagram for declaring the heading for a procedure that uses variable parameters is

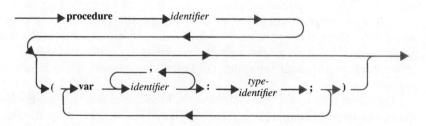

6.2 Procedures with variable parameters

Using this syntax diagram, you can create a heading for the procedure to exchange two real numbers:

```
procedure Exchange (var First, Second : real);
```

This heading says that the procedure **Exchange** has two variable parameters of data type **real.** The identifiers listed between the parentheses are called the *parameter list.* In this case, the parameters are **First** and **Second.** Notice in the diagram, however, that parameters are not required for correct Pascal syntax.

Note that variable parameters are declared the same way that variables are declared. Variable parameters of the same data type can be listed together and separated by commas just as they are in a **var** statement. Variable parameters of different types must be separated by semicolons, and the word **var** must be repeated after every semicolon. For example, a procedure that uses variable parameters to read values into an integer variable and a real variable might have the following heading:

```
procedure ReadSome (var HowMany : integer; var HowMuch : real);
```

Declaring a variable parameter is similar to declaring a local variable for the procedure.

This complete procedure for **Exchange** would be as follows:

```
procedure Exchange (var First, Second : real);
(*     Procedure Exchange exchanges the values stored in two real      *)
(*     variables that are passed into Exchange as First and Second.    *)

var    Save : real;

begin
  Save := First;
  First := Second;
  Second := Save
end; (* procedure Exchange *)
```

Procedure **Exchange** is the same as the two procedures **Exchange** and **Switch,** except that variable parameters have replaced the variables used in those procedures.

Inside the procedure, you can use the variable parameters declared in the parameter list just as you would use local variables. Once a variable parameter has been declared, you do not need to—and, in fact, cannot— declare it again in the declarative part of the procedure:

```
          procedure Exchange (var First, Second : real);

          var    First : real;
          ***error*** $101


          error messages :
          ***************

          101: identifier declared twice.
```

Invoking a procedure that uses variable parameters

Like any procedure declaration, the declaration for the procedure **Exchange** gives a definition. A procedure with a parameter list is invoked with the name of the procedure and the names of its arguments, which are the identifiers for the variables to be passed into the variable parameters. These identifiers are separated by commas, and the entire list is enclosed in parentheses.

The syntax for invoking a procedure that uses variable parameters is similar to the syntax for invoking the standard procedure **read.** The procedure **Exchange** above is invoked by

```
          Exchange (Highest, HighToday);
```

where **Highest** and **HighToday** are real variables that have previously been assigned values. **Lowest** and **LowToday** are variables in the same program that, when they have been assigned values, can be exchanged by invoking the procedure again as

```
          Exchange (Lowest, LowToday);
```

When the procedure is invoked, each argument is associated with the variable parameter that is in the same position in the parameter list as the argument is in the invocation. In the example above,

Lowest ↔ First
LowToday ↔ Second

Or   procedure **Exchange** (var **First, Second** : real);
                                ↑   ↑
                        Exchange (**Lowest, LowToday**);

Before the procedure is invoked, there are two storage locations reserved for the variables **Lowest** and **LowToday.** Assuming that **Lowest** and **LowToday** have previously been assigned values, you can visualize the storage locations:

6.2 Procedures with variable parameters

Lowest **LowToday**
  38.2   34.8

When the procedure is invoked, the storage locations are given second names that correspond to the names of the variable parameters:

Lowest ↔ First   **LowToday ↔ Second**
   38.2   34.8

The variable parameter **First** now has a value of 38.2, and the variable parameter **Second** has a value of 34.8. A storage location for the local variable **Save** is also created. When the procedure is invoked, **Save** does not have a value:

Lowest ↔ First   **LowToday ↔ Second**   **Save**
   38.2   34.8

While the procedure is being executed, any change made to the variable parameter **First** will change the variable **Lowest,** and any change made to the variable parameter **Second** will change the variable **LowToday.** The local variable **Save** will be assigned the value of **First.** When the last executable statement in the procedure **Exchange** is made, the storage locations are

Lowest ↔ First   **LowToday ↔ Second**   **Save**
   34.8   38.2   38.2

After the procedure has been executed, the double labels are removed and the local variables and the names of the variable parameters are erased:

Lowest **LowToday**
  34.8   38.2

Lowest, which had a value of 38.2 on entering the procedure, now has a value of 34.8, and **LowToday,** which had a value of 34.8, now has a value of 38.2.

Invoking the procedure as

```
Exchange (Highest, HighToday);
```

where **Highest** is a variable with a value of 75.6 and **HighToday** is a variable with a value of 83.1, results in the following:

Highest	HighToday
75.6	83.1

When the last statement in procedure **Exchange** is executed, the storage locations are

Highest ↔ First	HighToday ↔ Second	Save
83.1	75.6	75.6

When execution of the procedure has been completed, the double labels are removed and the variables have the following values:

Highest	HighToday
83.1	75.6

    As you can see, changing the value of a variable parameter in a procedure changes the value of the corresponding variable in the argument list. After the procedure has executed, the variables that were passed in as arguments have values different from those that they had prior to being passed through the parameters of the procedures. A variable parameter is a *two-way parameter:* values can be passed into the variable parameters in the procedure, and values can be returned to the variables.

    When data is passed from the main program into a procedure through a parameter list, the data must be passed into the correct parameter. The position of the argument, not its name, determines which values are passed to each parameter. In Program 6.1, the procedure **WriteIt** prints the variable parameters **Years** and **Pounds** from the parameter list. When the procedure is invoked, the first argument is always associated with the first parameter and the second argument with the second parameter. So, the first time the

---

REMINDER

Because a value is returned in a variable parameter, an argument that corresponds to a variable parameter must be a variable. Any other type of expression cannot be used to store a new value.

PROGRAM 6.1 AgeAndWeight

```
program AgeAndWeight (input, output);
(*      This program uses a procedure to print the age and the weight.   *)
(*      The arguments of the procedure are invoked in two different      *)
(*      orders to show that the order in which variables are passed to   *)
(*      a procedure is important.                                        *)

var   Age : integer;        (* global variable to be passed to procedure. *)
      Weight : integer;     (* global variable to be passed to procedure. *)

  procedure WriteIt (var Years, Pounds : integer);
  (*      This procedure prints two real variables passed as parameters.  *)

  begin
    writeln (' Years =', Years:4, '  Pounds =', Pounds:4)
  end; (* procedure WriteIt *)

(*-------------------------------------------------------------------------*)

begin (* main program *)
  Age := 18;
  Weight := 120;

  writeln (' Procedure invoked as: WriteIt (Age, Weight)');
  writeln;
  WriteIt (Age, Weight);
  writeln;
  writeln(' Procedure invoked as: WriteIt (Weight, Age)');
  writeln;
  WriteIt (Weight, Age)
end. (* program AgeAndWeight *)

=>
 Procedure invoked as: WriteIt (Age, Weight)

 Years =  18  Pounds = 120

 Procedure invoked as: WriteIt (Weight, Age)

 Years = 120  Pounds =  18
```

procedure is invoked, the variable **Age** is associated with the variable parameter **Years,** and the variable **Weight** is associated with the variable parameter **Pounds.** The second time the procedure is invoked, the variable **Weight** is associated with the variable parameter **Years,** and the variable **Age** is associated with the variable parameter **Pounds.** The output from the second call to the procedure is nonsense. The compiler cannot know that **Age** and **Years** are somehow associated and make the correct assignment of values. You must be sure to list the arguments in the correct order when you invoke a procedure.

> *REMINDER*
>
> Only the position of the variable in the argument list matters. The name is irrelevant.

When you invoke a procedure with a list of variable parameters, the data type of each argument must be the same as the data type of the corresponding variable parameter. It is not sufficient that the argument and the variable parameter be assignment compatible. If you try to pass a real variable into the procedure **WriteIt** in Program 6.1, you will get an error message telling you that the argument is not of the same data type as the parameter. For example, if **RealVar** is declared to be a real variable, invoking **WriteIt** with **RealVar** as an argument results in the compile-time error

```
        WriteIt (RealVar, Weight);
***error***              $142

error messages :
***************

  142: type conflict on parameters.
```

When you invoke a procedure, the number of arguments must be the same as the number of parameters. If you invoke the procedure **WriteIt** with more than two arguments or with fewer than two arguments, you will get an error message telling you that the number of arguments does not agree with the number of parameters in the procedure declaration, as in the following case:

```
        WriteIt (Age, Weight, Height);
***error***                    $126

        WriteIt;
***error***  $126

error messages :
***************

  126: number of parameters does not agree with declaration.
```

> *REMINDER*
>
> The number and type of the arguments must correspond to the number and type of the parameters in the procedure heading.

6.2 Procedures with variable parameters

Although you are not required to give your variable parameters names that are different from the argument variable names, doing so will clarify which identifier stands for the variable and which for the variable parameter. In Program 6.2, the variable parameters and the variables have the same identifiers:

PROGRAM 6.2 Nonsense

```
program Nonsense (input, output);
(*      This program uses a procedure to print the age and weight. The    *)
(*      parameters and the global variables have the same names. The      *)
(*      procedure is invoked with the variables in the wrong order.       *)

var   Age : integer;       (* global variable to be passed to procedure. *)
      Weight : integer;    (* global variable to be passed to procedure. *)

   procedure WriteAgeAndWeight (var Age, Weight : integer);
   (*     This procedure prints two real variables passed as parameters.  *)

   begin
      writeln (' Age =', Age:4, ' Weight =', Weight:4)
   end; (* procedure WriteAgeAndWeight *)


(*---------------------------------------------------------------------*)

begin (* main program *)
   Age := 18;
   Weight := 120;
   WriteAgeAndWeight (Weight, Age)
end. (* program Nonsense *)

=>
 Age = 120  Weight =  18
```

When the procedure **WriteAgeAndWeight** is invoked, the storage locations for the variables are given duplicate labels:

Age ↔ Weight   **Weight ↔ Age**
(argument) (parameter)   (argument) (parameter)

| 18 | | 120 |

Because the order of the arguments in the main program is the opposite of that of the parameters in the procedure, the storage location that has the name of the argument **Age** has the name of the parameter **Weight,** and vice versa. This situation is quite confusing. Once you have created a bug like the one in Program 6.2, you may never be able to track it down, because you would not question the fact that **Age** refers to the same quantity in the main program and in the procedure. Two quantities that are obviously the same to you may be different to the computer.

STYLE HINT

Give your variable parameters names that are different from your program variable names.

6.3 Procedures with value parameters

Variable parameters can pass only variables into procedures. Constants and other types of expressions whose values cannot be changed cannot be passed through variable parameters. They can only be passed through *value parameters.*

For some procedures in which it is important that the procedure not change the value of a variable, values of variables can be passed into the procedures through value parameters. When a value parameter is used, the value of the expression in the argument list is not changed when the procedure finishes execution. For example, in the procedure **ReportData** that follows, the values of **Mass** and **Force** should not be changed:

```
procedure ReportData (var Mass, Force : real);
(*      ReportData prints the variables passed through the parameters    *)
(*      Mass and Force.                                                  *)
begin
  writeln (' The current value for the Mass is ', Mass:10:2);
  writeln (' The current value for the Force is ', Force:10:2)
end; (* procedure ReportData *)
```

It would be rather disturbing to find that the value of **Mass** changed when a procedure to print its value was invoked. Value parameters can be used to prevent a procedure from changing the values of arguments.

The syntax for declaring a value parameter is the same as for declaring a variable parameter, except that the word **var** is omitted. The complete syntax diagram for a procedure heading is

6.3 Procedures with value parameters

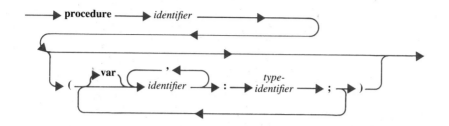

To change the variable parameters **Mass** and **Force** to value parameters in the procedure **ReportData,** you need only remove the word **var** from the declarations for **Mass** and **Force** in the procedure heading:

```
procedure ReportData (Mass, Force : real);
(*      ReportData prints the values passed through the parameters      *)
(*      Mass and Force.                                                 *)
```

When a value parameter is used, a new storage location is created and a *copy* of the value in the argument list is stored in the new location. The separate storage location, which can be referred to as a *dummy variable,* is used to store the value during execution of the procedure. The dummy variable can be used and changed by the procedure, but the original argument never enters the procedure so it cannot be changed. Because values can be passed into a procedure through value parameters but values cannot be returned from the procedure, value parameters are called *one-way parameters.*

Unlike a variable parameter, a value parameter and an expression passed into a value parameter need not be of the same data type, although the value parameter and the expression must be assignment compatible. An expression that is passed into a value parameter must have a value before it is passed into a procedure. If the expression does not result in a value when it is evaluated upon execution of the procedure, the dummy variable cannot be given a value. On most systems, passing an undefined expression into a value parameter will cause a run-time error. Indeed, passing an undefined expression into a value parameter is a logical error because, by definition, the procedure cannot change its value. It will remain undefined after the procedure execution, and thus there is no reason to pass it to the procedure.

Suppose that the procedure **ReportData** is invoked with the arguments **Grams** and **Newtons,** as follows:

```
                    ReportData (Grams, Newtons);
```

and that the variables **Grams** and **Newtons** have the following values:

Grams	Newtons
4.5	3.2

When the procedure is invoked, the values of the variables are passed into the dummy variables. The dummy variables are given the names of the corresponding value parameters in the parameter list:

Grams	→	Mass
4.5		4.5

Newtons	→	Force
3.2		3.2

While the procedure is executing, the dummy variables **Mass** and **Force** are used. The procedure has no effect on the values of the variables **Grams** and **Newtons;** they will be unchanged, as these variables are not used by the procedure. When the procedure finishes execution, the dummy variables are erased, and **Grams** and **Newtons** still have the same values that they had prior to the execution of the procedure.

Grams	Newtons
4.5	3.2

Program 6.3, **Parameters,** demonstrates passing a variable into one procedure as a variable parameter and into another procedure as a value parameter, in a simple program that reads and writes a variable. **NewVar** in the parameter list for the procedure **ReadIt** is a variable parameter, as it must be if the value read by the procedure is to be transferred to the global variable **PassIt.** A value parameter is used in the output procedure. This is sensible because a procedure whose stated purpose is to print values should not change values.

PROGRAM 6.3 Parameters

```
program Parameters (input, output);
   (*     This program passes a variable into the procedure ReadIt as a     *)
   (*     variable parameter and into WriteIt as a value parameter.         *)

   var    PassIt : integer;     (* global variable to be passed into a *)
                                (* variable parameter and into a value *)
                                (* parameter. *)
```

```
procedure WriteIt (OldVar : integer);
(*      This procedure prints the value of the value parameter OldVar.    *)

begin
  writeln (' The value of the variable in WriteIt is:', OldVar:5)
end; (* procedure WriteIt *)


procedure ReadIt (var NewVar : integer);
(*      This procedure assigns a value to the variable parameter NewVar.   *)

begin
  writeln;
  writeln (' Enter an integer');
  readln (NewVar)
end; (* procedure ReadIt *)



(*---------------------------------------------------------------------*)
begin (* main program *)
  ReadIt (Passit);
  WriteIt (PassIt)
end. (* program Parameters *)

=>
 Enter an integer
□5
 The value of the variable in WriteIt is:    5
```

If the purpose of a procedure is to change the value of a variable, the corresponding parameters must be declared as variable parameters. Suppose that in the heading of the procedure **Exchange** the word **var** is omitted. When the procedure is invoked as

$$\text{Exchange (Lowest, LowToday)}$$

a dummy variable is created to correspond to each value parameter and initialized with the value of the corresponding variable:

Lowest	→	First
38.2		38.2

LowToday	→	Second
34.8		34.8

When the last statement in the procedure is executed, the memory locations are

Lowest
| 38.2 |

First
| 34.8 |

LowToday
| 34.8 |

Second
| 38.2 |

Save
| 38.2 |

After execution of the program, the local variable **Save** and the dummy variables are erased:

Lowest
| 38.2 |

LowToday
| 34.8 |

and the values of **Lowest** and **LowToday** are unchanged. The procedure computed the results, but the results were blocked from being passed back to the variables. This is a common mistake. Whenever you get error messages for undefined variables, check your parameter lists for missing **var**s.

DEBUGGING HINT

Leaving out the **var** is as easy to do by mistake as on purpose. When you omit the word **var** when declaring a parameter, be sure you have done so on purpose. Undefined variables frequently arise because of missing **var**s in parameter lists.

Because value parameters can create so many bugs, you may be tempted to make all of your parameters variable parameters. There are two reasons not to do this; one is practical, the other is stylistic. Practically, not all parameters can be declared as variable parameters. Constants, values, and the control variables of the **for** statement, which you will learn in Chapter 7, must be passed into procedures as value parameters because their values cannot change. For example, if **Pi** is declared to be a constant, it cannot be passed as an argument into a variable parameter. The value of a constant cannot be changed, but the value of a variable parameter can be changed. Passing a constant, a value, or a **for** control variable into a variable parameter will result in a compile-time error.

Value parameters are used when a procedure does not change the value of a variable. Stylistically, value parameters are used as a signal to someone reading your program that a particular procedure uses a variable but does not change its value. This signal can be helpful to someone trying to understand or debug your program.

Now that you have learned to use parameters, you should write all of your procedures using parameter lists. Changing a variable declared in one program block within another program block can be confusing. If any procedure can modify any variable, tracing a bug is difficult. Referencing variables only in the program block in which they are declared enables you and anyone else reading your program to see how information is transmitted between the main program and the procedures and among procedures.

DEBUGGING HINT

Do not use global variables in your procedures. Use parameters to transfer information between the main program and the procedures.

Solution to Case Study 7

Program 6.4, **Weather,** is based on Program 5.2, the solution to the case study from Chapter 5. In **Weather,** the procedure **Exchange** replaces the two procedures used in Chapter 5. All of the procedures have been rewritten to use parameters to pass information between the main program and the procedures.

PROGRAM 6.4 Weather

```
program Weather (input, output);
(*      Program Weather is a revision of Program 5.2, Meteorology.   *)
(*      Weather uses procedures with parameters instead of global    *)
(*      variables to pass information within the program.            *)

var  Highest : real;       (* highest temperature in the observation *)
                           (* period so far - in degrees C. *)
     Lowest : real;        (* lowest temperature in the observation *)
                           (* period so far - in degrees C. *)
     TotalRain : real;     (* total rain that has fallen during the *)
                           (* observation period - in cm. *)
     Choice : char;        (* character entered by the user indicating *)
                           (* which procedure should be executed. *)
```

```
procedure Initialize (var HighestSoFar, LowestSoFar, RainSoFar : real);
(*      Initialize writes a message to the user and initializes the     *)
(*      parameters HighestSoFar, LowestSoFar, and RainSoFar to their    *)
(*      initial values.  In the final program, these values would be    *)
(*      passed to the main program from another procedure.              *)

begin
  writeln;
  writeln (' This program allows you to determine the highest',
           ' temperature, the');
  writeln (' lowest temperature, or the total amount of rainfall for',
           ' a given');
  writeln (' observation period.  You will be asked which you want to',
           ' determine.');
  writeln;

  HighestSoFar := -1000.0;
  LowestSoFar := 1000.0;
  RainSoFar := 0.0
end; (* procedure Initialize *)


procedure Exchange (var First, Second : real);
(*      Procedure Exchange exchanges the values stored in two real      *)
(*      variables that are passed into Exchange as First and Second.    *)

var   Save : real;

begin
  Save := First;
  First := Second;
  Second := Save
end; (* procedure Exchange *)


procedure HighCompare (var HighestSoFar : real);
(*      This procedure reads today's high temperature and compares it   *)
(*      to the highest temperature so far.  If today's temperature is   *)
(*      higher, then the values are exchanged by the procedure Exchange.*)

var  HighToday : real;        (* highest temperature of the current day *)
                              (* - in degrees Celsius. *)

begin
  writeln (' What was today''s highest temperature in degrees Celsius?');
  readln (HighToday);

  if HighToday > HighestSoFar then
    Exchange (HighestSoFar, HighToday)
end; (* procedure HighCompare *)
```

```
procedure LowCompare (var LowestSoFar : real);
  (*    This procedure reads today's low temperature and compares it    *)
  (*    to the lowest temperature so far.  If today's temperature is    *)
  (*    lower, then the values are exchanged using the procedure Switch. *)

  var   LowToday : real;       (* lowest temperature of the current day *)
                               (* - in degrees Celsius. *)

begin
  writeln (' What was today''s lowest temperature in degrees Celsius?');
  readln (LowToday);

  if LowToday < LowestSoFar then
    Exchange (LowestSoFar, LowToday)
end; (* procedure LowCompare *)



procedure Precipitation (var RainSoFar : real);
  (*    This procedure reads today's rainfall and adds it to the        *)
  (*    total rain that has fallen during the observation period.       *)

  var  RainToday : real;      (* rain fall today in cm. *)

begin
  writeln (' What was today''s rainfall in cm? ');
  readln (RainToday);

  RainSoFar := RainSoFar + RainToday
end; (* procedure Precipitation *)



procedure ReadChoice (var Selection : char);
  (*    ReadChoice presents the menu of choices to the user, asks which  *)
  (*    function is to be performed, and reads the response from the     *)
  (*    user.                                                            *)

begin
  writeln (' What do you want to do?');
  writeln (' Enter');
  writeln ('       h  to determine the highest temperature');
  writeln ('       l  to determine the lowest temperature');
  writeln ('       r  to determine the rainfall');
  writeln;

  readln (Selection)
end; (* procedure ReadChoice *)
```

```
      procedure Choose (Selection : char;
                        var HighestSoFar, LowestSoFar, RainSoFar : real);
      (*    Choose uses a case statement to make the choice among the      *)
      (*    alternatives presented to the user. One of three procedures    *)
      (*    will be executed depending on the value of the character       *)
      (*    variable Selection.                                            *)
      begin
        case Selection of
          'H', 'h' : HighCompare (HighestSoFar);
          'L', 'l' : LowCompare (LowestSoFar);
          'R', 'r' : Precipitation (RainSoFar)
        end (* case *)
      end; (* procedure Choose *)


      procedure Report (Selection : char;
                        HighestSoFar, LowestSoFar, RainSoFar : real);
      (*    This procedure reports the current values of HighestSoFar,     *)
      (*    LowestSoFar, or RainSoFar depending on the value of Choice.    *)
      begin
        writeln;
        writeln (' For the current observation period, ');

        case Selection of
          'H', 'h' : writeln (' the highest temperature so far is',
                              HighestSoFar:7:2, ' degrees C');
          'L', 'l' : writeln (' the lowest temperature so far is',
                              LowestSoFar:7:2, ' degrees C');
          'R', 'r' : writeln (' the total rain fall has been',
                              RainSoFar:7:2,' cm')
        end (* case Choice *)
      end; (* procedure Report *)

(*---------------------------------------------------------------------*)

begin (* main program *)
  Initialize (Highest, Lowest, TotalRain);
  ReadChoice (Choice);
  Choose (Choice, Highest, Lowest, TotalRain);
  Report (Choice, Highest, Lowest, TotalRain);
end. (* program Weather *)
```

=>

This program allows you to determine the highest temperature, the
lowest temperature, or the total amount of rainfall for a given
observation period. You will be asked which you want to determine.

```
What do you want to do?
Enter
      h  to determine the highest temperature
      l  to determine the lowest temperature
      r  to determine the rainfall
□h
What was today's highest temperature in degrees Celsius?
□23.0

For the current observation period,
the highest temperature so far is  23.00 degrees C
```

6.4 Functions

The rest of this chapter will cover the definition and use of functions and will present guidelines to help you determine whether a subprogram should be implemented as a procedure or as a function.

CASE STUDY 8
Minimum and maximum temperature solved with functions

A common task, related to the case study from the first part of this chapter, is to find the higher or lower of two numbers. Write a function to determine which temperature is higher and save the higher temperature in the variable called **Highest.** Do the same for the low temperatures, storing the lower temperature in the variable **Lowest.** Write a new program using the functions that return the higher and lower temperatures. How is the new program different from Program 6.4? Are the programs equivalent?

Algorithm development

Because most of the program for this case study is the same as for the previous case study, only the algorithms for the new functions will be developed. The algorithm to find the higher of two numbers can be stated as follows: If the first number is greater than the second number, the first number is higher; otherwise, the second number is higher. In algorithmic form, the function can be written as follows:

ALGORITHM 6.1 Algorithm to find the larger of two numbers

Find the **Maximum** of two numbers
  begin
    if **FirstNum** > **SecondNum** then
      Maximum ← **FirstNum**
    else
      Maximum ← **SecondNum**
  end

Similarly, the algorithm to find the smaller of two numbers can be written as follows:

ALGORITHM 6.2 Algorithm to find the smaller of two numbers

Find the **Minimum** of two numbers
  begin
    if **FirstNum** < **SecondNum** then
      Minimum ← **FirstNum**
    else
      Minimum ← **SecondNum**
  end

Definition of a function

The Pascal standard functions that you are familiar with correspond to mathematical functions of the form

$$y = f(x)$$

For example, the function for the absolute value

$$y = |x|$$

can be invoked in Pascal as

$$\mathtt{Y := abs(X)}$$

By definition, a mathematical function f returns a value of y for every value of x in the *range* of the function. And every function is by definition *one-to-one*. That is, for every value of x there is one and only one value of y; for example, the absolute value of -2 is always the number 2. The inverse, however, is not true; several different values of x may result in the same value of y. For example, the values $+2$ and -2 both have the same absolute value.

Functions can have more than one argument. Mathematically, functions of more than one argument are written as

$$y = f(x_1, x_2, \ldots, x_n)$$

The function that takes the average of *n* numbers has *n* arguments and returns a single number:

$$y = \frac{\sum_{i=1}^{n} x_i}{n}$$

Although no standard Pascal function has more than one argument, you will be able to write functions that have more than one argument.

Because Pascal procedures and functions have many similarities, they are both referred to as *subprograms*. Whether a subprogram is implemented as a function or as a procedure depends on the purpose and use of the subprogram. In general, a function is used when the purpose of the subprogram is to return a single value. More specific guidelines are given at the end of this section.

An important difference between a function and a procedure is that the name of the function acts like a variable; the effect of execution of a function is as if you had created a storage location whose identifier was the function name and had stored the result of the function in that location. So the function identifier must be assigned a data type; for example, the function **sqrt** takes a value of data type **real.** If you look at the standard Pascal functions, you will realize that they each have a data type: **sqrt** and **cos** are **real** functions, **trunc** and **ord** are **integer, chr** is **char,** and **eoln** and **eof** are **boolean.**

Defining a function

Pascal allows you to define your own functions. The syntax diagram for a function heading is quite similar to the syntax diagram for a procedure heading. As with procedures, the parameters are not necessary for correct Pascal syntax.

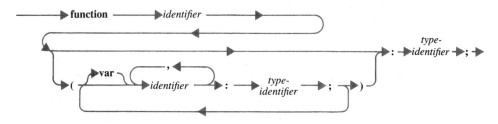

The *type identifier* of a function must be a simple data type.

A heading for the function that determines the higher temperature in the case study is

```
function Maximum (FirstNum, SecondNum : real) : real;
```

Like procedures, functions are defined in the declarative part of a program.

The heading indicates that a function with the identifier **Maximum** has two parameters of data type **real** and that the value returned by the function **Maximum** will be of data type **real.**

Assigning a value to a function

Only a few modifications are required to turn the algorithm for the function **Maximum** into a Pascal function. An important difference between a function and a procedure is that, within the body of the function, a value must be assigned to the function identifier. Thus, a value must be assigned to **Maximum** within the function.

```
function Maximum (FirstNum, SecondNum : real) : real;
(*      Function Maximum returns the larger of the two values FirstNum    *)
(*      and SecondNum.                                                     *)

begin
  if FirstNum > SecondNum then
    Maximum := FirstNum
  else
    Maximum := SecondNum
end; (* function Maximum *)
```

Notice that, within the function, the value of the higher parameter is assigned to the function identifier name, **Maximum.** Similarly, the function **Minimum** has the following definition:

```
function Minimum (FirstNum, SecondNum : real) : real;
(*      Function Minimum returns the smaller of the two values FirstNum   *)
(*      and SecondNum.                                                     *)

begin
  if FirstNum < SecondNum then
    Minimum := FirstNum
  else
    Minimum := SecondNum
end; (* function Minimum *)
```

If a value is not assigned to the name of the function within the function's body, a compile-time error will result:

```
function Maximum (FirstNum, SecondNum : real) : real;
(*      Function Maximum returns the larger of the two values FirstNum    *)
(*      and SecondNum.                                                     *)

var   Larger : real;     (* value of the larger of the two parameters. *)
```

```
begin
  if FirstNum > SecondNum then
    Larger := FirstNum
  else
    Larger := SecondNum
end; (* function Maximum *)
***error***   $179

error messages :
***************

  179: no assignment made to function identifier.
```

If the value assigned to the function identifier does not agree with the data type declared for the function in the heading, the compile-time error message for a type conflict of operands will be generated.

Invoking a function

A function is invoked by using the function identifier as a variable of the function's data type. As with a procedure, whenever a function is invoked, the number and type of the arguments must agree with the declaration in the parameter list. For example, if the function **Minimum** is declared to be a **real** function with two **real** parameters, it must be invoked with two **real** arguments:

```
LowTemp := Minimum (LowToday, Lowest)
```

This statement assigns the value returned by the function **Minimum** to the variable **LowTemp.** If the function **Minimum** were invoked without any arguments, a compile-time error would result, as follows:

```
LowTemp := Minimum;
***error***         $126

error messages :
***************

  126: number of parameters does not agree with declaration.
```

Notice that the parameters for the function **Minimum** and **Maximum** are value parameters rather than variable parameters. A Pascal function corresponds to a mathematical function $y = f(x)$ in which the value of y is returned for a given value of x. When the statement

```
          LowTemp := Minimum (LowToday, Lowest);
```

is invoked, **LowTemp** should be assigned a value, but **LowToday** and **Lowest** should remain unchanged. In order to ensure that the values do not change, use value parameters.

DEBUGGING HINT

Always use value parameters with functions.

Development of a boolean function

A common requirement in interactive programs is to determine whether the user entered a yes or a no from the terminal. As an example of a function that would have the data type **boolean,** you could write a function that returned a value of true or false depending on whether the user entered a yes or a no from the terminal. The basic algorithm for this function must determine whether the character 'y' or 'Y' was entered:

1. Write a message asking the user to enter a **Response** of yes or no
   Write 'Do you want to enter more data? Enter either y or n'
2. Read **Response**
3. if **Response** is 'y' or 'Y' then
       BoolVar ← true
   else
       BoolVar ← false

This algorithm assumes that any character other than a 'y' or a 'Y' implies that the user meant to type the word no.

As a general rule, functions should not include **write** or **read** statements. Within a program, the function should be able to be used like a variable without any side effects such as unwanted output. This rule, however, has exceptions. For a program in which it is often necessary to ask the user to enter yes or no and then to test the answer, the program will be easier to write and to understand if these two tasks are grouped in a single function. The heading for this function could be

```
          function YesEntered : boolean;
```

In the function **YesEntered,** a value of true or false must be assigned to the function identifier, **YesEntered.**

```
function YesEntered : boolean;
(*      YesEntered writes a message to the terminal asking the user to   *)
(*      enter yes or no.  If a 'y' or a 'Y' is entered, YesEntered       *)
(*      returns with a value of true; otherwise YesEntered is false.     *)

var Answer : char;      (* takes the value of the response from the user. *)

begin
  writeln (' Enter yes or no (y or n)');
  readln (Answer);
  if (Answer = 'y') or (Answer = 'Y') then
    YesEntered := true
  else
    YesEntered := false
end; (* function YesEntered *)
```

Notice that the variable **BoolVar** from the original algorithm was changed to **YesEntered,** the function identifier.

The following fragment invokes the function **YesEntered.** In the **if** statement, the expression **YesEntered** is used in exactly the same way as a **boolean** variable or **boolean** expression would be used.

```
            writeln (' Do you want to continue?');
            if YesEntered then PrintMainMenu

            =>
             Do you want to continue?
             Enter yes or no (y or n)
            □y
```

Every time a function identifier is used in a program, the statements in the function block are executed. Because a function looks and acts like a variable it is easy to forget this fact. For example, in the following program fragment, the function **AddOne** is invoked twice, once in the assignment statement and once in the output statement. To emphasize this fact, a **writeln** statement has been included in the function.

```
function AddOne (X : integer) : integer;
(*      The function AddOne adds 1 to the integer passed in the value    *)
(*      parameter X.                                                     *)

begin
  AddOne := X + 1;
  writeln (' End of function AddOne')
end; (* function AddOne *)
```

```
begin
  Number := 3;
  NewNumber := (AddOne (Number) * 3) mod 6;
  writeln (' (', Addone (Number):2, ' * 3) mod 6) = ', NewNumber:1)
end.

=>
End of function AddOne
( End of function AddOne
 4 * 3) mod 6) = 0
```

> **DEBUGGING HINT**
>
> Every time a function is used in an executable statement, the function block is executed. To avoid executing a function twice, you can assign the value of the function to a temporary variable. Then you can use the temporary variable in place of the function so that the function does not have to be executed again.

Functions versus procedures

A common question is, When do you use a function and when do you use a procedure? First, there are two general classes of subprograms for which a function cannot be used:

1. If a subprogram does not return any values, a function cannot be used. For example, a subprogram that reports a value is unsuitable for a function.

2. If a subprogram returns more than one value, a function should not be used. Sometimes an algorithm returns two values, and you may be tempted to write a function that returns one value in the function name and another value through a variable parameter. For example, suppose you want to compare a new temperature to both the previous low temperature and the previous high temperature to determine which temperature is the lowest of the temperatures and which is the highest. In the following function, the function identifier is set to the lower value and the higher value is returned through a variable parameter:

```
function Smaller (First, Second : real; var Larger : real) : real;
    (*    This is an improper function that returns the smaller value       *)
    (*    in the function identifier and the larger value through a         *)
    (*    parameter.  Also, deceptively, the function does not return       *)
    (*    the smallest of the three arguments, but the smaller of the       *)
    (*    first two arguments.                                              *)
```

```
begin
  if First > Second then
    begin
      Larger := First;
      Smaller := Second
    end   (* if *)
  else
    begin
      Larger := Second;
      Smaller := First
    end   (* else *)
end; (* function Smaller *)
```

The function could be invoked as:

```
LowTemp := Smaller (Lowest, NewTemp, Higher)
```

This function is deceptive and can lead to errors. Looking at the invoking statement, you would not expect the value of **Higher** to be changed after the function **Smaller** is executed.

DEBUGGING HINT

Do not write functions that return values through the parameter list.

Elimination of the two classes of subprograms for which functions are unsuitable leaves only one class of subprograms for which functions are suitable: those subprograms that return a single value. The rule dictating that a subprogram should perform one well-defined task must be strictly followed for functions. If a statement such as

```
Work := DotProduct (Force, Distance)
```

is encountered, you would not expect the function **DotProduct** to read the values into **Force** and **Distance** or to print the values of **Force** and **Distance**. A function should return a single value, and it should not change any values in the parameter list.

Solution to Case Study 8

The program for the case study is the same as Program 6.4, **Weather**, except that the statements

```
Exchange (HighestSoFar, HighToday);

Exchange (LowestSoFar, LowToday)
```

can be replaced by statements using the functions **Minimum** and **Maximum**:

```
Highest := Maximum (HighToday, Highest);

Lowest := Minimum (LowToday, Lowest)
```

When the function **Minimum** is invoked, two different tasks are performed. During execution of the function, the value of the smaller of the two variables **LowToday** and **Lowest** is assigned to the function identifier, **Minimum**. After the function has executed, the value returned by **Minimum** is assigned to the variable **Lowest**. This statement does *not* perform an exchange. If **LowToday** is the lowest temperature, **LowToday** and **Lowest** will both have the same value after the assignment statement is executed. Notice that the value of **Lowest** does not change when the function **Minimum** is executed. Its value changes when the assignment statement is executed.

Although both the algorithm that uses the procedure and the algorithm that uses functions solve the problem of determining the low temperature, the program that uses the procedure **Exchange** saves the values of both temperatures. Saving values is required in sorting procedures in which the values must be exchanged until they are arranged in order. Sorting will be covered in Chapter 11. If the problem requirement is only to determine the higher or lower of two temperatures, the functions **Minimum** and **Maximum** are better solutions.

Summary

Parameters that stand for expressions are used to pass variables or values from the main program to subprograms or from one subprogram to another. In keeping with modular programming, all references to a variable must occur in the program block in which the variable is originally declared. Pascal allows the use of both variable parameters, which are two-way parameters, and value parameters, which are one-way parameters.

Parameters are declared in the definition of the subprogram, and on execution of the subprogram the variables or values are passed as arguments into the parameters. Variable parameters are used when a subprogram may change the value of the variable in the main program or in another subprogram. Only variables can be passed through variable parameters. Value parameters are used when constants or other values that cannot be changed are passed into procedures or functions. Value parameters also can be used if the subprogram should not change the value of the variable, as in procedures that report values.

A function is the second type of subprogram. A value is returned in the identifier name of the function, and so a function can be used like a variable within an expression. Because a function can be used like a variable, it must have a declared data type and a value must be assigned to the identifier within the function block. The syntax for the parameter list of a function is the same as the syntax for the parameter list of a procedure. However, as a matter of style, functions should not have variable parameters.

Exercises

1. Correct the errors in each of the following fragments:

    (a)
    ```
    procedure Next (NewVar);
    begin
    end;
    ```

    (b)
    ```
    procedure (var Data : real);
      var Data : real;
    begin
      readln (Data)
    end;
    ```

    (c)
    ```
    procedure SquareIt(Top : integer);
    begin
      Top := Top * Top
    end;
    ```

    (d)
    ```
    function AddIt (X : real);
    begin
      X := X + 1
    end;
    ```

2. Given the following function, **Power**,

    ```
    function Power (V, R : real) : real;
    begin
      Power := V * V/R;
      writeln (' Call to Power')
    end; (* function Power *)
    ```

    determine the output generated by the fragment

    ```
    begin
      writeln (' Power of ', Power (Volts, Ohms):6:2);
      if (Power (Volts, Ohms) < Max ) and (Power (Volts, Ohms) > Min) then
        writeln (' acceptable')
      else
        writeln (' unacceptable')
    end.
    ```

for each of the following sets of data:
(a) **Max** = 4.0, **Min** = 0.5, **Volts** = 2.0, **Ohms** = 4.0
(b) **Max** = 2.0, **Min** = 1.5, **Volts** = 5.0, **Ohms** = 3.0

3. Using graph paper, graph each of the following functions over the domain specified.

   (a) $y = x^2$ $-2 \leq x \leq 2$ (e) $y = \text{trunc}(x)$ $-5 \leq x \leq 5$
   (b) $y = x^3 - 4x + 3$ $-10 \leq x \leq 10$ (f) $y = \ln(x)$ $x > 0$
   (c) $y = \sin(x)$ $-\pi/2 \leq x \leq \pi/2$ (g) $y = e^{-x}$ $x \geq 0$
   (d) $y = x!$ $x = 0, 1, 2, \ldots, 5$ (h) $y = 1 - e^{-x}$ $x \geq 0$

4. Identify and correct the syntax and logical error(s) in each of the following functions:

(a)
```
function Min (RealVar1, RealVar2 : real) : real;

begin
  if RealVar1 < RealVar2 then
    Min := RealVar1;
  else
    Min := RealVar2;
    writeln (' The minumum of ', RealVar1:6:2, ' and', RealVar2:6:2,
             ' is ', Min:6:2)
end; (* function Min *)
```

(b)
```
function Reciprocal (Number : real) : real;

const Small = 1.0e-8;

begin
  if abs (Number) > Small then
    Reciprocal := 1.0 / Number
end; (* function Reciprocal *)
```

Problems

PASCAL

1. Rewrite Program 4.12, **AnalyzeThreeCircuits3**, using procedures and functions with parameters.

2. The standard Pascal function **odd** returns a value of 0 if the argument is even and a value of 1 if it is odd. Write your own version of the function **odd**.

3. Write a boolean function for the logical operator XOR. XOR stands for "exclusive or." XOR operates on two logical expressions and returns a value of true if only one of the expressions is true. XOR is false if both values are true

or if both values are false. Write a test program that invokes the function for all possible combinations of values for the logical expressions.

4. Implement the Pascal operators **div** and **mod** as functions. Because **div** and **mod** are reserved words, you will have to give your functions different names. In your program, double check your functions using the standard **div** and **mod** functions. Write an error message if your functions do not return the same values as do **div** and **mod.** Test your program on the values

   -1 **mod** $\quad 3$
   10 **div** $\quad 5$
   5 **div** $\quad -2$
   9 **mod** $\quad -2$

5. A simple method for encrypting a message is to systematically replace each letter by a different letter. Write a modular program that reads a sentence from the terminal and then prints an encrypted message.

6. The lowest common divisor (LCD) of two integers is the smallest integer that evenly divides both numbers. Using modular subprograms, write a program that reads two integers from the terminal and determines their lowest common divisor.

7. Compute the derivative of the function $F(x) = \text{sqrt}(\sin(x))$ at 10 equally spaced values of x between $-\pi$ and π. Approximate the derivative with the formula

$$\frac{dy}{dx} \cong \frac{F(x) - F(x + \Delta x)}{\Delta x}$$

Use a function to evaluate the derivative. (In Chapter 15 a better formula will be developed.)

8. Write a program that computes the day of the week that your birthday will fall on next year, given today's day and date. Your program should take leap years into account and should work for anyone's birthday.

APPLICATION

9. Rewrite the program from Problem 6 in Chapter 4 so that the cost of producing electricity is computed by a function.

10. A bicycle wheel has a diameter, d, of 27 in. In one revolution, the wheel covers a distance equal to its circumference, πd. Using procedures and functions, write a program that computes the average revolutions per minute (rpm) of the wheel, given the average speed of the bicycle in miles per hour.

11. Rewrite the program for Problem 18 in Chapter 5 using a function that returns the fuel consumption, given the passenger-miles per gallon and the seating capacity.

CHAPTER 7

The For Statement

Computers are often used to solve problems in which the same computation must be repeated many times on different sets of data. For example, when grades are being computed for a class, the same steps are repeated to compute an individual grade for each student. The method for computing each grade is the same, but each student's grade must be calculated individually.

There are three Pascal constructions that allow you to repeat sections of your program as many times as you want: the **for** statement, the **repeat-until** statement, and the **while-do** statement. In this chapter you will learn to use the **for** statement. To use the **for** statement, you must know exactly how many times the section of code will be repeated. With a **for** statement, Program 4.11, **AnalyzeThreeCircuits,** could have been written much more compactly. The **repeat-until** and **while-do** statements are more powerful than the **for** statement because their repetition can be halted based on the outcome of a test. However, **repeat-until** and **while-do** are more complex constructions, so the concept of repetition will be taught with the **for** statement. The **while-do** and **repeat-until** statements will be covered in Chapter 8.

CASE STUDY 9
Average particle velocity

In an experiment, the velocity of a particle is measured five different times. The velocity may be either positive or negative. The average of the positive velocities, the average of the negative velocities, and the average of the absolute values of the velocities must all be determined. Write a program that reads in the experimental values and computes the average velocities.

Algorithm development

The problem requires that you read five values for the velocity of a particle. You must test whether each value is positive or negative, and you must find the three different averages. To solve the problem using the Pascal constructions learned so far, you would have to write subprograms to read a value for the velocity, to test its sign (+ or −) in order to include the velocity in the appropriate averages, to compute the averages, and to report the averages. The statement to invoke the first two procedures would have to be repeated five times, once for each experimental value. If the problem statement were then changed so that there were 100 experimental values for the velocity, you would have to edit the program and type the statement invoking the procedures 100 times. Although you could write the program this way, doing so would be extremely tedious. The **for** statement will be used in the final program for the problem solution to allow you to repeat the procedures without having to retype the same steps five times.

You know how to write procedures to read the data and to test whether each velocity is positive or negative. The problem does not state what is to be done if the velocity is equal to 0. Zero is usually included with the positive values, which are then, strictly speaking, referred to as nonnegative values. In the solution to the case study, the velocity will be positive if it is greater than or equal to 0.

You need to develop an algorithm for computing the averages of the positive velocities, the negative velocities, and the absolute values of the velocities. The average is the sum of the values divided by the number of values. For five values, the average can be written as

$$\overline{X} = \frac{V_1 + V_2 + V_3 + V_4 + V_5}{5} \qquad (7.1)$$

where $\overline{X}$ (read as "X bar") stands for the average, V_1 for the first measured value of the velocity, V_2 for the second value, and so on.

Equation 7.1 could be written in the alternative, mathematically equivalent form

$$\overline{X} = \frac{V_1}{5} + \frac{V_2}{5} + \frac{V_3}{5} + \frac{V_4}{5} + \frac{V_5}{5} \qquad (7.2)$$

This alternative form, however, has several computational disadvantages. One disadvantage is that the division operation must be performed five times instead of once as in Equation 7.1. Not only do the extra division operations waste time; more importantly, each division operation introduces some computational error into the calculation. (See the Application section in Chapter 5.) Thus the answer that results from using Equation 7.1 will be more accurate than the answer that results from using Equation 7.2. Another objection to Equation 7.2 is that it assumes that the number of values that will be in the sum is known when the first number is added.

To use Equation 7.1, you must keep a running total of the positive values, a running total of the negative values, and a running total of the absolute values. You know that the running total of the absolute values will be divided by five to find the average, but you do not know how many positive values or negative values there will be. You can declare variables that count the number of positive numbers and the number of negative numbers. Each time you add a positive value to the total, the counter for the number of positive values will be increased by 1. You must remember to initialize the values of the counters and of the running totals to 0 prior to using them.

A function can be written to compute the average. This function will be invoked three times, once to assign a value to each average that must be computed. After the function has been executed, a procedure to report the computed averages can be invoked.

ALGORITHM 7.1 Algorithm for Case Study 9

begin algorithm

1. Initialize the program
    begin
        a. Initialize the running totals for the absolute values, the positive values, the negative values, the counter for the positive values, and the counter for the negative values
            AbsoluteTotal ← 0
            PositiveTotal ← 0
            NegativeTotal ← 0
            PositiveCounter ← 0
            NegativeCounter ← 0
        b. Write a message to explain the program to the user
    end

2. Read the values and compute the running totals
    begin
        Repeat 5 times
            begin
                a. Read the velocities
                    begin
                        write 'Enter a real number for the velocity'
                        read **Velocity**
                    end

b. Compute the running totals
      begin
         AbsoluteTotal ← AbsoluteTotal + Abs(Velocity)
         if Velocity >= 0 then
            begin
               PositiveTotal ← PositiveTotal + Velocity
               PositiveCounter ← PositiveCounter + 1
            end if
         else
            begin
               NegativeTotal ← NegativeTotal + Velocity
               NegativeCounter ← NegativeCounter + 1
            end else
      end Step 2b
   end of repeat 5 times
end
3. Find the averages
   begin
      a. AbsoluteAverage ← AbsoluteTotal / 5
      b. PositiveAverage ← PositiveTotal / PositiveCounter
      c. NegativeAverage ← NegativeTotal / NegativeCounter
   end
4. Report the averages
   begin
      report AbsoluteAverage, PositiveAverage, and
         NegativeAverage
   end
end algorithm

7.1 The for statement

The **for** statement has a control variable that is automatically increased from a lower bound to an upper bound. The control variable serves as a counter to keep track of the number of times that a statement has been executed. The syntax diagram for a **for** statement is

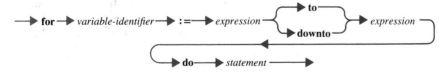

The words **for, to, downto,** and **do** are reserved words. Following the syntax diagram, this **for** statement can be created:

```
        for Counter := LowerBound to UpperBound do Action
```

where **Counter, LowerBound,** and **UpperBound** are integer variables and **Action** is a procedure. The statement that follows the **do** can be any executable statement: a procedure, an assignment statement, a compound statement, a **read** statement, or an output statement. The **for** statement above says: Initialize **Counter** to **LowerBound** and increment (increase) **Counter** by 1 until it equals **UpperBound**; each time the control variable **Counter** is incremented, **do** (execute) the procedure **Action.**

In the alternative form of the **for** statement that uses **downto** instead of **to,** the control variable is decremented (decreased) by 1 each time the statement following the **do** is executed. In the alternative form, the **for** statement above could be written as

```
        for Counter := UpperBound downto LowerBound do Action
```

The following discussion assumes use of the form of the **for** statement that uses **to.** The same rules apply if **downto** is used, but the control variable is decremented rather than incremented.

Because the control variable initially takes the value of the lower bound and is incremented until it equals the upper bound, the control variable, the lower bound, and the upper bound must all be of the same data type. Because the control variable is incremented by 1 each time the statement is executed, it must be an integer or an integer-like variable. Such data types are called *ordinal data types.* Of the data types discussed previously, **integer, char,** and **boolean** are ordinal data types that can be used for the control variable, the lower bound, and the upper bound in a **for** statement. If the data type of the control variable is **integer,** the control variable is incremented by 1 each time the loop executes. If the data type is **char,** the control variable is incremented to the next character, that is, from 'a' to 'b', from 'b' to 'c', etc. If the data type is **boolean,** the control variable can only be incremented from false to true.

Program 7.1 demonstrates use of a **for** statement to execute a **writeln** statement three times.

PROGRAM 7.1 SimpleFor

```
program SimpleFor (input, output);
(*      This program uses a for statement that prints the value of   *)
(*      the control variable in the for statement.                    *)

const Lower = 1;            (* lower bound of the for loop. *)
      Upper = 3;            (* upper bound of the for loop. *)

var   Index : integer;      (* for loop control variable. *)
```

7.1 The for statement

```
begin
  for Index := Lower to Upper do
    writeln (' The value of Index is', Index:3)
end. (* program SimpleFor *)
```

```
=>
 The value of Index is  1
 The value of Index is  2
 The value of Index is  3
```

In this **for** statement, the lower bound is the integer 1 stored in the constant **Lower,** and the upper bound is the integer 3 stored in the constant **Upper.** During execution of the **for** statement, the current value of the control variable determines whether or not the statement will be executed again. The control variable **Index** initially takes the value of the lower bound, 1. The control variable is incremented by 1 each time the **writeln** statement is executed until the control variable equals the upper bound, 3. The last time the statement is executed, the control variable equals the upper bound. When the **for** statement completes execution, the value of the control variable becomes undefined and the next statement after the **for** statement is executed. **Lower** and **Upper** are integer constants, so the control variable **Index** must be declared as an integer variable.

If the **for** statement is to execute, the lower bound must be less than or equal to the upper bound. If the lower bound is equal to the upper bound, the statement is executed once. If the lower bound is greater than the upper bound, the statement is not executed. For example, the following **for** statement has no output:

```
for Counter := 10 to 5 do
   writeln (' The value of Counter is', Counter:3)
```

If **downto** is used instead of **to,** the above statement will execute as follows:

```
for Counter := 10 downto 5 do
   writeln (' The value of Counter is', Counter:3)

=>
 The value of Counter is 10
 The value of Counter is  9
 The value of Counter is  8
 The value of Counter is  7
 The value of Counter is  6
 The value of Counter is  5
```

In standard Pascal, the value of the control variable is undefined before and after the execution of a **for** statement. Some compilers leave the value of the **for** control variable equal to the upper bound; however, you should assume that, at the completion of the **for** statement, the control variable is undefined.

DEBUGGING HINT

After a **for** statement has completed execution, the value of the control variable is *undefined*.

Look at Program 7.1 again. Suppose the main body of the program is changed as follows:

```
for Index := 1 to 3 do
   writeln (' The value of Index is', Index:3);
writeln (' The last value of Index is', Index:3)
```

This code will not execute correctly because the value of **Index** is undefined in the second **writeln** statement. The error that occurs will depend on the compiler, but essentially what happens is that the program compiles, the **for** statement executes three times, the character string "The last value of Index is" is printed, the undefined variable **Index** is encountered, and a run-time error results. If your compiler flags undefined variables, you will get an error message stating that the variable **Index** is undefined. If your compiler does not flag undefined variables, the value printed for **Index** could be any number, reasonable or not.

REMINDER

Do not put a ";" after the **do**. A semicolon after the **do** will cause the computer to do nothing (the null statement) as many times as you tell it to. The statement

for I := 1 to 1000 do;

will cause the computer to execute the loop, doing nothing 1000 times.

Program 7.2 illustrates legal **for** statements using different combinations of constants and variables for the upper and lower bounds. In each **for** statement, the upper and lower bounds and the control variable are all of the same data type, even though the upper and lower bounds may be constants, variables, or values. Note that, for integer control variables, the lower and upper bounds do not have to be positive integers. In the second **for** statement, the lower bound is a negative number. In the last two **for** statements, the control variable is a character. In these cases, the control variable is incremented alphabetically. Also note that a control variable can be used in more than one **for** statement: **CharCount** is used as the control variable for the last two statements.

PROGRAM 7.2 LegalForLoops

```
program LegalForLoops (input, output);
(*      This program shows the use of lower and upper bounds in for       *)
(*      statements; the upper and lower bounds of a for loop may be       *)
(*      constants, variables, or values. This program demonstrates        *)
(*      the variety of ways that for loops can be written.                *)

const Letter = 'a';        (* lower bound for a for statement. *)
      Gee = 'g';           (* upper bound for a for statement. *)

var   IntVar : integer;    (* integer variable for the lower bound of *)
                           (* a for statement. *)
      CharVar : char;      (* character variable for the upper bound *)
                           (* of a for statement. *)
      Counter, I : integer;(* for statement integer control *)
                           (* variables. *)
      CharCount : char;    (* for statement character control *)
                           (* variable. *)

begin
  (* This for statement increments the control variable from an integer *)
  (* variable to a constant value. *)

  IntVar := 1;
  write (' Integers:  ');
  for Counter := IntVar to 9 do write (Counter:2);
  writeln;

  (* The next for statement increments the control variable from a *)
  (* negative value to an integer variable. *)

  write (' Negative Integers:   ');
  for I := -10 to IntVar do write (I:4);
  writeln;
```

```
(* The next for statement increments the character control variable *)
(* from one character constant to another. *)

write (' Small Letters:   ');
for CharCount := Letter to Gee do write (CharCount:1);
writeln;

(* The next for statement increments the character control variable *)
(* from a character variable to a character value. *)

write (' Capital Letters:   ');
CharVar := 'I';
for CharCount := 'B' to CharVar do write (CharCount:1);
writeln
end. (* program LegalForLoops *)

=>
  Integers:    1 2 3 4 5 6 7 8 9
  Negative Integers:   -10  -9  -8  -7  -6  -5  -4  -3  -2  -1   0   1
  Small Letters:   abcdefg
  Capital Letters:   BCDEFGHI
```

7.2 Using a for statement with a compound statement

In the previous examples, only one statement has been executed each time the value of the control variable was incremented or decremented. In many problem solutions, more than one statement must be executed during each repetition. As with the **if** statement, the use of a compound statement with a **for** statement requires a **begin-end** pair.

The following program fragment is supposed to compute and print the squares of the first 20 integers:

```
for Counter := 1 to 20 do
   SquareVar := sqr (Counter);
   writeln (' The square of the counter is', SquareVar:4)

=>
The square of the counter is 400
```

But when the fragment is executed, **Counter** is incremented from the lower bound, 1, to the upper bound, 20, and the statement

SquareVar := sqr (Counter)

is executed twenty times. When the **for** statement finishes execution, the **writeln** statement is executed. The **writeln** statement is not part of the **for**

> **REMINDER**
>
> Whenever you compute a running total using a **for** statement, be sure that the running total is initialized before the **for** statement.

loop, even though it is indented as if it were a part. Remember that the compiler ignores indentation. The only way to indicate to the compiler that both statements should be executed each time the control variable is incremented is to use a compound statement with a **begin-end** pair, such as

```
for Counter := 1 to 20 do
   begin
      SquareVar := sqr (Counter);
      writeln (' The square of the counter is', SquareVar:4)
   end
```

When this statement is executed, each value of the counter will be squared, and each resulting value will be printed.

In order to solve the case study, more than one action must be performed for each of the five velocities. Each velocity must be read, and the procedure that determines whether it is positive or negative, adds it to the appropriate running totals, and increments the counter for either positive or negative values must be executed. You must declare a control variable for the **for** statement. If you call the control variable **ObservationNum,** the following fragment can be written:

```
for ObservationNum := 1 to 5 do
   begin
      ReadVelocity (Velocity);
      AddToTotals (Velocity, AbsoluteTotal, PositiveTotal, NegativeTotal,
                   PositiveCounter, NegativeCounter)
   end
```

where **ReadVelocity** is a procedure that asks the user for input and reads the velocity and **AddToTotals** is a procedure that performs the actions in Step 2b of the algorithm.

In standard Pascal, the value of the control variable cannot be changed within the loop. Once the **for** statement begins to execute, the control variable is incremented from the lower bound to the upper bound. Attempting to change the control variable's value within the body of the **for** statement will result in a compile-time error:

```
            for Index := 1 to 5 do
              begin
                Index := 3
***error***              $190

       error messages :
       ***************
       190: assignment to control variable forbidden
```

You can change the value of the upper or lower bound inside the **for** loop, but the change will have no effect on the number of times the loop is executed. For example, in the following program fragment, the value of **N,** the upper bound, is changed from 5 to 3 the first time the loop is executed. The **writeln** statement shows that, indeed, the value of **N** has been changed, but the loop still executes five times as the **Counter** is incremented from the lower bound of 1 to the original upper bound of 5.

```
N := 5;
for Counter := 1 to N do
  begin
    writeln (' The upper bound is', N:5,
             ' and this loop has executed', Counter:5, ' times.');
    N := 3
  end
```

```
=>
The upper bound is    5 and this loop has executed    1 times.
The upper bound is    3 and this loop has executed    2 times.
The upper bound is    3 and this loop has executed    3 times.
The upper bound is    3 and this loop has executed    4 times.
The upper bound is    3 and this loop has executed    5 times.
```

7.3 Using for statements with subprograms

Because the computer rather than the programmer has control over changing the values of **for** control variables, there are several restrictions on the use of these variables in subprograms.

A **for** control variable must be a local variable of the program or subprogram within which it is used. This rule prohibits the use of the same global variable as the control variable in a **for** statement in the main program and as the control variable in a **for** statement in a subprogram. If this usage were not prohibited, the **for** loop in the subprogram might redefine the value of the control variable in the main program. In the following example, declaring **Counter** as a global variable and not as a local variable results in a compile-time error:

7.3 Using for statements with subprograms

> *DEBUGGING HINT*
>
> In many of the examples in this chapter, numerical values are used as the lower and upper bounds in **for** statements. Numerical values are used to make the execution of the statements easier to see. Within programs, however, descriptive identifiers should be used for the bounds. For example, the statement
>
> > **for** StudentNum := 1 **to** MaxStudents **do**
> > ComputeGrade (StudentNum)
>
> is clearer than
>
> > **for** StudentNum := 1 **to** 25 **do**
> > ComputeGrade (StudentNum)
>
> Using descriptive identifiers also makes the program easier to modify. Suppose there are originally 25 students, 25 sets of homework, and 25 classes, and that all the **for** statements go from 1 to 25. Suppose then that the number of students is changed to 30, the number of sets of homework is changed to 20, and the number of classes is changed to 26. If numerical values are used for the upper and lower bounds, modification of the program will require that you find each occurrence of the number 25 and replace it with the corresponding new value. If descriptive identifiers are used, you need only assign new values to the identifiers.

```
program ComputeTotal (input, output);
(*      This program illustrates what happens if the for statement    *)
(*      control variable is declared as a global rather than a local  *)
(*      variable.                                                     *)

var     Counter : integer;      (* for statement control variable. *)

   procedure Total (DataPoint : real; var Sum : real);
   begin
     Sum := 0.0;
     for Counter := 1 to 5 do
***error***          $177

error messages :
****************

  177: for statement control variable must be locally declared.
```

A **for** loop control variable also cannot be declared in a parameter list:

```
procedure WriteLoop (var LoopCounter : integer );
  (*      In WriteLoop the control variable is passed to the procedure    *)
  (*      as a parameter resulting in an error message.  The control      *)
  (*      variable must be declared locally.  You cannot use a global     *)
  (*      variable or a parameter as the control variable.                *)
  begin
    for LoopCounter := 9 downto 3 do
***error***         $155

error messages :
***************

155: control variable may not be a formal parameter.
```

Within a **for** statement, the control variable can be passed as an argument into a subprogram, but it must be passed into a value parameter. Again, this restriction arises from the fact that the value of a **for** control variable cannot be changed. The following fragment shows the error message that results when the control variable is passed into a variable parameter:

```
procedure WriteIt (var N : integer):
  (*      This procedure shows that a for statement control variable      *)
  (*      declared in the main program cannot be passed to a procedure    *)
  (*      as a variable parameter.                                        *)

  begin
    writeln (' The value of the variable in WriteIt is', N:5)
  end;


begin (* main program *)
  for Index := 1 to 3 do
    WriteIt (Index)
***error***         $191

error messages :
***************

191: control variable may not be actual variable parameter.
```

If the **var** is removed from the parameter list in **WriteIt**, the preceding fragment will execute as follows:

```
procedure WriteIt (N : integer);

begin
   writeln (' The value of the variable in WriteIt is', N:5)
end;

begin (* main program *)
  for Index := 1 to 3 do
    WriteIt (Index)
end.

=>
  The value of the variable in WriteIt is    1
  The value of the variable in WriteIt is    2
  The value of the variable in WriteIt is    3
```

This last construction, passing the **for** loop control variable into a subprogram as a value parameter, is a common construction because the loop number is frequently required in a subprogram.

7.4 Using a for statement to perform exponentiation

Many scientific and engineering problems involve *exponentiation*, or raising a number to a power. For example, finding the volume of a sphere requires that the radius be cubed:

$$V = \frac{4}{3}\pi r^3 \tag{7.3}$$

Pascal does not have an arithmetic operator that performs exponentiation. In this section a **for** statement will be used to develop a function to raise real numbers to integer powers. In Chapter 8, in the application section, a general formula for exponentiation will be developed.

One way of raising a real number, z, to an integer power, n, is to multiply z by itself n times; for example,

$$z^4 = z \times z \times z \times z \tag{7.4}$$

Equation 7.4 is similar to a running total, except that it is a running product; instead of being added, the terms are multiplied.

Using a **for** loop, we can express exponentiation in Pascal as follows:

```
for Index := 1 to Power do
    Result := Z * Result
```

Assuming that **Z** and **Power** have been assigned values, to what value should

Result be initialized? If it is set to 0.0 as it would be if it were a running total, the value of **Result** will always be zero. After a little thought, you should be able to see that **Result** should be initialized to 1.0:

```
Result := 1.0;
for Index := 1 to Power do
    Result := Z * Result
```

The first time through the loop, when **Index** is equal to 1, the new value of **Result** is 1.0 times **Z**; the second time through the loop, **Result** is equal to its previous value of **Z** times **Z** (that is, Z^2); and so on. The preceding fragment can be incorporated into a general function that has a real number and an integer power as parameters.

```
function ZToPower (Z : real; Power : integer) : real;
(*      This function raises the real number Z to the positive power    *)
(*      Power.                                                          *)

var Index : integer;         (* for loop control variable. *)
    Result : real;           (* intermediate variable used to store the *)
                             (* running product. *)
begin
  Result := 1.0;             (* initialize the running product to 1.0 *)

  for Index := 1 to Power do (* multiply Z by itself Power times. *)
      Result := Z * Result;

  ZToPower := Result         (* assign the value of the running product to *)
                             (* the function name. *)
end; (* function ZToPower *)
```

After you have written the function **ZToPower,** you should test it using values of **Z** and **Power** for which you know the correct answer. For example, raising a number to the power of 1 results in the number itself; raising the number 1.0 to any power results in the number 1.0; raising the number 2.0 to the power of 3 results in the number 8.0. After you have tested your function for simple cases, you can test it by checking its response to extreme conditions. You might set all the integers to **maxint** and all the reals to large numbers. What happens if you raise a number to the power of **maxint?** What happens if you set **Z** to 0.0? What happens if you set **Z** to 0.0 and **Power** to 0? What happens if you set **Power** to a negative integer?

Program 7.3 uses the function **ZToPower** to determine the volume of a sphere.

PROGRAM 7.3 SphereVolume

```
program SphereVolume (input, output);
(*      This program uses the function ZToPower in the computation of    *)
(*      the volume of a sphere.                                          *)

const Cube = 3;          (* power to which radius of sphere is raised. *)
      Pi = 3.14159;      (* used in computation of sphere's volume. *)

var   Radius : real;     (* radius of sphere in inches entered by the user. *)
      Volume : real;     (* computed volume of sphere in cubic inches. *)

(*$i ztopower *)
(* Include function ZToPower (Z : real; Power : integer) : real; *)

begin
  writeln (' This program computes the volume of a sphere.');
  writeln (' Enter the radius of the sphere in inches');
  readln (Radius);

  Volume := 4/3 * Pi * ZToPower (Radius, Cube);

  writeln;
  writeln (' The volume of a sphere with radius', Radius:6:1, ' inches');
  writeln (' is', Volume:8:2, ' cubic inches.')
end. (* program SphereVolume *)
```

=>
 This program computes the volume of a sphere.
 Enter the radius of the sphere in inches
 □5.3
 The volume of a sphere with radius 5.30 inches
 is   623.6 cubic inches.

Solution to Case Study 9

The main program for the algorithm developed at the beginning of the chapter can be written with a **for** statement as follows:

```
begin (* main program *)
  Initialize (AbsoluteTotal, PositiveTotal, NegativeTotal,
              PositiveCounter, NegativeCounter);
```

```
    for ObservationNum := 1 to 5 do
      begin
        ReadVelocity (Velocity);
        AddToTotals (Velocity, AbsoluteTotal, PositiveTotal, NegativeTotal,
                     PositiveCounter, NegativeCounter)
      end; (* for *)

    AbsoluteAverage := Average (AbsoluteTotal, 5);
    PositiveAverage := Average (PositiveTotal, PositiveCounter);
    NegativeAverage := Average (NegativeTotal, NegativeCounter);

    ReportAverages (AbsoluteAverage, PositiveAverage, NegativeAverage,
                    PositiveCounter, NegativeCounter)
end. (* main program *)
```

The function **Average** is a general-purpose function that is invoked three times, once for the absolute values, once for the positive values, and once for the negative values.

In the main program above, the upper bound of the **for** statement is given as the integer 5, and the integer 5 is passed into the function **Average.** Suppose that, after having written the program, you were told that the number of experimental velocities had been changed to 100. You would have to change the value from 5 to 100 in both places. As a matter of style, although numbers can be passed through value parameters, identifiers that store the value should be passed into procedures and functions. You can either define a constant that takes the value of the number of observations or declare a variable and ask the user to supply the number of experimental observations that are to be read. Program 7.4 employs the first alternative. If the number of experimental values changes, you will need only to redefine the constant.

When the subprograms for the main program were written, the procedure **ReadVelocity** was modified to include the **ObservationNum** in the parameter list so that the message to the user could include the observation number to be entered next. Small modifications such as this one are often required during the final stages of converting an algorithm into a program.

In the procedure **AddToTotals,** a separate running total is kept for the sum of the absolute values of the velocities. How else could the sum of the absolute values have been found with fewer addition operations?

Notice that the function **Average** includes a check to be sure that **Counter** is not equal to 0. When this function was written a decision was made to set the value of **Average** to 0 when there were no values in the running total. How else could the problem have been handled?

In Program 7.4, if there were 100 values of the velocity to be entered instead of 5, running the program would be very tedious. All of the values to be read would have to be entered at the terminal during execution of the program. In Chapter 10 you will learn how to enter the data into a file that can be edited to eliminate typing mistakes and that does not have to be retyped every time a program is run.

PROGRAM 7.4   ParticleVelocity

```pascal
program ParticleVelocity (input, output);
   (*     This program reads five measured values of the velocity of a       *)
   (*     particle.  Averages of the absolute values, the positive values,   *)
   (*     and the negative values of the velocities are computed.  The       *)
   (*     values of the velocities are not saved.  Each time the for         *)
   (*     statement in the main program is executed, the newly entered       *)
   (*     value of the velocity replaces the previous value.                 *)

const NumberOfValues = 5;             (* number of experimental values. *)

var    ObservationNum : integer;      (* for statement control variable. *)
       Velocity : real;               (* user supplied value for velocity. *)

       AbsoluteTotal : real;          (* running total of absolute values. *)
       PositiveTotal : real;          (* running total of positive values. *)
       NegativeTotal : real;          (* running total of negative values. *)

       PositiveCounter : integer;     (* number of positive values. *)
       NegativeCounter : integer;     (* number of negative values. *)

       AbsoluteAverage : real;        (* average of the absolute values. *)
       PositiveAverage : real;        (* average of the positive values. *)
       NegativeAverage : real;        (* average of the negative values. *)

   procedure Initialize (var TotalAbs, TotalPos, TotalNeg : real;
                         var CounterPos, CounterNeg : integer);
      (*     Initialize writes a message to the user stating the purpose    *)
      (*     of the program and sets the values of the counters and the     *)
      (*     running totals to zero.                                        *)
   begin
     writeln (' This program reads the values for ', NumberOfValues:1);
     writeln (' experimental values of the velocity in meters per second.');
     writeln;
     writeln (' The averages of the absolute values, the positive values, ');
     writeln (' and the negative values are computed and reported.');
     writeln;

     TotalAbs := 0.0;
     TotalPos := 0.0;
     TotalNeg := 0.0;

     CounterPos := 0;
     CounterNeg := 0
   end; (* procedure Initialize *)
```

```
procedure ReadVelocity (N : integer; var Speed : real);
(*      ReadVelocity asks the user to enter a value for the velocity    *)
(*      and reads the value from the terminal.                          *)


begin
  writeln (' Enter value ', N:1,' of the velocity in meters per second');
  readln (Velocity)
end; (* procedure ReadVelocity *)




procedure AddToTotals (Speed : real;
                       var TotalAbs, TotalPos, TotalNeg : real;
                       var CounterPos, CounterNeg : integer);
(*      AddToTotals adds the absolute value of each velocity to the     *)
(*      running total of the absolute values and then tests whether     *)
(*      Speed is positive or negative and adds it to the proper         *)
(*      running total.                                                  *)



begin
  TotalAbs := TotalAbs + abs (Speed);
  if Speed >= 0 then
    begin
      TotalPos := TotalPos + Speed;
      CounterPos := CounterPos + 1
    end (* if *)
  else
    begin
      TotalNeg := TotalNeg + Speed;
      CounterNeg := CounterNeg + 1
    end (* else *)
end; (* procedure AddToTotals *)




function Average (Total : real; Counter : integer) : real;
(*      Average returns the value of the running total divided by       *)
(*      the value of Counter or the number of values in Total.          *)




begin
  if Counter > 0 then          (* test to prevent attempt to divide by zero *)
    Average := Total / Counter
  else
    Average := 0.0
end; (* function Average *)
```

```
    procedure ReportAverages (AverageAbs, AveragePos, AverageNeg : real;
                              CounterPos, CounterNeg : integer);
  (*       ReportAverages reports the computed averages of the absolute    *)
  (*       values, the positive values, and the negative values.           *)


    begin
      writeln;
      writeln (' The average of the absolute values of the observations');
      writeln (' is', AverageAbs:10:2, ' meters per second');
      writeln;
      writeln (' The average of the ', CounterPos:1, ' positive',
               ' observations is', AveragePos:10:2, ' meters per second');
      writeln (' The average of the ', CounterNeg:1, ' negative',
               ' observations is', AverageNeg:10:2, ' meters per second')
    end; (* procedure ReportAverages *)



(*---------------------------------------------------------------------------*)

begin (* main program *)
  Initialize (AbsoluteTotal, PositiveTotal, NegativeTotal,
              PositiveCounter, NegativeCounter);

  for ObservationNum := 1 to NumberOfValues do
    begin
      ReadVelocity (ObservationNum, Velocity);
      AddToTotals (Velocity, AbsoluteTotal, PositiveTotal, NegativeTotal,
                   PositiveCounter, NegativeCounter)
    end; (* for *)

  AbsoluteAverage := Average (AbsoluteTotal, NumberOfValues);
  PositiveAverage := Average (PositiveTotal, PositiveCounter);
  NegativeAverage := Average (NegativeTotal, NegativeCounter);

  ReportAverages (AbsoluteAverage, PositiveAverage, NegativeAverage,
                  PositiveCounter, NegativeCounter)
end. (* program ParticleVelocity *)


=>
 This program reads the values for 5
 experimental values of the velocity in meters per second.

 The averages of the absolute values, the positive values,
 and the negative values are computed and reported.
```

```
Enter value 1 of the velocity in meters per second
□-2.0
 Enter value 2 of the velocity in meters per second
□-1.0
 Enter value 3 of the velocity in meters per second
□0.0
 Enter value 4 of the velocity in meters per second
□1.0
 Enter value 5 of the velocity in meters per second
□2.0

The average of the absolute values of the observations
 is       1.20 meters per second

The average of the 3 positive observations is     1.00 meters per second
The average of the 2 negative observations is    -1.50 meters per second
```

APPLICATION
Sequences and Series

CASE STUDY 10
Bouncing ball

A ball is dropped from a height h. The ball has no forward motion; that is, it strikes the same spot on the ground with each bounce. The total distance that it travels is the sum of the up and down motions. The height of the bounce depends on the bounciness of the ball. The bounciness of the ball can be described by a number r that is less than 1.0. On every bounce, the ball rises to a height r times its previous height. Figure 7.1 shows the height of the ball as a function of bounce number. Write a Pascal program to determine how high the ball goes on each bounce and the total distance that the ball travels in n bounces.

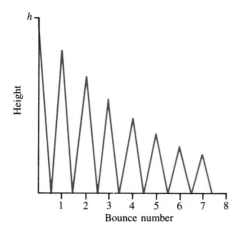

FIGURE 7.1
Distance traveled by a bouncing ball.

Algorithm development

Although you may not yet know the formula that you need to determine how high the ball bounces or how far it travels, you can write an algorithm in general terms and refine it later. You know that you will need an input procedure and an output procedure, and that you will need to determine the height of each bounce and the total distance traveled by the ball.

ALGORITHM 7.2 First iteration of the algorithm for Case Study 10

begin algorithm
1. Initialize the program
    begin
        a. Write a message to the user stating the purpose of the program
        b. Ask the user to supply **Height,** the height from which the ball is dropped, and **NumBounces,** the number of times that the ball bounces
    end
2. Compute the height and distance
    for **NumBounces**
    begin
        a. Compute the height of each bounce
        b. Add the new height to the total distance traveled by the ball
    end for
3. Report the heights and total distance traveled
end algorithm

Sequences

Many physical phenomena can be described by functions called *sequences*. A function is a sequence if the domain of the function is the positive integers—that is, if the function returns a value for every positive integer. Anyone who has taken a standardized math test has had some experience in determining what number comes next in a sequence. In this type of problem, you are usually given the first several elements in a sequence, say a_1, a_2, a_3, and asked to determine a_4. In order to solve this problem, you must find the function that converts the integer 1 into the value a_1, the integer 2 into the value a_2, and the integer 3 into the value a_3. If you were given the sequence

$$2 \quad 4 \quad 8 \quad 16$$

you could determine that the underlying sequence or function is 2^n and that the next number is 32. Formally, this sequence is written as

$$a_n = 2^n \tag{7.5}$$

where n is any positive integer. If you were asked for the fiftieth number in the sequence instead of the fourth, all you would have to do is to substitute 50 for n in the formula:

$$a_{50} = 2^{50}$$

In the case of the bouncing ball, the heights that the ball reaches as it bounces are elements in a sequence. After the first bounce, the height is equal to r, the bounciness factor, times the original height:

$$b_1 = hr \qquad (7.6)$$

Because the height at the end of the first bounce is the height from which the second bounce begins, the height at the end of the second bounce is

$$b_2 = b_1 r$$

or $\qquad (7.7)$

$$b_2 = (hr)r = hr^2$$

The height at the end of the nth bounce can be written as

$$b_n = hr^n \qquad (7.8)$$

This sequence, in which every term is the previous term multiplied by the same constant, is called a *geometric sequence*.

The first step of the algorithm must be refined so that a value is read for the **Bounciness** of the ball and for **NumBounces,** the number of bounces. The height at the end of each bounce, **NewHeight,** can be computed through use of the function **ZToPower** that was developed in Section 7.4. Steps 1 and 2 of the algorithm are now as follows:

ALGORITHM 7.3   Refinement of Steps 1 and 2 of Algorithm 7.2

1. Initialize the program
   begin
      a. Write a message to the user stating the purpose of the program
      b. Read **Height,** the height from which the ball is dropped
      c. Read **NumBounces,** the number of times that the ball bounces
      d. Read **Bounciness,** the bounciness factor of the ball
   end
2. Compute the height and distance traveled by the ball

for **N** := 1 to **NumBounces** do
  begin
    a. **NewHeight** ← **Height** * **ZToPower(Bounciness, N)**
    b. Add the new height to the total distance traveled by the ball
    c. Report **NewHeight**
  end for

Notice that the statement to report **NewHeight** must be included within the **for** statement. If it is not, the only value that will be reported will be for the last bounce. Step 2b that takes the running total will be completed in the next section.

One important question about any sequence that describes a physical process is, Where does it end? Intuitively, you know that the height decreases with each bounce and that eventually the ball stops bouncing. As n, the number of bounces, becomes large, the height of the bounce, b_n, becomes small. In mathematical terms, the question posed is, What is the behavior of the sequence as n approaches infinity? From experience, you would expect that the value of b_n will get closer and closer to 0 as n gets larger. The sequence b_n is said to *converge* to the *limit L* if, for every positive number epsilon, ϵ, there corresponds an integer N such that

$$|b_n - L| < \epsilon \quad \text{for all } n > N \tag{7.9}$$

The definition of the limit given in Equation 7.9 simply states that, if the limit exists, then after a while every element in the sequence is very close to the limit. Another way to make the same statement is to write

$$\lim_{n \to \infty} b_n = L \tag{7.10}$$

The left-hand side of this equation is read as "the limit of b_n as n approaches infinity." For the geometric sequence

$$b_n = hr^n$$

the limit is

$$\lim_{n \to \infty} b_n = 0 \quad \text{for } r < 1 \tag{7.11}$$

That is, the limit of the height of the bounce is 0. Any geometric sequence with $r < 1$ converges to 0. What will happen if r is greater than 1? For the bouncing ball, that would mean that on each bounce the ball would rise higher than before. Such a ball would continue to bounce until it went into orbit. If r is greater than 1, the geometric sequence does not have a limit. When a sequence does not have a limit, the sequence is said to *diverge*.

Look again at the sequence $a_n = 2^n$. As n becomes larger, a_n also becomes larger. The value of a_n does not get close to any limiting value N. So the sequence $a_n = 2^n$ diverges.

Two sequences that you may be familiar with are the *factorial* sequence and the *Fibonacci* sequence. The factorial sequence can be used to compute the number of *permutations* that can be made from n objects. The number of permutations is the number of ways that a given number of objects can be arranged. Suppose you want to know how many ways the cards in a 52-card deck can be arranged. The first card selected could be any of the 52 cards in the deck. Once the first card has been selected, the second card could be any of the remaining 51 cards in the deck. The third card could be any of the remaining 50 cards, and so on. The total number of ways of arranging the deck is

$$(52)(51)(50)\cdots(1)$$

or

$$(n)(n-1)(n-2)(n-3)\cdots(1)$$

where n is the number of cards in the deck. This is the factorial sequence, defined as

$$a_n = n(n-1)(n-2)\cdots(2)(1) \tag{7.12}$$

By definition,

$$a_0 = 1$$

The factorial sequence is often written as

$$a_n = n! \tag{7.13}$$

For example,

$$6! = (6)(5)(4)(3)(2)(1)$$

By definition, $0! = 1$. The factorial sequence will be used again in Chapter 8 for the exponential function and in Chapter 10 for probability and statistics.

The Fibonacci sequence is used to describe certain growth processes in physics and biology. The elements in the Fibonacci sequence are defined by the equation

$$a_{n+1} = a_n + a_{n-1} \quad \text{for } n > 2 \tag{7.14}$$

where

$$a_1 = 1 \quad \text{and} \quad a_2 = 1 \qquad (7.15)$$

In Equation 7.14, the Fibonacci sequence is defined in terms of itself; that is, each element is defined in terms of earlier terms in the sequence. Such a sequence is said to be *recursive*. In a recursive sequence, the initial conditions must always be stated, because the first term cannot be derived from any earlier terms. The factorial sequence can also be written recursively as

$$a_n = n(n-1)!$$

or $\qquad (7.16)$

$$a_n = na_{n-1} \quad \text{for } n > 0$$

where $0! = 1$.

Series

A *series* is the summation of the terms of a sequence. In some problems, such as those predicting the next element in a sequence of numbers, the summation of the terms in the sequence is of little interest. In other problems, such as that of the bouncing ball, the series is more useful. By summing the elements of this sequence, you can determine the total distance traveled by the ball.

To write an expression for the distance that the ball travels, you must realize that the ball first travels a distance h, the height from which it is dropped. On the first bounce the ball rises to a height of hr and then falls a distance hr. So on the first bounce the ball travels $2hr$. On the second bounce it travels $2hr^2$, and so on. The equation for the total distance traveled is given by

$$d = h + 2hr + 2hr^2 + 2hr^3 + \cdots \qquad (7.17)$$

This type of series is called an *infinite series*. In general, an infinite series is written in the form

$$\sum_{i=1}^{\infty} b_i = b_1 + b_2 + b_3 + \cdots + b_n + \cdots \qquad (7.18)$$

The Σ is a sigma, a capital s in Greek, which stands for summation. The left-hand side of Equation 7.18 is read as "the sum of b sub i from 1 to infinity."

The series does not have to go to infinity. A *partial sum* is the sum of the elements of a sequence up to a particular element; for example,

$$s_n = \sum_{i=1}^{n} b_i = b_1 + b_2 + \cdots + b_n \qquad (7.19)$$

If each element b_i has a finite value, the partial sum of always defined; that is, it always has a finite value. However, an infinite series may or may not have a finite sum. An infinite series is said to converge if the sequence of partial sums $\{s_n\}$ converges to a finite limit L. This test is more complicated than the test for convergent sequences, and it is beyond the scope of this text. You can refer to any calculus book to find out more about infinite series.

By now you should be aware that computers are finite machines. Because infinity cannot be represented in a digital computer, you will always be approximating an infinite series with a partial sum. However, a partial sum does not guarantee that you do not have to worry about divergent series. At some point, the value of a divergent series will become larger than the largest number that the computer can store. Unless you are warned otherwise, you can assume that any infinite series in this book will be a convergent series. Real life will not always be so considerate.

Suppose you want to write a program to approximate the area under the curve $y = 1/x$, for x between 1 and 100, as illustrated in Figure 7.2. To approximate the sum, you can find the sum of the areas of the shaded rectangles. Each rectangle has an area equal to its base, 1, times its height, $1/x$. The sum of the areas is given by the partial sum

$$s = \sum_{n=1}^{100} \frac{1}{n} \qquad (7.20)$$

This series, called the *harmonic series,* diverges even though the limit of the sequence $\{1/n\}$ is 0.

You can use a **for** statement to loop through the integers from 1 to 100, computing each element in the series and adding each element into a running total:

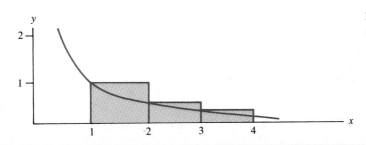

FIGURE 7.2
Area under the curve $1/x$.

```
Sum := 0.0;
for N := 1 to 100 do
  Sum := Sum + 1.0 / N;

writeln (' The sum of the first 100 terms of the harmonic series is',
         Sum:10:1)
```

=>
```
The sum of the first 100 terms of the harmonic series is    5.2
```

Notice that **Sum** must be a real variable because $1/n$ is always a fraction less than 1.0.

Now suppose you want to compute the first thousand terms of the sequence $1/n$. You will run into one of two problems. Either $1/n$ will become so small the computer will not be able to store it, or the sum will become so large the computer will not be able to store the sum. The first condition is called *underflow*, and the second condition is called *overflow*. For the harmonic series, the underflow condition will occur first. Try computing the harmonic series on your computer to see how many terms you can include and what the value of the series is before the underflow condition occurs.

Solution to Case Study 10

Going back to the bouncing ball, you may guess that the infinite series is convergent because a bouncing ball travels only a finite distance before it comes to rest. The sum of the distance is given by

$$d = h + \sum_{n=1}^{\infty} 2hr^n \tag{7.21}$$

From a branch of mathematics called analysis, it can be shown that the equation for the sum of a geometric series is

$$\sum_{n=1}^{\infty} r^n = \frac{r}{1-r} \tag{7.22}$$

So the total distance that the ball travels is

$$d = h + 2h\frac{r}{1-r} \quad \text{or} \quad d = h\frac{1+r}{1-r} \tag{7.23}$$

Given the original height, h, and the bounciness of the ball, r, you can compute directly the distance that it travels in an infinite number of bounces. If you only wanted to know how far it traveled in the first 5 bounces, you would use the partial sum:

$$d = h + \sum_{n=1}^{5} 2hr^n \qquad (7.24)$$

In refining Step 2 of the algorithm, you have already determined the height on each bounce. The distance traveled on each bounce is twice **NewHeight,** the height of the bounce. You can compute a running total for the distance by adding twice **NewHeight** to a running total. The **TotalDistance** must be initialized to the height from which the ball was originally dropped; otherwise this distance will not be included in the **TotalDistance.**

ALGORITHM 7.4 Completed algorithm for the case study

begin algorithm
1. Initialize the program
    begin
        a. Write a message to the user stating the purpose of the program
        b. Read **Height,** the height from which the ball is dropped
        c. Read **NumBounces,** the number of times that the ball bounces
        d. Read **Bounciness,** the bounciness factor of the ball
        e. **TotalDistance** ← **Height**
    end
2. Compute the height and distance traveled by the ball
    for **Num** := 1 to **NumBounces** do
    begin
        a. **NewHeight** ← **Height** ∗ **ZToPower(Bounciness, Num)**
        b. **TotalDistance** ← **TotalDistance** + 2 ∗ **NewHeight**
        c. Report **NewHeight**
    end for
3. Report **Total Distance**
end algorithm

Program 7.5, **Bounce,** is a translation of Algorithm 7.4 into Pascal. In program **Bounce,** each task in Step 2 is implemented as a separate subprogram. Function **BounceLength** computes the height of the *n*th bounce, procedure **ReportHeight** reports the value returned by **BounceLength,** and function **RunningTotal** computes the total distance traveled in the bounce and adds it to the running total. These three tasks could have been performed in a single procedure, but it would have needed a name such as **ComputeAndReportHeightAndAddDistanceToTotal.** A name such as this indicates that too many different tasks are being lumped into a single procedure. Even though the body of each subprogram is one statement, each subprogram performs a separate task.

PROGRAM 7.5 Bounce

```
program Bounce (input, output);
(*      Program Bounce computes the height that a ball bounces on each    *)
(*      of NumBounces and the total distance that the ball travels in     *)
(*      NumBounces.                                                       *)

var    Height : real;              (* height in inches from which ball drops. *)
       NumBounces : integer;       (* number of times the ball bounces. *)
       Bounciness : real;          (* bounciness factor of the ball. *)
       Num : integer;              (* for control variable in main program. *)
       NewHeight : real;           (* height ball bounces on current bounce. *)
       TotalDistance : real;       (* total distance traveled by the ball. *)

   procedure Initialize (var Hght, Rubberiness, Total : real;
                         var Number : integer);
   (*      Initialize writes a message to the user and reads values for   *)
   (*      the height, bounciness, and number of bounces. It also         *)
   (*      initializes the value of the total distance traveled by the    *)
   (*      ball to the value of the height so that the number of inches   *)
   (*      the ball initially drops is included in the total distance.    *)
   begin
     writeln (' This program computes the height that a ball bounces on');
     writeln (' each of N bounces and computes the total distance that');
     writeln (' the ball travels in N bounces.');
     writeln;
     writeln (' Enter the height in inches from which the ball is dropped');
     readln (Hght);
     writeln;

     writeln (' Enter the number of times that the ball bounces');
     readln (Number);
     writeln;
```

```pascal
      writeln (' Enter the bounciness coefficient of the ball.  It must be');
      writeln (' greater than 0.0 and less than 1.0 if the ball is to stop',
               ' bouncing');
      readln (Rubberiness);
      writeln;

      Total := Hght
   end; (* procedure Initialize *)



(*$i ztopower *)
(* Include function ZToPower (Z : real; Power : integer) : real; *)



   function BounceLength (Hght, Rubberiness : real;
                          Number : integer) : real;
   (*     BounceLength returns the Distance traveled by the ball on      *)
   (*     the nth bounce where Number equals n.                          *)

   begin
      BounceLength := Hght * ZToPower (Rubberiness, Number)
   end; (* function BounceLength *)



   procedure ReportHeight (NewHght : real; Number : integer);
   (*     ReportHeight reports the height of the nth bounce, where       *)
   (*     Number equals n.                                               *)

   begin
      writeln (' On bounce ', Number:1, ' the ball rises to a height of',
               NewHght:10:2, ' inches.')
   end; (* procedure ReportHeight *)



   function RunningTotal (Total, NewHght : real) : real;
   (*     Function RunningTotal adds the distance traveled on the Nth    *)
   (*     bounce (up and down) to the Total so far.                      *)

   begin
      RunningTotal := Total + 2 * NewHght
   end; (* function RunningTotal *)
```

```
    procedure ReportTotalDistance (Hght, Rubberiness, Total : real;
                                   Number : integer);
(*      ReportTotalDistance reports the total distance traveled by      *)
(*      the ball in a Number of bounces.                                *)

    begin
      writeln;
      writeln (' The total distance traveled by the ball with a bounciness',
               ' factor of ', Rubberiness:6:3);
      writeln (' when it is dropped from a height of ', Hght:6:1,
               ' inches');
      writeln (' and bounces ', Number:2, ' times is ', Total:10:2,' inches')
    end; (* procedure ReportTotalDistance *)

(*-------------------------------------------------------------------------*)

begin (* main program *)
  Initialize (Height, Bounciness, TotalDistance, NumBounces);

  for Num := 1 to NumBounces do
    begin
      NewHeight := BounceLength (Height, Bounciness, Num);
      ReportHeight (NewHeight, Num);
      TotalDistance := RunningTotal (TotalDistance, NewHeight)
    end; (* for *)

  ReportTotalDistance (Height, Bounciness, TotalDistance, NumBounces)
end. (* program Bounce *)



=>
 This program computes the height that a ball bounces on
 each of N bounces and computes the total distance that
 the ball travels in N bounces.

 Enter the height in inches from which the ball is dropped
 □12.0

 Enter the number of times that the ball bounces
 □5

 Enter the bounciness coefficient of the ball.  It must
 be between 0.0 and 1.0 if the ball is to stop bouncing
 □0.96
```

```
On bounce 1 the ball rises to a height of      11.52 inches
On bounce 2 the ball rises to a height of      11.06 inches
On bounce 3 the ball rises to a height of      10.62 inches
On bounce 4 the ball rises to a height of      10.19 inches
On bounce 5 the ball rises to a height of       9.78 inches

The total distance traveled by the ball with a bounciness factor of   0.960
when it is dropped from a height of    12.0 inches
and bounces  5 times is     118.35 inches
```

Summary

The **for** statement is used when an algorithm requires that a statement or sequence of statements be repeated a given number of times. In the **for** statement, a control variable is incremented from a lower bound to an upper bound or, alternatively, decremented from an upper bound to a lower bound, and the statement or statements in the loop are executed for each value of the control variable. The following list summarizes the restrictions on **for** statements:

1. The control variable, the upper bound, and the lower bound must have an ordinal data type and must all have the same data type.

2. If more than one statement is to be executed, the statements must be within a **begin-end** pair.

3. The control variable is undefined both prior to and after the execution of the **for** statement.

4. The value for a **for** control variable cannot be changed within the **for** statement.

5. The control variable of a **for** statement invoked within a subprogram must be declared within that subprogram as a local variable.

6. The value of either bound may be changed within the **for** statement, but this will not affect the number of times the loop executes.

Exercises

PASCAL

1. Correct all the syntax and run-time errors in each of the **for** statements below. Each **for** statement has at least one error. There are some subtle errors; for example, a reserved word is used as an identifier.

(a) `for Index = 1 to 5 do X := X + 1;`

(b)
```
Sum := 0;
for Number := -1 to 1 do
  Sum := Sum + Number;
writeln (' Sum up to number', Number:3, ' is ', Sum:5:1)
```

(c)
```
for Counter := Start to End do
  begin
    Backwards := End - Counter + 1;
    writeln (' For Counter =', Counter:4, ' Backwards =',
             Backwards:4);
    Counter := Backwards
  end
```

(d)
```
writeln (' The letters from a to z are:');
for I := 'a' to 'z' do;
  write (I:2);
writeln;
```

(e)
```
for Counter := 1 to 5 do
  begin
    Sum := 0.0;
    Sum := Sum + Counter;
    writeln (' The sum of the first', Counter:4,
             ' integers is', Sum:5)
  end;
```

2. Write a procedure that writes one line of your choosing at the terminal. Write a main program that invokes your procedure five times.

3. What is the output from the following statements?

(a)
```
for Index := 20 to 25 do
  case Index of
       20 : writeln (' things ');
    21,24 : begin
              writeln (' nonsense ');
              writeln (' or not ')
            end;
       23 : write (' worse ');
       22 : write (' could be ');
       25 : write (' the end ')
  end; (* case *)
writeln(' of all that')
```

(b)
```
for Counter := 1 to 3 do
   begin
      Index := 2 * Counter;
      case Index of
         2, 3 : writeln (' Counter =', Counter:4, ' Index =', Index:4);
         5, 6 : writeln (' Index =', Index:6, ' Counter =', Counter:6);
         4, 7 : writeln (' Counter =', Counter:5)
      end  (* case statement *)
   end  (* for statement *)
```

4. Write a program to compute and print the cube of each integer between 1 and 10.

5. In Program 7.4 **ParticleVelocity**, why is **AddToTotal** a procedure and why is **Average** a function?

APPLICATION

6. Write the first 5 terms and the 100th term for each of the following series:

   (a) $a_n = \dfrac{1}{n^2}$ (b) $a_n = \sin\left(\dfrac{\pi}{n}\right)$

   (c) $a_n = \dfrac{2n+1}{2n-1}$ (d) $a_n = \dfrac{2^n}{n+1}$

7. Write the first 5 terms for each of the following recursive series.

   (a) $a_n = \dfrac{1}{a_{n-1}}$, $a_0 = 2$ (b) $a_n = a_{n-1}^2$, $a_0 = 2$

   (c) $a_n = 2\ln(a_{n-1})$, $a_0 = 100$ (d) $a_n = a_{n-2}a_{n-1}$ $a_0 = 2$, $a_1 = 3$

8. Graph each of the sequences in Exercises 6 and 7.

9. Compute the partial sum from 1 to 5 for each of the series in Exercises 6 and 7.

Problems

PASCAL

1. The number of students in each of eight sections of a course is given in the list on the next page. Use a **for** statement to compute the total number of students enrolled in the course, and compute the average number of students per section.

Section	Number students
A1	52
A2	45
A3	48
A4	112
A5	10
A6	93
A7	57
A8	84

2 Using only the addition operator, **+**, and a **for** statement, write a program that multiplies two integers that are read from the terminal. The multiplication operator, $*$, should not appear anywhere in your program.

3. Write a program to find and report the smallest positive number in a list of integers. Ask the user how many numbers will be entered, and then read that many integers. You should print an error message if there are no positive numbers in the input. Test your program by entering the following seven numbers:

$$-3, 2, 1, 0, -1, 5, 0$$

Test it again by entering the following four numbers: $-1, -4, -6, -7$

4. Given an initial investment p (the principal) earning interest at an annual rate of $R\%$, the total amount of money P at the end of one year is

$$P = p\left(1.0 + \frac{R}{100.0}\right)$$

If the interest and the principal are not withdrawn from the account, the investment the next year is the total from the last year.

   (a) Write a function that computes the amount of money in a savings account at the end of one year.

   (b) Write a test program for your function that computes the interest for 10 years. For each year, write out a statement saying how much money is in the account at the end of that year and how much interest was earned that year.

   (c) Run your program with the following two sets of data:

   $100.00 at 4.25% per year for 10 years
   $100.00 at 10.00% per year for 10 years

5. Calculate all the ways that a dollar bill can be converted into nickels, dimes, quarters, and half-dollars. For example, one way uses three quarters, one dime, and three nickels. Another way uses a half-dollar, a quarter, two dimes, and a nickel. The output of your program should be a chart in the following (or similar) format:

	Half-dollars	Quarters	Dimes	Nickels
1.	2	0	0	0
2.				
3.				
(etc.)				

6. A company produces bolts on four different machines. Each machine produces bolts at a different rate. The company runs one eight-hour shift per day. The machines break down frequently and need to be repaired. At the end of each day, each operator fills out a form listing production totals and breakdown time and gives it to the supervisor, who gives it to data processing. Write a program that computes the breakdown time in hours for each machine. Also compute the average number of bolts produced per hour by each machine. In computing the average, include only the time that the machine is working. The following form is taken from a typical day:

Machine #	Bolts produced	Breakdown time in minutes
1	1239	10.2
2	463	230.0
3	897	79.3
4	106	400.9

7. Each part manufactured in a small factory is assigned a one-letter identification code (a–z). The parts are manufactured by sophisticated robots that perform their tasks in an apparently random order, so the parts coming off the end of the line are all mixed together. The last robot reads the one-letter character code and puts the part in a bin with other parts of the same type.

You have been hired to write the sorting robot a program that will count the number of parts that the robot puts in each bin. You must make a well-formatted table, to be read by the president of the company, that lists the part identification codes and the number of parts with each code. For data, enter a line of 80 (or more) letters from the terminal.

8. A test program is run on 10 different computers, and the running time of the program on each computer is recorded. Write a program that (a) reads an identifier for each computer (a single character) and the running time for the test program on that machine, (b) finds the computer with the shortest running time, and (c) writes the identifier and running time for the fastest computer. Test your program on the following data:

Computer	Running time
a	10
b	15
c	22
d	9
e	8
f	20
g	6
h	24
i	26
j	13

9. You are in charge of quality control for an assembly line that produces widgets. You must select a random sample of 10 widgets out of each batch of widgets that comes off the assembly line. Each widget in the sample is tested by a robot that measures the diameter of the widget. The robot sends the data on the diameter of each widget in the random sample to a Pascal program that helps you analyze this data.

   Widgets are labeled good or bad according to the following criteria:

   $3.25 \leq$ Diameter ≤ 3.75 good widget
   otherwise   bad widget

   Batches are labeled rejected, substandard, or acceptable according to the following criteria:

   $20.0\% \leq$ percent of bad widgets in the sample   REJECTED
   $10.0\% \leq$ percent of bad widgets in the sample $< 20.0\%$ SUBSTANDARD
   percent of bad widgets in the sample $< 10.0\%$ ACCEPTABLE

   Write a program that reports the number of bad widgets and decides whether the batch should be labeled rejected, substandard, or acceptable. Test your program for the following batch:

   3.24   3.50   3.51   3.75   3.76   3.52   3.56   3.76   3.25   3.26

10. You are given the hourly temperatures for one day, and you want to know the first hour in which the temperature drops below the temperature for the previous hour. Write a program that prints the hour in which the temperature first drops and the temperatures for that hour and the previous hour. Test your program on the following data:

    65.2   65.2   65.2   65.2   65.3   66.0   66.5   67.1   68.2   70.3   72.4   75.7

    77.8   77.9   80.6   80.7   78.6   77.5   77.2   76.4   74.3   73.9   73.8   72.5

11. You must decide which of three motors should be used in a new product, based on the results of 10 tests administered to each motor. Because the tests measure very different physical phenomena, the numerical results of the test are not relevant. There is no possibility of a tie on the individual test scores.

    The following system is used to assign points to each motor for each test:

	Highest score	2 points
	Second highest score	1 point
	Lowest score	0 points

The motor with the greatest number of points will be considered the best motor and will be put into production. The results of each of the 10 tests are stored in sets of three scores.

	Motor #1	Motor #2	Motor #3
Test 1	7.56	34.21	22.09
Test 2	123.45	50.41	75.65
Test 3	1.01	1.22	2.03
Test 4	5.98	6.23	3.73
Test 5	1000.0	1500.0	1750.0
Test 6	9.03	9.01	9.04
Test 7	65.1	59.2	49.3
Test 8	543.22	578.99	567.98
Test 9	32.88	31.43	30.11
Test 10	3456.78	3344.55	327.98

Write a program that reads in the test data, reports the total score for each motor, and decides which motor is the best motor.

12. Function **ZToPower** is valid only for positive values of **Power**. Rewrite the procedure so that **Power** can be positive or negative. Remember that

$$x^{-n} = \frac{1}{x^n}$$

Be sure that your function gives the correct answers for 0^{-1} and 0^0.

APPLICATION

13. The heights and weights of five students are read from the terminal. The height is in feet and inches, and the weight is in pounds. Convert the height to centimeters and the weight to kilograms. Compute the average height in centimeters and the average mass in kilograms. Test your program on the following data:

	Feet	Inches	Pounds
Pascal	6	2	160
Curie	5	1	110
Milton	5	8	135
Gauss	5	10	200
Anthony	5	3	115

Notice that each line starts with the name of the student. Just enter the numerical data.

Chapter 7 □ The For Statement

14. Write a program that generates and reports the first 20 Fibonacci numbers.

15. If *n* coins are tossed each one can come up a head or a tail, so the total number of possible outcomes, or permutations, is 2^n. Write a program to generate a table of the number of permutations when *n* coins are tossed. Make a table of the values of the function

$$f(n) = 2^n$$

for $n = 1, 2, 3, \ldots, 25$. Use the fact that $2^2 = 2 * 2$ and $2^3 = 2^2 * 2$; that is, each term in the series is a multiple of the previous term.

16. Generalize the program from Problem 15 to compute the number of permutations when there are more than two possible outcomes. Test your program first by computing the number of possible outcomes when three dice are tossed. Test your program again by computing the number of possible outcomes when 2 cards are drawn from a 52-card deck. Assume that the first card is replaced in the deck before the second card is drawn.

17. Write a program that computes the distance traveled in the first 100 bounces of a ball. Compare this value to the distance traveled in an infinite number of bounces according to Equation 7.23.

18. Write a program that computes the terms in the series

$$f(m) = m^n, \quad \text{for } n = 1, 2, \ldots, N$$

where *m* is a real number and *N* is a positive integer. Print each term of the series in a table.

19. A *p*-series in a mathematical series of the form

$$f(p) = \sum_{n=1}^{\infty} \frac{1}{n^p}$$

A *p*-series converges if $p > 1$ and diverges if $p \leq 1$.

(a) Use a calculator to compute the first five terms in the *p*-series where $p = 1/2$.

(b) Write a program that computes and sums the first 50 terms of the *p*-series for $p = 1/2$. Print out a table giving n, $1/n^{1/2}$, and the partial sum for each value of *n*.

20. Modify the *p*-series program from Problem 19 to compute the first 50 terms for the *p*-series with $p = 2$.

21. Write a program that approximates the sine function using a Taylor series. The Taylor series for $\sin(x)$ is given by

$$\sin(x) = x - \frac{x^3}{3!} + \frac{x^5}{5!} - \frac{x^7}{7!} + \frac{x^9}{9!} - \cdots$$

where the angle *x* is in radians. Compute the first 10 terms of the series and compare the result with the value returned by the standard Pascal function **sin**.

Make the comparison for $x = 0$ radians, $x = \pi/4$ radians, $x = \pi/2$ radians, $x = 3\pi/4$ radians, and $x = \pi$ radians. Hint: If **maxint** is smaller than 21! on your computer, you can declare the factorial as a **real** function.

22. Using the *p*-series program from Problem 19, write a program that computes the first N terms of the *p*-series with $p = 1$. The program should print every Mth term. For example, the user could request that the first 1000 terms be computed and that every 100th term be printed.

CHAPTER 8

Conditional Loops

Now that you are familiar with writing input/output procedures, we will often omit listing them in the remaining chapters in order to make the programs more compact and easier to read.

In this chapter, you will learn Pascal constructions that allow you to write loops that terminate based on the outcome of a test, rather than after a specified number of repetitions as in a **for** statement. The **while-do** and **repeat-until** statements require that you specify the conditions under which the loop continues to be executed or stops execution.

CASE STUDY 11
Population growth

A colony of rabbits has an initial population of 10 rabbits. At the end of the first time period, each of the 5 pairs of rabbits has 3 surviving offspring. At the end of one time period, the number of rabbits will be

$$P_1 = 10 + \frac{10}{2}(3) = 25$$

The population at the end of the first time period can be written mathematically as

$$P_1 = p + pk \qquad (8.1)$$

or

$$P_1 = p(1 + k) \qquad (8.2)$$

where p = the initial population
P_1 = the total number of rabbits at the end of the time period 1
k = the reproductive rate of the rabbits

For the case study, k is 3/2. That is, for every 2 existing rabbits, 3 new rabbits come into being. To simplify the problem, assume that no rabbits die.

At the beginning of the next time period, there are $p(1 + k)$ rabbits that can reproduce. So, the number of rabbits at the end of the second time period is

$$P_2 = p(1 + k) + kp(1 + k) \tag{8.3}$$

or

$$P_2 = p(1 + k)^2 \tag{8.4}$$

And at the end of a third time period, the total would be

$$P_3 = p(1 + k)^3 \tag{8.5}$$

This pattern continues for each time period and leads to the following conclusion: If p rabbits reproduce at a rate of k per time period, after n time periods the total population is

$$P_n = p(1 + k)^n \tag{8.6}$$

Write a Pascal program that computes the number of time periods required for the rabbit population to more than quintuple (be multiplied by 5). Report the population at the end of each time period and report the number of time periods for the population to quintuple.

Algorithm development

The stopping criterion for the problem is based on a condition—that the population has become 5 times the original population—rather than on the number of times the loop is executed. The problem can be restated as follows: Find the value of n such that $P_n \geq 5p$. Equation 8.6 gives the formula for P_n as a function of p, k, and n. Since both k and p are known, one method for solving the problem of finding the number of time periods would be to substitute $5p$ for P_n and solve directly for n. But you also must determine the population in the time periods between 1 and n, and the use of a loop will allow you to solve both parts of the problem.

Algorithm 8.1 has two alternative but equivalent solutions for the third step; the first alternative leads to a problem solution that uses the **while-do** statement and the second to a solution that uses the **repeat-until** statement. The first version of Step 3 solves the problem by computing the population at the end of each time period *while* the population is *less than* 5 times the original population. The other version computes the population at the end of each time period *until* the population is *greater than or equal to* 5 times the original population.

ALGORITHM 8.1   First iteration of the algorithm for Case Study 11
begin algorithm
 1. Initialize the program
 2. Read values for the original population and the growth rate
 3. While the population is less than 5 times the original population
    begin
       a. Compute the population at the end of each time period
       b. Report the population at the end of each time period
       c. Add 1 to a running total of the number of time periods
    end
    Or
 3. Repeat
    begin
       a. Compute the population at the end of each time period
       b. Report the population at the end of each time period
       c. Add 1 to a running total of the number of time periods
    end
    until the population is greater than or equal to 5 times the original population
 4. Report the number of time periods from Step 3
end algorithm

Equation 8.6 will be used to compute the population at the end of each time period. Since this equation involves exponentiation, you can use the function **ZToPower** that was developed in Chapter 7 to find the new population. The value of $(1 + k)$ can first be computed and assigned to a variable so that it can be passed as a parameter to the function. Step 3a thus becomes

ALGORITHM 8.2   Refinement of Step 3a of Algorithm 8.1

3a. Compute the new population, **NewPop**
    begin
       RatePlus1 ← (1.0 + **Rate**)
       RateToN ← **ZToPower**(**RatePlus1**, **TimePeriod**)
       NewPop ← **OriginalPop** ∗ **RateToN**
    end

The values of the variables **TimePeriod, Rate,** and **OriginalPop** must be initialized prior to Step 3a.

8.1 The while-do statement

The **while-do** statement is used for loops that are to continue while a condition is true. When the condition becomes false, the execution of the loop ends. As in the **if-then** statement, the **while-do** statement uses the value of a boolean expression to test whether the condition is true or false. The syntax diagram for a **while-do** loop is

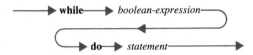

The **while-do** statement below can be constructed from the syntax diagram:

```
while Sum < 5 do Sum := Sum + 1
```

In this **while-do** statement, the *boolean expression* is

```
Sum < 5
```

and the *statement* is

```
Sum := Sum + 1
```

The **while-do** statement is equivalent to the following sequence of statements: If the current value of **Sum** is less than 5, then execute the statement that increments the **Sum.** After the statement is executed once, test the boolean expression again to determine if the current value of **Sum** is still less than 5. If it is, then increment the sum again. Keep executing the statement as long as **Sum** is less than 5.

A **while-do** loop begins execution only if the initial value of the boolean expression is true. If the initial value of the boolean expression is false, the program executes the next statement after the loop and the statement in the **while-do** is not executed.

The **while-do** statement above is not complete because the initial value of **Sum** is unspecified. If the initial value of **Sum** is less than 5, the loop will execute until **Sum** is greater than or equal to 5. If the initial value of **Sum** is greater than or equal to 5, the loop will not execute because the first time the boolean **Sum < 5** is evaluated, the result will be false. Without knowing the value of **Sum** before the loop is entered, it is not possible to know how many times, if any, the loop will execute. The fragment could be written with the initial value for **Sum** explicitly stated, as follows:

```
Sum := 0;
while Sum < 5 do Sum := Sum + 1
```

In this loop, the statement

```
Sum := Sum + 1
```

will execute 5 times. This **while-do** is not very interesting, because all it does is count until it finishes execution. In this respect, it is like a **for** statement with a null statement after the **do**.

This example points up a major difference between the **for** statement and the **while-do** statement. In the **for** statement, you do not increment the control variable. If you use a **for** statement such as

```
for Sum := 1 to 5 do
  NewNum := NewNum * 2
```

Sum is incremented by 1 each time the **for** loop is repeated. The **while-do** statement does not keep track of the number of times the loop has been repeated nor does it automatically stop the execution of the loop after a certain number of repetitions.

To confirm that the **while-do** statement above repeats 5 times, an output statement can be included as part of the loop. To execute more than one statement as part of the **while-do** loop, you must use a compound statement with a **begin-end** pair, just as you did for the **for** and the **if** statements:

```
Sum := 0;
while Sum < 5 do
  begin
    Sum := Sum + 1;
    writeln (' The current value of the sum is ', Sum:2)
  end

=>
  The current value of the sum is  1
  The current value of the sum is  2
  The current value of the sum is  3
  The current value of the sum is  4
  The current value of the sum is  5
```

For a **while-do** loop to begin execution, the boolean expression must initially be true. Then, inside the loop, the value of the boolean expression must eventually become false or the loop will never terminate. A loop that does not terminate is called an *infinite loop*. The following fragment is an infinite loop because the semicolon after the **do** is interpreted as meaning that the **while-do** statement is complete and the next statement is about to begin. If you do not include a statement to be executed when the boolean expression has a value of true, the null statement will be executed. The boolean expression **Number** < 2 is true the first time the loop executes, and it remains true because the null statement is executed each time the loop executes. The statement that increments **Number** will never execute.

```
Number := 1;
while Number < 2 do;
   Number := Number + 1
```

DEBUGGING HINTS

Whenever you write a **while-do** loop, be sure that the boolean expression is defined when the loop begins execution and that the boolean expression will eventually become false during execution of the loop. Do not put a semicolon after the **do**.

The next fragment is also an infinite loop. Because the value of **Counter** never changes, the value of the boolean expression **Counter** < 5 is always true.

```
Counter := 1;
Sum := 0;
while Counter < 5 do Sum := Sum + 1
```

Unless a statement is added to increment the value of **Counter,** the boolean will never become false and the loop will not terminate. The following loop will execute correctly:

```
Counter := 1;
Sum := 0;
while Counter < 5 do
   begin
      Sum := Sum + 1;
      Counter := Counter + 1
   end
```

In the preceding example, a variable has acted like the control variable in a **for** statement to count the repetitions of the loop. But because the execution of a **while-do** loop does not have to depend on a counter, the **while-do** is more versatile than a **for** loop. In Program 8.1, **WhileDo,** the program continues to execute as long as integers between 1 and 5 are entered. As soon as a value outside the range is entered, the program halts execution.

PROGRAM 8.1 WhileDo

```
program WhileDo (input, output);
(*      This program uses a while-do statement so that execution of the   *)
(*      loop continues as long as a Number between 5 and 10 is entered    *)
(*      from the terminal.  Notice that Number must be initialized to     *)
(*      a value between 5 and 10 in order for the while-do statement to   *)
(*      be executed when it is first encountered during program execution. *)

var    Number : integer;       (* variable to be initialized to 5 and then *)
                               (* to be assigned subsequent values by the  *)
                               (* user. *)
begin
  Number := 5;                 (* initialize Number so that the boolean *)
                               (* expression is true. *)

  while (Number >= 5) and (Number <= 10) do
    begin
      writeln;
      writeln (' Enter a number between 5 and 10 to continue');
      readln (Number)
    end; (* while *)

  writeln (' That''s all')
end. (* program WhileDo *)
```

=>

 Enter a number between 5 and 10 to continue
▫5

 Enter a number between 5 and 10 to continue
▫7

 Enter a number between 5 and 10 to continue
▫3
 That's all

8.2 The repeat-until statement

A **repeat-until** statement is similar to a **while-do** statement except that a **repeat-until** loop repeats as long as a boolean expression is false and stops when it becomes true. The syntax diagram for a **repeat-until** statement is

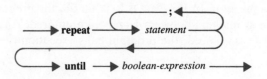

The first example of the **while-do** statement can be rewritten as a **repeat-until** statement as follows:

```
Sum := 0;
repeat
   Sum := Sum + 1
until Sum >= 5
```

In this fragment, the initial value of **Sum** is 0. When the loop is entered, the statement that causes **Sum** to be incremented is executed. After this statement is executed, the boolean expression is evaluated. If the boolean expression has a value of false, the loop will be executed again; if it has a value of true, the execution of the loop will stop and the next program statement will be executed. Comparing the **repeat-until** loop with the equivalent **while-do** loop, you will notice that the stopping condition for the **repeat-until** is the complement of the continue condition for the **while-do.** That is, the loop repeats **until Sum >= 5** or continues **while Sum < 5.**

A major difference between the **repeat-until** and the **while-do** statements is that the body of the **repeat-until** statement is always executed at least once because the boolean expression is evaluated at the end of the loop. No matter what the value of the boolean expression is when the **repeat-until** is entered, the expression is not tested until the end of the loop.

A minor difference between the two statements is that the **repeat-until** does not need a **begin-end** pair; the words **repeat-until** act as the **begin-end**. You can put in a **begin-end** pair, but, like an extra set of parentheses, it has no effect on the execution of the program.

Repeat-until has the same potential as **while-do** for infinite loops. Be sure that the value of the boolean expression eventually becomes true somewhere inside the loop.

Three types of errors commonly arise with **while-do** and **repeat-until** statements: (1) The entry conditions for the loop may be improperly stated, (2) the loop may not be exited properly or may not be exited at all, or (3) the loop may not perform the correct number of iterations.

When the loop is entered under improper conditions, usually either the variable that controls the loop is being computed incorrectly or the boolean

---

DEBUGGING HINT

It is a good idea to put a **writeln** statement inside the first **repeat-until** and **while-do** loops that you write. When your program executes you will be able to tell if you have written an infinite loop. As soon as your **writeln** statement prints more times than the loop should execute, you should cancel your program's execution.

expression is incorrect. If the boolean has an initial value of false, a **while-do** statement will not execute. For example, the logical bug in the following fragment is due to an incorrect boolean expression:

```
Number := 5;
while (Number < 5) and (Number > 10) do
   begin
      writeln (' Enter an integer between 5 and 10 to continue');
      readln (Number)
   end;  (* while *)
writeln (' That''s all')
```

=>
That's all

The boolean expression (**Number < 5**) **and** (**Number > 10**) will always be false because no number can be both less than 5 and greater than 10. Therefore the **while-do** loop will never execute.

The **repeat-until** statement that is equivalent to the **while-do** statement above illustrates the second type of error, that of incorrect exit from a loop:

```
repeat
   writeln (' Enter an integer between 5 and 10 to continue ');
   readln (Number)
until (Number >= 5) or (Number <= 10);
writeln (' That''s all')
```

=>
 Enter an integer between 5 and 10 to continue
□7
 That's all

This fragment also has a logical bug—it will stop when any number greater than or equal to 5 is entered or when any number less than or equal to 10 is entered. Since every number is greater than or equal to 5 and/or less than or equal to 10, any number that is entered will stop the loop. If the boolean expression in the **repeat-until** statement is written as

$$(\text{Number} < 5) \text{ and } (\text{Number} > 10)$$

then an infinite loop will result because the value of the boolean expression will always be false.

In the third type of error, the loop does not perform the correct number of iterations. This error often occurs when a **repeat-until** or a **while-do** is being used like a **for** statement; usually the number of iterations is off by one. In the following fragment, the programmer intended to compute the sum of the first five integers:

8.2 The repeat-until statement

```
X := 0;
Sum := 0;
while X < 5 do
  begin
    Sum := Sum + X;
    writeln (' The sum of the first ', X:2, ' integers is ', Sum:2);
    X := X + 1
  end (* while *)
```
```
=>
The sum of the first  0 integers is  0
The sum of the first  1 integers is  1
The sum of the first  2 integers is  3
The sum of the first  3 integers is  6
The sum of the first  4 integers is 10
```

But, because the boolean condition in the **while-do** is incorrect, the loop executes four times instead of five times. If the number of iterations is known, a **for** statement is usually the better construction to use.

Problems in which the number of iterations of the loop is unknown are suitable to **repeat-until** and **while-do.** One such problem would be to find the number of times that e can be raised to the power e before the computed value is greater than 100.0. You can use the standard Pascal function **exp** to compute e. The function **exp** is discussed in more detail in the application section of this chapter.

```
X := 1.0;
while X < 100.0 do
  begin
    writeln (' The current value of X is', X:10:1);
    X := exp(X)
  end; (* while *)
writeln;
writeln (' The final value of X is', X:12:2)
```
```
=>
The current value of X is       1.0
The current value of X is       2.7
The current value of X is      15.2

The final value of X is   3814279.10
```

DEBUGGING HINT

Always walk through parts of your code that contain conditional loops. Check the values during the first and last iterations to make sure that the values of the variables on entering and on completing the loop are what you intend.

Notice that when a counter is used to keep track of the number of iterations in **repeat-until** and **while-do** loops, the counter is not part of the Pascal construction, so it still has a value after the loop terminates, just like any other variable.

STYLE HINT

If you know the number of times that a loop is to be repeated, a **for** loop is usually a better solution than a **while-do** or **repeat-until** loop. If you know the condition under which repetition of the loop is to terminate, but not the number of times that the loop will be repeated prior to the condition's being reached, you must use a **while-do** or **repeat-until** loop.

8.3 Simple data checking

The ability to repeat statements until a condition is met gives you more control over the execution of your program. In this section, we will discuss simple data checks that you can perform using **while-do** and **repeat-until** loops. A more complete discussion of data checking is found in Chapter 12.

The easiest data checks to perform are those that test the range of input data. For example, if you request a person's birth date, you can test if the birth year entered is before the current year but not more than 100 years before the current year. Using a conditional loop, you can request that the user enter the birth year until it falls within the correct range.

Another example of a place where a conditional loop could be used for range checking is the procedure **ZToPower,** developed in Chapter 7, which requires that the power to which **Z** is raised be a positive integer. As part of the interactive input procedure, you could use a **while-do** loop to prevent the user from entering a negative value, as follows:

```
procedure ReadPower (var X : real; var N : integer);
(*     This procedure reads a value X to be raised to the power N.       *)
begin
  writeln (' Enter a real number to be raised to a power');
  readln (X);

  N := -1;
  while N < 0 do       (* if N is negative, keep requesting another integer. *)
    begin
      writeln (' Enter a positive integer for the power to which the number',
               ' is to be raised');
      readln (N)
    end (* while *)
end;   (* procedure ReadPower *)
```

```
=>
 Enter a real number to be raised to a power
□2.5
 Enter a positive integer for the power to which the number is to be raised
□-5
 Enter a positive integer for the power to which the number is to be raised
□5
```

Procedure **ReadPower** requests another value if a negative value is entered for **Power.**

In the preceding examples, the conditional loops can prevent data that is out of range from being entered, but cannot prevent entry of a character instead of an integer. If you have never typed a character in response to a prompt for numerical data, you should try doing so. On most systems, you will immediately get a run-time error stating that the data is not of the appropriate type, and the program will halt execution. This response can be annoying if you have already entered a lot of data and have to start over.

Entering a character in response to a prompt for numerical data, whether real or integer, will result in a run-time error. The following discussion will be limited to using conditional loops to prevent run-time errors when a single digit is requested from the user. In Chapter 12, complete algorithms for trapping illegal data will be developed. The key to solving the problem of preventing illegal data from being entered is to realize that, since anything on your keyboard is a valid character, any input from the terminal must be a valid character—that is, it can be read into a Pascal variable of data type **char.** No data that is not of the appropriate data type can be entered if all types of data are read as characters. But if numerical data is read into character variables, it must somehow be converted into numbers because you cannot perform arithmetic operations on characters. The character '3' somehow must be converted to the integer 3 so that it can be used in arithmetic expressions.

The key to the conversion from character to integer is knowing that the characters '0' through '9' are sequentially ordered within the character sets. For example, in the ASCII character set, the characters '0' to '9' have ordinal values between 48 and 57. The standard Pascal function **ord** can be used to find the position of a character within the character set that your computer uses. **Ord('3')** returns the ordinal position of the character '3' within a character set. You are interested in the ordinal value of the character relative to the ordinal value of '0'. **Ord('0')** is evaluated as 48 and **Ord('3')** is evaluated as 51 on a computer that uses ASCII; thus **Ord('3')** − **Ord('0')** would be evaluated as the integer 3. If the difference between the ordinal value of a character and the ordinal value of '0' is between 0 and 9, the character represents a digit. Appendix A contains a table of the ordinal values for all the characters in the ASCII and EBCDIC character sets.

The following fragment uses the algorithm outlined above to convert the character input from the terminal into numerical data. The fragment uses a **repeat-until** loop to continue to request new data until a character that corresponds to an integer is entered. **Answer** is declared as a character variable.

```
OffSet := ord('0');
repeat
  writeln (' Please enter a digit between 0 and 9');
  readln (Answer);
  Digit := ord (Answer) - OffSet
until (Digit >= 0) and (Digit <= 9)
```

=>
 Please enter a digit between 0 and 9
□B
 Please enter a digit between 0 and 9
□0

Solution to Case Study 11

To write the complete program to solve Case Study 11, you must develop at least four subprograms: a procedure to initialize the program, a procedure to read the values for the initial population and the growth rate, a procedure or a function to compute the number of time periods until the population quintuples, and a procedure to report the results. By now, the procedures to initialize, read, and report should be easy to write, so only the subprogram for Step 3, computing the time until the population quintuples, will be discussed in detail.

Step 3 of the algorithm can be solved with either a **repeat-until** or a **while-do** statement, so Algorithm 8.2 can be translated directly into a Pascal procedure. This procedure will require some revisions, but the first iteration is

```
procedure NumberOfPeriods (OrigPop, Rate, Factor : real; var N : integer);
(*     This is a first attempt at a procedure to compute the number of      *)
(*     time periods, N, until the size of a population has multiplied by    *)
(*     a Factor.  This procedure has a bug in that N is off by one at       *)
(*     the end of execution of the repeat-until statement.                  *)

var    RatePlus1 : real;      (* growth rate plus 1. *)
       NewPop : real;         (* population at end of time period N. *)

begin
  repeat
    RatePlus1 := 1 + Rate ;
    NewPop := OrigPop * ZToPower (RatePlus1, N);     (* use Equation 8.8 *)
                                                     (* to compute NewPop. *)
    ReportPop (NewPop, N);    (* report the NewPop at the end of each time *)
                              (* period. *)
    N := N + 1                (* increment the counter for the time periods. *)
  until NewPop >= Factor * OrigPop
end; (* procedure NumberOfPeriods *)
```

In the procedure **NumberOfPeriods,** the problem of reporting the new population has been relegated to another procedure. **Rate** and **OrigPop** are passed through the parameter list rather than being local constants because at some future time you might be interested in rates and populations other than those given in the original problem. **RatePlus1** and **NewPop** are locally declared because they are not used in any other part of the program and their values do not need to be initialized before the computation begins.

N is the time-period counter that must initialized before the loop is executed. Suppose that the value of **N** is initialized to 0 before the **repeat-until** statement. The first time the function **ZToPower** is invoked, **RatePlus1** will be raised to the zero power and **NewPop** will equal **OrigPop.** So, the first iteration of the loop does nothing but restate the initial conditions. Initializing **N** to 1 will eliminate this problem; however, the final value of **N** will be off by 1. In the final version of the procedure **NumberOfPeriods** the order in which the computations are performed is rearranged so that the loop executes the correct number of times and **N** equals the correct value at the end of the loop. In this version, **N** is initialized to 0 outside the procedure.

```
procedure NumberOfPeriods (OrigPop, Rate, Factor : real; var N : integer);
  (*      This is a corrected version of NumberOfPeriods that computes      *)
  (*      the number of time periods until the population multiplies by     *)
  (*      a Factor.                                                         *)

var    RatePlus1 : real;      (* growth rate plus 1. *)
       NewPop : real;         (* population at end of time period N. *)
begin
  repeat
    N := N + 1;                    (* increment the counter for the time periods. *)
    RatePlus1 := 1 + Rate;
    NewPop := OrigPop * ZToPower (RatePlus1, N);   (* use Equation 8.8 *)
                                                   (* to compute NewPop. *)

    ReportPop (NewPop, N)     (* report the NewPop at the end of each time *)
                              (* period. *)
  until NewPop >= Factor * OrigPop
end; (* procedure NumberOfPeriods *)
```

As an exercise, you could try rewriting the procedure **NumberOfPeriods** with a **while-do** instead of a **repeat-until.** Why was **NumberOfPeriods** implemented as a procedure instead of as a function?

Program 8.2, **Rabbits,** is a complete program that solves Case Study 11. When the program was written, it was decided that the factor by which the population grew should be an input variable rather than a constant so that the user could experiment with different values of **GrowthRate, OriginalPop,** and **GrowthFactor.**

PROGRAM 8.2 Rabbits

```pascal
program Rabbits (input, output);
(*      This program computes the number of years a colony of rabbits   *)
(*      takes to multiply by the given factor.                          *)

var     GrowthRate : real;        (* annual growth rate per rabbit. *)
        OriginalPop : real;       (* number of rabbits in the colony in *)
                                  (* the beginning. *)
        GrowthFactor : real;      (* factor by which the rabbits will *)
                                  (* multiply. *)
        TimePeriods : integer;    (* number of time periods so far. *)


(*$i initialize *)
(*  procedure Initialize (var N : integer);                             *)
(*      Initialize initializes the current time period, N, to zero     *)
(*      and writes an opening message to the user.                     *)


(*$i readdata *)
(*  procedure ReadData (var OrigPop, Rate, Factor : real);              *)
(*      This procedure reads values for the initial population, the    *)
(*      growth rate per year, and the multiplication factor.           *)


(*$i reportpop *)
(*  procedure ReportPop (PopSoFar : real; Time : integer);              *)
(*      ReportPop reports the number of rabbits at the end of time     *)
(*      period Time and is used in the procedure NumberOfPeriods.      *)


(*$i ztopower *)
(*$i numberofperiods *)


(*$i report *)
(*  procedure Report (OrigPop, Rate, Factor : real; N : integer);       *)
(*      Procedure Report prints the final report for program           *)
(*      Rabbits.  The original data is echo-printed and the            *)
(*      computed number of time periods is reported.                   *)


(*---------------------------------------------------------------------*)
begin (* main program *)
  Initialize (TimePeriods);

  ReadData (OriginalPop, GrowthRate, GrowthFactor);
```

```
        NumberOfPeriods (OriginalPop, GrowthRate, GrowthFactor, TimePeriods);

        Report (OriginalPop, GrowthRate, GrowthFactor, TimePeriods)
end. (* program Rabbits *)


=>
This program computes the number of years that a colony of
rabbits takes to multiply in size by a given factor.

You will be asked to supply real numbers for the size of
the original population, the growth rate per year, and the
factor by which the population is to increase.

What is the initial population?
□50.0

Enter the growth rate per year
□1.5

By what factor do you want the population to grow?
□5.0

After   1 time periods, there are         125 rabbits
After   2 time periods, there are         313 rabbits

It takes 2 time periods for a colony of   50 rabbits growing at a
rate of   150.00 percent per time period to grow by a factor of    5.0
```

The procedure for computing the number of time periods can be made more efficient in several ways. One way to make it more efficient involves changing the way $(1 + k)^n$ is computed. Using the **ZToPower** function saves programming time because **ZToPower** has already been written and is known to work. If you look at **ZToPower,** you can see that every time it is invoked, it starts at 1 and multiples z times itself n times. But if you have already computed z^{n-1}, then z^n is just $z * z^{n-1}$. In the procedure **NumberofPeriods, RatePlus1** is first raised to the first power, then to the second power, and so on. The alternative is to incorporate the algorithm from **ZToPower** into the loop in **NumberOfPeriods** to eliminate the duplication. Another way is to store **Factor * OrigPop** in a variable so that the multiplication does not have to be repeated each time the loop is executed. The procedure could then be written more efficiently as follows:

```
procedure NumberOfPeriods (OrigPop, Rate, Factor : real; var N : integer);
(*      The final version of NumberOfPeriods computes the number of       *)
(*      time periods the population takes to multiply by a factor of      *)
(*      Factor.  The final version is more efficient because it           *)
(*      incorporates the algorithm from the function ZToPower within      *)
(*      its repeat-until loop.  It also saves computation by per-         *)
(*      forming other arithmetic operations outside the loop.             *)

var   RatePlus1 : real;         (* growth rate plus 1. *)
      RateToN : real;           (* intermediate variable -- running product *)
                                (* for RatePlus1 ** N. *)
      NewPop : real;            (* population at the current time period. *)
      TargetPop: real;          (* final population target. *)
begin
  TargetPop := Factor * OrigPop;

  RatePlus1 := 1 + Rate;
  RateToN := 1.0;                       (* initialize the running product. *)

  repeat
    N := N + 1;                         (* increment the time counter. *)

    RateToN := RateToN * RatePlus1;     (* x**n = x*(x**n-1), so this *)
                                        (* period's RateToN equals last *)
                                        (* period's RateToN multiplied *)
                                        (* by RatePlus1. *)
    NewPop := OrigPop * RateToN;        (* use Equation 8.8 to compute *)
                                        (* the new population. *)
    ReportPop (NewPop, N)
  until NewPop >= TargetPop
end; (* procedure NumberOfPeriods *)
```

APPLICATION
The Exponential Function

This application assumes that the application from Chapter 7 on sequences and series has been covered.

CASE STUDY 12
Exponential growth

Many bacteria show resistance to the antibiotics commonly used to treat the diseases that the bacteria cause. Suppose that you have developed a new antibiotic drug that you think can be used against bacteria for which an antibiotic already exists. You must test the drug to determine whether it is effective in stopping the growth of bacteria. If it is effective, you want to compare its effectiveness to that of the standard antibiotic. To make the comparison you need to determine the growth rate of the bacteria and the change in the growth rate when the standard antibiotic or your new antibiotic is used to stop the growth of the bacteria. For the experiment, three colonies of bacteria are studied: one untreated, one treated with the standard antibiotic, and one treated with the new antibiotic. The time each bacteria colony takes to reach either half or double its original size is determined. Write a program that will evaluate the three growth rates and determine the lowest growth rate.

Algorithm development

To solve the problem presented in Case Study 12, you must determine the growth rate of the bacteria when they are not being killed by an antibiotic. You must then determine the growth rate of the bacteria in the presence of each antibiotic. An actual comparison of the drugs would require that the growth rates be determined for many experimental trials and a statistical analysis made of the experimental results. Statistics will be covered in the application section of Chapter 10.

ALGORITHM 8.3   Initial algorithm for Case Study 12

begin algorithm
1. Initialize the program
2. Read the doubling time or halving time for each colony. Include with the input a flag that indicates whether the input is a doubling or halving time
3. Determine the growth rate for each colony
4. Determine the smallest growth rate
5. Report the results

end algorithm

It is easiest to begin study of the exponential function by studying the growth that occurs at discrete time increments. We will look first at two situations in which growth occurs during discrete time periods at a constant rate. Then we will look at what happens as the time period becomes shorter and shorter so that time can be represented as a continuous (real) variable rather than a discrete (integer) variable.

Discrete time growth

Population growth in discrete time

In the case study on the population growth of rabbits, we assumed that the rabbits reproduced once per time period at the constant rate of 3/2 per rabbit each time period. The statement that the population grows at a constant rate can be written mathematically as

$$\frac{\Delta P}{\Delta t} = kP \qquad (8.7)$$

where k = the constant denoting the growth rate
P = the number of members in the population
ΔP = the change in the number of members in the population
Δt = the time increment

Equation 8.7 is called a finite-difference equation; it states that the change in the population, ΔP, for a finite time increment, Δt, is equal to a constant, k

Application: The Exponential Function

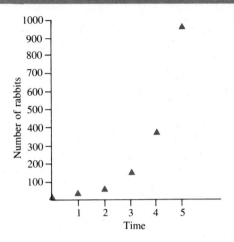

FIGURE 8.1
Growth of rabbits.

per rabbit. Thus, the change in the population is proportional to the time increment, and this ratio is equal to a constant, k.

From Equation 8.6 at the beginning of the chapter, you also know that the total number of rabbits at the end of n time periods is given by

$$P_n = p(1 + k)^n \tag{8.8}$$

This equation gives the total population at the end of a time period, and Equation 8.7 gives the change in the total population in a small time period. Either equation can be used to create the curve in Figure 8.1, which plots the number of rabbits versus time. You should notice that even though the growth rate is constant, the total population grows rapidly.

Compound interest in discrete time

A money-market fund allows investors either to withdraw the interest earned on their principal or to use their earned interest to buy more shares in the fund. If an investor chooses to reinvest the interest, the interest is added to the principal and begins to earn its own interest. Because each dollar earns interest at the same rate during each time period, the equation expressing the increase in the investment per unit time period is

$$\frac{\Delta M}{\Delta t} = rM \tag{8.9}$$

where r = the interest rate
M = the total amount of money invested
ΔM = the change in the total investment
Δt = the time increment

If m dollars of principal is invested at the beginning of the first time period, by the end of the first time period the m dollars of principal has grown to

$$M_1 = m + mr \text{ dollars}$$

or (8.10)

$$M_1 = m(1 + r) \text{ dollars}$$

If the new total, $m(1 + r)$ dollars, is left in the fund for a second time period, the total at the end of the second time period is the new total plus the interest this amount has earned:

$$M_2 = m(1 + r) + rm(1 + r)$$

or (8.11)

$$M_2 = m(1 + r)^2$$

And at the end of a third time period, the total is

$$M_3 = m(1 + r)^2 + rm(1 + r)^2$$

or (8.12)

$$M_3 = m(1 + r)^3$$

You can conclude that if m dollars is invested in an interest rate of r per time period, the total investment after n time periods, M_n, is

$$M_n = m(1 + r)^n \text{ dollars} \qquad (8.13)$$

Again, Equations 8.9 and 8.13 describe the same phenomenon, but Equation 8.9 describes the change in the total investment in a short time period, and Equation 8.13 describes the total investment at the end of a time period.

Continuous time and the exponential function

In the examples above, the time increment is assumed to be discrete. That is, rabbits are born only once a time period, and interest is added only once every time period. For rabbits, this assumption might be reasonable; however, for other populations, such as the bacteria colonies, this assumption is not reasonable, because bacteria reproduce on very short time scales. For bacteria, the time increment is *continuous*. The transition from discrete time to continuous time is illustrated below, as the time increment for compound interest gets smaller and smaller, becoming continuous in the limit.

Compound interest in continuous time

As a general rule, the more often the interest on money that you have invested is compounded, the better off you are. But there is a limit to how well you can do. For example, assume that you are investing $1.00 at an interest rate of 100% ($r = 1.00$). From Equation 8.13, the total investment at the end of n years is

$$M_n = 2^n \tag{8.14}$$

Now assume that, instead of being compounded annually, the interest is posted to your account j times per year. For a subperiod that is $1/j$th of a year, the interest rate is r/j. When $r = 1.00$, the interest rate in each of the j subperiods is $(1/j)$, and the number of subperiods in one year is j. From Equation 8.13, the total investment at the end of the jth subperiod is

$$M_j = \left(1 + \frac{1}{j}\right)^j \tag{8.15}$$

M_j is the amount to which $1.00 will grow in one year if 100% interest is compounded j times during the year. As shown in Table 8.1, once j is greater than 730, increasing j does not significantly increase the amount of interest that is earned in one year. The investment is worth $2.72 at the end of the year no matter how big j is.

The calculations in Table 8.1 suggest that as j approaches infinity, the value of $(1 + 1/j)^j$ approaches a specific real number: 2.7182. . . . The real number 2.7182 . . . is denoted by e. The limit of the sequence can be written mathematically as

$$\lim_{j \to \infty} \left(1 + \frac{1}{j}\right)^j = e \tag{8.16}$$

Equation 8.16 can be written in a more general form for e^x:

$$\lim_{j \to \infty} \left(1 + \frac{x}{j}\right)^j = e^x \tag{8.17}$$

Equations 8.17 and 8.13 can be combined to express the total investment. M_1 represents the total investment after 1 time period if an initial investment of m

TABLE 8.1
Interest Compounded at Decreasing Time Intervals

Interest Is Compounded	j	$\left(1 + \frac{1}{j}\right)^j$
Annually	1	$\left(1 + \frac{1}{1}\right)^1 = 2.00$
Semiannually	2	$\left(1 + \frac{1}{2}\right)^2 = 2.25$
Quarterly	4	$\left(1 + \frac{1}{4}\right)^4 = 2.4414$
Monthly	12	$\left(1 + \frac{1}{12}\right)^{12} = 2.6130$
Daily	365	$\left(1 + \frac{1}{365}\right)^{365} = 2.71457$
Twice Daily	730	$\left(1 + \frac{1}{730}\right)^{730} = 2.71642$
Hourly	8760	$\left(1 + \frac{1}{8760}\right)^{8760} = 2.718127$
Every second	31,536,000	$\left(1 + \frac{1}{31536000}\right)^{31536000} = 2.7182828$

dollars grows at an interest rate r which is compounded continuously

$$M_1 = me^r \qquad (8.18)$$

As an alternative to Equation 8.17, the value of e^x can also be expressed as an infinite sum

$$\sum_{i=0}^{\infty} \frac{x^i}{i!} = e^x \qquad (8.19)$$

where $i!$ is i factorial as defined by Equation 7.12. Although we will not use the equation in this chapter, it is a useful formula to have in your back pocket.

Application: The Exponential Function

The exponential function and the natural log are inverse functions; that is,

$$e^x = y \quad \text{if and only if} \quad x = \ln(y) \qquad (8.20)$$

where $\ln(y)$ is the natural log, or the logarithm base e, of y. Using Equation 8.20 and the fact that $e^1 = e$, it follows that

$$\begin{aligned} \ln(e) &= 1 \\ e^{\ln(1)} &= 1 \\ \ln(e^x) &= x \\ e^{\ln(x)} &= x \end{aligned} \qquad (8.21)$$

The exponential function is frequently written as exp:

$$e^x = \exp(x)$$

exp is the exponential function in Pascal. Both **exp** and **ln** are standard functions. Given the fact, from Equation 8.20, that exp and ln are inverse functions, the following Pascal statements leave the value of **X** unchanged:

```
X := 5.2;
Y := ln (exp(X));
writeln (' The value of ln (exp(', X:6:2, ') is', Y:6:2);

X := 8.3;
Y := exp (ln(X));
writeln;
writeln (' The value of exp (ln(', X:6:2, ') is', Y:6:2)
```

=>
```
The value of ln (exp(  5.20) is   5.20

The value of exp (ln(  8.30) is   8.30
```

Program 8.3, **Exponent,** approximates the value of e^x using Equation 8.19. The program uses the function **ZToPower** to compute the value of x^i. A new function **Factorial** computes the values of $i!$

A complete derivation and discussion of the exponential function is beyond the scope of this book. However, several common applications of the exponential function will be discussed.

PROGRAM 8.3 Exponent

```pascal
program Exponent (input, output);
(*      This program compares the value of the transcendental number e    *)
(*      raised to the Xth power computed using the series approximation   *)
(*      to the value computed using the Pascal function exp(X).           *)

var   X : real;            (* value for power input by the user. *)
      Num : integer;       (* number of terms in the series -- input by the *)
                           (* user. *)
      ApproxE : real;      (* computed approximate value of e using series *)
                           (* approximation. *)
      TrueE : real;        (* value of e using standard Pascal function exp. *)

(*$i ztopower *)

  function Factorial (I : integer) : real;
  (*      This function computes and returns the value of I!              *)

  var   Count : integer;      (* for loop control variable. *)
        NFac : real;          (* intermediate variable to store partially *)
                              (* computed I!. *)
  begin
    NFac := 1.0;              (* Initialize the running product. *)
    for Count := 1 to I do
      NFac := NFac * Count;   (* I! = I * (I - 1) * (I - 2) * ...* 2 * 1. *)
    Factorial := NFac         (* assign the value of I! to function *)
                              (* identifier. *)
  end; (* function Factorial *)


  function ApproximateE (Y : real; N : integer) : real;
  (*      function ApproximateE returns the series approximation of the   *)
  (*      value of e raised to a power using the first N terms in the     *)
  (*      series.                                                         *)

  var   SeriesE : real;    (* intermediate variable to store the sum of *)
                           (* the terms in the series. *)
        K : integer;       (* for statement control variable. *)

  begin
    SeriesE := Y;            (* Y/0! = Y. *)
    for K := 1 to N do
      SeriesE:= SeriesE + ZToPower (Y, K) / Factorial (K);
    ApproximateE := SeriesE
  end; (* function ApproximateE *)
```

Application: The Exponential Function

```pascal
procedure ShowDifference (Y : real; N : integer; FuncE, SeriesE : real );
(*      ShowDifference compares the functional value of e raised to a    *)
(*      power with the series approximation of the value.                *)

   var   Diff : real;       (* difference between the true value of e raised *)
                            (* to a power and the computed approximation. *)

   begin
     writeln;
     writeln (' The series approximation for exp(', Y:10:2, ')',
              ' is', SeriesE:15:12);
     writeln (' using', N:3, ' terms in the series');
     writeln;
     writeln (' The function value of exp(', Y:10:2, ') is', FuncE:15:12);

     Diff:= abs (FuncE - SeriesE);

     writeln;
     writeln (' The difference between them is', Diff)
   end; (* procedure ShowDifference *)


(*---------------------------------------------------------------------*)



begin (* main program *)
  writeln (' This program computes the value of e to the x using a series',
           ' approximation.');
  writeln;
  writeln (' What value of x do you want to use?');
  readln (X);

  writeln (' How many terms do you want in the series?');
  readln (Num);

  TrueE   := exp(X);
  ApproxE := ApproximateE (X, Num);

  ShowDifference (X, Num, TrueE, ApproxE)
end. (* program Exponent *)
```

```
=>
This program computes the value of e to the x using a series approximation.

What value of x do you want to use?
□1.0
How many terms do you want in the series?
□5

The series approximation for exp(     1.00) is 2.716666666667
using  5 terms in the series

The function value of exp(     1.00) is 2.718281828459

The difference between them is  1.6151617923787497e-03


=>
This program computes the value of e to the x using a series approximation.

What value of x do you want to use?
□1.0
How many terms do you want in the series?
□10

The series approximation for exp(     1.00) is 2.718281801146
using 10 terms in the series

The function value of exp(     1.00) is 2.718281828459

The difference between them is  2.73126614658281O8e-08
```

Exponential growth

Examples of exponential growth can be observed in many fields ranging from biology to business. Population growth, whether human or bacterial, can be described by an exponential function. The accumulation of money in a continuously compounded account also can be described by the exponential function.

For the earlier compound interest example, if the interest is compounded continuously, Equation 8.18 can be used to compute the amount of money in the account at the end of one time period. If $m = \$1.00$, $t = 1$ year, and $r = 100\%$ or 1.0, then

$$M_1 = \$1.00e^{1.0} = \$2.72 \tag{8.22}$$

If the \$2.72 is left in the account to accrue interest, the amount of money at the end of the second year is

$$M_2 = \$2.72e^{1.0}$$

or, substituting Equation 8.22 for \$2.72,

$$M_2 = (\$1.00e^{1.0})e^{1.0} = \$1.00e^{2.0} \tag{8.23}$$

In general, for exponential growth the equation for the population at time t is given by

$$M_t = me^{kt} \tag{8.24}$$

By convention, the variable t is used to represent continuous time and the variable n is used to indicate discrete time.

Equation 8.24 can be derived mathematically by making the assumption that the instantaneous growth rate of a population is proportional to the number of members of the population. This assumption is similar to the earlier assumption that the growth rate of rabbits during a time period is proportional to the total number of rabbits in the colony, except that in Equation 8.24 the time period over which the growth occurs is infinitesimally small. The instantaneous growth rate can be described in terms of a derivative, as follows:

$$\frac{dP}{dt} = kP \tag{8.25}$$

where k = the constant denoting the growth rate
t = the unit of time
P = the number of members in the population

The derivative in Equation 8.25 states that the change in the population, dP, over an infinitesimally small time increment, dt, is a constant equal to k. Solving this differential equation results in the equation

$$P_t = pe^{kt} \tag{8.26}$$

where p is the initial size of the population. Equation 8.26 is the equation for *exponential growth*. When the instantaneous growth rate is constant, exponential growth is said to occur.

Equation 8.25 describes the increase in population over an infinitesimal time period. Equation 8.25 is the continous time equivalent of Equation 8.7, which describes the population increase in a discrete time period. Similarly,

Equation 8.26, the solution to Equation 8.25, is the continuous time equivalent of Equation 8.8, the solution to Equation 8.7.

Population growth

Going back to the original problem of population growth, suppose that the colonies of bacteria are growing continuously. By analogy with compound interest, the original population is the principal and the growth rate is the interest rate. So

$$P_t = pe^{kt} \tag{8.27}$$

where P_t = the size of the population at time t
k = the constant growth rate
p = the original population

Again, t is used to indicate that the time variable is continous rather than discrete.

One interesting statistic is the *doubling time* of a population. The doubling time is the time that a population takes to reach double its original size, or

$$pe^{kd} = 2p \tag{8.28}$$

where p is the original size and d is the doubling time. You can see that the p's cancel and you are left with

$$e^{kd} = 2 \tag{8.29}$$

The doubling time does not depend on the original population, only on the growth rate, k. The easiest way to solve for d, the doubling time, is to take the natural log of both sides of the equation and to use the properties given in Equation 8.21:

$$\ln(e^{kd}) = \ln(2)$$

or (8.30)

$$d = \frac{\ln(2)}{k}$$

From Equation 8.30 you can see that the doubling time depends only on the growth rate, since $\ln(2)$ is a constant. This equation says that the larger the growth rate, k, the shorter the doubling time, d.

For the case study, the doubling time, d, is given and the problem is to find the growth rate, k. Equation 8.30 can be rearranged to give the growth rate as a function of the doubling time:

$$k = \frac{\ln(2)}{d} \tag{8.31}$$

Exponential decay

As with exponential growth, when exponential decay occurs, the change in a population is proportional to the change in time and equal to a constant. However, in exponential decay, the population is decreasing rather than increasing.

Population decline

Rabbits were used in the example of population growth because they are known to be prolific breeders. In the case of decay, think instead of an animal on the endangered species list, such as the California condor. If each pair of condors produced only one offspring before they died, the condor population would die out, because every two condors would be replaced by one condor until none were left. (In the case of declining animal populations, the reason for the decline is often predation, disease, or some other factor other than lack of reproduction; however, here we will take into account only the lack of reproduction.) The formula for population decline is the same as that for population growth, except that, in case of decline, k is less than zero and thus the population becomes smaller with successive time periods rather than larger. For decline in the condor population, the time interval would be discrete, so Equation 8.6 would be used.

The growth of a bacteria colony whose growth is inhibited by an effective antibiotic would be more likely to exhibit exponential decay. The mathematical formula for exponential decay is the same as the formula for exponential growth, except that the growth rate is less than zero:

$$P_t = pe^{-kt} \tag{8.32}$$

To make clearer that a population exhibits exponential decay, we will include the minus sign in the exponent. Using this convention means that k, the growth rate, will always be a positive number.

For populations in decline, the *half-life* gives the time required for a population to reach half its original size. The equation for the time that has elapsed when half the original population is left is

$$\frac{p}{2} = pe^{-kh} \tag{8.33}$$

where h is the half-life. Canceling and taking natural logs,

$$-kh = \ln(1/2)$$

or (8.34)

$$h = \frac{\ln(2)}{k}$$

In Case Study 12, you are given the time that has elapsed when half the original bacteria population is left, h, so the growth rate k can be determined.

Radioactive decay

Another example of exponential decay is radioactive decay. Radioactive substances have the property that the rate at which radioactivity decreases is proportional to the amount of radioactivity. That is, they exhibit exponential decay. For radioactive decay, the formula is

$$M_t = me^{-kt} \tag{8.35}$$

where M = the mass of time t
m = the original mass
k = the decay rate

The decay rate of radioactive substances is usually stated in terms of the half-life. The half-life of a radioactive element is the time a given quantity takes to decay to one-half of its original mass:

$$me^{-kh} = \frac{m}{2} \tag{8.36}$$

Using Equation 8.34 to eliminate k from the general equation yields

$$M_t = me^{-(\ln(2)/h)t} = m(e^{-\ln(2)})^{t/h}$$

or (8.37)

$$M_t = m\left(\frac{1}{2}\right)^{t/h}$$

where h is the half-life and m is the original mass of the element.

Voltage decay in an electrical circuit

In electrical circuits with resistors and capacitors, called RC circuits, the equation describing the decrease of the charge or voltage is an exponential decay equation. In an RC circuit, a resistor and a capacitor are connected in series, as illustrated in Figure 8.2.

A capacitor consists of two separated charged plates, one of which is positive and the other negative, so that a potential difference or voltage exists between them. When the switch is open, the current cannot flow. As soon as the switch is closed, electrons flow from the negatively charged plate to the positively charged plate through the resistor. A current, I, is created by the electrons. The current flows from the positively charged plate to the negatively charged plate until there is no charge left in the capacitor. (By convention, the current is positive in the direction opposite that of the electron flow.) In an RC circuit, the relationship among the voltage, the capacitance, and the resistance is given by

$$\frac{dV}{dt} = -\frac{1}{RC} V \qquad (8.38)$$

where V = the voltage in volts
R = the resistance in ohms (Ω)
C = the capacitance in farads (F)

This equation describes the *natural response* of the circuit as energy stored in the capacitor is dissipated through the resistance.

Equation 8.38 has the same form as Equation 8.25 that describes the increase in the rabbit population in an infinitesimally small time increment. Equation 8.38 can be solved in the same way that Equation 8.25 was solved.

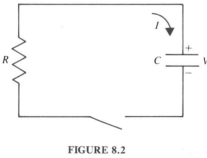

FIGURE 8.2
An RC circuit.

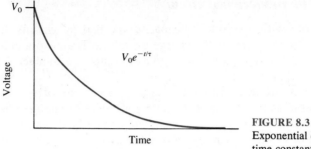

FIGURE 8.3
Exponential decay. Tau is called the time constant of the decay.

Equation 8.39, the solution to Equation 8.38, gives the voltage remaining at time t

$$V_t = V_0 e^{-t/RC} \qquad (8.39)$$

where V_0 is the initial voltage stored in the capacitor. Notice that Equation 8.39 is the same as Equation 8.26 except that the variable names are different.

One difference between Equations 8.25 and 8.38 is that in Equation 8.38 the constant $-1/RC$ is negative. In Equation 8.26 the constant k is positive. Thus, Equation 8.38 describes exponential decay in continuous time, whereas Equation 8.25 describes exponential growth in continuous time.

The quantity RC is called the *time constant* of the circuit and is referred to as tau (τ). Figure 8.3 shows the exponential function in standard form.

Exponentiation in Pascal

Pascal does not have a standard operator or function for taking exponents. The procedure **ZToPower** will not work if **Power** is not an integer. In developing an algorithm for raising a number to a real power, the following properties of y^x will be useful:

$$\begin{aligned} y^{a+b} &= y^a y^b \\ y^{a-b} &= \frac{y^a}{y^b} \\ y^{ab} &= (y^a)^b \end{aligned} \qquad (8.40)$$

To write a general algorithm to raise a number to a power, you can use the properties of the exponential function given in Equation 8.21 and the properties of y^x given in Equations 8.40 to derive the relationship

Application: The Exponential Function

$$x^y = e^{y \ln(x)} \qquad (8.41)$$

or

```
XToY := exp (Y * ln(X))
```

You can write a function with two parameters **X** and **Y** that takes the natural log **ln** of **X,** multiplies it by **Y,** and returns the result as the argument of the exponential function **exp.** X and Y can be either real or integer numbers, so the parameters of the function must be declared as **real** value parameters. Since integers are assignment compatible with real variables, the function will be able to accept reals or integers in the argument list.

Before completing the function to raise a number to a real power, you must be aware that x^y is *undefined* when x is negative. That is, $(-2)^{5.2}$ has no mathematical definition. If you try to compute its value using Equation 8.41, the argument of the function **ln** will be negative and a run-time error will result. Note, however, that negative numbers can be raised to integer powers because integer exponentiation can be performed through multiplication. For example, $(-4)^3$ is equal to $(-4)*(-4)*(-4)$, but $(-4)^{3.0}$ is undefined.

The function **XtoY** is a general function that raises a number to a real power.

```
function XToY (X, Y : real) : real;
(*      XToY returns the value of X raised to the power of Y.  If X is      *)
(*      less than zero an error message is written to the terminal, and     *)
(*      the function returns with a value of -1.0e30.                       *)
const Negative = -1.0e30;        (* value to which XToY is set if X < 0.0. *)
begin
  if X < 0.0 then
    begin
      writeln (' (-x)**y is undefined.  Value set to ', Negative);
      XToY := Negative
    end
  else
    XToY := exp(Y * ln(X))
end; (* function XToY *)
```

In writing the function **XToY,** it was necessary to make a decision on how errors would be handled. In this case, the decision was made to write an error message and to return a large negative number as the value of **XToY**. How else could the error have been handled?

Some compilers, as an extension to standard Pascal, include the exponentiation operator ******. **X**Y,** usually read as "x star star y," means x raised to the power y. Because ****** is not standard, the error that results from taking a negative number to a power depends on the particular implementation. The ****** notation is common and convenient, so it will be used to refer to exponentiation in algorithms within this text, but the programs will use **XToY**. If your compiler includes the ****** operator, you can use it instead of the function **XToY**. But you should be aware that if you use nonstandard Pascal, your programs may not be portable to other Pascal compilers.

Solution to Case Study 12

The solution to Case Study 12 involves solving a series of smaller problems. One problem arises in Step 2, where you must keep track of whether the input time is the doubling time or the half-life. The problem in Step 3 turns out to be simple to solve—Equation 8.31 can be used to compute the growth rate given the doubling time, and Equation 8.34 can be used to compute the growth rate given the half-life. The final problem occurs in Step 4, where the smallest of three growth rates must be found. In Chapter 6, a function was developed that returned the smaller of two numbers. That algorithm can be extended or modified to return the smallest of three values.

Several methods are available for entering the initial data. One method is to ask the user to enter the character 'h' if a half-life is to be entered next or a 'd' if a doubling time is to be entered next. Another method is to ask the user to enter a negative time if the time is a half-life and a positive time if it is a doubling time. Still another method is to ask the user if a doubling time is going to be entered and then to use the boolean function **YesEntered** from Chapter 6 to determine if the user entered yes.

Coupled with the problem of determining what type of number will be entered is the problem of deciding how the information will be stored. For example, if you decide to use the first technique for entering data, each colony needs to have a character variable associated with it. Using this technique, you can create the following data structure:

```
var     TimeType1 : char;       (* indicates whether the time entered for *)
                                (* culture 1 is a half-life or doubling time. *)
        TimeType2 : char;       (* indicates whether the time entered for *)
                                (* culture 2 is a half-life or doubling time. *)
        TimeType3 : char;       (* indicates whether the time entered for *)
                                (* culture 3 is a half-life or doubling time. *)

        Life1 : real;           (* doubling time or half-life for culture 1. *)
        Life2 : real;           (* doubling time or half-life for culture 2. *)
        Life3 : real;           (* doubling time or half-life for culture 3. *)
```

What data structures would you use for the other methods?

For Step 3, the same equation can be used for both the doubling time and the half-life, the difference being whether the growth rate k is reported as a positive or a negative growth rate:

```
Rate1 := ln(2) / Life1;
if TimeType1 = 'h' then
   Rate1 := - Rate1
```

Finally, you must determine which of the three rates is the smallest. Since the function **Minimum** has already been written, you can use it to find the minimum of three numbers by means of the following algorithm:

ALGORITHM 8.4 Minimum of three numbers

SmallestSoFar ← **Minimum (Num1, Num2)**
Smallest ← **Minimum (Num3, SmallestSoFar)**

This logic can easily be incorporated into the program. The completed program for the case study is given in Program 8.4, **Bacteria.**

In Program 8.4, two of the procedures are invoked three times. However, a **for** statement is not used because the growth rates have to be given distinct identifiers in order to be passed into the function **Smallest.** In the next chapter, the **array** data structure will be covered. Among other things, arrays allow similar variables to be given distinct names within **for** statements.

PROGRAM 8.4 Bacteria

```
program Bacteria (input, output);
(*      This program computes the growth rate for three different    *)
(*      bacterial cultures given either the doubling time or the half- *)
(*      life.  The program also reports which culture has the smallest *)
(*      growth rate.                                                  *)

var     TimeType1 : char;       (* indicates whether the time entered for *)
                                (* culture 1 is a half-life or doubling time. *)
        TimeType2 : char;       (* indicates whether the time entered for *)
                                (* culture 2 is a half-life or doubling time. *)
        TimeType3 : char;       (* indicates whether the time entered for *)
                                (* culture 3 is a half-life or doubling time. *)

        Life1 : real;           (* doubling time or half-life for culture 1. *)
        Life2 : real;           (* doubling time or half-life for culture 2. *)
        Life3 : real;           (* doubling time or half-life for culture 3. *)

        Rate1 : real;           (* rate of growth or decay for culture 1. *)
        Rate2 : real;           (* rate of growth or decay for culture 2. *)
        Rate3 : real;           (* rate of growth or decay for culture 3. *)

        Smallest : real;        (* smallest growth rate. *)


(*$i minimum *)


(*$i writeintro *)
(*   procedure WriteIntro;                                            *)
(*      This procedure writes the introductory message to the user.   *)


   procedure ReadLife (var TimeType : char; var Life : real);         *)
   (*      This procedure reads a character variable that indicates   *)
   (*      whether the doubling time or the half-life will be entered *)
   (*      entered and then reads the doubling time or half-life.     *)

   begin
     repeat
       writeln;
       writeln (' Will you enter the doubling time or the half-life?');
       writeln (' Enter d or h');
       readln (TimeType)
     until (TimeType = 'd') or (TimeType = 'h');
```

```
      if TimeType = 'd' then
        writeln (' Enter the doubling time for the colony')
      else
        writeln (' Enter the half life for the colony');
      readln (Life)
    end; (* procedure ReadLife *)


    function GrowthRate (TimeType : char; Life : real) : real;
    (*     GrowthRate returns the growth rate of the bacterial culture.    *)

    var   Rate : real;             (* intermediate variable for GrowthRate. *)

    begin
      Rate := ln(2) / Life;        (* compute the growth rate using Equation *)
                                   (* 8.31. *)
      if TimeType = 'h' then
        Rate := - Rate;            (* for exponential decay, rate is negative. *)
      GrowthRate := Rate
    end; (* function GrowthRate *)


    function SmallestOfThree (Num1, Num2, Num3 : real) : real;
    (*     SmallestOfThree finds the smallest of three numbers using the   *)
    (*     function Minimum which finds the smaller of two numbers.        *)

    var   SmallestSoFar : real;    (* intermediate variable for finding *)
                                   (* SmallestOfThree. *)
    begin
      SmallestSoFar := Minimum (Num1, Num2);
      SmallestOfThree := Minimum (Num3, SmallestSoFar)
    end; (* function SmallestOfThree *)


(*$i reportrates *)
(*  procedure ReportRates (Growth1, Growth2, Growth3 : real);              *)
(*      This procedure reports the growth rates for each of the three      *)
(*      bacterial cultures.                                                *)



(*--------------------------------------------------------------------------*)

begin (* main program *)
  WriteIntro;
```

```
    ReadLife (TimeType1, Life1);
    ReadLife (TimeType2, Life2);
    ReadLife (TimeType3, Life3);

    Rate1 := GrowthRate (TimeType1, Life1);
    Rate2 := GrowthRate (TimeType2, Life2);
    Rate3 := GrowthRate (TimeType3, Life3);

    ReportRates (Rate1, Rate2, Rate3);

    Smallest := SmallestOfThree (Rate1, Rate2, Rate3);

    writeln (' The smallest growth rate is', Smallest * 100:10:1, ' percent')
end. (* program Bacteria *)
```

```
=>
  This program reads the halving or doubling times for
  three different bacterial cultures.

  It computes the growth rate for each of the cultures and
  reports which culture has the smallest growth rate.

  Will you enter the doubling time or the half-life?
  Enter d or h
□d
  Enter the doubling time for the colony
□3.0

  Will you enter the doubling time or the half-life?
  Enter d or h
□d
  Enter the doubling time for the colony
□6.0

  Will you enter the doubling time or the half-life?
  Enter d or h
□h
  Enter the half life for the colony
□3.0

  The growth rate of the first colony is      23.1 percent per time period
  The growth rate of the second colony is     11.6 percent per time period
  The growth rate of the third colony is     -23.1 percent per time period

  The smallest growth rate is      -23.1 percent
```

Summary

The **while-do** and **repeat-until** statements allow the repetition of action to be initiated and terminated depending on the value of a **boolean** expression. A **while-do** statement is repeated *while* a **boolean** expression is true. A **repeat-until** loop is repeated *until* a **boolean** expression becomes true. In a **while-do** statement, the **boolean** expression is evaluated prior to the execution of the action. If the **boolean** expression is false, the action of the **while-do** statement is not executed. In a **repeat-until** statement, the **boolean** expression is evaluated after the execution of the action. Hence, a **repeat-until** always executes at least once. A **while-do** or a **repeat-until** statement is used when the conditions under which the loop is to terminate are known but the number of iterations of the loop is unknown.

One use of a **while-do** or **repeat-until** is to check data entered from the terminal. The user can be asked to enter data until a specified criterion is met. If digits are read into character variables, the ordinal value of the character can be used to determine whether the character represents a digit. If the character does not represent a digit, the user can be asked to continue entering data until a digit is entered.

Exercises

PASCAL

1. Correct the errors in each of the following fragments:

    (a)
    ```
    IntVar := 1
    while IntVar < 50 do;
      begin
        IntVar := IntVar + 2.5;
        if odd (IntVar) then
          writeln (' IntVar is odd ')
        else
          writeln (' IntVar is even ')
      end
    ```

    (b)
    ```
    Offset := ord('0');
    repeat
       writeln (' Enter a digit ');
       readln (Digit);
    until (chr (Digit) > Offset) or (chr (Digit) < Offset + 9)
    ```

(c) ```
RealNum := 5.0;
while RealNum < 10.0 do
  begin
    RealNum := 1/RealNum;
    writeln(' RealNum = ', RealNum:10:2)
  end
```

(d) ```
C := true;
repeat;
  A := not C
until A and C
```

2. What is the output from each fragment below?

(a) ```
RealVar := 1.4;
while RealVar < 5.5 do
  begin
    writeln (' RealVar = ', RealVar:5:1);
    RealVar := RealVar + 1.2
  end
```

(b) ```
Number := 1.4;
repeat
  writeln (' Number = ', Number:5:1);
  Number := Number + 1.2
until Number >= 5.5
```

(c) ```
Num1 := 100.0;
while Num1 > 1.0 do
  begin
    Num1 := ln (Num1);
    writeln (' Num1 = ', Num1:5:1)
  end
```

(d) ```
Num2 := 100.0;
repeat
  Num2 := ln (Num2);
  writeln (' Num2 = ', Num2:5:1)
until Num2 <= 1.0
```

3. Rewrite the following program using a **while-do** statement. Your program should produce the same output as does program **DuplicateOutput**.

```
program DuplicateOutput (input, output);
   (*    DuplicateOutput works in a strange way.  Determine how it works,   *)
   (*    and create another program that uses a while-do statement instead  *)
   (*    of a repeat-until statement and prints the same output.  Your      *)
   (*    program should work for any value of test.                         *)
```

```
const Test = 3;

var   Number : integer;
      A : integer;

begin
  Number := 3;
  A := 3;
  repeat
    Number := (A * Number) mod 5;
    if (Number = 4) or (Number = 2) then
      writeln (' Number is even.')
    else
      if Number <> 3 then
        writeln (' Number =', Number:2)
      else
        writeln (Number:2, ' again!')
  until Number = Test
end. (* program DuplicateOutput *)
```

4. Write a program that reads an arbitrary number of integers and counts the number of positive values and the number of negative values. Use the number -9999 as a flag to signal the end of the data. At the end of the program, print the number of positive and the number of negative values read. Do not count the -9999 in with the negative numbers. Why can't you use a **for** statement?

   (a) Write the program using a **while-do** loop.

   (b) Write the program using a **repeat-until** loop.

5. Write a program that reads real numbers one at a time. The program must stop when a number less than -10.0 is read. Be sure to include a message telling the user how to stop the program.

   (a) Write the program using a **while-do** loop.

   (b) Write the program using a **repeat-until** loop.

6. Write a program that reads real numbers until the number entered is between 1.0 and 2.0. Be sure to write a message to the user stating the stopping criteria.

7. Write an interactive program that reads characters one at a time from the terminal and echoes the input value. Stop the program when the character 's' and 'S' is entered.

   APPLICATION

8. Use a calculator to evaluate the following expressions:

   (a) $\sum_{i=1}^{4} \frac{3i}{i!}$ (b) $\exp(3)$ (c) $\ln(\exp(3.2) + \exp(-2))$
   (d) $\sin(\ln(\pi))$ (e) $\exp(\pi)$ (f) $\exp(5^2)$

9. Simplify the following expressions:
   (a) $\ln(e^y)$
   (b) $\ln(ye^5)$
   (c) $e^{[7 + \ln(z)]}$
   (d) $\ln\left(\dfrac{a}{b}\right)$
   (e) $\dfrac{(e^{3x})^2}{e^x}$
   (f) $e^x(e^{-y} + e^{-x})$

10. Show how Equations 8.21 and 8.40 can be used to derive Equation 8.41.

Problems

PASCAL

1. Write a program that reads integer numbers from the terminal until the user guesses the secret number that stops the program. Each time a number is entered, tell the user whether it is larger or smaller than the secret number. Have someone else try out your program. Hint: Use an **if-then-else** inside a **repeat-until**.

2. Write a program that reads integers one at a time and echoes the input value. When either a negative number is entered or the thirtieth integer is entered, the program stops.

3. (a) Write a program that, given a specific interest rate, finds the year in which the total invested becomes more than twice the initial investment. Write the program using a **repeat-until** statement and the procedures from Program 8.2, **Rabbits.**
   (b) Write the program using a **while-do** loop.

4. Write a program that reads four-letter words from the terminal until the user enters the word stop.

5. Using Problem 9 from Chapter 7, write a program for a robot that reads the diameters of the widgets in a lot. If the diameter of a widget is less than 3.25 or greater than 3.75, the widget is a bad widget. The robot (and the program) stops and rejects the lot when the second bad widget is encountered. Report the position and diameter of the bad widgets that are encountered. For example: Bad widget #1 is widget #1 and has a diameter of 3.24. Be sure your program terminates correctly if there are fewer than two bad widgets in the lot.

6. In Problem 10 of Chapter 7 you were asked to write a program to find the first hour in which the temperature was lower than the temperature in the previous hour. Rewrite the program using a conditional loop. Be sure that your program will terminate correctly if the temperature does not drop during the day.

7. On page 277, a **repeat-until** loop was used to compute the number of times that e can be raised to the power e before the value exceeds 100.0. The following algorithm is a more general version of this problem:

begin algorithm
   1. Initialize the program
      Read **X, Max**
      NumberofPowers $\leftarrow 0$
   2. repeat
      $Y \leftarrow e^X$
      $X \leftarrow Y$
      NumberofPowers $\leftarrow$ **NumberofPowers** $+ 1$
     until **Y** $>=$ **Max**
   3. **Report NumberofPowers** and **Y**
end algorithm

Write a program to implement the algorithm. Are there initial values of **X** for which this loop will always be infinite? Are there initial values of **Max** for which this loop will always be infinite? Are there particular combinations of **X** and **Max** for which this loop will be infinite? Include checks in your program to prevent infinite loops. Would the checks be different if the algorithm were implemented using a **while-do** loop? You may want to put the reporting step inside the loop so that you can tell if your program is executing correctly.

APPLICATION _____

8. Rewrite the bouncing ball program, Program 7.5, so that the program stops when the new height of the ball is less than 0.1 m.

9. Write a program that computes the time it takes a baseball, dropped from a specified height, to hit the ground. Compute the speed and distance every second (that is, for $t = 1, t = 2, \ldots$) until the distance from the ground is less than 0.0 m. Use a **while-do** or a **repeat-until** statement.

10. In the application section of Chapter 5 on computational error, it was stated that the order in which numbers were added could affect the answer. Write a program that approximates the value of e^x first by adding the terms in the series in ascending order, then by adding them in descending order. Which is more accurate and why? Does the magnitude of x make a difference?

11. Write a Pascal program that computes the remaining radioactive mass of carbon -14 at the end of every 500 years over a time period of 10^4 years. The formula for the radioactive mass remaining at time t is

$$M_t = m\left(\tfrac{1}{2}\right)^{t/h}$$

where $t =$ the time in years
       $M_t =$ the mass remaining at time t
       $m =$ the original mass
       $h =$ the half-life in years

For carbon -14, the half-life, h, is 5700 years. If the original radioactive mass is 300 grams, print out a (good-looking) table for the mass at $t = 500, t = 1000, t = 1500$, etc., up to $t = 10,000$.

12. A rope is wrapped around a cleat to make a sail tight. The tope is held by the friction of the rope against the cleat. The formula for the force (tension) in a rope that is just about to slip is given by

$$T_2/T_1 = e^{f\theta}$$

where T_1 = the force exerted on the rope in newtons
T_2 = the tension on the rope from the sail in newtons
f = the coefficient of friction (dimensionless)
θ = the angle in radians

The rope is wrapped twice around the cleat, making an angle of 4π. If the coefficient of friction between the rope and the cleat is 0.29, and if you can exert a force of 150 N, what is the maximum force on the sail that you can resist before the rope starts to slip?

13. Write a program that uses Equation 8.17 to approximate the value of e^x for three different values of j for a given value of x. The values of x and j are entered interactively. Your program should prevent the user from entering negative values for j. Compare the computed values with the value of **exp(x)**. How does the accuracy of Equation 8.17 compare to the accuracy of the series expression, Equation 8.19, for different values of i?

14. The Taylor series expansion for the cosine function is given by

$$\cos(x) = 1 - \frac{x^2}{2!} + \frac{x^4}{4!} - \frac{x^6}{6!} + \cdots$$

where the angle x is in radians. Write a program that evaluates the first 10 terms of the Taylor series for $\cos(x)$.

15. Rewrite the sine program from Problem 21 in Chapter 7 or the cosine program from Problem 14 above, so that the number of terms in the series is determined by the size of the last term. Stop the series when the absolute value of a term is less than 0.0005. What is the error in the calculation?

16. In an RC circuit, the voltage across a capacitor at time t is given by

$$V = V_0 e^{-t/RC}$$

where V = the voltage across the capacitor in volts
V_0 = the initial voltage across the capacitor at time $t = 0$
R = the resistance in ohms
C = the capacitance in farads

Write a program that computes the voltage across a 0.1-microfarad capacitor with a resistance of 5.0 kilo-ohms and an initial voltage of 5.0 V. Pay attention to units in this problem. Compute the voltage across the capacitor for 10 time increments that are multiples of the time constant RC. That is, compute the voltage when $t = RC$, when $t = 2RC$, $t = 3RC$, etc., up to $t = 10RC$.

CHAPTER 9

Arrays

In this chapter, you will learn to use an **array,** which is a structured data type. **Array** variables allow you to handle large amounts of data easily.

CASE STUDY 13
Sample averages of experimental data

A professor has set up an experiment to study how students do on different types of tests. The test types are multiple choice, short answer, and program writing. There are 6 tests during the semester, so each student takes 2 tests of each type. There are 12 students in the class. The professor has arranged the test scores in the following table:

TABLE 9.1
Scores for Different Types of Tests, (MC = multiple choice, SA = short answer, PW = program writing)

Student Number	Test Number					
	1 MC	2 SA	3 PW	4 MC	5 SA	6 PW
1	99	93	88	95	78	95
2	75	80	83	70	77	80
3	92	95	85	88	92	89
4	50	62	50	65	70	50
5	76	70	60	78	77	55
6	81	85	78	87	82	70
7	72	74	90	73	81	92
8	78	70	50	67	65	50
9	85	87	80	89	85	78
10	67	80	65	70	75	67
11	77	78	70	72	78	68
12	80	85	82	78	87	83

Write a program that will find the average score for each test, the average score for each student, and the average score for each type of test. Assume that none of the students misses a test.

Algorithm development

The problem for the case study can be stated as follows: Read the data for all of the students for all 6 tests. Compute the average for each test, the average for each student, and the average for each type of test. Report the averages.

ALGORITHM 9.1 First iteration of the algorithm for Case Study 13

begin algorithm
  1. Read the 72 scores for all the students on all the tests
  2. Compute the average score on each test
     for each test
       begin
         Sum the test scores in each column of the table
         Divide the sum by the number of students
       end
  3. Compute the average score for each student
     for each student
       begin
         Sum the test scores in each row of the table
         Divide the sum by the number of tests
       end
  4. Compute the average score for each type of test
     for each test type
       begin
         Sum the test scores in the two columns for that test type
         Divide the sum by 2 * number of students
       end
  5. Report all the averages
end algorithm

The major problem to be solved is how to handle the data for all the students on all the tests so that 72 separate variable names do not have to be created and so that the function to compute the average can operate on any row or column of the data. In Chapter 7, the algorithm to compute an average was developed. That algorithm required that each number be added to the running total as the number was read. For the case study, each number must be added to several different running totals. A possible strategy would be to read a test score and then use **if** statements to decide in which test average and in which student average the score should be included. With the data structure of an **array**, however, this approach is unnecessary. With an **array** variable, all the test scores can be read at the same time. Once the test scores are stored in an **array** variable, any score can be accessed at any time.

9.1 Structured data types

A *structured data type* is a data type that is constructed by the user from the basic data types. As an analogy, consider a house made up of the basic building blocks of nails and boards. When you want to describe a house, you can describe how the nails and boards are put together, you can describe how the rooms are put together, or you can just use the word *house* to describe the entire construction. Similarly, a structured data type can refer to a structure made up of the basic data types. Structured data types are *user-defined data types* because you, the user, define what structure the data type is to have. Once the structure has been defined, any variable can be declared to be of the user-defined type. In this chapter, you will learn the syntax for the structured data type **array.** Other structured data types that will be covered later are **sets, records,** and **files.**

9.2 One-dimensional arrays

Arrays can be used to store pieces of data that are similar but that have distinct values. For example, the scores of each of the 12 students on the first exam could be stored in an array. Instead of having to name 12 different variables (one for the score for each student), if you use an array you can give all 12 pieces of data a single name such as **Scores.** The name refers to the entire array made up of 12 individual *elements,* in this case the students' scores. For you to be able to refer to any particular element, an *index* is required. The score of the first student is stored in the first element of the array, and the index is 1. The score of the last student is stored in the last element, and the index is 12.

A one-dimensional array can be thought of as a number of boxes in a column. Each box (element) in the array contains one piece of data. A one-dimensional array can be visualized as follows:

Student Index	Scores
1	99
2	75
.	.
.	.
.	.
12	80

In this array, $Scores_2 = 75$.
The *element* stored at *index* 2 of the array **Scores** is the integer number 75.

In mathematics, the elements of an array are referred to by subscripts. For example,

$$\text{Scores}_i, \quad i = 1, 2, \ldots, 12$$

refers to the *i*th score in the array. In the array above, Scores_2 is 75.

In Pascal, the syntax

$$\text{Scores[I]}$$

refers to the *i*th score in an **array** variable. **Scores[2]** is 75.

Once a Pascal **array** variable has been created, any element within the array can be referenced with the array name followed by the index number which is put between square brackets. In the following example, the variable **StudentNum** is used as the index for the **array** variable. Execution of the statements will cause the score of the second student to be written to the terminal:

```
StudentNum := 2;
writeln (' Student #', StudentNum:3, ' received a',
         Scores [StudentNum] :4)
```

```
=>
Student #  2 received a  75
```

In the statements above, the index is 2 and the element stored in the **array** variable **Scores** at index 2 is the integer number 75. In the following statements, the index is 1 and the element stored at index location 1 is the integer number 99:

```
StudentNum := 1;
writeln (' Student #', StudentNum:3, ' received a',
         Scores [StudentNum] :4)
```

```
=>
Student #  1 received a  99
```

Before going on, you should be sure that you understand the difference between the value of an element and the index that points to the element. Look at **Scores** and verify that **Scores[2]** is 75 and **Scores[1]** is 99.

Some terminals do not have the characters '[' and ']'. If you are working on such a terminal, you can use '(.' for '[', and ').' for ']'. In almost all implementations of Pascal, the following two statements are equivalent:

```
Scores [StudentNum] := 100;
```

```
Scores (.StudentNum.) := 100
```

9.2 One-dimensional arrays

All of the elements in an **array** data type must be of the same base data type. In the example above, all the elements are integers. You can also create arrays in which the elements are all of the data type **real**, all of the data type **char**, or even all of another **array** data type. However, you cannot make a single **array** data type that mixes data types. There is a data type called a **record** that will allow you to create a data structure with more than one base data type. Records will be covered in Chapter 13.

Declaring arrays

There are two methods for creating the data structure of an **array** in Pascal. The first method uses the new concept of a user-defined data type. The second method of declaring an **array** variable is shorter and simpler, but it limits the ways that the **array** variable can be used. Declaring an **array** with a user-defined data type will be covered first because this method must be used if the array appears in a parameter list, and therefore is the preferred method.

A simplified syntax diagram for creating an **array** data type with a *type definition* is

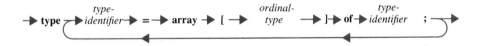

where an *ordinal-type* is defined by the simplified syntax diagram

The constants must be of an ordinal data type.

Interpreting this syntax diagram to create an **array** data type with the **type** statement gives a generic definition of an **array** data type:

```
type  NewType = array [LowerBound..UpperBound] of datatype;

var   ArrayName : NewType;
```

where

    array   is the reserved Pascal word that creates the **array** data structure.

    NewType   is the name you have given to your new data type. **NewType** can be used as a data type in variable declarations.

    LowerBound   is the index of the first element in the **array** data type.

UpperBound is the index of the last element in the **array** data type. **UpperBound** should be greater than or equal to **LowerBound.**

datatype is a valid data type—**real, integer, char, boolean,** or any previously defined data type.

ArrayName is the name you give your **array** variable.

With this syntax, you have defined a new data type. The data type, **NewType,** that you define can be used just like the standard data types **real, integer, char,** and **boolean.** When you use the **type** syntax, you must obey the following rules:

1. The **type** statement appears in the declarative part of a program or subprogram and precedes the **var** statement. You are defining your own data type, and it must be defined before you can use it.

2. Once you have defined the name **NewType,** it stands for a data type, *not* a variable. You *cannot* use **NewType** as a variable name in your program. It is treated just like the reserved words **real, integer, char,** and **boolean.**

3. **LowerBound** and **UpperBound** must be of an ordinal data type—for example, **integer** or **char.**

The syntax **LowerBound..UpperBound** creates a *subrange*. The index values that are used with the **array** variable must be within the subrange created by specifying the upper and lower bounds of the array; for example, for **Scores,** a reference to **Scores[55]** would be an error because 55 is not within the range 1..12. Subranges are covered in more detail in Chapter 11.

More specifically, you could create an **array** variable to store the students' scores using the following declaration:

```
type  ScoreArray = array[1..12] of integer;

var   Scores : ScoreArray;
```

The **type** definition creates a new data type, **ScoreArray.** The **var** declaration declares the variable **Scores** to be of the data type **ScoreArray.** A variable, such as **Scores,** that is declared to be of a structured data type is sometimes referred to as a *structured variable* in order to distinguish it from variables that are of a simple data type such as **real** or **integer.**

The shorter method for declaring an **array** allows you to implicitly declare a new type in the **var** statement. This method does not define a new data type. A syntax diagram for declaring an array by this method is

→ **var** → *identifier* → : → **array** → [→ *ordinal-type* →] — **of** → *type-identifier* → ; →

For example,

```
var   Scores : array [1..12] of integer;
```

This statement looks like the longer declaration with the right-hand side of the **type** statement placed on the right-hand side of the **var** statement. Although this statement is shorter and thus requires less typing, there are many constructions in which it cannot be used.

Using arrays

For statements are frequently used with arrays because they allow you to perform the same operation on every element in the **array** variable. For example, suppose you want to compute the average score for 12 students. If the scores are stored in an **array** variable, a loop can be used to read the 12 scores into the array, and another loop can be used to compute a running total of the array.

The **for** statement to read values into the array can be written as follows:

```
for StudentNum := 1 to NumStudents do
  begin
    writeln;
    writeln (' Enter the score for student number ', StudentNum:1);
    read (Scores [StudentNum])
  end
```

Each time the loop executes, one value is read into the array. After the **for** statement has completed execution, all 12 values will be stored in the array variable.

The operation of finding the average of the sequence of 12 numbers can be written in the same form as Equation 7.1, which was used to compute the average velocity. The average of the scores, $\overline{X}$, is given by

$$\overline{X} = \frac{x_1 + x_2 + \cdots + x_{12}}{12} \tag{9.1}$$

where x_1 is the first score, x_2 is the second score, and so on.

The following program fragment computes the average score using Equation 9.1. In this fragment, the average score of all the students is found by adding every element in **Scores** to the running total stored in **Sum**, and then dividing the result by the number of students.

```
const  NumStudents = 20;          (* total number of students. *)

type   ScoreArray = array [1..NumStudents] of integer;

var    Scores : ScoreArray;       (* array of test scores -- given. *)
       AvgScore : real;           (* average of the test scores -- computed. *)
       Sum : real;                (* running total -- intermediate variable. *)
       StudentNum : integer;      (* for-statement control variable. *)
         .
         .
         .
begin
  Sum := 0.0;                     (* initialize the running total to zero. *)

  for StudentNum := 1 to NumStudents do     (* add each students score to *)
      Sum := Sum + Scores [StudentNum];     (* the running total. *)

  AvgScore := Sum / NumStudents             (* the average is the total *)
                                            (* divided by the number. *)
end
```

When using structured variables, you must always be aware that they are constructed from simple data types. Many constructions require that structured variables be broken down into their simple data types. For example, in standard Pascal, when an **array** is read, each element must be read individually. The exception to this rule is character arrays, which are covered in Chapter 11. To print the elements of an **array** variable, you must print the elements one at a time. The following statement, which tries to print the **array** variable **Scores** as a single entity, will produce an error message:

```
                    writeln (' The scores are: ', Scores);
  ***error***                                                      $116

    116: error in type of standard procedure parameter.
```

The following fragment will produce 12 lines of output, one for each student, giving the student number and the student's score:

```
    for StudentNum := 1 to NumStudents do
        writeln (' Student #', StudentNum:3, ' received a',
                  Scores [StudentNum]:3)
```

You must individually initialize every element in an **array** variable if you use the array for running totals. You cannot set the **array** identifier to zero. This code sets each element of the **array** variable **Scores** to 0:

```
    for StudentNum := 1 to NumStudents do Scores [StudentNum] := 0
```

Array types

If you use the **type** definition to create a new data type and then declare two arrays of the new data type, these arrays are said to be of *identical type* and are therefore assignment compatible. That two **array** variables are of the identical type does not mean that all their elements must be equal, just as two **real** variables do not require equal values to be of the identical type.

In standard Pascal, if two **array** variables are of the identical data type, one **array** variable can be set equal to the other **array** variable:

```
type  VectorType = array [1..4] of real;

var   VectorA, VectorB : VectorType;
         .
         .
      VectorA := VectorB
```

VectorA and **VectorB** are of identical type because they are both of type **VectorType.** The effect of the assignment statement is to set the four elements in **VectorA** equal to the corresponding elements in **VectorB.** Although the assignment statement above is standard Pascal, not all compilers support it.

If you create two **array** variables using the syntax

```
var   VectorC, VectorD : array[1..4] of real;
```

the **array** variables are said to be of a *compatible data type.* **VectorC** and **VectorD** are stored the same way, but because they do not have the same data type name, the Pascal compiler does not recognize them as the same type.

9.3 Arrays as parameters

So far, the examples discussed have been program fragments that did not involve using arrays in parameter lists. To pass an **array** variable as an argument into a subprogram, you must use the **type** syntax to define the **array** structure, and you must declare the parameter to be an array of the same type. For example,

```
function AverageScore (Data : ScoreArray; Num : integer) : real;
(*      Function AverageScore returns the average of the Num integer   *)
(*      numbers in the array Data.                                     *)

var   Sum : integer;          (* running total -- intermediate variable. *)
      StudentNum : integer;   (* for control variable *)
```

```
begin
  Sum := 0;                        (* initialize the running total to zero. *)

  for StudentNum := 1 to Num do
    Sum := Sum + Data [StudentNum];  (* add the score to the total. *)

  AverageScore := Sum / Num        (* assign the value of the average to the *)
                                   (* function identifier. *)
end; (* function AverageScore *)
```

Suppose the function **AverageScore** is invoked in a main program as follows:

```
TestAvg := AverageScore (Scores, NumStudents)
```

If **Scores** has been declared to be of type **ScoreArray** and **NumStudents** has been declared to be **integer,** when the function **AverageScore** is invoked, the **array** variable **Scores** is copied into the local **array** variable **Data** and the **integer** variable **NumStudents** is copied into the local **integer** variable **Num.**

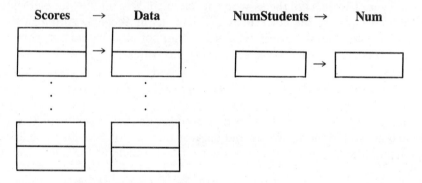

The function **AverageScore,** which performs the general task of finding the average, can be used to compute the averages of many different arrays through use of a parameter list, as in the following fragment

```
Test1Avg := AverageScore (FirstScores, NumStudents);
Test2Avg := AverageScore (SecondScores, NumStudents)
```

The first time **AverageScore** is invoked, the values stored in **FirstScores** are passed into the function and are copied into the local array called **Data.** Within the function, the array of grades is called **Data,** but it contains the values of **FirstScores.** When control returns to the main program, **Test1Avg** has the value of the average of **FirstScores.** The second time the procedure is called, the values stored in **SecondScores** are passed into the function and are copied into the local array called **Data.** The average is computed and returned to **Test2Avg.**

9.3 Arrays as parameters

When you pass an **array** variable into a subprogram, normally the entire array is passed by means of the array name, as illustrated in the previous example. However, you can also pass individual elements of an **array** variable into a parameter whose data type is the same as the base data type of the elements of the array, as illustrated in the procedure **OneAtATime:**

```
procedure OneAtATime (OneScore, Student, Num : integer; var Avg: real);
   (*      procedure OneAtATime computes the average of an array by      *)
   (*      adding each element from 1 to Num as it is passed into        *)
   (*      the procedure in the parameter OneScore. The parameter Avg    *)
   (*      is used to store the running total because a local variable   *)
   (*      would be deleted at the end of each execution of the function. *)

begin
   if Student = 1 then         (* if it is the first student, set the *)
      Avg := 0;                (* running total (stored in Avg) to zero. *)

   Avg := Avg + OneScore;      (* add the current score to the running *)
                               (* total. *)

   if Student = Num then       (* if it is the last student, compute the *)
      Avg := Avg / Num         (* average. *)
end;   (* procedure OneAtATime *)
```

For the procedure **OneAtATime** to compute a student's average, the variable passed from the main program into the parameter **OneScore** would have to be a single **integer** variable:

```
for StudentNum := 1 to NumStudents do
   OneAtATime (Score [StudentNum], StudentNum, NumStudents, TestAvg)
```

Notice how awkward procedure **OneAtATime** is compared to the function **AverageScore.** In procedure **OneAtATime,** the loop must be in the main program, and two tests are required within **OneAtATime** to see if the beginning or the end of the loop has been reached. As a general rule, place a loop in the innermost possible block to avoid unnecessary computation and to make the program easier to read.

REMINDER

The data type for a parameter can be a user-defined data type such as an array. To pass a variable that is not of a simple data type, you *must* explicitly define a data type. Both the global variable and the parameter must be of the identical type.

9.4 Multidimensional arrays

Array data structures are not limited to one column. Because each student has taken 6 tests, this information could be stored in a two-dimensional **array** variable with 12 rows (one row for each student) and 6 columns (one column for each test), as illustrated in the array **AllScores.**

AllScores

	Test Index					
Student Index	*1*	*2*	*3*	*4*	*5*	*6*
1	99	93	88	95	78	95
2	75	80	83	70	77	80
⋮						
12	80	85	82	78	87	83

By convention, a reference to an element in a two-dimensional array gives the row number first and the column number second. The row number tells you how far down to go to locate the element, and the column number tells you how far over to go. You can think of the row and column numbers as street and avenue addresses. To get to a particular block, you need to know the avenue and the cross street. In the array **AllScores,** the element **AllScores[12, 6]** is equal to 83. That is, the twelfth student received an 83 on the sixth test. Notice that each box contains only one number.

All the elements in the two-dimensional array variable **AllScores** have the data type **integer.** As with one-dimensional arrays, multidimensional arrays cannot store data of different data types. You cannot store integers in one dimension of the array and characters in another. For example, if you wanted to store the students' final letter grades of A, B, C, D, and F in an array, you would need to create a separate array of data type **char.** You could not store the characters as a third dimension of the array **AllScores.**

Although in theory you can have as many subscripts (indices) in your **array** data structure as you want, in practice you are limited by the size of the computer's memory. The total number of elements is the product of the number of elements in each dimension. For example, the array **TestScore[Student, Test, Class, Dept, Year]** has **Student** × **Test** × **Class** × **Dept** × **Year** elements. The total number of elements may be larger than the storage

9.4 Multidimensional arrays

capacity of the computer. Three dimensions is generally a reasonable limit. Another name for a multi-dimensional array is a *matrix*.

A simplified syntax diagram for declaring a data type for a multidimensional array is

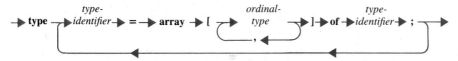

To define a multidimensional **array** data structure like the two-dimensional array of test scores, give the lower and upper bounds for the first index, followed by a comma, then the lower and upper bounds for the second index, as in

```
type ScoreMatrix = array [1..12, 1..6] of integer;
```

> *STYLE HINT*
>
> Many of the examples for declaring and using arrays have used actual numbers for the upper and lower bounds. Numbers are used so that you can see the indexing of arrays. Using constants for upper and lower bounds is a better solution once you understand how the indexing works. If it is necessary to change a bound in a long program, it is then necessary to make only a change in the constant definitions.

The **type** statement sets up the data structure for the test scores. After the **type** has been defined, a variable of that **type** can be declared:

```
var   AllScores : ScoreMatrix;
```

To reference an element in a two-dimensional **array** variable, you must use two index values separated by a comma. The following statements will print the twelfth student's score on the sixth test and the first student's score on the second test:

```
StudentNum := 12;
TestNum := 6;
writeln (' Student #', StudentNum:3, ' received a ',
         AllScores [StudentNum, TestNum]:4, ' on test', TestNum:4);
writeln;
StudentNum := 1;
TestNum := 2;
writeln (' Student #', StudentNum:3, ' received a ',
         AllScores [StudentNum, TestNum]:4, ' on test', TestNum:4)
```

```
=>
Student #  12 received a  83 on test   6

Student #   1 received a  93 on test   2
```

> *REMINDER*
>
> By convention in mathematics, the row of a two-dimensional array is referenced by the index *i,* and the column of a two-dimensional array is referenced by the index *j.* Although you do not have to follow this convention, many programmers do follow it, so knowing the convention makes other people's programs easier to read.

Notice that two numbers are needed to reference an element of the **array** variable **AllScores,** but only one number will be printed. You need both index values to uniquely determine the location of a particular element within the array.

Function **TestAverage** computes the average of all the students on the first test. **TestAverage** performs the same task as the function **AverageScore.** The difference is that the data structure for the function **TestAverage** contains the scores for 6 tests. Therefore, the test number must be specified for the average to be computed.

```
function TestAverage (Grades : ScoreMatrix; Num, TestNum : integer) : real;
(*     Function TestAverage computes the average score of all the    *)
(*     students on the test TestNum.  NumStudents is the number of   *)
(*     students in the class.  TestNum is the number of the test     *)
(*     for which the average is to be found.                         *)

var    Student : integer;       (* for statement control variable. *)
       Sum : integer;           (* running total -- intermediate variable. *)

begin
  Sum := 0;                     (* initialize the running total to zero. *)

  for Student := 1 to Num do
    Sum := Sum + Grades [Student, TestNum];     (* add the test to the *)
                                                (* running total *)

  TestAverage := Sum / Num
end;   (* function TestAverage *)
```

Solution to Case Study 13

The basic functions needed to solve Case Study 13 have been developed, but several tasks remain. First, the data structure for the program must be decided upon. Second, procedures to read the test scores and report the averages are required. Third, because the function **TestAverage** developed in the last section takes an average down a column, but not across a row, another function or a more general function must be developed to compute the average for each student.

The data structure using the data type **ScoreMatrix** has been created for the **array** variable that stores the input grades for each student on each test. The output data, each student's average and each test average, can also be stored in **array** variables. The students' averages can be stored in a one-dimensional **array** variable:

```
type  StudentArray = array [1..NumStudents] of real;

var   StudentAvg : StudentArray;
```

and the test averages can be stored in a one-dimensional **array** variable:

```
type  TestArray = array [1..NumTests] of real;

var   TestAvg : TestArray;
```

Finally, an **array** can be created for the average for each test type:

```
type  TestTypeArray = array [1..3] of real;

var   TestTypeAvg : TestTypeArray;
```

Next, the input procedure requires that the grades be read from the terminal in a logical order. Assume that all the tests have been taken when the grades are entered, so that all the grades from a single student can be entered at once. Procedure **ReadGrades** reads the students' scores in this format.

```
procedure ReadGrades (var Grades : ScoreMatrix; Num, NumTests : integer);
(*     Procedure ReadGrades prompts the user at the terminal for      *)
(*     the grades for each student for all the tests.                 *)

var   Student, Test : integer;      (* for statement control variables. *)
```

```
begin
  for Student := 1 to Num do
    begin
      writeln;
      writeln (' Enter the ', NumTests:1, ' test scores for student',
               Student:3);

      for Test := 1 to NumTests do
        read (Grades [Student, Test])
    end    (* for Student *)
end;    (* procedure ReadGrades *)
```

Notice that **Grades** in the parameter list must be a variable parameter; otherwise the grades that were entered would not be passed back to the global variable. In the procedure **ReadGrades,** there are two loops, one nested inside the other. The outer loop controls the index for the student number, and the inner loop controls the index for the test number. A walkthrough of this procedure will show you how the loops interact. The student index starts at 1 when the procedure begins execution:

Student	Test	AllScores[Student, Test]
1	—	—

Then the compound statement

```
begin
  writeln;
  writeln (' Enter the ', NumTests:1, ' test scores for student',
           Student:3);

  for Test := 1 to NumTests do
    read (Grades [Student, Test])
end
```

is executed with the student index equal to 1. At the terminal, the execution of this statement looks like

        Enter the 6 test scores for student 1
        □99 93 88 95 78 95

As the grades are read from the terminal, the test-loop counter is incremented from 1 to 6. So the walkthrough chart becomes

Student	Test	AllScores[Student,Test]
1	—	—
1	1	99
1	2	93
1	3	88
1	4	95
1	5	78
1	6	95

The compound statement has been completely executed when the last test score of the first student has been read. Control then returns to the outer loop, and the student index is incremented to 2. With the student index set at 2, the compound statement is repeated and the test index is incremented from 1 to 6 again. As you can see from the walkthrough, the inside loop moves faster than the outside loop.

The next task is to develop an algorithm that will compute the average for any student, for any test, and for each type of test. The same basic algorithm can be used to compute each average, but the algorithm must be able to sum one row, sum one column, or sum two columns. Although it might be possible to write a general function to perform these tasks, the program will be easier to write and easier to read if the tasks are performed in different functions. **TestAverage** has already been written, and it can easily be modified for the function **StudentAverage.** The difference between the two functions is that in **StudentAverage** the sum is taken across a row, instead of down a column as was done in **TestAverage.**

```
function StudentAverage (Grades : ScoreMatrix;
                         StudentNum, Num : integer) : real;
   (*     Function StudentAverage computes the grade for StudentNum    *)
   (*     on tests 1 through Num.  The scores are stored in the array  *)
   (*     Grades.                                                      *)
var    Test : integer;      (* for statement control variable. *)
       Sum : integer;       (* running total -- intermediate variable. *)
begin
  Sum := 0;                 (* initialize the running total to zero. *)

  (* The following for statement adds the grade on  each test to Sum *)

  for Test := 1 to Num do
    Sum := Sum + Grades [StudentNum, Test];

  StudentAverage := Sum / Num
end;   (* function StudentAverage *)
```

A function to take the average for each type of test could be written; however, a mathematical shortcut can be used instead. The average of the 12 students on test 1 can be written as

$$\overline{X}_1 = \frac{x_{1,1} + x_{2,1} + \cdots + x_{12,1}}{12}$$

and the average of the 12 students on test 4 can be written as

$$\overline{X}_4 = \frac{x_{1,4} + x_{2,4} + \cdots + x_{12,4}}{12}$$

where, for example, $x_{1,4}$ is the score of the first student on the fourth test. The average of all 24 grades on tests 1 and 4 is

$$\overline{X}_{1+4} = \frac{x_{1,1} + x_{2,1} + \cdots + x_{12,1} + x_{1,4} + x_{2,4} + \cdots + x_{12,4}}{24}$$

which means that the average $\overline{X}_{1+4}$ is equivalent to

$$\overline{X}_{1+4} = \frac{1}{2}\overline{X}_1 + \frac{1}{2}\overline{X}_4$$

Therefore, once the averages for each test are known, the average for each type of test can be found.

Program 9.1 is a complete program that reads the test scores for the students and prints the original scores, the average for each student, and the average for each test. Study the procedure **ReportGrades** to be sure that you understand why the grades are printed in the order in which they are. Also notice how the **write** and **writeln** statements are used so that the output is in rows and columns.

STYLE HINT

Arranging and labeling output can be time consuming, but unlabeled or unformatted output can be confusing both to you and to someone else reading the output. Always spend enough time to ensure that your output will be neat, well labeled, and easy to read.

PROGRAM 9.1   TestScores

```
program TestScores (input, output);
(*      This program reads the grades of 12 students on 6 tests and    *)
(*      computes the average grade for each student and the average    *)
(*      on each test.  In addition, it computes the average for each   *)
(*      type of test: multiple choice, short answer, and program       *)
(*      writing.                                                       *)
```

```
const NumStudents = 12;      (* number of students. *)
      NumTests = 6;          (* number of tests that were given. *)
      NumTestTypes = 3;      (* number of different types of tests that *)
                             (* were given. *)


type  ScoreMatrix = array [1..NumStudents, 1..NumTests] of integer;
      StudentArray = array [1..NumStudents] of real;
      TestArray = array [1..NumTests] of real;
      TestTypeArray = array [1..NumTestTypes] of real;


var   AllScores : ScoreMatrix;           (* array of the scores of all the *)
                                         (* students on all the tests. *)
      TestAvg : TestArray;               (* array of the test averages. *)
      StudentAvg : StudentArray;         (* array of the student averages. *)
      TestTypeAvg : TestTypeArray;       (* array of test type averages. *)

      Test : integer;                    (* for statement control variable. *)
      Student : integer;                 (* for statement control variable. *)



(*$i readgrades *)
(*$i testaverage *)
(*$i studentaverage *)



  procedure ComputeTestTypeAverages (TAvg : TestArray;
                                     var TTypeAvg : TestTypeArray);
  (*     ComputeTestTypeAverages computes the average for each type    *)
  (*     of test that was given.  The same type of test was given      *)
  (*     for the first and fourth, second and fifth, and third and     *)
  (*     sixth tests.                                                  *)
  begin
    TTypeAvg[1] := 0.5 * TAvg[1] + 0.5 * TAvg[4];   (* multiple choice. *)
    TTypeAvg[2] := 0.5 * TAvg[2] + 0.5 * TAvg[5];   (* short answer. *)
    TTypeAvg[3] := 0.5 * TAvg[3] + 0.5 * TAvg[6]    (* program writing. *)
  end;  (* procedure ComputeTestTypeAverages *)


  procedure ReportGrades (Grades : ScoreMatrix; SAvg : StudentArray;
                          TAvg : TestArray; TTypeAvg : TestTypeArray);
  (*     Procedure ReportGrades reports the scores for each student    *)
  (*     on each test and reports the test averages and the student    *)
  (*     averages.                                                     *)

  var   Student, Test : integer;     (* for statement control variables. *)
```

```
    begin
      writeln;
      writeln (' Student', 'Test Number':24, 'Student Average':29);
      writeln;
      write (' ':8);
      for Test := 1 to NumTests do
        write (Test:6);
      writeln;

      for Student := 1 to NumStudents do
        begin
          write (Student:4,'    ');
          for Test := 1 to NumTests do
            write (Grades [Student, Test]:6);
          writeln (SAvg [Student]:8:1)
        end;

      writeln;
      writeln (' Test');
      write (' Average   ');
      for Test := 1 to NumTests do
        write (TAvg [Test]:6:1);
      writeln;
      writeln;
      writeln (' The average on the multiple choice tests was':45,
               TTypeAvg[1]:8:1);
      writeln (' The average on the short answer tests was':45,
               TTypeAvg[2]:8:1);
      writeln (' The average on the program writing tests was':45,
               TTypeAvg[3]:8:1)
    end;   (* procedure ReportGrades *)

(*--------------------------------------------------------------------------*)

  begin  (* main *)
    ReadGrades (AllScores, NumStudents, NumTests);

    for Test := 1 to NumTests do
      TestAvg [Test] := TestAverage (AllScores, NumStudents, Test);

    ComputeTestTypeAverages (TestAvg, TestTypeAvg);

    for Student := 1 to NumStudents do
      StudentAvg [Student] := StudentAverage (AllScores, Student, NumTests);

    ReportGrades (AllScores, StudentAvg, TestAvg, TestTypeAvg)
  end.   (* program TestScores *)
```

=>

 Enter the 6 test scores for student 1
□99 93 88 95 78 95
 Enter the 6 test scores for student 2
□75 80 83 70 77 80
 Enter the 6 test scores for student 3
□92 95 85 88 92 89
 Enter the 6 test scores for student 4
□50 62 50 65 70 50
 Enter the 6 test scores for student 5
□76 70 60 78 77 55
 Enter the 6 test scores for student 6
□81 85 78 87 82 70
 Enter the 6 test scores for student 7
□72 74 90 73 81 92
 Enter the 6 test scores for student 8
□78 70 50 67 65 50
 Enter the 6 test scores for student 9
□85 87 80 89 85 78
 Enter the 6 test scores for student 10
□67 80 65 70 75 67
 Enter the 6 test scores for student 11
□77 78 70 72 78 68
 Enter the 6 test scores for student 12
□80 85 82 78 87 83

Student	Test Number						Student Average
	1	2	3	4	5	6	
1	99	93	88	95	78	95	91.3
2	75	80	83	70	77	80	77.5
3	92	95	85	88	92	89	90.2
4	50	62	50	65	70	50	57.8
5	76	70	60	78	77	55	69.3
6	81	85	78	87	82	70	80.5
7	72	74	90	73	81	92	80.3
8	78	70	50	67	65	50	63.3
9	85	87	80	89	85	78	84.0
10	67	80	65	70	75	67	70.7
11	77	78	70	72	78	68	73.8
12	80	85	82	78	87	83	82.5
Test Average	77.7	79.9	73.4	77.7	78.9	73.1	

The average on the multiple choice tests was   77.7
   The average on the short answer tests was   79.4
The average on the program writing tests was   73.2

APPLICATION
Vectors and Matrices

Vectors and matrices are used in physics and in every branch of engineering. This section will discuss vector and matrix notation and illustrate some of the basic operations used with vectors and matrices.

CASE STUDY 14
Torque of a bicycle wheel

When a bicycle is ridden, a frictional force is exerted on the front wheel which makes the wheel rotate. How much work must be done to rotate the wheel? What are the magnitude and the direction of the torque created by rotating the wheel?

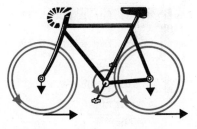

FIGURE 9.1
Forces on a bicycle wheel.

Algorithm development

The solution to the case study requires a knowledge of the concepts of work and torque. These concepts in turn require a knowledge of vectors, dot products, and cross products. Using top-down design, you can write a general algorithm for the case study, assuming that the subprograms to compute the work and the torque will be developed later.

ALGORITHM 9.2   First iteration of algorithm for Case Study 14

begin algorithm
   1. Initialize the program

2. Compute **Work**

3. Compute **Torque**

4. Report the values for **Work** and **Torque**

end algorithm

Algorithm 9.2 is distressingly vague. Without knowing the definitions of work and torque, you cannot specify even the input variables. However, if you look at the main program of **BicycleTorque,** the solution to the case study at the end of the chapter, you will see that it follows the algorithm above.

Vector operations

Vectors

A *vector* is a physical quantity that has magnitude and direction. A vector can be visualized as an arrow with a specified length pointing in a particular direction. Quantities that can be represented by vectors include the velocity of a particle, specified by its speed (magnitude) and its direction of motion; the force on a wheel, specified by the strength and direction of the force; the position of a planet in space, specified by its distance and direction from the sun. The last vector is illustrated in Figure 9.2.

Vectors are drawn as arrows because an arrow conveys the sense of direction. The *tail* of the vector refers to that end of the vector where the feathers of the arrow would be, and the *head* of the vector refers to that end of the vector where the point of the arrow would be.

The *magnitude* of a vector is the absolute value of its length. If the vector is oriented with its tail at the origin, the *direction* is given by the angle, in degrees or radians, from the positive x-axis to the vector. A vector that is

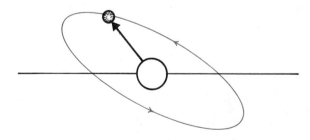

FIGURE 9.2
A vector specifying the distance and direction of a planet from the sun.

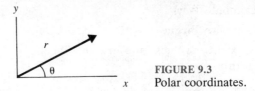

FIGURE 9.3
Polar coordinates.

specified by its magnitude and angle is said to be in *polar coordinates* as illustrated in Figure 9.3

A vector in polar coordinates is represented by an ordered pair of numbers (r, θ). The first number is the magnitude, and the second number is the angle. The vector in Figure 9.3 is represented as $(5, 36.8°)$; that is, you can construct the vector by moving 36.8 degrees from the x-axis and then moving out a distance of 5 units.

A vector that emanates from the origin of a coordinate system can also be represented in *Cartesian coordinates,* also known as *rectangular coordinates.* Cartesian coordinates are named for René Descartes (1596–1650), a French philosopher and mathematician and a contemporary of Pascal. In Cartesian coordinates a two-dimensional vector is represented by an ordered pair of numbers (x, y). By convention, the x-coordinate is given first and the y-coordinate is given second. The vector in Figure 9.4 is represented as $(4, 3)$; that is, moving 4 units in the x direction and 3 units in the y direction will bring you to the head of the vector that originates at $(0, 0)$ and has the cartesian coordinates $(4, 3)$.

Not all vectors have positive coordinates. Figure 9.5 illustrates the following vectors: $(4, 3), (-4, 3), (-4, -3)$, and $(4, -3)$. You can see that the x-axis and the y-axis divide the x,y-plane into four regions. These regions are referred to as *quadrants*. The standard names for the quadrants are given in Figure 9.5. The first vector, $(4, 3)$, is in quadrant I; $(-4, 3)$ is in quadrant II; $(-4, -3)$ is in quadrant III; and $(4, -3)$ is in quadrant IV.

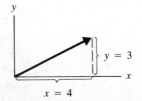

FIGURE 9.4
Cartesian coordinates.

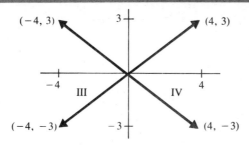

FIGURE 9.5
Vectors in each quadrant.

Vector notation

Because vectors are used frequently in mathematics, science, and engineering, a compact standard notation for representing vectors has been developed. The *basis* of a vector—that is, the coordinate system in which the components are given—must always be specified. The most common basis is the Cartesian coordinate system. The basis for this system in two dimensions is a vector with a length of one unit parallel to the x-axis and a vector with a length of one unit parallel to the y-axis, as shown in Figure 9.6.

Any two-dimensional vector can be written in terms of the unit basis. That is, any point in the x,y-plane can be represented by its x-coordinate, which is a multiple of the unit vector **i**, and its y-coordinate, which is a multiple of the unit vector **j**. By convention, the vector parallel to the x-axis is called **i**, and the vector parallel to the y-axis is called **j**. For example, a two-dimensional vector **D** can be represented by its components:

$$\mathbf{D} = (d_1, d_2)$$

where d_1 is the component along the x-axis and d_2 is the component along the y-axis.

To store a two-dimensional vector in Cartesian coordinates in a Pascal program, you could declare variables for the x- and y-coordinates for the point, or you could use a more general solution and store the coordinates in

FIGURE 9.6
The Unit Basis.

an array. The two-dimensional vector *D* can be represented by an array with a subrange of 2:

```
const NumDimensions = 2;

type  Rectangular = array [1..NumDimensions] of real;

var   RectD : Rectangular;
```

The vector *D* can be visualized as

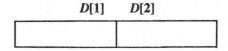

The first element of the array *D* is used to store the *x*-coordinate of the vector, and the second element is used to store the *y*-coordinate.

Similarly, if a vector is represented in polar coordinates, the same Pascal data structure can be used to represent the vector:

```
const NumDimensions = 2;

type  Polar = array [1..NumDimensions] of real;

var   PolarD : Polar;
```

The vector *D* can still be visualized as

except that *D*[1] now represents *r*, the magnitude of the vector, and *D*[2] now represents θ, the angle that the vector makes with the *x*-axis. This example emphasizes the importance of specifying the basis of a vector. The vector (1, 1) in rectangular coordinates is not the same vector as (1, 1) in polar coordinates.

Converting between coordinate systems

Because some vector operations are performed more easily in polar coordinates and some are performed more easily in Cartesian coordinates,

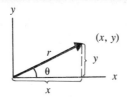

FIGURE 9.7
Cartesian and polar coordinates.

converting between coordinate systems is frequently required within a program. In your library of subprograms, it is useful to have procedures or functions that convert between coordinate systems.

Finding the polar coordinates r and θ of a vector from its Cartesian coordinates x and y is a problem very similar to the one solved for the case study in Chapter 2. Referring to Figure 9.7, you can see that a line drawn from the head of the vector to the x-axis forms a right angle. A right triangle is formed by the line, the vector, and the x-axis. As before, the Pythagorean theorem can be used to write the formula for the hypotenuse, r, which in this case is the length of the vector:

$$r = (x^2 + y^2)^{1/2} \tag{9.2}$$

and θ can be found from the **arctan** function:

$$\tan \theta = \frac{y}{x} \quad \text{or} \quad \theta = \tan^{-1}\left(\frac{y}{x}\right) \tag{9.3}$$

The function **arctan(Z)** returns the *principal* angle whose tangent is Z. The principal angle is the angle in the first or fourth quadrant whose tangent is Z. The vectors (4, 3) and (−4, −3) will have the same arctan because y/x is equal to 4/3 in both cases. To get the correct angle for vectors in the second or third quadrant, you must add π to the angle. This is equivalent to adding 180° to an angle expressed in degrees.

In writing the procedure to convert from rectangular to polar coordinates, you must not only compute which quadrant a point is in, but also consider what will happen if the point lies on the y-axis. If the point lies on the y-axis, x is equal to zero and the angle is $\pi/2$ radians. In Equation 9.3, what will happen if x is equal to zero? In the procedure, to prevent division by zero, you must test the value of x before the angle is computed. If x is zero, then the angle must be set to $\pi/2$. Procedure **RecToPole** converts rectangular to polar coordinates.

```
procedure RecToPole (X, Y : real; var R, Theta : real);
(*      RecToPole returns the polar coordinates, R and Theta, of a      *)
(*      vector given the rectangular coordinates, X and Y.              *)
const Small = 1.0e-6;           (* a real value close to small is considered *)
                                (* to be equal to zero. *)
      Pi = 3.1415926;           (* used to compute the angle when X is zero. *)
begin
  R := sqrt (sqr (X) + sqr (Y));    (* the magnitude is the length of *)
                                    (* hypotenuse. *)

  Theta := 0.0;                     (* initialize Theta to zero. *)

  if abs (Y) <= Small then
    writeln (' The y coordinate equals zero, so Theta is undefined. ',
             ' Theta has been set to 0.0')
  else
    begin
      if abs (X) > Small then       (* if X is not close to zero, *)
        Theta := arctan (Y/X)       (* Theta is the angle whose tangent *)
                                    (* is Y/X. *)
      else
        begin                       (* otherwise if X is close to zero, *)
          if Y > Small then         (* and if Y is greater than zero, *)
            Theta := Pi / 2.0;      (* then Theta is +90 degrees. *)
          if Y < Small then         (* or, if Y is less than zero, *)
            Theta := -Pi / 2.0      (* then Theta is -90 degrees. *)
        end; (* else *)

      if X < -Small then            (* if X is negative, then Theta is *)
        Theta := Theta + Pi         (* in the 2nd or 3rd quadrant. *)
    end (* else Theta defined. *)
end;  (* procedure RecToPole *)
```

To convert a vector in polar coordinates to Cartesian coordinates, refer to Figure 9.7 again. Using the trigonometric identities, you can rewrite x and y directly in terms of r and θ:

$$x = r \cos \theta \quad \text{and} \quad y = r \sin \theta \qquad (9.4)$$

These two equations can be incorporated into a procedure that converts polar coordinates to rectangular coordinates.

```
procedure PoleToRec (R, Theta : real; var X, Y : real);
(*      Procedure PoleToRec returns the rectangular coordinates, X    *)
(*      and Y, of a vector given the polar coordinates, R and Theta.  *)
begin
  X := R * cos (Theta);
  Y := R * sin (Theta)
end;   (* procedure PoleToRec *)
```

In the procedures **PoleToRec** and **RecToPole,** the variables **X, Y, R,** and **Theta** are used in the parameter lists instead of the **array** data structures developed earlier. The simple variables are used to make the procedures easier to read. There is, however, a tradeoff between simplicity in declaring the procedure and simplicity in invoking the procedure. Suppose that a vector **RectD** in rectangular coordinates is to be converted to a vector **PolarD** in polar coordinates and that both are **array** data types. As defined above, the procedure **RecToPole** is invoked as

```
RecToPole (RectD[1], RectD[2], PolarD[1], PolarD[2])
```

The argument list for **RecToPole** is difficult to write. On the other hand, if the declaration of **RecToPole** were

```
procedure RecToPole (RectVec : Rectangular; var PolVec : Polar);
```

then **RecToPole** could be invoked as

```
RecToPole (RectD, PolarD);
```

This argument list is easier to write and understand. The tradeoff is that the procedure itself would be harder to read, because every reference to **R** would become **PolVec[1],** **Theta** would become **PolVec[2],** etc.

Multidimensional vectors

Vectors are not restricted to two dimensions. There are three-dimensional vectors, *n*-dimensional vectors, and infinite-dimensional vectors. Although vectors of more than three dimensions cannot be visualized spatially, multidimensional vectors are used in many mathematical formulations of physical systems.

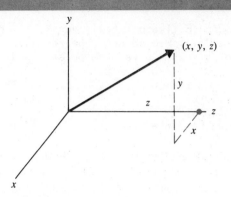

FIGURE 9.8
Cartesian coordinates for a three-dimensional vector.

In standard notation for multidimensional vectors, the components of an n-dimensional vector D are represented by $d_1, d_2, \ldots, d_n$:

$$D = [d_1, d_2, \ldots, d_n]$$

Again, the basis of the vector must be specified. The standard basis is the rectangular coordinate system in which the basis vectors are mutually perpendicular vectors of unit length. There are alternative bases, such as spherical and cylindrical bases, but these bases are harder to generalize in higher dimensions and are difficult to work with even in three dimensions.

You can construct three-dimensional vectors that have three elements: one for the x-axis, one for the y-axis, and one for the z-axis. (See Figure 9.8.) In Cartesian coordinates, three-dimensional vectors are given by their x-, y-, and z-coordinates, (x, y, z). An n-dimensional vector can be stored in a data structure such as the following:

```
type  Vector = array [1..N] of real;
```

where **N** is a constant that has been declared to be the number of dimensions for the vector. Note that in Pascal a three-dimensional vector can be represented by a one-dimensional array with three components.

Scalar multiplication of vectors

In scalar multiplication, each element of the vector is multiplied by a constant. The result of scalar multiplication is another vector that has the

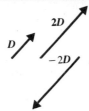

FIGURE 9.9
The vector D scaled by 2 and by -2.

same direction as the original vector, but that has been shortened or lengthened by the value of the constant multiplier. Figure 9.9 shows a vector that has been scaled first by a factor of 2, and then by a factor of -2. The vectors have been offset so that you can see them more clearly.

If a vector D is scaled in the x and y directions by a constant c, the components of the new vector, D' (read as "D prime"), are

$$d'_1 = cd_1 \quad \text{and} \quad d'_2 = cd_2 \qquad (9.5)$$

Equations 9.5 can be written more compactly as

$$d'_i = cd_i \quad \text{for } i = 1, 2 \qquad (9.6)$$

Equation 9.6 is equivalent to stating "Take every component of the vector and multiply it by the constant c." Equation 9.6 can be written even more compactly as

$$\boldsymbol{D'} = c\boldsymbol{D} \qquad (9.7)$$

When you see vector notation such as that in Equation 9.7, you must realize that it does not represent a single multiplication operation. The notation in Equation 9.7 means that every component of the vector D must be multiplied by the constant c to create the scaled vector D'. A Pascal procedure to scale a vector can be written as follows:

```
procedure ScaleVector (N : integer; OldD : Vector; Scale : real;
                 var NewD : Vector);
(*     Procedure ScaleVector multiplies each element of the N-           *)
(*     dimensional vector OldD by the real number, Scale, to create       *)
(*     the new N-dimensional vector, NewD.                                *)

var    Index : integer;      (* for statement control variable. *)

begin
  for Index := 1 to N do
    NewD [Index] := Scale * OldD [Index]
end;  (* procedure ScaleVector *)
```

Vector addition

Graphically, two vectors can be added as follows. One vector is drawn, and the second vector is drawn so that its tail coincides with the head of the first vector. The resultant vector is then drawn between the tail of the first vector and the head of the second vector, as in Figure 9.10.

Mathematically, vector addition can be performed by adding the corresponding components of each vector together. That is, the x-components of the two vectors are added together, and their sum becomes the x-component of the new vector. If the two vectors are

$$\mathbf{D} = (d_1, d_2) \quad \text{and} \quad \mathbf{T} = (t_1, t_2)$$

then

$$\mathbf{D'} = (d'_1, d'_2) = (d_1 + t_1, d_2 + t_2) \tag{9.8}$$

The operation of addition can be written for each component of the vector:

$$d'_i = d_i + t_i \quad \text{for } i = 1, 2 \tag{9.9}$$

Or the addition can be written in terms of the vectors themselves:

$$\mathbf{D'} = \mathbf{D} + \mathbf{T} \tag{9.10}$$

Again, you must remember that Equation 9.10 does not represent a single addition operation, but indicates that each element of vector $\mathbf{T}$ should be added to the corresponding element of vector $\mathbf{D}$. Notice that vector addition can be performed only if the vectors have the same number of elements.

A procedure to add two vectors can be written with the same data structure as was used for **ScaleVector.** Notice how the **for** statement is used so that each element in the second vector is added to the corresponding element of the first vector.

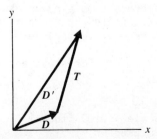

FIGURE 9.10
Vector addition: $\mathbf{D'}$ is the sum of the vectors $\mathbf{D}$ and $\mathbf{T}$.

```
procedure AddVectors (N : integer; First, Second : Vector;
                      var NewD : Vector);
(*      This procedure adds the coordinates of the first N-dimen-    *)
(*      sional vector to the corresponding coordinates of the second *)
(*      N-dimensional vector to create the resultant vector, NewD.   *)

var   Index : integer;      (* for statement control variable. *)
begin
  for Index := 1 to N do
    NewD [Index] := First [Index] + Second [Index]
end;  (* procedure AddVectors *)
```

The dot product of two vectors

The dot and cross products are functions that operate on vectors in a multiplicative fashion. The definition of these functions are, in some sense, arbitrary formulas for combining two vectors; however, both the dot product and the cross product arise naturally in the solution of many problems involving vectors and thus are widely used in physics and engineering.

The *dot product* of two vectors D and F is defined to be the real number

$$D \cdot F = |D||F| \cos \phi \qquad (9.11)$$

where $|D|$ = the magnitude of the vector D

$|F|$ = the magnitude of the vector F

ϕ = the angle between the vectors D and F

The expression $D \cdot F$ is read as "D dot F." The dot product is a scalar function of the two vectors; the magnitudes of the vectors are scalars and so is cos ϕ, so the dot product of two vectors is a single real number. For this reason, the dot product is also referred to as the *scalar product*.

The dot product is used in computing the work in physics. Work is defined as a force applied over a distance. To pull a massive block across a floor, you can attach a rope to the block and pull on the rope. The work that is done to move the block is the product of the force times the distance. Only the force applied in the direction of motion contributes to the work. In Figure 9.11, the force is diagonal, with both a horizontal and a vertical component. The force applied in the upward direction does not contribute to the work that is performed in moving the block forward. As will be shown below, work can be expressed as the dot product of the force and the distance.

To compute the amount of force that contributes to the motion, you can

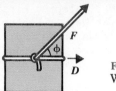

FIGURE 9.11
Work in progress.

take the *projection* of the force vector onto the distance vector. The projection of vector **F** onto vector **D** is found by dropping a line from the head of **F** perpendicular to **D**, creating a right triangle. One side of the new triangle lies along **D**. The new vector is the projection of **F** onto **D**. This vector represents the force that is useful in moving the block in the direction **D**. From Figure 9.12, you can see that the length of the projection is $|F| \cos \phi$.

If the work is defined to be the dot product of the force times the distance,

$$W = D \cdot F$$

or
$$W = |D||F| \cos \phi \tag{9.12}$$

then only the force that acts in the direction of motion is included in the work. The work done in moving the block is the product of the distance that the block moves and the force that acts in the direction in which it moves.

An alternative definition for the dot product of two vectors uses the rectangular coordinates, x and y, of the vectors. For example, the three-dimensional vector **D** can be represented by its components in the x, y, and z directions:

$$D = (d_1, d_2, d_3)$$

In rectangular coordinates, the dot product is defined to be the sum of the products of the corresponding elements of two vectors. For a three-dimensional vector, the dot product is

$$D \cdot F = d_1 f_1 + d_2 f_2 + d_3 f_3$$

or, in general, for an n-dimensional vector, the dot product is defined as

$$D \cdot F = \sum_{i=1}^{n} d_i f_i \tag{9.13}$$

In exercise 7 at the end of the chapter, you are asked to verify that the definition in rectangular coordinates is equivalent to the definition in polar coordinates. Verifying that two equations give the same result for several data

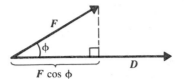

FIGURE 9.12
Projection of vector F onto vector D.

points is not the same as proving that the equations are equivalent. You might also think about how to prove that the definitions are equivalent.

From Equation 9.13, it is easy to show that the dot product is commutative—that is,

$$D \cdot F = F \cdot D \tag{9.14}$$

The work can be thought of either as (1) the force applied in the direction of motion times the distance moved or (2) the distance moved in the direction of the force times the force applied. The second interpretation involves taking the projection of the distance vector onto the force vector, as in Figure 9.13. Multiplying the magnitude of the projection by the magnitude of the force results in the same value for the work as was obtained before.

From Equation 9.11, it is easy to show that the dot product of two perpendicular vectors is zero. This result follows directly from the fact that $\cos 90°$ is equal to zero. Applied to the problem of determining the work done by a force, this result means that if you push down (apply a downward force) on the block, the block will not move forward, no matter how hard you push. The downward force is perpendicular to the direction of motion, so no work is performed.

Another property of the dot product is that the dot product of a vector with itself is the square of the magnitude of the vector. This property can be

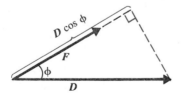

FIGURE 9.13
Projection of vector D onto vector F.

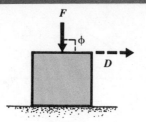

FIGURE 9.14
Force perpendicular to the direction of motion.

derived directly from either Equation 9.11 or Equation 9.13. Using Equation 9.11, $D \cdot D$ is

$$D \cdot D = |D||D| \cos \phi \qquad (9.15)$$

Since a vector makes an angle of 0° with itself, Equation 9.15 becomes

$$D \cdot D = |D|^2 \qquad (9.16)$$

Equation 9.16 can be used as an alternative definition of the magnitude of a vector:

$$|D| = \sqrt{D \cdot D} \qquad (9.17)$$

The dot product as a Pascal function

To write a subprogram to compute the dot product of two vectors, you must make several decisions. One decision is whether to write a function or a procedure. Another decision is whether to use Equation 9.11 or Equation 9.13 to compute the dot product.

To decide whether to write a procedure or a function, you should consider that the dot product does not change the values of the original vectors and that the result of the dot-product operation is a single real number. So the dot product is more appropriately implemented as a function than as a procedure. Having made that decision, you must now decide which definition of the dot product to use. Comparing the two definitions, you can see that Equation 9.11 uses the polar coordinates of the vectors, and Equation 9.13 uses the rectangular coordinates. Since vectors are more commonly stored in their rectangular coordinates, Equation 9.13 is the more natural equation to use. Another argument against using Equation 9.11 is that in the dot product, ϕ is the angle between the vectors, but in polar coordinates, θ is the angle the vector makes with the x-axis. For these reasons, Equation 9.13 will be easier to implement than Equation 9.11. If the vectors are given in polar coordinates, the procedure **RecToPole** can be used to find the rectangular coordinates before the dot product is computed.

The last step before implementing the dot-product function is to decide on the data structure. The vectors that will be passed into the function must have a defined data type. Given the decision to use Equation 9.13, the vectors must be stored in rectangular coordinates. Since the dot product is used in physics problems involving three-dimensional space, it is logical to use a three-dimensional vector, which can be stored in the following one-dimensional array:

```
type   Vector = array [1..3] of real;

var    Distance, Force : Vector;
```

With this definition of **Force, Force[1]** stores the *x*-component of the force vector, **Force[2]** stores the *y*-component, and **Force[3]** stores the *z*-component.

The function **DotProduct** uses the data structure above and Equation 9.13 to compute the dot product for any pair of *n*-dimensional vectors:

```
function DotProduct (N : integer; Vector1, Vector2 : Vector) : real;
(*       Function DotProduct returns the dot (scalar) product of the   *)
(*       N-dimensional vectors Vector1 and Vector2.                    *)

var    Sum : real;           (* running total for the dot product --  *)
                             (* intermediate variable. *)
       Coord : integer;      (* for statement control variable. *)
begin
  Sum := 0.0;                (* initialize the running total to zero. *)

  (* The next statement uses Equation 9.13 to compute the dot product. *)

  for Coord := 1 to N do
    Sum := Sum + Vector1 [Coord] * Vector2 [Coord];

  DotProduct := Sum
end; (* function DotProduct *)
```

The cross product

The *cross product,* or *vector product,* of two vectors **R** and **F** is a vector that is mutually perpendicular to the two vectors, with a magnitude of $|R||F| \sin \phi$, where ϕ is the angle between **R** and **F**. (See Figure 9.15.) The

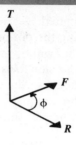

FIGURE 9.15
T equals the cross product of R and F.

cross product operation is represented by a ×, and the expression $R \times F$ is read as "R cross F."

The direction of the vector product T can be found through use of the right-hand rule. For $R \times F$, curl your fingers from the vector R to the vector F and your thumb will point in the direction of their cross product. (See Figure 9.16.)

The right-hand rule has practical applications because most screws and threads are right-handed. To illustrate this rule, take a lightbulb, point your right thumb in the direction in which you want the bulb to go. Curl your fingers, as illustrated in Figure 9.17. Turn the bulb in the direction in which your fingers curl, and the bulb will move in the direction your thumb points.

The magnitude and direction of the cross-product vector given above are sufficient to specify the vector; however, the vector can also be defined in terms of rectangular coordinates. If T is the cross product of R and F—that is, if $T = R \times F$—the elements of the vector T are defined to be

$$\begin{aligned} t_1 &= r_2 f_3 - r_3 f_2 \\ t_2 &= r_3 f_1 - r_1 f_3 \\ t_3 &= r_1 f_2 - r_2 f_1 \end{aligned} \qquad (9.18)$$

From the definition, it should be clear that the cross product is defined *only* for three-dimensional vectors.

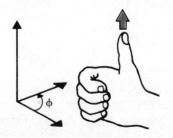

FIGURE 9.16
The right-hand rule.

FIGURE 9.17
Using the right-hand rule.

The cross product is not commutative, that is, $R \times F \neq F \times R$. Reversing the order of the vectors changes the sign of the resulting vector. You can find the direction of the cross product $F \times R$ by curling your fingers from F to R, as illustrated in Figure 9.18. The resulting vector points in the direction opposite that of vector $R \times F$ in Figure 9.15. Mathematically, $F \times R = -(R \times F)$.

A common phenomenon that can be described by the cross product is *torque*. When a wrench is used to remove a bolt, one end of the wrench is fixed to the bolt and a force is applied to the other end of the wrench. This force creates a torque, that is, a force that tends to produce a rotating or twisting motion, as illustrated in Figure 9.19. By definition, the torque τ of a force about a point in the vector $R \times F$, where R is the distance vector from the point to the force:

$$\tau = R \times F \tag{9.19}$$

For the bicycle wheel in the case study, when the force is applied on the rim at a distance r from the wheel hub, a torque results. The torque has a magnitude of $|R||F| \sin \phi$ and, as illustrated in Figure 9.20, points directly into the page.

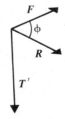

FIGURE 9.18
The vector T' equals the cross product of F and R.

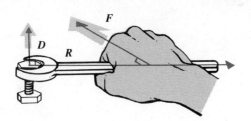

FIGURE 9.19
Force applied through a wrench to a bolt, creating a torque.

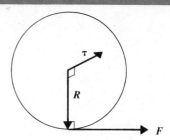

FIGURE 9.20
Torque of a bicycle wheel.

The cross product as a Pascal subprogram

To write a subprogram to compute the cross product of two vectors, you again must decide whether to write a procedure or a function. The cross product meets several of the criteria for a function: the operation does not change the values of the original vectors, and the result is a single vector. A function, however, cannot return a structured data type; it must return a scalar data type. Therefore the cross product must be implemented as a procedure, because the following function declaration is illegal:

```
function CrossProduct (Vec1, Vec2 : Vector) : Vector;
***error***                                              $120

error messages :
***************

   120: function result type must be scalar, subrange or pointer.
```

The body of the procedure **CrossProduct** consists of Equations 9.18, translated into Pascal using the same data structure that was used for the function **DotProduct.**

```
procedure CrossProduct (Vec1, Vec2 : Vector; var Vec3 : Vector);
   (*     CrossProduct computes the cross product of the 3-dimensional   *)
   (*     vectors, Vec1 and Vec2.  It returns the cross product in the   *)
   (*     vector Vec3.                                                   *)
```

```
begin
  Vec3 [1] := Vec1 [2] * Vec2 [3] - Vec1 [3] * Vec2 [2];
  Vec3 [2] := Vec1 [3] * Vec2 [1] - Vec1 [1] * Vec2 [3];
  Vec3 [3] := Vec1 [1] * Vec2 [2] - Vec1 [2] * Vec2 [1]
end;  (* procedure CrossProduct *)
```

Matrix operations

Matrices

A *matrix* can be used to represent several vectors in a single mathematical structure. A matrix is a multidimensional array of numbers. For example, the following two-dimensional matrix is a 3 by 4 array of real numbers:

$$A = \begin{bmatrix} 2.2 & 4.1 & 7.4 & 0.2 \\ 3.8 & 8.6 & -2.1 & 9.4 \\ 6.3 & 9.0 & 5.5 & 1.7 \end{bmatrix}$$

By convention, a matrix that has m rows and n columns is called an m by n matrix. So the matrix above is a 3 by 4 matrix.

You can think of the matrix above as a stack of three four-dimensional vectors; in the section on simultaneous linear equations in the Advanced Applications supplement, you will see that a matrix can also be used to represent the coefficients of equations. A matrix is a very general mathematical tool.

The elements of a matrix are specified by their row and column number, $a_{i,j}$. As with arrays, the row number is designated as i and the column number as j. For example, in the matrix above, the element $a_{3,2}$ has a value of 9.0. You can find this value by going down 3 rows and over 2 columns. The general form for an m by n matrix is

$$A = \begin{bmatrix} a_{1,1} & a_{1,2} & a_{1,3} & \cdots & a_{1,n} \\ a_{2,1} & a_{2,2} & a_{2,3} & \cdots & a_{2,n} \\ a_{3,1} & a_{3,2} & a_{3,3} & \cdots & a_{3,n} \\ \cdot & \cdot & \cdot & \cdots & \cdot \\ \cdot & \cdot & \cdot & \cdots & \cdot \\ \cdot & \cdot & \cdot & \cdots & \cdot \\ a_{m,1} & a_{m,2} & a_{m,3} & \cdots & a_{m,n} \end{bmatrix}$$

Notice that a vector is just a special case of a matrix. A *row vector* is a 1 by n matrix and a *column vector* is an m by 1 matrix.

A matrix, like a vector, can be stored in a Pascal **array.** The difference between the data structure for a matrix and for a vector is that the array in which a matrix is stored is multidimensional. The following data structure can be used to store an *m* by *n* matrix, where *M* and *N* have been declared as constants equal to the number of rows and columns.

```
type  Matrix = array [1..M, 1..N] of real;
```

Scalar multiplication of matrices

In scalar multiplication of a matrix, each element is multiplied by a single constant value. If the matrix is **R** and the scale factor is *c*, the elements in the new matrix **R'** are given by

$$r'_{i,j} = c r_{i,j} \quad \text{for } i = 1, \ldots, m \text{ and } j = 1, \ldots, n \tag{9.20}$$

or, in matrix notation,

$$\mathbf{R'} = c\mathbf{R} \tag{9.21}$$

The notation in Equation 9.21 indicates that *n* by *m* multiplications are to be performed, one for each element in the matrix. Notice that Equation 9.21 looks exactly the same as Equation 9.7, the equation for scalar multiplication for vectors. Since a vector is a one-dimensional matrix, it is reasonable that scalar multiplication is expressed the same way for both vectors and matrices.

Procedure **ScaleMatrix** scales a matrix stored in an array of the type **Matrix.**

```
procedure ScaleMatrix (M, N : integer; OldM : Matrix; Scale : real;
                      var NewM : Matrix);
(*      ScaleMatrix multiplies the elements of the M by N matrix    *)
(*      OldM by the real factor Scale to create the scaled          *)
(*      matrix NewM.                                                *)

var   Row : integer;      (* control variable for the rows. *)
      Col : integer;      (* control variable for the columns. *)
begin
  for Row := 1 to M do
    for Col := 1 to N do
      NewM [Row, Col] := Scale * OldM [Row, Col]
end;  (* procedure ScaleMatrix *)
```

Notice that scaling an entire matrix does not take much more programming effort that scaling a vector. The ease with which the procedure **ScaleVector** can be modified to **ScaleMatrix** is due to the compactness and power of the array data structure.

Matrix addition

In matrix addition the corresponding elements of matrices are added together. Matrix addition is defined only for matrices that have the same dimensions. You cannot add a 2 by 2 matrix to a 10 by 10 matrix. If the elements of the matrix X are added to the matrix R to yield a new matrix R', the elements of R' are

$$r'_{i,j} = r_{i,j} + x_{i,j} \quad \text{for } i = 1, \ldots, m \text{ and } j = 1, \ldots, n \quad (9.22)$$

or, in matrix notation,

$$R' = R + X \quad (9.23)$$

Procedure **AddMatrices** is an extension of the procedure **AddVectors**.

```
procedure AddMatrices (M, N : integer; First, Second : Matrix;
                      var NewM : Matrix);
(*      This procedure adds the coordinates of the first M by N       *)
(*      matrix to the coordinates of the second M by N matrix         *)
(*      to create the M by N matrix, NewM.                            *)

var   Row : integer;     (* control variable for the rows. *)
      Col : integer;     (* control variable for the columns. *)
begin
  for Row := 1 to M do
    for Col := 1 to N do
      NewM [Row, Col] := First [Row, Col] + Second [Row, Col]
end;  (* procedure AddMatrices *)
```

Matrix multiplication

Matrix multiplication is a very powerful tool. If you have never seen matrix multiplication before, it may take a while before it makes sense, but because matrix multiplication is used so frequently in science and engineering, it is worth taking the time to learn.

In matrix notation, multiplication is written as

$$B = AD \quad (9.24)$$

However, matrix multiplication is defined only for pairs of matrices in which the number of columns in the first matrix, A, is equal to the number of rows in the second matrix, D.

In matrix multiplication, the elements of one row of a matrix are multiplied by the elements of one column of the other matrix. The products are summed and the result becomes the row, column element of the new matrix. If A is m by n and D is n by l, mathematically the elements of the new matrix AD, are given by

$$b_{i,j} = \sum_{k=1}^{n} a_{i,k} \, d_{k,j} \quad \text{for } i = 1, \ldots, m \text{ and } j = 1, \ldots, l \quad (9.25)$$

To find the product of the two matrices

$$A = \begin{bmatrix} 1.0 & 2.0 \\ 3.0 & 4.0 \\ 5.0 & 6.0 \end{bmatrix} \quad \text{and} \quad D = \begin{bmatrix} 7.0 & 8.0 & 9.0 & 6.0 \\ 10.0 & 11.0 & 12.0 & 15.0 \end{bmatrix}$$

you start by finding the value for $b_{1,1}$. Equation 9.25 can be rewritten with 1 substituted for i and 1 for j. The upper bound of the sum, n, has a value of 2 because there are two columns in matrix A and 2 rows in matrix D:

$$b_{1,1} = \sum_{k=1}^{2} a_{1,k} \, d_{k,1}$$

When k is equal to 1, the element $a_{1,1}$, which has a value of 1.0, is multiplied by the element $d_{1,1}$, which has a value of 7. The resulting value of 7 is included in the running total. When k is 2, the element $a_{1,2}$, which has a value of 2.0, is multiplied by $d_{2,1}$, which has a value of 10.0. The resulting value of 20 is added to the running total. When k is equal to 2, all the terms in the summation have been calculated and the resulting value, 27, is assigned to $b_{1,1}$.

$$b_{1,2} = a_{1,1}d_{1,2} + a_{1,2}d_{2,2} = (1.0)(8.0) + (2.0)(11.0) = 30$$
$$b_{1,3} = a_{1,1}d_{1,3} + a_{1,2}d_{2,3} = (1.0)(9.0) + (2.0)(12.0) = 33$$

If you locate each number in the matrices as you multiply and add, you will notice a pattern of moving across the rows of the first matrix and down the columns of the second matrix. To be sure that you understand the algorithm, you can verify the remainder of the elements in the B matrix.

$$\begin{bmatrix} 1.0 & 2.0 \\ 3.0 & 4.0 \\ 5.0 & 6.0 \end{bmatrix} \begin{bmatrix} 7.0 & 8.0 & 9.0 & 6.0 \\ 10.0 & 11.0 & 12.0 & 15.0 \end{bmatrix} = \begin{bmatrix} 27.0 & 30.0 & 33.0 & 36.0 \\ 61.0 & 68.0 & 75.0 & 78.0 \\ 95.0 & 106.0 & 117.0 & 120.0 \end{bmatrix}$$

Notice that a 3 by 2 matrix multiplied by a 2 by 4 matrix results in a 3 by 4 matrix. In more general terms, multiplying an m by n matrix by an n by q matrix results in an m by q matrix.

Suppose the matrix **S**, which is 2 by 2, is to be multiplied by the matrix **D**, which is 2 by 3:

$$S = \begin{bmatrix} 2.0 & 1.0 \\ 4.0 & 3.0 \end{bmatrix} \quad \text{and} \quad D = \begin{bmatrix} 5.0 & 4.0 & 3.0 \\ 9.0 & 8.0 & 2.0 \end{bmatrix}$$

The resulting matrix **T** will be a 2 by 3 matrix with the following elements:

$$T = \begin{bmatrix} 19.0 & 16.0 & 8.0 \\ 47.0 & 30.0 & 18.0 \end{bmatrix}$$

Procedure **MultiplyMatrices** multiplies two matrices and returns their product. If you are not sure how matrix multiplication works, you can make your own version of **MultiplyMatrices** and include write statements to print the variables each time the **for** loop is executed. Watching someone else perform matrix multiplication may also prove helpful.

```
procedure MultiplyMatrices (Num : integer; A, D : Matrix; var B : Matrix);
(*      Procedure MultiplyMatrices multiplies the Num by Num square    *)
(*      matrices A and D and returns the result in the matrix B.       *)
(*      The procedure uses Equation 9.25 to compute the value for      *)
(*      each element in B.  Note that this version of Multiply-        *)
(*      Matrices works only if both matrices are square.               *)
var     Row : integer;          (* control variable for the rows. *)
        Col : integer;          (* control variable for the columns. *)
        InnerLoop : integer;    (* control variable for the inner loop. *)
        Sum : real;             (* intermediate variable to hold the *)
                                (* running total for each element. *)
begin
  for Row := 1 to Num do
    for Col := 1 to Num do
      begin
        Sum := 0.0;                     (* for element Row, Col set *)
                                        (* the sum to zero. *)
        for InnerLoop := 1 to Num do
          Sum := Sum + A [Row, InnerLoop] * D [InnerLoop, Col];

        B [Row, Col] := Sum             (* assign the running total *)
                                        (* to element Row, Col. *)
      end
end;    (* procedure MultiplyMatrices *)
```

Solution to Case Study 14

In the solution to the case study, the computation of the work is complicated because the force is applied at the point of contact of the wheel with the road. As the wheel rotates, the point of contact changes. Therefore, when the wheel is analyzed, the force at the point of contact is assumed to rotate the wheel a small distance Δd, as illustrated in Figure 9.21. When the wheel has rotated the distance Δd, the point of contact changes and the force rotates the wheel another Δd.

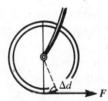

FIGURE 9.21
Wheel rotated a distance Δd.

Therefore, the force moves the wheel a distance Δd before the point of contact changes. If you assume that force is applied parallel to the road surface in the direction of motion of the wheel, the work performed to move the wheel a distance Δd is just the force times Δd. This result follows from the definition of the dot product. The vectors for the wheel are illustrated in Figure 9.22. (See References 1 and 2 at the end of this chapter for a more detailed description of the forces on a wheel.)

In the analysis above, only the work involved in rotating the wheel a small distance, Δd, was computed. The total work that is done in rotating the wheel is found by adding together the work from each small rotation:

$$W = \sum F \, \Delta d \qquad (9.26)$$

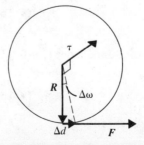

FIGURE 9.22
Vector diagram for the bicycle wheel.

Because a constant force has been assumed, the total work can be written as
$$W = F \sum \Delta d \qquad (9.27)$$
The summation can be simplified again, since it represents the total distance that the wheel travels. If the wheel makes one rotation, the distance that the wheel travels is its circumference. The work can be expressed as
$$W = Fr\omega \qquad (9.28)$$
where r is the radius of the wheel and ω is the number of radians through which the wheel rotates. For a ten-speed bike with a 27-in-diameter wheel, going at a speed of 10 mi/hr, a typical force is 1.1 lb. In metric units, these values are approximately equivalent to a radius of 0.33 m, a speed of 4.47 m/s, and a force of 7.6 N. The work in joules required to rotate the wheel once is equal to (0.33 m)(2π radians)(7.6 N), or 15.76 J.

The torque on the wheel is given by the formula
$$\tau = \mathbf{R} \times \mathbf{F} \qquad (9.29)$$
Making the same assumptions that were made in computing the work and assuming that the applied force is perpendicular to the radius, the magnitude of the torque is $|\mathbf{R}|$ times $|\mathbf{F}|$, or (0.33 m)(7.6 N). This result follows from the definition of the cross products. The direction of the torque is perpendicular to the plane of the bicycle wheel. In Figure 9.22, the torque goes into the page, following the right-hand rule.

In analysis of rotational motion, the definition of work is generally based on the torque rather than the force. The formula based on the torque is
$$W = \sum \tau \Delta \omega \quad \text{or} \quad W = \int \tau \, d\omega \qquad (9.30)$$
Equation 9.30 can be derived from the definition of work:
$$W = \mathbf{F} \cdot \mathbf{D} \qquad (9.31)$$
and the relationship among the radius of a circle, the angle of rotation, and the distance that a point on the circumference moves:
$$\Delta d = r \, \Delta \omega \qquad (9.32)$$
Equation 9.32 is the formula for the length of an arc of a circle, as can be seen in Figure 9.22. If Equation 9.31 is expressed as a difference equation and geometric identities are employed, Equations 9.31 and 9.32 can be combined to yield Equation 9.30.

Program 9.2, **BicycleWheel,** computes the torque and the work for the bicycle wheel. The program makes the same assumptions as were made above about the angles at which the force is applied to the wheel. This restriction means that the user can enter only the radius of the wheel, the magnitude of the force, and the number of rotations that the wheel makes. Given these values, the program uses the subprograms **Dot-Product** and **CrossProduct** to compute the torque vector, its magnitude, and the work that is done in rotating the wheel.

PROGRAM 9.2   BicycleWheel

```
program BicycleWheel (input, output);
(*      Program BicycleWheel computes the work that is done to rotate   *)
(*      a wheel. It also computes the torque that is created when       *)
(*      the wheel is rotated.                                           *)

const Pi = 3.14159;        (* pi is used to compute the distance traveled by *)
                           (* the wheel. *)
      Dim = 3;             (* number of dimensions for the vectors. *)

type  Vector = array [1..Dim] of real;

var   Force : Vector;      (* force vector -- in newtons. *)
      Radius : Vector;     (* radius vector -- in meters. *)
      Turns : real;        (* number of times the wheel turns. *)
      Work : real;         (* work done to rotate the wheel -- in joules. *)
      Torque : Vector;     (* torque vector -- in newton-meters. *)
      MagTorque : real;    (* magnitude of the torque -- in newton-meters. *)


(*$i dotproduct *)
(*$i crossproduct *)
(*$i initialmessage *)

  procedure ReadAndSetVectors (var WheelRadius, WheelForce : Vector;
                               var Rotations : real);
  (*      Procedure ReadAndSetVectors reads the radius of the wheel in    *)
  (*      meters, the magnitude of the force applied to the wheel in     *)
  (*      newtons, and the number of rotations that the wheel makes.     *)

  var MagRad : real;         (* magnitude of the radius vector. *)
      MagForce : real;       (* magnitude of the force vector. *)
```

Application: Vectors and Matrices

```pascal
  begin
    writeln (' What is the radius of the wheel in meters?');
    readln (MagRad);

    WheelRadius[1] := 0.0;         (* set the values for the distance *)
    WheelRadius[2] := -MagRad;     (* vector that points straight up in *)
    WheelRadius[3] := 0.0;         (* the y direction. *)

    writeln (' What is the magnitude of the force applied to the wheel',
             ' in newtons?');
    readln (MagForce);

    WheelForce[1] := MagForce;     (* set the values for the force vector *)
    WheelForce[2] := 0.0;          (* that points forward in the x *)
    WheelForce[3] := 0.0;          (* direction. *)

    writeln (' How many times does the wheel rotate?  Enter a real number');
    readln (Rotations)
  end;   (* procedure ReadAndSetVectors *)

function ComputeWork (WheelForce : Vector; Rotations : real) : real;
(*      Function ComputeWork returns the work that is done when a       *)
(*      wheel is turned Rotations number of times using a force of      *)
(*      WheelForce.  The work is the product of the force times the     *)
(*      distance over which the force is applied.                       *)

  var  MagForce : real;      (* magnitude of the force - computed using *)
                             (* square root of the dot product. *)
       Distance : real;      (* distance the wheel travels -- in meters. *)
  begin
    MagForce := sqrt (DotProduct (Dim, WheelForce, WheelForce));
    Distance := 2.0 * Pi * Rotations;
    ComputeWork := MagForce * Distance
  end;   (* function ComputeWork *)


procedure ComputeTorque (WheelRadius, WheelForce : Vector;
                         var WheelTorque : Vector; var Mag : real);
(*      ComputeTorque computes the torque vector using the procedure    *)
(*      CrossProduct.  Torque = Radius x Force.  The function           *)
(*      DotProduct is used to compute the magnitude of the torque.      *)

  begin
    CrossProduct (WheelRadius, WheelForce, WheelTorque);

    Mag := sqrt (DotProduct (Dim, WheelTorque, WheelTorque))
  end;   (* procedure ComputeTorque *)
```

```
(*$i reportvalues *)
(*   procedure ReportValues (WheelRadius, WheelForce, WheelTorque : Vector;  *)
(*                           Rotations, MagTorque, WorkDone : real);         *)
(*            Procedure Report prints the initial and computed values.       *)


(*---------------------------------------------------------------------------*)
begin    (* main program *)
  InitialMessage;

  ReadAndSetVectors (Radius, Force, Turns);

  Work := ComputeWork (Force, Turns);                 (* compute the work. *)

  ComputeTorque (Radius, Force, Torque, MagTorque);   (* find the torque *)
                                                      (* vector. *)
  ReportValues (Radius, Force, Torque, Turns, MagTorque, Work)
end.   (* program BicycleWheel *)

=>
 This program computes the work performed and the
 torque created when a wheel is rotated.

 What is the radius of the wheel in meters?
□0.33
 What is the magnitude of the force applied to the wheel in newtons?
□7.6
 How many times does the wheel rotate?  Enter a real number
□6.0

 If a force of       7.60 newtons is applied to
 a wheel with a radius of    0.33 meters,
 the work performed to rotate it    6.0 times
 is    286.51 joules

 The magnitude of the torque created is    2.51 newton-meters
 The components of the torque vector are  (     0.00      0.00      2.51)
```

Summary

An **array** can be thought of as a structure that organizes data in a table format. An **array** variable can be used to store data that are similar but have distinct values. For example, a table of results from different trials in an experiment can be stored in an **array** variable. Each piece of data that is stored in an array is called an *element* of the array and is referenced by an index (in a one-dimensional array) or indices (in a multidimensional array). All the elements in an array must have the same data type. The elements of the array can be treated like variables of the base data type of the array. That is, **Scores[2],** the second element in the **array** variable **Scores,** is an **integer** variable and can be treated as such in assignment statements and argument lists.

With a **type** definition, a *user-defined* **array** data type can be created. The user-defined data type can then be used just like the standard Pascal data types **real, integer, char,** and **boolean.** The **array** definition defines an upper and lower bound for each index and defines the base data type of the elements in the array. An **array** variable that has been declared to be of a user-defined **array** data type can be passed as an argument into an **array** parameter of the same data type.

References

1. F. R. Whitt and D. G. Wilson, *Bicycling Science,* Cambridge, Mass.: M. I. T. Press, 1982.

2. A. Sharp, *Bicycles and Tricycles, An Elementary Treatise on Their Design and Construction,* Cambridge Mass.: M. I. T. Press, 1979. Reprinted from the 1896 edition.

Exercises

PASCAL

1. Correct the syntax in each of the following statements:

   (a) ```type NewType : array of [1..10] of real;```

   (b) ```var NewArray : array[1,10] of integer;```

   (c) ```for Index := 1 to N do
          Score (Index) := 0```

(d) ```
    type Matrix = array[1..10, 1..20] of real;
    var OldResults : Matrix;
         Row, Col : integer;
  begin
    for Col := 1 to 10
      for Row := 1 to 20 do
        begin
          OldResults[Row, Col] := 0.0;
          Matrix[Row, Col] := 0.0
        end
```

2. What is the output from the following fragments? (Assume that all arrays and variables have been declared correctly.)

   (a) ```
   Value [0] := 2;
   for Index := 1 to 4 do
     begin
       Value [Index] := Value [Index - 1] * 2;
       writeln (' For Index =', Index:4, ' Value equals',
                Value [Index]:5)
     end
   ```

   (b) ```
   for Index := 1 to 5 do
     Number [Index] := Index;
   for Index := 2 to 5 do
     Number [Index] := Number [Index] + Number [Index - 1];
   for Index := 1 to 5 do
     writeln (' Number [', Index:1, '] =', Number [Index]:3);
   ```

3. Write a program to read five social security numbers in the format 999-99-9999 into an array and then print them at the terminal. Do not forget that social security numbers may begin with a 0 and that, for integers, leading 0s are not printed.

4. Write a program to read a five-character word from the terminal and then print it backwards at the terminal.

## APPLICATION

5. Given the two vectors $A = (5.3, 6.2, 7.8)$ and $B = (11.1, -3.2, 16.5)$, using a calculator,

   (a) compute the polar coordinates for $A$ and $B$.
   (b) compute $C = A \cdot B$ using Equation 9.11. Remember that $\phi$ is the angle between the vectors. Compute the dot product again using Equation 9.13.
   (c) compute the cross products $D = A \times B$ and $E = B \times A$. Show that the vectors $D$ and $E$ have the same magnitude and opposite direction.
   ```

6. Given

$$A = \begin{bmatrix} 1 & 3 \\ 2 & -5 \end{bmatrix} \quad B = \begin{bmatrix} 1 & 2 & 1 \\ 1 & 0 & 2 \end{bmatrix}$$

$$C = \begin{bmatrix} 6 & 4 \\ 3 & 2 \end{bmatrix} \quad D = \begin{bmatrix} 9 & 4 \\ 6 & 8 \\ 1 & 3 \end{bmatrix}$$

compute the elements of the following matrices:

(a) 2.5 A   (b) A + C   (c) AB   (d) BD   (e) AC   (f) A + C + AC

Problems

PASCAL

1. Rewrite the program to find the fastest computer from Problem 8 in Chapter 7 so that the computer identification codes and the run times are read into arrays. Find the average time for the 10 computers. Report the average time, the fastest time, and the slowest time.

2. Write a program that reads four integer numbers into an array, computes the average of the numbers, and reports which numbers in the array are greater than the average. Both the values of the numbers and their positions in the array should be reported.

3. In order to determine with accuracy the weight and density of a block of metal, you have measured each dimension 5 times. The weight is in kilograms and the density is in kilograms per meter cubed. Write a program that finds the average of the measurements for the weight and the density as given below:

Weight in kgs	Density in kgs/m^3
24.97	3.22×10^3
24.89	3.23×10^3
25.00	3.31×10^3
24.83	3.22×10^3
24.92	3.25×10^3

4. The record high temperatures for each day of July in Some City are stored in an array of 31 integer numbers. The high temperatures for each day of July for this year are stored in another array. The first array must be corrected upward by the second array. That is, if the fifth of July this year was hotter than the record high for the fifth of July, the fifth temperature in the first array must be replaced by the fifth temperature in the second array. Report the date, the old record temperature, and the new record temperature for any changes that are made.

Date	1	2	3	4	5	6	7	8	9	10	11	12	13	14	15	
Record	98	99	94	96	100	98	95	93	95	95	97	96	98	94	97	
This year	88	85	90	92	95	97	98	95	90	89	89	87	90	79	80	
Date	16	17	18	19	20	21	22	23	24	25	26	27	28	29	30	31
Record	101	100	98	97	98	95	95	98	98	99	98	97	98	96	96	97
This year	82	85	91	93	97	97	97	96	95	92	92	95	94	94	93	90

5. Suppose you need to add positive integers that can have up to 50 digits in them. These values are much greater than **maxint**, the largest integer that can be stored in an integer variable. Write a program that will add two such large integers.

   You will not be able to add these integers in the conventional way. First, you must figure out how to store them in pieces (use an array); then you must devise an algorithm that adds them column by column, carrying the appropriate value to the next column, from right to left. Run your program with two sets of data, one in which the two integers each have fewer than 50 digits and one in which the two integers each have exactly 50 digits.

6. Suppose that the professor in Case Study 13 has decided to redesign the experiment to account for the fact that some topics are easier than others. Instead of giving the entire class one test, the professor has divided the class into thirds and given each third of the class a different type of test.

   Modify the program from the case study to compute the average for each type of test each time a test is given and for each type of test over all 6 tests.

APPLICATION

7. The *transpose* of a matrix is the matrix with its rows and columns interchanged. The transpose of the matrix

$$A = \begin{bmatrix} a_{11} & a_{12} & a_{13} \\ a_{21} & a_{22} & a_{23} \end{bmatrix} \text{ is } A^T = \begin{bmatrix} a_{11} & a_{21} \\ a_{12} & a_{22} \\ a_{13} & a_{23} \end{bmatrix}$$

   Write a program that reads a matrix, prints it, and prints its transpose.

8. The voltage in a three-loop circuit is the current times the resistance,

$$V = RI$$

   V and I are column vectors with three elements and R is a 3 by 3 matrix.

   Write a program that reads the elements for the current vector and the elements for the resistance matrix and then, using matrix multiplication, computes the three elements of the voltage vector. The program should print the input values and the resulting voltage vector.

9. Write a general procedure that multiplies matrix A, which is an m by n matrix, times matrix B, which is a k by l matrix. If the number of columns in the first matrix is not equal to the number of rows in the second matrix, the matrices

cannot be multiplied. Your program should halt execution with an appropriate message if the two matrices cannot be multiplied.

Test your program by multiplying the 3 by 3 identity matrix by the 3 × 1 vector $(-1, 0, 0)$:

$$\begin{bmatrix} 1 & 0 & 0 \\ 0 & 1 & 0 \\ 0 & 0 & 1 \end{bmatrix} \begin{bmatrix} -1 \\ 0 \\ 0 \end{bmatrix}$$

10. Two vectors are linearly dependent if one vector is a scalar multiple of the other. Write a program to read a pair of three-dimensional vectors, test whether the pair is linearly dependent, and then write a message stating whether or not the pair is linearly dependent.

11. The *determinant* of the 2 by 2 matrix

$$A = \begin{bmatrix} a_{11} & a_{12} \\ a_{21} & a_{22} \end{bmatrix}$$

is

$$\det(A) = a_{11}a_{22} - a_{12}a_{21}$$

Given n consecutive points on the Euclidean plane with coordinates (x_i, y_i), where $i = 1, 2, \ldots, n$, the enclosed area is one half the sum of the determinants of the adjacent points, or

$$\text{area} = \frac{1}{2} \left[\det \begin{bmatrix} x_1 & y_1 \\ x_2 & y_2 \end{bmatrix} + \det \begin{bmatrix} x_2 & y_2 \\ x_3 & y_3 \end{bmatrix} + \cdots + \det \begin{bmatrix} x_n & y_n \\ x_1 & y_1 \end{bmatrix} \right]$$

If the points are taken in a counterclockwise direction, the area will have a positive value. If they are taken in a clockwise direction, the area will have a negative value.

Write a program that will compute the area for each polygon and report the following information:

(a) the type of polygon—triangle, quadrilateral, pentagon, hexagon, heptagon, octagon, nonagon, or decagon

(b) the coordinates of the vertices

(c) the area of the polygon

12. In nodal analysis of linear time-invariant electrical networks, the node admittance matrix Y_n is equal to the product of three matrices:

$$Y_n = AGA^T$$

where A^T is the transpose of the A matrix. (See problem 7 for the definition of the transpose.)

Write a program that evaluates Y_n for an A matrix and a G matrix. Include a procedure that computes the transpose of a matrix and a procedure that multiplies two matrices. The multiplication procedure should test whether the matrices can be multiplied.

13. When a small amount of liquid is introduced into a closed system, a fixed percentage of the liquid will change into a vapor state and a fixed percentage of the vapor will change back into a liquid state. The process will be repeated indefinitely. Let us assume that 1/4 of the liquid present at the beginning of the day turns into vapor during the day, and that an amount equal to 1/10 of the vapor present at the beginning of the day turns into liquid during the day. This information can be represented in matrix form as follows:

	Portion into liquid	Portion into vapor
Liquid	3/4	1/4
Vapor	1/10	9/10

At first glance, you might think that eventually all the substance will turn into vapor; however, this is not the case.

Let L_o be the proportion of the substance originally in liquid state and V_o the proportion originally in vapor state. The state is represented by the vector (L_o, V_o). In a similar fashion (L_i, V_i) represents the proportion in liquid and vapor states at the end of the ith day. At the end of day 1,

$$(L_1, V_1) = (L_o, V_o)\begin{bmatrix} 3/4 & 1/4 \\ 1/10 & 9/10 \end{bmatrix}$$

At the end of day 2,

$$(L_2, V_2) = (L_1, V_1)\begin{bmatrix} 3/4 & 1/4 \\ 1/10 & 9/10 \end{bmatrix} = (L_o, V_o)\begin{bmatrix} 3/4 & 1/4 \\ 1/10 & 9/10 \end{bmatrix}^2$$

At the end of day 3,

$$(L_3, L_3) = (L_2, V_2)\begin{bmatrix} 3/4 & 1/4 \\ 1/10 & 9/10 \end{bmatrix} = (L_o, V_o)\begin{bmatrix} 3/4 & 1/4 \\ 1/10 & 9/10 \end{bmatrix}^3$$

No matter what the original proportions (L_o, V_o) may be, (L_k, V_k) approaches the same vector for large values of k. After a long time, what proportion of the substance will be liquid and what proportion will be vapor?

Write a program that computes the proportion of liquid and vapor on days 2, 3, 5, 10, and 20, for any input values of L_o and V_o.

CHAPTER 10

Textfiles

In this chapter, you will learn a new method for entering data into a Pascal program and for storing the output from a program. When you have a program that requires entry of many lines of data, you are likely to make typing errors. If there is even one incorrect entry the output will be wrong, and there is no way to correct an entry after a line has been entered. The structured Pascal data type **file** makes it possible to use the computer's editor to create a file containing the data, in the same way that the editor is used to create a file containing the program. You can then correct mistakes with the editor instead of retyping all the data. The program then reads the data from the file instead of from the terminal.

CASE STUDY 15
Saving experiment data

The professor who gave the ten quizzes in Chapter 9 would like to have a program that will allow the grades to be entered into a computer file after each quiz, so that all ten quiz grades do not have to be entered at the same time at the terminal. She would also like to be able to save the results of the program in a computer file so that she has a permanent record of the final grades. Write a program that reads the quiz grades from a file that has been created using the computer's editor and that writes the report into a file in the computer's long-term memory.

Algorithm development

The program developed in the last chapter will be used to determine the averages of the quizzes. The only difference is that the grades will be read from a file stored in the computer's long-term memory and the results will be saved in a file in long-term memory. So the algorithm can be written as follows:

ALGORITHM 10.1 First iteration for the solution to Case Study 15

begin algorithm
1. Read the quiz scores from a computer file in long-term memory
2. Use the algorithm from Chapter 9 to compute the averages
3. Write the report on the averages to a file in long-term memory rather than to the printer or the user's terminal

end algorithm

By the end of the chapter, you will be able to write the procedures for Steps 1 and 3.

10.1 Declaring textfiles

A **file** is a Pascal data structure that allows you to manipulate and store large amounts of data simply and efficiently. In this chapter, only one type of **file**, the **textfile**, will be discussed. The more general structure of the **file** will be covered in Chapter 14.

A **textfile** is a structured data type that consists of a sequence of elements all of the type **char**. Any computer file that you create using an editor can be thought of as a **textfile** because, when you type, you are entering a sequence of characters into the computer file. All types of data, including numbers, can be represented using characters.

You can also create **textfiles** with a Pascal program. All of the output from a program can be represented by character data, so it too can be stored in a **textfile**. The output from a **write** statement can be directed to the computer's long-term memory rather than to the terminal or the printer. You can store the output of a program in a computer file so that you can refer to it later or so that it can be used as the input for another program.

The word *file* is used to refer both to the Pascal data structure **file** and to a file stored in the computer's memory. To avoid confusion, we will always refer to the Pascal data structure as a **file** or a **textfile**; a file that is stored in the computer's long-term memory will be referred to as a computer file and will not be printed in bold face.

Text is a standard, predefined data type in Pascal like **real** or **integer**. Files declared to be of type **text** (that is, **textfile**) have special attributes that the **file** types you will learn in Chapter 14 do not have. Like any other Pascal variable, a **textfile** variable must be declared, as in the statement.

```
var    newfile : text;
```

The elements of a **textfile** variable are characters, and the characters are organized into lines. Thus the structure of a **textfile** variable is the same as the structure of a computer file that you create using an editor.

Suppose that you created the following computer file using the editor for your system:

```
g   1   12.5
Hello    202%
```

If this computer file is assigned to the **textfile** variable **newfile**,* you can visualize the **newfile** variable as having the structure

```
|g|  |1|  |1|2|.|5|                                        <eoln>
|H|e|l|l|o|  |  |2|0|2|%|                                  <eoln>
<eof>
```

The symbols ⟨eoln⟩ and ⟨eof⟩ stand for the end-of-line and end-of-file markers. When you create a computer file using your system's editor, an end-of-line marker is automatically placed at the end of the line when you press the carriage return. Depending on how computer files are created on the system you are using, the end-of-line marker may be placed where the carriage return is entered or in a particular column. An end-of-file marker is automatically placed after the last line in the computer file. You cannot see the markers in your file, but their presence allows you to test whether you are at the end of a line or at the end of a computer file when you access a computer file as a **textfile.** Within the text, the presence of an end-of-line marker is indicated by ⟨eoln⟩ and an end-of-file marker is indicated by ⟨eof⟩.

In the example above, the length of the line in the computer file is fixed at 80 characters, so the ⟨eoln⟩ marker is placed in column 80. If the computer file had been created with a variable line length, the ⟨eoln⟩ marker would have been placed after the last character in each line.

The structure of a **textfile** variable is similar to the structure of a variable that is an **array** of characters. Two major differences are that the length of a **textfile** variable does not have to be declared and that the elements of a **textfile**

---

*__File__ identifiers, unlike other other user-declared identifiers, will be written all in lowercase letters. This one exception to the convention that we have used until now will make input/output statements for files easier to comprehend.

variable can only be accessed sequentially. When you use an **array** variable, you can access any element of the array simply by giving the index number. In a **textfile** you must begin with the first element of the **textfile** and go through the elements of the **textfile** in the order in which they were entered.

To use a **textfile** variable in a Pascal program, you must declare the **textfile** in your program header. In writing program headers, you have been using the format

```
program ProgramName (input, output);
```

The identifiers **input** and **output** are included because these are the **files** that you have been using to read from the terminal and write to the terminal. Standard Pascal recognizes these two **files**, and they do not need to be declared in your programs. Other **files** must be declared in the program header and defined within the program. The complete syntax diagram for a program header is

where each *identifier* within the parentheses is the name of a Pascal **file**. Normally, the standard file identifiers, **input** and **output**, are the first identifiers in the list.

Suppose you want to declare a **textfile** called **infile** that will contain the data on scores for all the students on the tests, and suppose you want to save the output in a **textfile** called **report**. The program header would look like

```
program FileScores (input, output, infile, report);
```

Putting **infile** and **report** in the header allows **files** called **infile** and **report** to access data outside the program. **Infile** and **report** must also be declared in the **var** statement of the program:

```
var   infile : text;     (* file of student scores -- input file. *)
      report : text;     (* file for grade report -- output file. *)
```

10.2 Reading from textfiles

When reading from the terminal, the Pascal **file** called **input** is assigned to your computer terminal. Any data read from the terminal is automatically read from **input,** the default file for the **read** statement. Assume that you have

10.2 Reading from textfiles

used the editor to create a computer file containing the grades that were entered for Program 9.1. Also assume that the computer file has been assigned to the Pascal **textfile** named **infile.** After the computer file is assigned to **infile** and the **read** procedure is directed to read data from **infile,** the data is retrieved from the computer file. This process is not very different from reading data from the terminal. Exactly how the assignment of a Pascal **file** to a computer file is achieved is covered in the next section. But once the assignment has been made, the Pascal **textfile** and the computer file are essentially the same thing. You can think of the Pascal **textfile** variable as an image of the computer file.

To visualize the process of reading from a **textfile,** imagine a pointer moving through the **textfile** as data is read. When your program begins execution, the pointer must be set at the beginning of the **textfile.** When the **read** procedure is executed, it converts the character data in the **textfile** into the data type appropriate for the variable specified in the read-argument list. The value is assigned to the current read-argument variable, and then the pointer moves to the next position in the **textfile.** Thus the **read** procedure advances the read pointer sequentially through the **textfile.**

Before reading from any **file** except **input,** you must move the read pointer to the beginning of the **file** by executing the standard procedure **reset:**

```
reset (infile)
```

You can visualize the **reset** procedure as moving the read pointer so that it points to the first character in the **textfile:**

```
|9|9|  |9|3|  |8|8|  |9|5|  |9|5|  |7|8|                    <eoln>
 ↑
|7|5|  |8|0|  |8|3|  |7|0|  |7|7|  |8|0|                    <eoln>
```

If you do not **reset** the **textfile,** you will get an error message the first time you try to read from the **textfile.** The error message will state that the **reset** procedure must be executed before the **read** statement. If you execute the **reset** procedure after you have started to read, the read pointer will be moved back to the beginning of the **textfile.**

To read from a **textfile,** you must specify the **textfile** name in the first field in the **read** statement. The complete syntax for a **read** statement is

The statement

```
read (infile, OneGrade)
```

reads a value into a variable called **OneGrade** from the **textfile, infile.**

If you do not specify a **file** name, the program will request data from **input** because that is the default device for the **read** statement. That is, the two statements

```
read (RealVar);
read (input, RealVar)
```

are equivalent.

When you use an editor to create a computer file that is to be read by a Pascal program, you should keep several things in mind. One is that all the rules for entering data from the terminal apply to entering data in a file: For real and integer data, each value must be separated from the following value by at least one blank space. For character data, blanks should be entered only when the character ' ' is required. Remember not to use commas to separate your data.

Another point to keep in mind is that the data that you enter must be in the right order so that the correct values are assigned to variables. When you are entering information into a file, there are no prompts to tell you what data is required next by the program. You must put the right values in the correct locations in the file. You must also be sure that the values are of the same data type as the variables in the read-argument list.

DEBUGGING HINT

Give your file identifiers descriptive names. Depending on the variable declarations, the statement

```
read (Data, FirstValue)
```

might read two values from **input** and assign the values to variables called **Data** and **FirstValue,** or it might read a value from a **file** variable called **Data** and assign it to a variable **FirstValue.** If the purpose of the statement is to read from a **file** variable, the **file** variable should be given a more descriptive identifier such as **infile** or **indata.** We have written **file** identifiers all in lower case letters so that you will not confuse them with other variables. The purpose of the statement

```
read (datafile, FirstValue)
```

is clearer than the purpose of the previous statement.

10.2 Reading from textfiles

For the **TestScores** program to read the scores from a computer file, the body of the procedure **ReadGrades** must be changed to

```
reset (infile);                          (* move the read pointer to *)
                                         (* the beginning of infile. *)

for Student := 1 to NumStudents do
  for Test := 1 to NumTests do
    read (infile, Scores [Student, Test])   (* read the score of each *)
                                            (* student on each test from *)
                                            (* infile. *)
```

In the fragment above, the **reset** procedure positions the read pointer at the beginning of **infile**. The first time the **read** statement is issued, it reads characters from the **textfile** variable **infile** until the first blank character is encountered; the characters '9' and '9' are converted to the integer number 99, and then the number 99 is assigned to the integer variable **Scores[1,1]**. After one score has been read, the pointer is positioned at the next blank character:

|9|9| |9|3| |8|8| |9|5| |9|5| |7|8| ⟨eoln⟩
 ↑
|7|5| |8|0| |8|3|7| |7|0| |7|7| |8|0| ⟨eoln⟩

Remember that a **textfile** is *sequential*. Sequential data is processed in the order in which it is given. You cannot jump around. The pointer only moves forward.

The following fragment is logically incorrect. It will compile and execute without any errors, but the results will be nonsense:

```
for Student := 1 to NumStudents do
  for Test := 1 to NumTests do
    begin
      reset (infile);
      read (infile, Scores [Student, Test])
    end
```

REMINDER

Do not **reset** a file more than once unless you are certain that you want to start reading the data from the beginning of the file again.

The problem in this fragment is that the **textfile, infile,** is **reset** every time the loop is executed. Each element of the **array** variable **Scores** will have the same value: the grade of the first student on the first test.

The format of each input value in the **textfile** must correspond to the type of the variable that it is assigned to in the **read** statement. That is, if you declare an **integer** variable, you cannot read a value of 'a' into it. Suppose a **textfile,** created using the editor, consisted of the following line:

b 1.0 2

To assign these input values to a character, real, and integer variable respectively, you could use the following program:

PROGRAM 10.1 ReadData

```
program ReadData (input, output, newfile);
(*      Program ReadData reads three values from newfile, which is a    *)
(*      textfile, and echo prints the values at the terminal.           *)

var     newfile : text;         (* textfile that stores the values to be read. *)
        CharVar : char;         (* character variable read from newfile. *)
        RealVar : real;         (* real variable read from newfile. *)
        IntVar : integer;       (* integer variable read from newfile. *)

begin (* main program *)
  reset (newfile);

  read (newfile, CharVar, RealVar, IntVar);

  writeln (' The character is ', CharVar, ', the real number is',
           RealVar:5:1, ', and the integer is', IntVar:4)
end. (* program ReadData *)
```

```
=>
 The character is b, the real number is  1.0, and the integer is   2
```

So far, reading from **textfiles** has not seemed very different from reading from the terminal. However, there is a major difference: Input from a **textfile** can be on different lines. The **readln** statement can be used to indicate to the computer that the read pointer should move to the next line in the input **textfile** after all the variables in the argument list of the statement have been assigned values.

Assume that **voltdata** has been declared as a **textfile** and assigned the values in a computer file that contains the following data:

1 2 3
4 5
6

The following fragment uses the **read** statement to read the data stored in **voltdata:**

```
reset (voltdata);

read (voltdata, Amps, Ohms);
read (voltdata, Farads);
read (voltdata, Henrys);

writeln (' Amps =', Amps:2, ' Ohms =', Ohms:2);
writeln (' Farads =', Farads:2, ' Henrys =', Henrys:2)
```
=>
```
 Amps = 1  Ohms = 2
 Farads = 3  Henrys = 4
```

The **reset** statement at the beginning of the fragment puts the read pointer at the beginning of **voltdata**. The first **read** statement assigns the value 1 to the variable **Amps** and the value 2 to the variable **Ohms**. When the first **read** statement has completed execution, the read pointer is positioned at the next blank character. When the second **read** statement is executed, the value 3 is assigned to the variable **Farads**. Now the read pointer is positioned at the next blank character—that is, immediately after the 3. When the last **read** statement is executed, the read pointer advances to the next nonblank character—that is, the number 4 on the next line, and the value 4 is assigned to the variable **Henrys**. After the last **read** statement has been executed, the read pointer is positioned immediately after the number 4. If another **read** statement is executed, the pointer will move forward from its current location. If no other **read** statement is executed, the program will never process the data remaining in the file.

In the next fragment, the same data

```
        1 2 3
        4 5
        6
```

is read with **readln** statements:

```
reset (voltdata);

readln (voltdata, Amps, Ohms);
readln (voltdata, Farads);
readln (voltdata, Henrys);

writeln (' Amps =', Amps:2, ' Ohms =', Ohms:2);
writeln (' Farads =', Farads:2, ' Henrys =', Henrys:2)
```
=>
```
 Amps = 1  Ohms = 2
 Farads = 4  Henrys = 6
```

Again, the **reset** statement moves the read pointer to the beginning of **voltdata.** The first **readln** statement assigns the value 1 to the variable **Amps** and the value 2 to **Ohms,** and then moves the read pointer to the beginning of the next line. The pointer is now positioned at the number 4. The number 3 was skipped. When the second **readln** statement is executed, the value 4 is assigned to **Farads,** and again the read pointer is moved to the beginning of the next line. The number 5 is skipped. When the last **readln** is executed, the value 6 is assigned to the variable **Henrys,** and the read pointer is advanced to the next line. The next line is the invisible ⟨eof⟩ marker.

The markers ⟨eof⟩ and ⟨eoln⟩ are useful when reading from **textfile**s. Two standard Pascal **boolean** functions can be used to test if the end of a line or the end of a **textfile** has been reached:

- **eof**(filename)   returns a value of true when the next character to be read is the special character ⟨eof⟩. If no file is specified, the default file is **input.**
- **eoln**(filename)   returns a value of true when the next character to be read is the special character ⟨eoln⟩. If no file is specified, the default file is **input.**

The function **eoln** is particularly useful in reading character data from **textfile**s. Because blanks are valid characters, when reading character data the read pointer advances not to the next blank character but to the next character, whether or not it is a blank. For some text processing, it is important to know when the end of a line has been reached. The following fragment reads the first line of the **textfile voltdata** character by character until the end of the line is reached:

```
reset (voltdata);
NumChars := 0;

repeat
  read (voltdata, CharVar);
  write (CharVar);
  NumChars := NumChars + 1
until eoln (voltdata);

writeln;
writeln (' There were ', NumChars:1, ' characters read in the line')
```

=>
1 2 3
There were 80 characters read in the line

When this fragment has completed execution, **eoln(voltdata)** is true and **CharVar** contains the value of the last character in the line. Note that

10.2 Reading from textfiles

eoln(voltdata) is automatically reassigned the value false as soon as the read pointer is not at the end of the line.

The standard Pascal function **eof** can be used when the amount of data to be read is not known before the program begins execution. The following fragment reads data from **voltdata** until the end of the **textfile** is reached:

```
reset (voltdata);

while not eof (voltdata) do
  begin
    read (voltdata, NewData);
    write (NewData:5)
  end; (* while not end of file voltdata *)
writeln
```

```
=>
     1    2    3    4    5    6
```

eof(voltdata) has a value of false unless the read pointer is beyond the end of the **file**.

An error people commonly make when using **files** is to forget to include the **file** name in the **read, eof,** or **eoln** statement. If you forget the **file** name in the **read** statement, the computer will look for input from the terminal because the default **file** for **read** is the terminal. If you forget the **file** name in **eof** or **eoln,** you will eventually get a run-time error when you run out of data in the input **file**. The following fragment produces a run-time error because **voltdata** was omitted from **eof:**

```
while not eof do
  begin
    read (voltdata, NewData);
    write (NewData:5)
  end; (* while not end of file voltdata *)
writeln
```

```
=>
*** error ***           file "voltdata", get or read issued after end of data
```

REMINDER

Be sure to include the correct **file** name when you test for the end of a **file** or the end of a line. If you are reading from a **file** called **infile,** and you must test **eof(infile)**. A common bug is to test **eof,** which is the end of the default input **file**.

10.3 Writing textfiles

As was mentioned at the beginning of the chapter, you can also create computer files using Pascal **textfiles**. The two major uses of this feature are to save output reports so that they can be reproduced when needed and to save the results of calculations so that they can be used as the input to another program.

Declaring an output **textfile** is exactly the same as declaring an input **textfile**. The Pascal identifier must be included in the header, and the identifier must be declared as a variable of type **text**.

```
program ProgramName (input, output, report);
(*     This program writes an output file to the textfile report.        *)
var    report : text;
```

Before you write a Pascal **file,** you must set up the **file** so that data can be written to it. The **rewrite** statement is used to move the write pointer to the beginning of the **file**. The **rewrite** statement also moves the ⟨eof⟩ marker to the beginning of the **file**. The statement to **rewrite** the **file** called **report** is

```
                    rewrite (report)
```

After the **rewrite** statement has been issued, the **report file** can be visualized as

```
                         ⟨eof⟩
                           ↑
```

If you issue the **rewrite** statement after you have written to the **file,** everything that you have written will essentially be erased: the ⟨eof⟩ marker will be placed at the beginning of the **file,** and, since you cannot read beyond ⟨eof⟩, you will not be able to access your data.

For all **files** except **output,** if you do not **rewrite** the **file** before the first write statement, you will get a message similar to the one that you would get if an input **file** were not **reset.**

Once the **textfile** has been rewritten, you can write to it just as you would to the terminal, except that you must specify the **file** to be written to. The complete syntax diagram for a **write** statement is

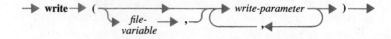

And the complete syntax diagram for a **writeln** statement is

10.4 Assigning disk files to textfiles

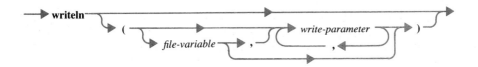

where *write-parameter* is

For example, the following fragment writes a blank line and then the word "Test" in the **textfile** called **report**:

```
writeln (report);
writeln (report, 'Test')
```

Note that if you want a blank line in your output **file,** you must specify the **file** where the blank line is to be written. If you are using a **textfile** for output and the only output at your terminal is a few carriage returns, you will know that you have forgotten to include the **file** name in some of your **writeln** statements.

10.4 Assigning disk files to textfiles

If your computer data file is stored on a floppy disk or a hard disk, how do you tell the Pascal program which computer file to read from when it reads a **textfile**? A Pascal **file** exists only within the Pascal program. In order for data to be read from a computer file, you must include a system command in your program that tells the operating system of your computer which computer file you want to use and what it will be called inside the Pascal program. The syntax for this statement will be different for every operating system. For some systems, all that is required is a statement within the program, such as

```
assign (report, 'output.dat');
```

or

```
open (report, output.dat);
```

where **report** is the Pascal **textfile** variable and output.dat is the name of the computer file. Some systems also require that an output file be closed for it to be saved in long-term memory. Other systems require that a system command be issued to create an output file before the program is executed. You will have to consult the documentation for your system, your instructor, or someone in your computer center in order to find out what commands are required for your system.

On any system, a computer file associated with an input **textfile** *must* exist before the program is executed; otherwise data cannot be read from it. A computer file that is associated with an output **textfile** may or may not exist before a program begins execution, but any information in a previously existing computer file assigned to an output **textfile** will be erased when the program begins execution. The computer file in which the output from a program is stored can then be assigned as an input **file** for another program. (See Exercise 5.)

Solution to Case Study 15

Program 10.2 is the same as Program 9.1, **TestScores,** except that the input is read from a **file** and the output is written to a **file.** When the program is executed at a terminal, the only response at the terminal is the statement that the output is in the computer file output.dat. Whenever you send the output of a program to a computer file, it is a good idea to include a statement that tells the user what file the output is in.

PROGRAM 10.2   ReadWriteGrades

```
program ReadWriteGrades (input, output, infile, report);
(*      This program reads the scores of 12 students on 6 tests from      *)
(*      the Pascal textfile infile.  The program computes the average     *)
(*      grade for each student and the average on each test.  In          *)
(*      addition, it computes the average for each type of test:          *)
(*      multiple choice, short answer, and program writing.  The output   *)
(*      is written to the textfile report.                                *)



var   infile : text;           (* input textfile for the scores. *)
      report : text;           (* output textfile for the averages. *)




(*$i testaverage *)
(*$i studentaverage *)
(*$i computetesttypeaverages *)

   procedure ReadGrades (var Grades : ScoreMatrix; Num, NumTests : integer);
      (*      Procedure ReadGrades reads the student scores on each test       *)
      (*      from the textfile infile.                                        *)
      var   Student, Test : integer;      (* for statement control variables. *)
```

```
  begin
    for Student := 1 to Num do
        for Test := 1 to NumTests do
          read (infile, Grades [Student, Test])
  end; (* procedure ReadGrades *)
  procedure ReportGrades (Grades : ScoreMatrix; SAvg : StudentArray;
                          TAvg : TestArray; TTypeAvg : TestTypeArray);
  (*      Procedure ReportGrades reports the scores for each student    *)
  (*      on each test and reports the test averages and the student    *)
  (*      averages.  The output is written to the textfile report.      *)
  var   Student, Test : integer;      (* for statement control variables. *)
  begin
    writeln (report);
    writeln (report, ' Student', 'Test Number':24, 'Student Average':29);
    writeln (report);
    write (report, ' ':8);

    for Test := 1 to NumTests do write (report, Test:6);
    writeln (report);

    for Student := 1 to NumStudents do

      begin
        write (report, Student:4, '    ');
        for Test := 1 to NumTests do
          write (report, Grades [Student, Test]:6);
        writeln (report, SAvg [Student]:8:1)
      end;   (* for Student *)

    writeln (report);
    writeln (report, ' Test');
    write (report, ' Average ');

    for Test := 1 to NumTests do
      write (report,  TAvg [Test]:6:1);

    writeln (report);
    writeln (report);
    writeln (report, ' The average on the multiple-choice tests was':45,
                     TTypeAvg [1]:8:1);
    writeln (report, ' The average on the short-answer tests was':45,
                     TTypeAvg [2]:8:1);
    writeln (report, ' The average on the program-writing tests was':45,
                     TTypeAvg [3]:8:1)
  end; (* procedure ReportGrades *)

(*-----------------------------------------------------------------------*)

begin (* main program *)
  reset (infile);
  rewrite (report);
```

```
    ReadGrades (AllScores, NumStudents, NumTests);
              .
              .
              .
    ReportGrades (AllScores, StudentAvg, TestAvg, TestTypeAvg);
    writeln (' The output report is in the file output.dat')
end. (* program ReadWriteGrades *)

=>
The output report is in the file output.dat
```

Before the program **ReadWriteGrades** was executed, the following data was entered in the computer file assigned to **infile**:

```
99 93 88 95 78 95
75 80 83 70 77 80
92 95 85 88 92 89
50 62 50 65 70 50
76 70 60 78 77 55
81 85 78 87 82 70
72 74 90 73 81 92
78 70 50 67 65 50
85 87 80 89 85 78
67 80 65 70 75 67
77 78 70 72 78 68
80 85 82 78 87 83
```

The following is a listing of the computer file assigned to **report** after the program **ReadWriteGrades** was run:

Student	Test Number						Student Average
	1	2	3	4	5	6	
1	99	93	88	95	78	95	91.3
2	75	80	83	70	77	80	77.5
3	92	95	85	88	92	89	90.2
4	50	62	50	65	70	50	57.8
5	76	70	60	78	77	55	69.3
6	81	85	78	87	82	70	80.5
7	72	74	90	73	81	92	80.3
8	78	70	50	67	65	50	63.3
9	85	87	80	89	85	78	84.0
10	67	80	65	70	75	67	70.7
11	77	78	70	72	78	68	73.8
12	80	85	82	78	87	83	82.5
Test Average	77.7	79.9	73.4	77.7	78.9	73.1	

```
The average on the multiple-choice tests was      77.7
   The average on the short-answer tests was      79.4
The average on the program-writing tests was      73.2
```

Remember that the system-specific commands that associate the computer data files with the Pascal **texfiles, infile** and **report,** must be included in the program **ReadWriteGrades.**

REMINDERS AND WARNINGS ON FILES

1. If you get an error message that a **read** was issued after the end of the data,

    (a) be sure every **read** statement has the correct **file** variable as the first argument;

    (b) be sure every test of **eoln** and **eof** has the correct **file** variable;

    (c) be sure the computer file exists;

    (d) be sure the computer file has the correct amount of data in it;

    (e) be sure the system file command assigns the correct computer file to the Pascal **file.**

2. If you get an error message that the input data is not of the appropriate type, check the computer file against the program to be sure that the data is in the correct order. Also double check to be sure that **read** and **readln** are used correctly.

3. If you get a message to the effect that the program cannot open or close a computer file, check to be sure that the system file command was issued and that it was issued with the correct computer file names.

4. If you get an error message stating that a **file** must be reset or rewritten, check to be sure that you included the **reset** or **rewrite** command and that you included the correct **file** name.

5. Whenever you write an output **file** from a Pascal program, double check the name of the computer file that you are writing to. Be sure that it is not the name of a computer file containing information that you want to save, because as soon as you issue the **rewrite** command, the data in the computer file is erased.

APPLICATION
Introduction to Probability

This section introduces a branch of mathematics that is important in the experimental and applied sciences. Because all experimental measurements are subject to random errors, an engineer or scientist never takes just one measurement of the outcome of an experiment. Experiments are repeated many times. Then, through use of *statistics,* the experimental value can be estimated within a stated degree of accuracy.

The statistical analysis of scientific experimentation is based on *probability* theory. The study of probability involves the analysis of chance situations through mathematics. Probability theory can be used to make predictions about the likelihood that an event will occur, and to determine the expected outcome of an experiment. By comparing the actual outcome of an experiment with the expected outcome, the likelihood that the difference is due to chance alone or to some other factor can be computed.

Probability theory and some common probability functions will be discussed first. Then the use of this theory in statistical analysis will be covered.

CASE STUDY 16
Arrival of phone calls

Phone calls arrive at a switchboard in a random fashion at an average rate of 50 calls per hour. In order to plan for new equipment, the phone company wants to know the probability that fewer than 5 calls will arrive within a ten-minute period. They would also like to know the probability that more than 5 and fewer than 10 calls will arrive within a ten-minute period and the probability that more than 10 calls will arrive within a ten-minute period. Write a program that computes these values.

Algorithm development

To solve the problem in the case study, you need to know how the average number of calls per hour can be used to compute the probability that a given number of calls will arrive within a time interval. The mathematics needed to compute these probabilities from the expected number of calls will be developed in this section. Knowing that the probabilities can be derived from the average, you can write the following algorithm:

ALGORITHM 10.2 First iteration for Case Study 16

begin algorithm
1. Read the average number of calls per hour
2. Compute the probabilities
   a. that there are fewer than 5 calls in 10 minutes
   b. that there are between 5 and 10 calls in 10 minutes
   c. that there are more than 10 calls in 10 minutes
3. Report the probabilities

end algorithm

Random variables*

You are familiar with many random processes. For example, the outcome of the roll of a die, the time until a light bulb burns out, and the number of cars that are stopped at a red light can all be described using probability theory.

A *random variable* is a numerical quantity that is determined from the outcome of an experiment. It is usually denoted by X. The nature of the experiment determines the quantity chosen to be the random variable. If a die is to be rolled, the random variable, X, might be the number appearing on the die:

$X = 1$ if a 1 is rolled
$X = 2$ if a 2 is rolled
$X = 3$ if a 3 is rolled
$X = 4$ if a 4 is rolled
$X = 5$ if a 5 is rolled
$X = 6$ if a 6 is rolled

If you are interested only in whether the number appearing on the die is odd or even, the number 0 might be assigned to X if the value is even and the number 1 might be assigned to X if the value is odd:

$X = 0$ if a 2, 4, or 6 is rolled
$X = 1$ if a 1, 3, or 5 is rolled

---

*From Finger, *Pascal Programming for Engineers Using VPS*. Copyright © 1983 by Kendall/Hunt Publishing Company.

Notice that in the examples X is assigned a value for all possible outcomes of the experiment.

Also notice that the outcomes of the experiment are *mutually exclusive*. In the first experiment, if one die is rolled once, it is not possible for the events $X = 1$ and $X = 2$ to both occur, so $X = 1$ and $X = 2$ are mutually exclusive events. An example of two events that are *not* mutually exclusive is the event that X is odd and the event that X is greater than 4. X is odd when X equals 1, 3, or 5. X is greater than 4 when X equals 5 or 6. Since $X = 5$ is in both events, the events are not mutually exclusive.

Another important property that events can have is *independence*. Events are said to be independent if the knowledge that one event occurred gives you no information about whether the other event occurred. A precise mathematical statement of independence is beyond the scope of this text. An example of independent events is the outcomes of successive rolls of a die. The outcome of the first roll of the die does not affect the outcome of the second roll of the die. An example of dependent events is the outcomes of drawing successive cards from a deck without replacement. The outcome of each draw depends on which cards have been drawn previously from the deck.

Distributions of random variables

The *probability mass function* or *distribution* of a random variable gives the probability that the random variable will assume a certain value. For example, you might want to determine the probability that the number appearing on a die will be 2. This probability is denoted $P(X = 2)$ and is read as "the probability that X equals 2." When a die is rolled, there are six possible outcomes. If the die is fair, each outcome is equally likely, or each has a probability of occurrence of 1/6. So

$$P(X = 1) = 1/6$$
$$P(X = 2) = 1/6$$
$$P(X = 3) = 1/6$$
$$P(X = 4) = 1/6$$
$$P(X = 5) = 1/6$$
$$P(X = 6) = 1/6$$

Consider the random variable that represents the sum of the faces of two dice. There is only one way that the random variable can assume the value 2 (if each die has one dot appearing); however, there are six possible outcomes for the value 7. There are a total of 36 possible outcomes; $X = 2$ occurs in 1/36 of the outcomes and $X = 7$ occurs in 6/36 of the outcomes. Table 10.1 and

TABLE 10.1
Tabulation of the probability distribution for the sum of the numbers on two dice

x	2	3	4	5	6	7	8	9	10	11	12
$P(X = x)$	$\frac{1}{36}$	$\frac{2}{36}$	$\frac{3}{36}$	$\frac{4}{36}$	$\frac{5}{36}$	$\frac{6}{36}$	$\frac{5}{36}$	$\frac{4}{36}$	$\frac{3}{36}$	$\frac{2}{36}$	$\frac{1}{36}$

Figure 10.1 show the mass function of a discrete random variable X representing the sum of the numbers on a pair of dice. Lowercase x represents the possible outcomes of the sum X. The equation $P(X = 3) = 2/36$ is read as "The probability that X equals 3 is two-thirty-sixths."

Suppose you were asked to write a program that printed Table 10.1. The sums of all possible combinations of two dice can be computed using two nested **for** statements:

```
for Die1 := 1 to 6 do
    for Die2 := 1 to 6 do
        SumOfFaces := Die1 + Die2
```

The problem, however, is how to count how many times each sum occurs. One solution is to set up an **array** variable that uses the sum of the faces as its index. That is, the array looks just like Table 10.1: **X** is the index, and the value stored in element **NumberOfTimes[X]** is the number of times that the sum x has occurred.

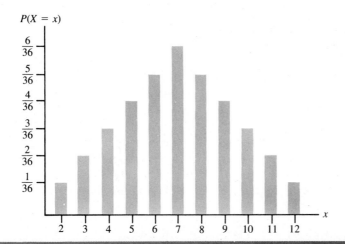

FIGURE 10.1
Graph of the probability distribution for the sum of the numbers on two dice.

```
type  NumberArray = array[1..12] of integer;
var   NumberOfTimes : NumberArray;
```

If the elements of the array **NumberOfTimes** are initialized to 0, the array can be used to keep a running total of the number of times that each value of **SumOfFaces** (that is, x) occurs.

```
for Die1 := 1 to 6 do
  for Die2 := 1 to 6 do
    begin
      SumOfFaces := Die1 + Die2;
      NumberOfTimes [SumOfFaces] := NumberOfTimes [SumOfFaces] + 1
    end
```

In Exercise 6, you are asked to use the fragments above to write a program that prints a copy of Table 10.1.

Properties of distribution functions of random variables

Two of the basic properties of probability mass functions are

$$\sum_{i=0}^{\infty} P(x_i) = 1 \tag{10.1}$$

and

$$0 \le P(x_i) \le 1 \quad \text{for all } i \tag{10.2}$$

The first property states that the sum of the probabilities of all possible outcomes of an experiment is equal to 1. Notice that in each of the previous examples the sum of the probabilities is 1.0. The die must show a 1, 2, 3, 4, 5, or 6; no other outcomes can occur. A probability of 0 is implicitly assigned to the outcome in which the die balances on a corner or an edge. The impossible outcome is not part of the *state space*. The state space contains all possible outcomes of an experiment.

The second property states that the probability of the occurrence of any specific value of x lies between 0 and 1. This property can be easily derived by considering that each probability must be greater than 0 and that the sum of all probabilities is 1; thus the probability of any specific value must be less than or equal to 1.

If two events are mutually exclusive, their probabilities can be added; that is, if

$$P(X = 1 \text{ and } X = 2) = 0$$

then

$$P(X = 1 \text{ or } X = 2) = P(X = 1) + P(X = 2) \quad (10.3)$$

When two events are independent, the probability that both events will occur is the product of their probabilities. If X and Y are independent, then

$$P(X \text{ and } Y) = P(X)P(Y) \quad (10.4)$$

Distributions can be described in terms of other basic properties. The mean, median, and mode are ways of measuring the average behavior of a random variable. The variance and standard deviation measure the dispersion of the distribution about the center. If the values of x are clustered near the center, the variance will be small. If the values of x are widely dispersed, the variance will be larger.

The *mean* of a mass distribution is the average of the values of the outcomes weighted by their probability of occurrence:

$$\mu = \sum_{i=0}^{\infty} x_i P(x_i) \quad (10.5)$$

The mean, or expected value, is usually represented by the Greek letter mu, μ. In the example involving two dice, the mean of the distribution of the sum of the dice is 7.

The function **Mean** computes the mean of a probability distribution stored in an **array** variable that is of the data type **ProbArray**. The arguments **MinX** and **MaxX** are used so that the function can be used for a range of x other than the one from 2 to 12.

```
function Mean (Probability : ProbArray; MinX, MaxX : integer) : real;
(*      Function Mean returns the mean of the probability              *)
(*      distribution where x ranges from MinX to MaxX.                  *)
var    Temp : real;         (* temporary variable to store the functional *)
                            (* value during computation. *)
       X : integer;         (* for-statement control variable -- equal to *)
                            (* the values of x, the random variable. *)
```

```
begin
  Temp := 0.0;              (* initialize the running total to zero. *)

  for X := MinX to MaxX do
    Temp := Temp + X * Probability[X];   (* add the value of X * P(X = x) *)
                                         (* to the running total. *)
  Mean := Temp
end; (* function Mean *)
```

How is the function **Mean** different from the functions developed in Chapter 9 to compute averages?

The *variance*, σ^2, or *standard deviation*, σ, is used to measure the variability of the random variable from its mean. The variance is an average of the squared differences between the mean and the values of the random variable, weighted by the probability. The variance can be thought of as a measure of the average distance of all outcomes from the expected outcome:

$$\sigma^2 = \sum_{i=0}^{\infty} (x_i - \mu)^2 P(x_i) \tag{10.6}$$

In the dice example, $\sigma^2 = 5.8$ and $\sigma = 2.4$. Because the units of the variance are the units of the random variable squared, the standard deviation is usually used. If there are units attached to the random variable, the standard deviation is in the same units as the original values.

The function **Variance** computes the variance of the distribution **Probability**. Note that one of the parameters is the mean of the distribution, which must be known before the variance can be computed using Equation 10.6.

```
function Variance (Probability : ProbArray; MinX, MaxX : integer;
                   MeanX : real) : real;
(*    Function Variance returns the variance of the probability       *)
(*    distribution where x ranges from MinX to MaxX.  The mean of     *)
(*    the distribution must have been previously computed and         *)
(*    passed through the parameter MeanX.                             *)

var   Temp : real;         (* temporary variable for the functional value. *)
      X : integer;         (* for-statement control variable -- equal to *)
                           (* the values of x, the random variable. *)
```

```
begin
  Temp := 0.0;            (* initialize the running total to zero. *)

  (* The following for statement adds the value of P(X) * (X - MeanX) ** 2 *)
  (* to the running total. *)

  for X := MinX to MaxX do
    Temp := Temp + sqr (X - MeanX) * Probability[X];

  Variance := Temp
end; (* function Variance *)
```

Realization of a random variable

The probability distribution is an idealization used to describe a random process. However, when an experiment is actually performed, there is only one outcome. So, for example, before you flip a coin, you can say that the probability of a head is 1/2 and the probability of a tail is 1/2. After the coin has been tossed, you can say with certainty what the outcome was. Tossing the coin and recording the result is the *realization* of the random process. When you toss a coin exactly twice, you do not expect always to get 1 head and 1 tail. (If you could count on always getting 1 head and 1 tail, the process would not be random.) However, if the coin were tossed 1000 times, you would expect the results to be close to 1/2 heads and 1/2 tails. The more times the coin is tossed, the closer you expect the results to be to the random distribution. Because these are random events, some deviation from the underlying distribution function is to be expected.

Discrete distributions

Several patterns of distributions of random variables occur frequently in engineering and science. Three discrete distributions (Bernoulli, binomial, and Poisson) and two continuous distributions (uniform and Gaussian) are described below. Several others are used in the homework problems at the end of the chapter.

There are two types of random variables: *discrete* and *continuous*. The examples so far have been of discrete random variables. Discrete random variables take specific values within a range of values. The sum of numbers appearing on two dice is an example of a discrete random variable. The variable can take values between 2 and 12, but cannot take the value 3.5. A continuous random variable can take any value within a range. Measurements of the heights or weights of people would be continuous variables.

Bernoulli distribution

James Bernoulli (1654–1705) and Blaise Pascal laid the foundations of probability theory. By using mathematics to study games of chance, these mathematicians developed the idea that a probability could be associated with the outcome of an experiment. From this idea, or model of the physical world, has developed the field of probability and statistics.

In a Bernoulli experiment, there are two and only two possible outcomes. The outcomes are usually expressed in terms of success or failure. Tossing a coin is a Bernoulli experiment. One side can be designated a success, the other a failure. Many other types of experiments have two outcomes—a student passes or fails a course, a basketball goes into the basket or does not, a randomly selected person is female or male, and so on. In each case, either outcome can be designated as a success and the other a failure. The probability of a success is designated p and that of failure is designated q. If the random variable X is set to 0 if there is a failure and to 1 if there is a success, the *Bernoulli distribution* has the following properties:

$$P(X = 1) = p$$
$$P(X = 0) = q$$

and (10.7)

$$p + q = 1$$

In a coin toss, if heads is a success and tails a failure, then

$$P(X = 0) = 0.5$$
$$P(X = 1) = 0.5$$

Heads and tails are equally likely, and one or the other must occur.

For the Bernoulli distribution,

$$\mu = p \quad \text{and} \quad \sigma^2 = pq \quad (10.8)$$

Program 10.3, **Bernoulli,** computes the mean and the variance of a Bernoulli distribution using the functions **Mean** and **Variance.**

PROGRAM 10.3 Bernoulli

```
program Bernoulli (input, output);
(*      This program computes the mean and the variance of a Bernoulli  *)
(*      random variable.                                                *)

const Failure = 0;          (* if the outcome of a trial is a failure, the *)
                            (* random variable x is assigned a value of zero. *)
      Success = 1;          (* if the outcome of a trial is a success, the *)
                            (* random variable x is assigned a value of one. *)

type  ProbArray = array [Failure..Success] of real;

var   Prob : ProbArray;     (* array to store the probability density *)
                            (* function for the Bernoulli distribution. *)
      MeanB : real;         (* mean of the Bernoulli distribution -- *)
                            (* computed. *)
      VarB : real;          (* variance of the Bernoulli distribution -- *)
                            (* computed. *)
(*$i mean *)
(*$i variance *)

(*-------------------------------------------------------------------------*)

begin
   writeln (' This program computes the mean and variance of a Bernoulli',
            ' random variable');
   writeln (' What is the probability of success?');
   readln (Prob [Success]);                 (* read probability of suc- *)
                                            (* cess, p. *)

   Prob [Failure] := 1 - Prob [Success];    (* compute probability of *)
                                            (* failure, q; for the *)
                                            (* Bernoulli distribution, *)
                                            (* p + q = 1. *)
   MeanB := Mean (Prob, Failure, Success);  (* compute the mean over the *)
                                            (* range of X from Failure to *)
                                            (* Success (0 to 1). *)
   VarB := Variance (Prob, Failure,         (* compute the variance over *)
                     Success, MeanB);       (* the range of X from Failure *)
                                            (* to Success (0 to 1) given *)
                                            (* the mean, MeanB. *)
   writeln;
   writeln (' The mean is ', MeanB:10:2, ' and the variance is', VarB:10:2)
end. (* program Bernoulli *)
```

```
=>
This program computes the mean and variance of a Bernoulli random variable
What is the probability of success?
□0.75

The mean is       0.75 and the variance is       0.19
```

You can verify that the values for the mean and variance using the functions **Mean** and **Variance,** which are based on Equations 10.5 and 10.6, are the same values as those that result from Equation 10.8.

Binomial distribution

In a *binomial distribution,* the random variable X represents the number of successes in a series of Bernoulli trials. For example, suppose you wanted to know the probability of getting exactly three tails in ten tosses of a coin. Since each trial is independent, you could write down the probability of getting a tail in any one toss using the Bernoulli distribution. Then you could write down all the possible outcomes of ten tosses of a coin and count the number of ways to get exactly three tails. If all your calculations were correct, you would finally arrive at the binomial distribution

$$P(X = x) = \binom{n}{x} p^x q^{n-x} \tag{10.9}$$

where n = the number of trials

p = the probability of success

q = the probability of failure

$$\binom{n}{x} = \frac{n!}{x!(n-x)!} \tag{10.10}$$

$n!$ is the factorial function:

$$n! = (n)(n-1)(n-2) \cdots 1$$

$\binom{n}{x}$ is read as "n choose x." The function **NChooseX** uses the function **Factorial** from Chapter 7 to compute the number of combinations of n things chosen x at a time

```
function NChooseX (N, X : integer) : integer;
(*      Function NChooseX computes the number of different ways that    *)
(*      N things can be arranged using X of them at a time.  The        *)
(*      function Factorial must already be defined.                     *)
begin
  NChooseX := Factorial(N) / (Factorial(X) * Factorial(N - X))
end; (* function NChooseX *)
```

How could the function **NChooseX** be made more efficient?

In a binomial distribution, the mean and variance are given by

$$\mu = np \quad \text{and} \quad \sigma^2 = npq \quad (10.11)$$

If a coin is tossed ten times, the probability of tossing exactly three heads is

$$P(X = 3) = \binom{10}{3}(0.5)^3(0.5)^7 = 0.117 \quad (10.12)$$

The expected number of heads in ten tosses is n times p, or 5. The Bernoulli distribution is a binomial distribution with $n = 1$. In Exercise 8, you are asked to write a program that computes the mean and variance of a binomial distribution using the functions **NChooseX**, **Mean**, and **Variance**.

Poisson distribution

The *Poisson distribution* is used as a model for many phenomena in which discrete events occur in a continuous interval. The interval is usually a time interval. In a Poisson distribution, the probability that the phenomena will occur is proportional to the length of the interval. The probability that two or more events will occur in a very small subinterval must be 0, and the occurrence of an event in one subinterval must be independent of an occurrence in another subinterval. Some events that are typically modeled using the Poisson process are the number of calls coming into a switchboard in a given time interval (as in Case Study 16), the number of radioactive particles that decay in a given time interval, and the number of people entering a supermarket in a given time interval.

The Poisson random variable X represents the number of events that occur in a given time interval t, and its distribution function is

$$P(X = x) = \frac{e^{-\lambda t}(\lambda t)^x}{x!} \quad (10.13)$$

where x is the number of events, t is the length of time, and the Greek letter lambda, λ, is the average number of events per unit of time. For the Poisson distribution,

$$\mu = \lambda t \quad \text{and} \quad \sigma^2 = \lambda t \qquad (10.14)$$

The program at the end of this section uses the Poisson distribution to compute the probability that a given number of phone calls will arrive in a ten-minute interval.

Continuous distributions

For a continuous random variable, it is not possible to specify the probability that a random variable will take on a particular value. All possible outcomes of a random variable and the probability of each outcome cannot be enumerated. An example of a continuous random variable is the time between the arrival of telephone calls. Because every possible interarrival time cannot be given, a range of values and a function called the *probability density function* (pdf) are specified instead.

A probability density function has the following properties:

$$f(x) \geq 0 \quad \text{for all } x$$

$$\int_{-\infty}^{\infty} f(t) \, dt = 1 \qquad (10.15)$$

$$P(a \leq X \leq b) = \int_{a}^{b} f(t) \, dt$$

Because the total probability must be equal to 1.0, and because within any specified range an infinity of values exist for a continuous variable, any particular value of the random variable has a vanishingly small probability of occurrence. Therefore, only the probability that a random variable is within a range can be found, using the third property given in Equation 10.15.

Uniform distribution

If all possible values of a random variable within an interval are equally likely to occur, the random variable X has a *uniform distribution*. For example, if you randomly pick a point on a line that has a length of 1.0 ft, any point is as likely to be chosen as any other. The probability that X, the distance of the point from the end of the line, will take a value within a subinterval is proportional to the length of the subinterval. The density

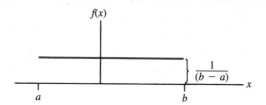

FIGURE 10.2
The uniform distribution.

function of a uniform random variable over an interval from a to b is

$$f(x) = \frac{1}{b-a} \quad \text{for } a < x < b \tag{10.16}$$

The graph of the uniform distribution in Figure 10.2 shows that the density of the function over an interval is a constant.

The mean and the variance of the uniform distribution are

$$\mu = \frac{b+a}{2} \quad \text{and} \quad \sigma^2 = \frac{(b-a)^2}{12} \tag{10.17}$$

Why can't you use the functions **Mean** and **Variance** to compute the mean and the variance of the uniform distribution?

Gaussian distribution

The *Gaussian* or *normal distribution* is used in describing large numbers of observations. The distribution is named after Karl Gauss (1777–1855), a German mathematician who made major contributions to the field of electricity and magnetism as well as to the field of probability and statistics.

Under specified conditions, the behavior of a large number of random variables can be shown to be Gaussian. Gaussian distributions are used frequently in statistics and are also used to approximate other distributions such as the binomial and Poisson distributions. The density function is a bell-shaped curve. In Gaussian distributions, most values are near the mean. Heights and weights of people, IQ test scores, SAT scores, and the amounts of time many individuals take to read the same book are random variables that can be approximated with the normal distribution.

Figure 10.3 shows three density functions of normal distributions with the same mean but different variances. The variance would be greatest for curve C, in which the random variable is widely dispersed about the mean. The variance would be smallest for curve A. In a normal curve, 68% of the area

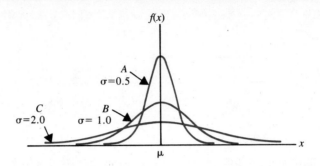

FIGURE 10.3
The Gaussian distribution with three values of σ.

under the curve is contained within $x \pm \sigma$, and 95% of the area under the curve is contained within $x \pm 2\sigma$.

The density function of a normal distribution is

$$f(x) = \frac{1}{(2\pi\sigma^2)^{1/2}} e^{-1/2[(x-\mu)/\sigma]^2} \tag{10.18}$$

where μ is the mean and σ^2 is the variance. The probability that a normally distributed random variable lies between two values can be calculated by combining Equations 10.15 and 10.18:

$$P(a < X < b) = \int_a^b \frac{1}{(2\pi\sigma^2)^{1/2}} e^{-1/2[(t-\mu)/\sigma]^2} \, dt \tag{10.19}$$

Equation 10.19 can be interpreted as follows: the probability that a random variable has a value between two numbers a and b is equal to the area

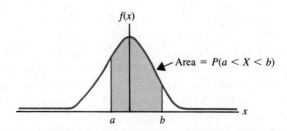

FIGURE 10.4
$P(a < X < b)$ is equal to the area under the density function between a and b.

under the curve of the probability density function between the points a and b. This interpretation is illustrated in Figure 10.4.

The integral in Equation 10.19 cannot be written in a simple form. But, because the Gaussian distribution is used often in statistics, some method for evaluating the integral quickly is required. The integral can be evaluated by means of numerical methods and the resulting values stored in tables; however, the value of the integral depends on a, b, μ, and x, so it is difficult to create tables for the Gaussian distribution in the form given in Equation 10.19.

Because of certain properties of the Gaussian distribution, any random variable X that has a Gaussian distribution with mean μ and variance σ^2 can be transformed into a *standardized normal* random variable by subtracting the mean and dividing by the standard deviation:

$$Z = \frac{X - \mu}{\sigma} \tag{10.20}$$

The standardized normal random variable Z has a mean of 0.0 and a variance of 1.0. Its distribution is given by

$$P(-\infty < Z < z) = \int_{-\infty}^{z} \frac{1}{(2\pi)^{1/2}} e^{-t^2/2} \, dt \tag{10.21}$$

The integral in Equation 10.21 is often represented by the Greek letter Φ:

$$P(Z < z) = \int_{-\infty}^{z} \frac{1}{(2\pi)^{1/2}} e^{-t^2/2} \, dt = \Phi(z) \tag{10.22}$$

From Equation 10.20, the probability that a Gaussian random variable X with mean μ and variance σ^2 is less than some value x can be written as

$$P(X < x) = P\left(\frac{X - \mu}{\sigma} < \frac{x - \mu}{\sigma}\right) = P\left(Z < \frac{x - \mu}{\sigma}\right) \tag{10.23}$$

Then, since the distribution for Z, the standardized normal variable, is tabulated as $\Phi(z)$, the probability that $X < x$ is given by

$$P(X < x) = \Phi\left(\frac{x - \mu}{\sigma}\right) \tag{10.24}$$

Table 10.2 is a short table of values for $\Phi(z)$. To see how to use this table, suppose that the heights of freshman college students are normally distributed with a mean of 165 cm and a standard deviation of 10 cm. The probability that

TABLE 10.2
Tabulation of the Gaussian distribution

$$\Phi(z) = \int_{-\infty}^{z} \frac{1}{(2\pi)^{1/2}} e^{-x^2/2} \, dx = P(Z < z)$$

z	$\Phi(z)$	z	$\Phi(z)$	z	$\Phi(z)$	z	$\Phi(z)$
0.0	0.5000	1.0	0.8413	2.0	0.9772	3.0	0.9986
0.1	0.5398	1.1	0.8643	2.1	0.9821	3.1	0.9990
0.2	0.5793	1.2	0.8849	2.2	0.9861	3.2	0.9993
0.3	0.6179	1.3	0.9032	2.3	0.9893	3.3	0.9995
0.4	0.6554	1.4	0.9192	2.4	0.9918	3.4	0.9997
0.5	0.6915	1.5	0.9332	2.5	0.9938	3.5	0.99977
0.6	0.7257	1.6	0.9452	2.6	0.9953	3.6	0.99984
0.7	0.7580	1.7	0.9554	2.7	0.9965	3.7	0.99989
0.8	0.7881	1.8	0.9641	2.8	0.9974	3.8	0.99993
0.9	0.8159	1.9	0.9713	2.9	0.9981	3.9	0.99995

a freshman is shorter than 180 cm (6 ft) is given by

$$P(X < 180) = P\left(Z < \frac{180 - \mu}{\sigma}\right) = P(Z < 1.5) = \Phi(1.5) \quad (10.25)$$

From Table 10.2, the value of $\Phi(z)$ for $z = 1.5$ is 0.9332. Thus,

$$P(X < 180) = 0.9332 \quad (10.26)$$

So, the probability that a freshman is shorter than 180 cm is 0.9332.

Solution to Case Study 16

Since you know that the arrival of phone calls can be modeled with a Poisson distribution, the missing steps in the case study can be solved.

Telephone calls arrive at an average rate of 50 per hour, so $\lambda = 50$ calls per hour. The time interval t is equal to ten minutes or one-sixth of an hour. The probability that exactly one call will arrive in the next one-sixth of an hour is given by

$$P(X = 1) = \frac{[50(1/6)]^1 e^{-[50(1/6)]}}{1!} \quad (10.27)$$

For the case study, you need to know the probability that fewer than 5 calls arrive in ten minutes. This probability is equal to the probability that 0, 1, 2, 3, or 4 calls arrive in the next one-sixth of an hour:

$$P(X < 5) = \sum_{i=0}^{4} \frac{(\lambda t)^i e^{-\lambda t}}{i!} \qquad (10.28)$$

And the probability that between 5 and 10 calls arrive is equal to the probability that 5, 6, 7, 8, 9, or 10 calls arrive:

$$P(5 \leq X \leq 10) = \sum_{i=5}^{10} \frac{(\lambda t)^i e^{-\lambda t}}{i!} \qquad (10.29)$$

Finally, the probability that more than 10 calls arrive in one-sixth of an hour is given by

$$P(X > 10) = \sum_{i=11}^{\infty} \frac{(\lambda t)^i e^{-\lambda t}}{i!} \qquad (10.30)$$

If you look at Equation 10.30 and think about how you would compute the value of $P(X > 10)$ on a computer, you can see that the infinite sum presents a problem. To circumvent the problem, you can use the fact that the events $X < 5$, $5 \leq X \leq 10$, and $X > 10$ are mutually exclusive (Equation 10.3) and that their probabilities must add up to 1.0. (Equation 10.1). Using these properties, you can find $P(X > 10)$:

$$P(X \leq 10) + P(X > 10) = 1$$

or

$$P(X > 10) = 1 - P(X \leq 10) \qquad (10.31)$$

or

$$P(X > 10) = 1 - [P(X < 5) + P(5 \leq X \leq 10)]$$

Program 10.4 computes the probability distribution for the arrival of phone calls using Equations 10.28 through 10.31, plus the function **Factorial** from Chapter 7 to compute $i!$ and the function **ZToPower** from Chapter 8 to compute $(\lambda t)^i$.

PROGRAM 10.4 PhoneCalls

```pascal
program PhoneCalls (input, output);
(*      This program computes the probability that fewer than 5,          *)
(*      between 5 and 10, and more than 10 phone calls will arrive at     *)
(*      a switchboard in a given interval.  The arrival of phone          *)
(*      calls is assumed to have a Poisson distribution.                  *)

const MinutesPerHour = 60;      (* conversion factor for the arrival rate. *)
      LowerBound = 0;           (* lowest value for the random variable. *)
      MiddleBound = 5;          (* middle value for the random variable. *)
      UpperBound = 10;          (* largest value for the random variable. *)

var   TimeInMinutes : real;           (* length of time over which random *)
                                      (* events occur. *)
      Lambda : real;                  (* average arrival rate in calls *)
                                      (* per minute. *)
      ProbFewerThanMid : real;        (* probability that there are be- *)
                                      (* tween LowerBound and MiddleBound *)
                                      (* arrivals in TimeInMinutes. *)
      ProbBetweenMidAndUp : real;     (* probability that there are more *)
                                      (* than MiddleBound and less than *)
                                      (* or equal to UpperBound arrivals. *)
      ProbMoreThanUp : real;          (* probability that there are more *)
                                      (* than UpperBound arrivals. *)

(*$i ztopower *)
(*$i factorial *)
(*$i writemessage *)


(*$i readlambdaandtime *)
(*   procedure ReadLambdaAndTime (var AverageRate, Minutes : real);       *)
(*         Procedure ReadLambdaAndTime reads the average arrival rate     *)
(*      of phone calls per hour and the length of time from the user.    *)
(*      The rate per hour is converted to the rate per minute.           *)


  function Poisson (X : integer;  AverageRate, Minutes : real) : real;
  (*      Function Poisson returns the probability that there are X      *)
  (*      Poisson arrivals in a time length of Minutes when the average  *)
  (*      arrival rate is AverageRate.  The probability is computed      *)
  (*      using Equation 10.13.                                          *)

  var  Arrivals : real;     (* local variable to store rate times time. *)
```

```
  begin
    Arrivals := AverageRate * Minutes;

    Poisson := (ZToPower (Arrivals, X) * exp (-Arrivals)) / Factorial(X)
  end; (* function Poisson *)

  function SumPoisson (Min, Max : integer;
                       AverageRate, Minutes : real) : real;
  (*     Function SumPoisson returns the probability that between      *)
  (*     Min and Max arrivals occur in a time length of Minutes.       *)
  (*     This cumulative probability is computed by summing the        *)
  (*     probability that the number of arrivals equals each value     *)
  (*     between Min and Max.                                          *)

    var  X : integer;      (* for statement control variable. *)
         Sum : real;       (* intermediate variable to store the running *)
                           (* total. *)

  begin
    Sum := 0.0;          (* initialize the running total to zero. *)

    (* add the probability that there are X arrivals in Minutes given *)
    (* that the average arrival rate is AverageRate per minute. *)

    for X := Min to Max do
      Sum := Sum + Poisson (X, AverageRate, Minutes);

    SumPoisson := Sum
  end; (* function SumPoisson *)
(*$i reportdistrib *)
(*   procedure ReportDistrib (Prob1, Prob2, Prob3 : real;              *)
(*                            AverageRate, Minutes : real);            *)
(*       This procedure reports the final values of the distribution.  *)


(*---------------------------------------------------------------------*)
begin (* main *)
  WriteMessage;

  ReadLambdaAndTime (Lambda, TimeInMinutes);

  ProbFewerThanMid := SumPoisson (LowerBound, (MiddleBound - 1),
                                  Lambda, TimeInMinutes);
```

```
      ProbBetweenMidAndUp := SumPoisson (MiddleBound, UpperBound,
                                        Lambda, TimeInMinutes);

      ProbMoreThanUp := 1.0 - ProbFewerThanMid - ProbBetweenMidAndUp;

      ReportDistrib (ProbFewerThanMid, ProbBetweenMidAndUp, ProbMoreThanUp,
                    Lambda, TimeInMinutes)
end. (* program PhoneCalls *)

=>
 This program computes the probability that
 more than 0 and fewer than 5 phone calls,
 more than 5 and fewer than 10 phone calls,
 and more than 10 phone calls
 arrive at a switch board in a given time interval

 What is the average number of calls that arrive per hour?
 □50.0
 How long a time interval do you want to study (in minutes)?
 □10.0

 At a switch board that receives calls at an average rate
 of      50.00 calls per hour
 In a period of      10.0 minutes

 P { 0 <   number of phone calls <  5} =    0.0821
 P { 5 <=  number of phone calls <=10} =    0.6994
 P {       number of phone calls >  10} =    0.2185
```

APPLICATION
Introduction to Statistics

CASE STUDY 17
Variance of experimental data

The professor in Case Study 15 at the beginning of this chapter now has a program that reads the grades from a computer file, computes the average for each student and the average for each test, and writes the report to the computer file. Next she would like a program that computes and reports the variance for each test and for the student's final averages.

Algorithm development

Knowing the variance or the standard deviation of test scores can give you additional information about how well you have done. Suppose you took a test and got an 85. If you know that the mean on the test is 75, you know that you did better than average. If you also know that the standard deviation is 5, you know that your score is higher above average than most of the other scores. If the standard deviation is 20, your score is less outstanding.

Most of the algorithm for the case study has already been developed in the previous sections. All that needs to be done is to add steps to Algorithm 9.1 to compute the variance of the test scores and the variance of the student averages. Again, the formula for the variance will be developed in the chapter, so the new steps for the algorithm can just be written as follows:

ALGORITHM 10.3 Additional steps for Algorithm 9.1

5. Compute the variance of the average score for each student
6. Compute the variance of the average score for each test
7. Report the averages and the variances

Statistics and experiments

In statistics, the underlying probability function that describes the way a process such as the tossing of a die behaves is estimated on the basis of experimental results. Then, using statistical analysis, you can test the probability that the particular die is fair. For example, the die can be rolled 1000 times, and the frequency with which each number occurs can be compared to the known frequency of each number for a fair die. Scientific experimentation often involves the use of experimental and control groups in which all factors are the same except for the one factor under study. The purpose of the control group is to provide an estimated probability mass function against which an estimated probability mass function from the experimental group can be compared. The probability that the differences are due to chance can then be determined.

Statistics enables you to infer the properties of a population from a small number of measurements. Because it is usually not possible to study an entire population, a *sample* of the population is selected. The sample taken from the population is used to infer properties of the entire population.

Statistics can also be used to analyze the relationship between two or more variables, through a technique called *regression*. Some aspects of regression are covered in the *Advanced Applications* supplement.

Using regression, a power-systems engineer can find an equation that relates the age of existing power plants to their reliability. Then the engineer can use the resulting equation to predict the reliability of a different power plant at any age. Regression can be used to determine the correlation coefficient between two variables. For example, the power-systems engineer might find that as the age of the power plant increases, so does its reliability. In this case, the correlation between age and reliability is positive. If the reliability decreases with age, the correlation is negative. Correlation coefficients range from -1.0 to $+1.0$. A coefficient of 0.0 means that knowing one variable does not give you any information about the other variable. For example, the reliability of a power plant and your grade on the last test are uncorrelated variables. Correlation coefficients close to either -1.0 or $+1.0$ mean that the variables are highly correlated; that is, they are good predictors of one another. For example, in most cases you would expect that your grade on a final exam would be highly correlated with your final grade. However, a high correlation coefficient does not necessarily mean that one variable causes the other. For example, although the height of the sun in the sky may be highly correlated with the number of traffic jams in a city, one does not necessarily cause the other.

Sample population

The way in which the sample is selected and the number of samples taken are both important issues in statistics. For a poll for a municipal election to be accurate, for example, a representative sample of voters must be reached. If the poll is taken on a college campus, the result will almost certainly be different from the election result. A college campus is not a representative sample of a community.

An engineer, a scientist, or a pollster must determine a way to select a representative sample from the population to be studied. A pollster, for example, might contact every twentieth person on a list of registered voters to obtain a representative sample.

The size of the sample is also important. Theoretically, to be as certain as possible you would always take the largest possible sample. However, this is frequently too expensive or impractical. In quality control, the test may require that the part be stressed until it breaks. If all parts were tested, there would be no parts. In an election poll, it would be quite expensive to contact everyone who was going to vote. Research in biology often involves the use of live animals. For any population, taking each new sample has a cost, be it time, money, or ethical considerations. Thus statistics is used to find the smallest sample size that still allows a valid conclusion to be reached.

Statistics is a branch of mathematics, but a *statistic* is any function of the observed sample that does not contain any unknowns. Suppose that an engineer has made the following 20 measurements of the current through a diode:

0.56 0.55 0.54 0.55 0.55 0.54 0.57 0.54 0.56 0.56
0.56 0.53 0.55 0.57 0.54 0.55 0.55 0.53 0.55 0.58

The first measurement is labeled x_1, the second one is labeled x_2, and so on up to the last measurement, x_{20}. The following are all statistics of the observations:

$$x_{10}, \quad \frac{x_2}{2}, \quad x_{18} + 3, \quad \frac{\sum x_i}{20}$$

Each statistic can be determined directly from the measurements and contains only known values.

The *median* and the *mode* are also statistics. The *median* is the value of x_i that divides the ordered sample in half: the values of half of the observations are above it and the values of the other half are below it. Ordered from highest to lowest, the sample above would be

0.58 0.57 0.57 0.56 0.56 0.56 0.56 0.55 0.55 0.55
0.55 0.55 0.55 0.55 0.54 0.54 0.54 0.54 0.53 0.53

The median for this sample is 0.55. The *mode* is the value of x that appears most often. The mode in the sample above is also 0.55. The median is useful where a population has a few very large or very small values. For example, the median is often used instead of the mean in reporting average incomes. If the incomes of a few millionaires were averaged in with a sample of incomes, the mean would be much higher than the income of the average person. The median thus provides a better measure of typical income.

Some statistics are more commonly used than others. The sample mean and variance are useful statistics that describe the general shape of the underlying distribution. The formula for the sample mean is

$$\overline{X} = \frac{\sum\limits_{i}^{N} x_i}{N} \qquad (10.32)$$

where $\overline{X}$ (read as X bar) is the average or *sample mean* of the N observations. Each x_i stands for a single observation of the data. If you look at the formula for the sample mean, you will see that it is exactly the same as the formula for the average of the observations. It can be shown that, in most cases, taking the average of the observations gives the best estimate for the mean of the underlying distribution of the population.

The mean of the sample is denoted by $\overline{X}$, and the mean of the population is denoted by μ. The sample mean, $\overline{X}$, serves as an estimator of the population mean, μ, which can never be known with certainty.

Another useful statistic is the sample variance. The formula for the sample variance is

$$\text{Var}(X) = \frac{\sum\limits_{i}^{N} (x_i - \overline{X})^2}{N} \qquad (10.33)$$

If you are familiar with statistics, Equation 10.33 may be slightly different from the one you are used to. The formula above is for the sample variance. However, it can be shown that the sample variance $\text{Var}(X)$ is not an unbiased estimator for the variance σ^2 of the underlying distribution of the population. The formula for the unbiased estimator of the population variance has $N - 1$ instead of N in the denominator.

The function **SampleVariance** computes the sample variance of the values stored in an array. Note that Equation 10.33 can be used to compute the variance only if the sample mean is known.

Application: Introduction to Statistics

```
function SampleVariance (Data : DataArray; Num : integer;
                    MeanData : real) : real;
(*      Function SampleVariance computes the sample average of an      *)
(*      array given the sample mean, MeanData and the number of        *)
(*      pieces of data, Num.                                            *)

var   Index : integer;       (* for statement control variable. *)
      Sum : real;            (* running total for the variance. *)
begin
   Sum := 0.0;               (* initialize the running total to zero. *)

   for Index := 1 to Num do
      Sum := Sum + sqr (Data [Index] - MeanData);   (* Equation 10.33 *)

   SampleVariance := Sum / Num
end; (* function SampleVariance *)
```

As was explained earlier with reference to probability distributions, the variance is a measure of the spread of a distribution. The factor $(x_i - \overline{X})$ in Equation 10.33 is a measure of the distance of an individual observation from the overall average. The sample variance is the average of the *square* of these distances. So, you can think of the variance as being a measure of the average distance of the observations from the sample mean.

Notice the units of the sample mean and variance. If the engineer's observations for the current through the diode were in amperes, the sample mean would also be in amperes, but the variance would be in amperes squared. Because the squared units are awkward, the standard deviation, which is the square root of the variance, is frequently used instead.

Hypothesis testing

An important topic in statistics is *hypothesis testing*. The following examples should give you a feel for hypothesis testing. A quality control engineer tests the hypothesis that a batch of parts meets quality specifications; only a small number of parts out of the batch are tested, but the test results are used to accept or reject the hypothesis for the entire batch. A biologist hypothesizes that a newly developed drug is more effective than the currently used drug; based on test results from a limited sample of animals, the biologist accepts or rejects the hypothesis that the new drug is more effective.

By its nature, a hypothesis test assumes that some errors in experiments are due to random occurrences that cannot be controlled. To perform a hypothesis test, you must have some knowledge of the sources and magnitudes of errors. A common assumption is that the results of experiments have Gaussian (normal) distributions. This assumption is based on a powerful mathematical theorem called the central limit theorem.

Suppose that the professor in the case study knows from previous experience that the students' final grades are normally distributed with a mean of 72 points and a standard deviation of 4 points. She thinks that perhaps this year's students were better than average. She decides to test her hypothesis using statistics. She sets up two mutually exclusive hypotheses. The first hypothesis is that this year's higher averages were due to random fluctuations and the student's abilities are the same as any other year. This hypothesis is called the *null hypothesis*. Her alternative hypothesis is that this year's students come from a different population with a higher mean. The two hypotheses can be stated as

$H_0: \mu = 75$ The null hypothesis
$H_1: \mu > 75$ The alternative hypothesis

The hypotheses are set up so that the null hypothesis will be accepted unless there is strong evidence that it is false. The professor would like to be certain that the higher scores were not due to random fluctuations before she accepts the alternative hypothesis. Of course she cannot be absolutely certain, but if the null hypothesis has only a five percent chance of being correct, she will reject it and accept the alternative hypothesis.

Figure 10.5 is a graph of the probability distribution of the final grades of the students (not the distribution for a single test). As illustrated in Figure 10.5, there must be a number c, the critical value, such that if the null hypothesis is true, the sample mean, $\overline{X}$, is less than c at least 95 percent of the time. The critical value, c, is a cutoff value. If the class average is greater than this cutoff value, there is only a five percent chance that the null hypothesis is true. Thus, the professor can be 95 percent certain that the null hypothesis is false and therefore that the alternative hypothesis is true. If the class average is less than the cutoff value, the professor must accept the null hypothesis that the higher grades are due to random fluctuation.

To find the value c such that

$$P(\overline{X} < c) = 0.95 \qquad (10.34)$$

you can use the standardized normal random variable given in Equation 10.20. Equation 10.34 can then be written as

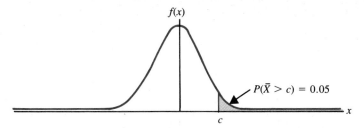

FIGURE 10.5
Critical region such that $P(\overline{X} > c) = 0.05$.

$$P(\overline{X} < c) = P\left(Z < \frac{c - \mu}{\sigma}\right) = \Phi\left(\frac{c - \mu}{\sigma}\right) = 0.95 \qquad (10.35)$$

To determine c, look in Table 10.2 to find the value of z for which $P(Z < z) = 0.95$. From the table, the value of z is approximately 1.65. So

$$z = \frac{c - \mu}{\sigma} = 1.65 \qquad (10.36)$$

Given that $\mu = 72$ and $\sigma = 4$, solving for c yields

$$c = 78.6$$

Because the final average of all the students was 76.8, the professor must accept the hypothesis that this year's students are no better than the average class.

Solution to Case Study 17

In the original version of Program 9.1, two different **array** data types were created: one for the student averages and one for the test averages. The use of two different data types required that two different averaging functions be written, because the arrays must be passed by data type. Continuing to use two different **array** data types would mean that two sample variance functions would have to be written: one for the student averages and one for the test averages. An alternative solution is to write, for the sample mean and sample variance, general functions that have the matrix data structure in their parameter lists.

Let us start by looking at the function to find the sample mean. If the entire matrix is passed into the function, you must specify whether the average is to be taken across a row or down a column. The simplest method for specifying which elements are to be included in the average is to declare four parameters: the index of the first row to be included, the index of the last row to be included, the index of the first column to be included, and the index of the last column to be included. For example, for the average for the third student, whose six test grades are stored in the third row of the matrix, the first and last row indices would both be 3, the first column index would be 1, and the last column index would be 6. For the average for the second test, whose 12 scores are stored in the second column of the matrix, the first row index would be 1, the last row index would be 12, and the first and last column indices would both be 2. Function **SampleMean** uses this method to compute the sample mean of either a row or a column:

```
function SampleMean (Data : DataArray; MinRow, MaxRow,
                    MinCol, MaxCol : integer) : real;
(*    Function SampleMean computes the sample average of a matrix.    *)
(*    The values in the rows from MinRow to MaxRow and from MinCol    *)
(*    to MaxCol are added to the running total.  To take the          *)
(*    average of one column, set MinCol and MaxCol to the column      *)
(*    number, and set MinRow to the first row index, and MaxRow to    *)
(*    the last row index.  To take the average of a single row, set   *)
(*    MinRow and MaxRow to the row and set MinCol to the first        *)
(*    column index and MaxCol to the index of the last column.        *)

var   Row, Col : integer;     (* for statement control variables. *)
      Sum : real;             (* running total. *)
      Num : integer;          (* number of elements in the running total. *)
begin
  Sum := 0.0;
  Num := (MaxRow - MinRow + 1) * (MaxCol - MinCol + 1);

  for Col := MinCol to MaxCol do
    for Row := MinRow to MaxRow do
      Sum := Sum + Data [Row, Col];
  SampleMean := Sum / Num
end; (* function SampleMean *)
```

The function **SampleMean** is versatile because it can compute the mean of either a column or of a row; however, it is harder to read than the functions that have been developed previously to compute means and averages. Tradeoffs must always be made among clarity, versatility, and efficiency. If you choose a more efficient but less obvious algorithm be sure to document your program thoroughly.

All that is necessary to complete the case study is to modify the data structure, add the **SampleVariance** function to the end of Program 10.4, and add the variance to the report procedure. Program 10.5 completes the case study problem:

PROGRAM 10.5   FinalGrades

```
program FinalGrades (input, output, infile, report);
(*      This program is the same as Program 10.2, ReadWriteGrades,   *)
(*      except that in addition to computing the averages, it computes *)
(*      the variances in the student averages and the test averages. *)
(*      The output is written to the textfile report.                *)

const NumStudents = 12;
      NumTests = 6;

type  DataArray = array [1..NumStudents, 1..NumTests] of integer;
      StudentArray = array [1..NumStudents] of real;
      TestArray = array [1..NumTests] of real;
      TestTypeArray = array [1..3] of real;

var   AllScores : DataArray;           (* array of the original student *)
                                       (* scores. *)
      TestAvg : TestArray;             (* array of the test averages. *)
      TestVar : TestArray;             (* array of the test variances. *)
      StudentAvg : StudentArray;       (* array of the student averages. *)
      StudentVar : StudentArray;       (* array of the student variances. *)
      TestTypeAvg : TestTypeArray;     (* array of the test-type averages. *)

      OverAllMean : real;              (* final average of the class. *)
      OverallVar : real;               (* variance of the final average. *)

      Test, Student : integer;         (* for statement control variables. *)

      infile : text;                   (* textfile of the input scores. *)
      report : text;                   (* output textfile for the report. *)

(*$i computetesttypeaverages *)
(*$i samplemean *)
(*$i samplevariance *)
(*$i readgrades *)
```

```
(*      The subprograms above were written for ReadWriteGrades and used    *)
(*      the array data structure ScoreMatrix.  In the subprograms for      *)
(*      the more general problem of finding means and variances the        *)
(*      name of the data structure is DataArray.  Even though the          *)
(*      two data structures are the same except for their identifiers,     *)
(*      a variable of type ScoreMatrix cannot be passed into a parameter   *)
(*      of type DataArray.  This is a drawback to the strong data typing   *)
(*      in Pascal.  To circumvent this problem, the data types of the      *)
(*      parameters in the previous subprograms were changed to DataArray.  *)
(*$i reportgrades *)
(*  procedure ReportGrades (Grades : DataArray; SAvg, SVar : StudentArray; *)
(*                          TAvg, TVar : TestArray;                        *)
(*                          TTypeAvg : TestTypeArray;                      *)
(*                          TotalMean, TotalVar : real);                   *)
(*      Procedure ReportGrades reports the scores for each student         *)
(*      on each test.  It reports the test averages, the student           *)
(*      averages, the test variances and the student variances.            *)
(*      The output is written to the textfile, report.                     *)


(*-------------------------------------------------------------------------*)
begin (* main program *)
  reset (infile);
  rewrite (report);

  ReadGrades (AllScores, NumStudents, NumTests);

  for Test := 1 to NumTests do
    begin
      TestAvg [Test] := SampleMean (AllScores, 1, NumStudents,
                              Test, Test);
      TestVar [Test] := SampleVariance (AllScores, 1, NumStudents,
                              Test, Test)
    end;

  ComputeTestTypeAverages (TestAvg, TestTypeAvg);

  for Student := 1 to NumStudents do
    begin
      StudentAvg [Student] := SampleMean (AllScores, Student, Student,
                                  1, NumTests);
      StudentVar [Student] := SampleVariance (AllScores, Student, Student,
                                  1, NumTests)
    end;
```

```
    OverallMean := SampleMean (AllScores, 1, NumStudents, 1, NumTests);
    OverallVar  := SampleVariance (AllScores, 1, NumStudents, 1, NumTests);

    ReportGrades (AllScores, StudentAvg, StudentVar, TestAvg, TestVar,
                  TestTypeAvg, OverallMean, OverAllVar);

    writeln (' The output report is in the file output.dat');
end. (* program FinalGrades *)

=>
The output report is in the file output.dat
```

In program **FinalGrades,** the output procedure is a large part of the program. This situation is not unusual in programs that require a formatted table in the output. The following is a listing of the computer file containing the output from the program **FinalGrades**:

Student	Test Number						Student Average	Student Standard Deviation
	1	2	3	4	5	6		
1	99	93	88	95	78	95	91.3	6.8
2	75	80	83	70	77	80	77.5	4.2
3	92	95	85	88	92	89	90.2	3.2
4	50	62	50	65	70	50	57.8	8.2
5	76	70	60	78	77	55	69.3	8.9
6	81	85	78	87	82	70	80.5	5.5
7	72	74	90	73	81	92	80.3	8.1
8	78	70	50	67	65	50	63.3	10.3
9	85	87	80	89	85	78	84.0	3.8
10	67	80	65	70	75	67	70.7	5.2
11	77	78	70	72	78	68	73.8	4.0
12	80	85	82	78	87	83	82.5	3.0
Test Average	77.7	79.9	73.4	77.7	78.9	73.1		
Test Standard Deviation	11.7	9.4	13.6	9.4	7.0	15.1		

```
The average on the multiple-choice tests was    77.7
   The average on the short-answer tests was    79.4
The average on the program-writing tests was    73.2

The class average for the test was  76.78
with a standard deviation of  11.66
```

Summary

Pascal **files** can be used for storage of information that is to be used as input to a program or information that a program produces as output. In this chapter only **files** of type **text** were covered. **Textfiles** are composed of elements of the data type **char**; the more general **file** structure is covered in Chapter 14.

Text is a standard, predefined data type in Pascal and a **file** must be declared as a **file** of type **text** prior to its being used. The file must also be declared in the program heading, as are the standard files **input** and **output**. The structure of a textfile is similar to that of an **array** of **char**acters, with two important differences: the length of a textfile is not declared and the information stored within a textfile can only be accessed sequentially. When information is stored in textfiles, the compiler places an end-of-line marker, ⟨eoln⟩, at the end of each line and an end-of-file marker, ⟨eof⟩, at the end of the file.

In previous chapters, **read** and **readln** were used to read from the standard **input** file. To read from a file other than **input**, the file must first be **reset** to the beginning of the file. After execution of the **reset** procedure, information can be read from the file. The name of the file from which data is to be read must precede the names of the variables to be read. When **read** is used, the information in the file is read sequentially and assigned to the variables in the **read**-argument list. A pointer moves through the file and points to the next element in the file to be read. When **readln** is used, the pointer moves to the beginning of the next line after assigning values to the variables in the argument list.

The standard Pascal functions **eoln** and **eof** can be used to determine whether the end of a line within a **textfile** or the end of a **file** has been reached, respectively. The file name must be included as the argument of **eoln** or **eof**.

To create a **textfile**, the file name identifier must be declared both in the heading and as a variable of type **text**, as it was to read from a previously existing file. The write pointer must be moved to the beginning of the file with a **rewrite** statement. The file name is included as the argument of the **rewrite** statement. The statements to write to the **textfile** can then be issued. As with the **read** statement, the file name must precede the arguments of the write statement. If the file name is not included, the data will be written to the standard **output** file.

Information that is stored in computer files can be assigned to Pascal **textfiles** so that it can be read by a program. The commands for assigning a computer file to a Pascal **file** are system dependent. The use of textfiles eliminates the need to type input data while a program is running. The information can be stored in a **textfile** and the data can be read from the file instead of from the terminal. If the output from one program is to be used as input to another program, it can be saved in a **textfile** that is associated with a computer file.

Exercises

PASCAL

1. What is the output from each of the fragments below? Show the output for each **file** that is used in the fragment. Indicate where the ⟨eoln⟩ and ⟨eof⟩ markers are placed.

    (a) ```
    rewrite (outfile);
    for Index := 1 to 10 do
       write (outfile, sqr(Index):4);
    writeln (outfile)
    ```

    (b)  ```
    for Index := 1 to 10 do
       begin
          rewrite (outfile);
          write (outfile, sqr(Index):4)
       end;
    writeln (outfile)
    ```

2. What are the values of **A, B, C,** and **D** after each fragment below is executed? The computer file assigned to **datafile** contains the following data:

    ```
    12 15 90 50
    20 31 53 30 31
    2000
    ```

    (a) ```
    var A, B : char;
        C, D : integer;
        datafile : text;

    begin
       reset (datafile);
       read (datafile, A);
       readln (datafile, B);
       read (datafile, C, D)
    end
    ```

    (b) ```
    var A, B, C, D : integer;
        datafile : text;

    begin
       reset (datafile);
       readln (datafile, A, B);
       read (datafile, C, D)
    end
    ```

3. Each of the following programs or fragments has a bug. Find the problem in each fragment, then execute each fragment as part of a program to see what error messages your compiler issues. Be sure that you have included the necessary system commands for each **file** that you use. Create a computer file that contains the following data and assign it to **infile**.

    ```
    12   12.5
    a
    665
    ```

(a) ```
program FirstBug (input, output, infile);
var infile : text;
    FirstVar: integer;
begin
  read (infile, FirstVar)
end.
```

(b) ```
program SecondBug (input, output, infile);
var infile : text;
    FirstVar : integer;
begin
  reset (infile);
  write (infile, FirstVar)
end.
```

(c) ```
program ThirdBug (input, output, infile);
var infile : text;
    FirstVar : integer;
begin
  reset (infile);
  while not eof (infile) do
    readln (infile, FirstVar)
end.
```

4. Rewrite any one of the programs that you have already written so that the input data is read from a Pascal **textfile** and the output is written to a Pascal **textfile**.

5. Write a program that writes the even integers from $-10$ to $10$ into a **textfile**. Then write a program that reads the same file and prints the contents of the file.

**APPLICATION**

6. Write a program that prints Table 10.1.

7. Write a program that computes the expected value, the variance, and the standard deviation for the outcome of the roll of two dice.

8. Using the functions **NChooseX**, **Mean**, and **Variance**, write a program that computes the probability distribution of the number of heads in $N$ coin tosses and also computes the mean and the variance. Read $N$ from the terminal. Assume that the coin is fair—that is, that $p = 0.5$ and $q = 0.5$.

9. For a normal random variable $X$, with $\mu = 5.0$ and $\sigma^2 = 4.0$, compute the following probabilities:
   (a) $P(X > 5.0)$
   (b) $P(1.0 < X < 2.0)$
   (c) $P(X > -1.5)$
   (d) $P(X < -1.5)$
   (e) $P(-2.0 < X < -1.0)$
   (f) $P(X < 25)$
```

Problems

PASCAL

1. Write a program that compresses a paragraph stored in a computer file by removing all the blanks and punctuation from it. Enter the paragraph using your computer's editor. Read the text from the computer file, remove the blanks and punctuation, and write the compressed text to the terminal.

2. You have measured the current through one of 16 samples of a 5.1-kΩ resistor manufactured by your company. Your value for the current is 0.012 A. The other 15 resistors have been tested by a technician who measured the voltage across each resistor and stored the results in a computer file with the following format:

62.0001	60.7790	61.9742	61.0591	59.9994
60.0133	62.0007	59.5960	60.0133	61.2003
61.0000	62.8420	62.3922	61.3113	61.9987

Write a program that reads your value of the current from the terminal and the 15 values for the voltages from the computer file. The program should compute the current for each resistor, and then write a report stating the average value of the current for the 16 samples tested.

3. To assemble a particular gadget, you need the largest possible widget. Write a program that finds the largest of 100 widgets. The data on the widgets is stored in a computer file. Test your program on the following data:

3.68	3.12	3.12	3.66	3.80	4.08	3.00	3.84	3.12	3.99
3.87	3.75	3.62	3.42	3.02	3.17	3.94	3.17	3.00	3.16
3.40	3.87	3.57	3.03	3.17	3.97	3.48	3.00	3.11	3.81
3.69	3.61	3.05	3.55	3.17	3.88	3.65	3.68	3.62	3.69
3.52	4.15	3.41	3.91	4.14	3.28	3.11	3.50	2.74	3.19
3.59	3.00	3.16	3.96	3.91	3.66	4.16	3.08	3.00	3.11
3.55	3.99	3.16	4.09	3.20	4.15	3.72	3.76	3.58	3.91
3.91	3.11	3.55	3.04	3.04	3.11	3.11	3.52	3.01	3.42
3.73	3.74	3.22	3.59	3.12	3.98	2.76	3.47	3.74	3.20
3.12	2.78	4.11	3.52	2.70	3.59	3.16	3.41	3.07	3.27

4. In analyzing a diode, you need to find the smallest positive voltage for which current flows. Your data is stored in a computer file and organized as follows: the number of measurements, N, is given on the first line and N real measurements for voltage follow on the next line(s). The program should determine and report the smallest of the positive (that is, greater than 0) N voltages. Create a computer file that contains the following data:

5
0.55   −1.5   0.3   0.43   0.61

5. The bolt-manufacturing company from Problem 6 in Chapter 7 has expanded and now has 20 machines. The data on the machines is automatically stored in a computer file that constantly monitors the machines. Write a program that reads the data from the computer file and computes the breakdown time in hours for each machine. It should also compute the number of bolts produced per hour by each machine. Only the time that the machine is working should be considered when the average is computed. The following data is taken from a typical day.

Machine number	Bolts produced	Breakdown time in minutes
1	1239	10.2
2	463	230.0
3	897	79.3
4	106	400.9
5	1313	0.0
6	815	42.8
7	725	135.2
8	444	199.8
9	213	313.5
10	913	72.1
11	1111	15.6
12	1212	3.7
13	209	200.4
14	250	111.4
15	521	219.0
16	937	22.3
17	822	96.7
18	157	375.0
19	753	100.6
20	1017	2.8

APPLICATION

6. Rewrite the program from Problem 9 in Chapter 9 so that the two matrices are read from a **textfile** and the output report is written to a **textfile.**

7. As head engineer, you have been collecting data on the number of defects per widget produced in your factory. From your data, you have deduced that the probability that a widget has X defects is given by

$$P(\text{defects} = 0) = 0.50$$
$$P(\text{defects} = 1) = 0.20$$
$$P(\text{defects} = 2) = 0.15$$

$$P(\text{defects} = 3) = 0.10$$
$$P(\text{defects} = 4) = 0.05$$

Write a program that computes the expected number of defects in a widget.

8. Rewrite the Pascal program from Problem 3 in Chapter 9 to find the sample mean and sample variance of the density and weight of the five blocks of metal. Read the data from a **textfile.**

9. A five-inch-long string is cut in two. The cut is made at a random location along the string. The probability that the cut occurs at a distance less than x from the beginning of the string is given by

$$P(X < x) = \frac{x - a}{b - a}$$

For the five-inch string, $a = 0$ and $b = 5$. Write a program that reads x from the terminal, computes $P(X < x)$, and writes the output report to a **textfile.**

10. A basketball player misses the basket with a probability of 0.30. The player decides to keep shooting until he gets a basket. The probability that it takes x shots up to and including the first basket is described by the *geometric distribution*. If x is the number of tosses until there is a basket, there must have been $x - 1$ misses before the basket. If a miss occurs with probability q and a basket occurs with probability p and each toss is independent of the others, the probability of $x - 1$ misses and then a basket is

$$P(X = x) = pq^{x-1}$$

For the geometric distribution,

$$\mu = \frac{1}{p} \quad \text{and} \quad \sigma^2 = \frac{1-p}{p^2}$$

Write a program that computes the probability that $x = 1$, $x = 2$, and $x = 3$. The expected value and the variance of the number of shots until the player gets a basket should also be computed.

11. Suppose you live in Boston and take the Green Line trolley to work. If you assume that the arrival of trolleys is totally random—that is, that the trolleys run independently of one another and that one is as likely to arrive this minute as the next minute—you can use the Poisson distribution to describe the arrival of trolleys. From past experience, you know that you can expect four trolleys to arrive in one hour. Write a program to calculate the probability that a trolley will arrive in the next n minutes, where $n = 1, 2, 3, 4, 5, 6, 7, 8, 9$, and 10.

12. For Problem 11, suppose that you wanted to know not how many trolleys would arrive in the next n minutes, but how long you could expect to wait for the next trolley. This problem is analogous to the problem of finding the number of shots until there is a success (see Problem 10). Each second (or arbitrarily short time interval) that a trolley does not arrive is a failure, and the first second that one does arrive is a success. Using this analogy and the Poisson distribution, one can derive the *exponential distribution*

$$f(x) = \lambda e^{-\lambda x}$$

and

$$P(0 < X < t) = \int_0^t \lambda e^{-\lambda x}\, dx$$
$$= 1 - e^{-\lambda t}$$

where X is the time interval and lambda, λ, is the average number of events per unit time. For the exponential distribution,

$$\mu = \frac{1}{\lambda} \quad \text{and} \quad \sigma^2 = \frac{1}{\lambda^2}$$

The exponential distribution is used to describe such things as the survival time of experimental animals, the lifetime of electronic components, and the time between machinery breakdowns.

Given that trolleys arrive at an average rate of four per hour, write a program to compute the probability that you will wait less than t minutes for a trolley. Values should be computed and reported for $t = 5, 10, 15, \ldots, 60$ minutes.

13. A sample of the time in minutes between arrivals of 30 trolleys on the Green Line is

    1.0 4.2   2.3 11.5   5.4 20.1   1.3 11.5   4.3 0.2
    14.5 0.2 12.0   2.4   7.7   6.9 13.5   0.4   1.5 6.8
    0.7 3.7 26.3   8.6 17.7   0.5   5.6   8.3   1.4 0.2

    Enter this data in a computer file. Write a program that computes the mean and variance of the waiting time between trolleys and then computes the average number of trolleys per minute.

14. The speed of a trolley is related to the number of passengers on the trolley. Twenty paired measurements of the speed and the number of passengers are made and stored in a computer file. The speed of the ith trolley is x_i, and the number of passengers on the ith trolley is y_i. The *sample covariance* is

$$\text{Cov}(X, Y) = \frac{\sum_{i}^{N} (x_i - \overline{X})(y_i - \overline{Y})}{N}$$

where N is the number of values and x_i and y_i are the paired measurements of the speed and number of passengers.

Write a program that computes and reports the sample covariance between the speed of a trolley and the number of passengers on the trolley.

CHAPTER 11

Enumerated Types and Character Arrays

Many programs require that ordinal, nonnumerical data, such as the days of the week or the months of the year, be used in an understandable manner. Pascal has a data type, the *enumerated type,* that can be used for this purpose. An ordinal data type is any data type that can be mapped onto a subset of the integers. The data type **boolean** is an ordinal data type because its values, true and false, can be mapped onto the integers 0 and 1. The data type **char** is an ordinal data type because its values can be mapped onto the integers from 1 to the number of characters in the character set, and, of course, the data type **integer** is ordinal. In Pascal, you could, therefore, declare a data type called **Months** that could be assigned the values January, February, March, etc., where the value of January would be less than that of February, the value of February less than that of March, and so on. In addition, Pascal has special functions and data types for character variables so that words or phrases, rather than single characters, can be represented.

> CASE STUDY 18
> **Fuel types for power plants**
>
> A Pascal program currently in use simulates the operation of a large interconnected electric power pool. There are many different power plants, and each power plant has characteristics that depend on the type

of fuel burned by the plant. For convenience, the plants have been broken down into classes by their fuel type. Currently, the classes are labeled 1, 2, 3, . . .,. The class numbers are used to input the type of fuel the plant uses and as subscripts in arrays. The program is large, has been expanded several times, and is used by many people. Because there are so many classes, the people who use the program cannot remember which number stands for which class of power plants. You have been asked to edit and modify the program so that the power plants can be referred to by the type of fuel that they burn rather than by numbers.

Algorithm development

For this case study, you do not need to write an algorithm for the program, as it is assumed that you have been given a working program. The modifications that you make will involve changing the data type of the variable that stores the type of fuel used by a plant so that a name rather than a number can be used within the program.

The following steps must be taken to solve the problem given in the case study:

1. Decide which fuel names you will need to use.

2. Specify the fuel names as the values to be taken by an enumerated data type.

3. Edit the program so that the numbering system now in use is replaced by the enumerated data type.

4. Convert the enumerated fuel types to character strings for output.

The use of enumerated data types and character strings in making these modifications will be discussed in the solution to the case study.

11.1 Enumerated data types

In Pascal, you can create user-defined enumerated types in which the integers are essentially given constant identifiers. You could create an enumerated data type for the plant fuels, for the days of the week, or the members of your family in order of their ages. Within the data type, each identifier is associated with an integer. For example, fuel type 24 might be bituminous coal, but rather than having to remember which fuel is type 24, with enumerated types you can use the identifier **BituminousCoal** instead of the number 24.

The syntax for creating an enumerated type is

→ **type** → *type-identifier* → = → (→ *identifier* →) → ; →

You could create the data type **FuelType** to store the fuel types:

```
type  FuelType = (Oil, Gas, Nuclear, Coal, Hydro, Sun, Wind);
```

The constant identifiers **Oil, Gas, Nuclear, Coal, Hydro, Sun,** and **Wind** are each associated with an integer. The assignment starts from 0 and is made in the order in which the identifiers are listed. In this example, **Oil** is assigned a value of 0, **Gas** a value of 1, **Nuclear** a value of 2, etc.

Notice that there is no reserved word that defines a type as an enumerated type. The enumerated type is declared by listing the constant identifiers separated by commas and enclosed in parentheses.

Each of the constant identifiers declared previously belongs exclusively to the data type name **FuelType**. Once **Sun** is defined in the data type **FuelType, Sun** cannot appear in a list of the days of the week. But the identifier **Sunday** can appear in a list of the days of the week since **Sun** and **Sunday** are different identifiers.

The Pascal statement above declares a data type, **FuelType**. In order to use the data type, you must declare a variable to be of that type. For example:

```
var   PlantFuel : FuelType;
```

This variable declaration creates a variable that can take on any of the values listed in the type **FuelType.**

Internally, Pascal stores the enumerated types as integers: in the data type **FuelType, Oil** is stored as 0 and **Wind** is stored as 6. This internal storage scheme results in several important properties for user-defined data types.

One common use of variables of enumerated data types is as **for** statement control variables. The following **for** statement executes the procedure **ConventionalFuels** for all the fuels from **Oil** to **Coal:**

```
for PlantFuel := Oil to Coal do
   ConventionalFuels (PlantFuel)
```

This statement is clearer than one using an integer variable as the **for** statement control variable even though the effect is the same. Notice that you must be aware of the order in which the identifiers have been listed in the declaration. If your enumerated type is for days of the week, you know the order in which they should be, but for power plants, there is no predetermined order.

Enumerated types are also easy to use as the control expression in a **case** statement. If all the values that are declared in the type statement are included in the case constant list, the control expression never takes a value that is not in the list:

```
case PlantFuel of
    Oil, Gas, Coal  : HydroCarbon (PlantFuel);
            Nuclear : Uranium (PlantFuel);
    Sun, Wind, Hydro : TimeDependent (PlantFuel)
end    (* case PlantFuel *)
```

Enumerated type variables can be compared using the relational operators. The result of the comparison will depend on the order in which the identifiers are declared in the type declaration. Given the previous declaration for **FuelType, Oil < Coal** has a value of true.

The standard functions **pred, succ,** and **ord** can be used with enumerated types.

Function	Function Data Type	Argument Data Type	
pred	scalar	scalar	Returns the constant prior to the argument in the enumerated list
succ	scalar	scalar	Returns the constant after the argument in the enumerated list
ord	integer	scalar	Returns the position of the argument in the enumerated list. The **ord** of the first constant in the enumerated list is 0

Creating an enumerated type has some advantages: enumerated types take up less memory and are faster than character arrays, are often more convenient to use than integer constants, and can make program structures clearer. Program 11.1 demonstrates a program that uses the functions **pred, succ,** and **ord** for a scalar data type declared for the days of the week.

PROGRAM 11.1 Days

```
program Days (input, output);
(*     This program demonstrates a user-defined data type, DayType,   *)
(*     that is defined to take values of the days of the week.        *)

type   DayType = (Monday, Tuesday, Wednesday, Thursday, Friday, Saturday,
                  Sunday);

var    Day : DayType;          (* variable of the user-defined data type. *)
       DayNum : integer;       (* integer conversion of variable the Day. *)
```

```
begin (* main program *)
  DayNum := ord (Monday) + 1;
  writeln;
  writeln (' Monday is day', DayNum:2, ' in the week.');

  Day := succ (Tuesday);
  writeln;
  if Day = Wednesday then
    writeln (' The day after Tuesday is Wednesday.');

  Day := pred (Tuesday);
  if Day = Monday then
    writeln (' The day before Tuesday is Monday.');

  if Tuesday < Saturday then
    writeln (' Tuesday comes before Saturday.')
end. (* program Days *)
```

=>

Monday is day 1 in the week.

The day after Tuesday is Wednesday.
The day before Tuesday is Monday.
Tuesday comes before Saturday.

Enumerated variables have one major disadvantage: although many Pascal implementations have an extension that allows you to read and write enumerated variables, in standard Pascal they cannot be read or written. They have a representation only inside the Pascal program. So, for the declaration of **PlantFuel** given previously, the following **read** and **write** statements result in error messages:

```
    readln (PlantFuel);
***error***         $153

    writeln (PlantFuel);
***error***         $116

error messages :
***************

 116: error in type of standard procedure parameter.
 153: illegal parameter type for read.
```

This restriction is a severe drawback to the use of enumerated data types. For a program like the one in the case study, a solution is to write functions or procedures that convert enumerated types to and from character arrays. For this program, the fuel type would have to be read either as a number or a character array, converted to an enumerated type for internal use, and then converted to a character array for output. The procedure **WriteName** prints the name that corresponds to the enumerated identifier. The conversion from characters to enumerated identifiers is covered in the solution to the case study.

```
procedure WriteName (Name : FuelType);
(*      WriteName prints the word corresponding to the enumerated     *)
(*      constant identifier.                                          *)
begin
  case Name of
    Oil : write (' oil ');
    Gas : write (' gas ');
    Nuclear : write (' nuclear ');
    Coal : write (' coal ');
    Hydro : write (' hydro ');
    Sun : write (' sun ');
    Wind : write (' wind ')
  end (* case *)
end; (* procedure WriteName *)
```

11.2 Subrange types

In some situations, it is necessary to restrict the range of values that can be assigned to a variable. For example, the number of fuel classes in any given run might be declared as an integer variable but must be an integer between 1 and **MaxClass,** a constant declared as the maximum number of classes. In Pascal, you can define new data types that consist of subranges of ordinal types to restrict the range of values that can be assigned to a variable. For example, when you declare an **integer** variable, it can take on values between the smallest and largest integer numbers. With a subrange, you can restrict the variable to values between 0 and 100. A subrange can be defined for any ordinal data type. The only predefined data type you have learned so far that is not ordinal is the **real** data type. So you cannot assign subranges for **real** data types or variables. The syntax diagram for declaring a subrange is

type ⟶ *type-identifier* ⟶ = ⟶ *constant* ⟶ .. ⟶ *constant* ⟶ ; ⟶

Just as with any Pascal type declaration, the new data type can be defined explicitly with an identifier, or the type can be implicitly defined in a **var** statement as the following statements illustrate:

11.2 Subrange types

```
const MaxClass = 300;              (* maximum number of fuel classes. *)

type  FuelType = (Oil, Gas, Nuclear, Coal, Hydro, Sun, Wind);

      ClassRange = 1..MaxClass;    (* range of valid class numbers. *)

var   NumClasses : ClassRange;     (* number of classes of fuel. *)
      NonConventional : Sun..Wind; (* nonconventional fuel types. *)
      Conventional : Oil..Coal;    (* range of conventional fuels *)
```

The data type of the subrange is the same as the data type of the constants used for the lower and upper bounds of the subrange. You may recognize the syntax for the subrange as the syntax for declaring the bounds of an array. When you declare an array, you use the subrange syntax to limit the values that can be assigned to the array index.

Subranges can be useful when a program is being written and debugged because they can prevent the bugs that occur when a subprogram and/or parameter list are used incorrectly:

```
type  FuelType = (Oil, Gas, Nuclear, Coal, Hydro, Sun, Wind);
      ThermalFuels = Oil..Coal;


   procedure ComputeThermalEfficiency (Fuel : ThermalFuels);
   begin
   end;


begin (* main program *)
   ComputeThermalEfficiency (Wind);
***error***                     $303

error messages :
***************

  303: value to be assigned is out of bounds.
```

However, subranges are not so useful for data checking as a program is run. When an out-of-range value is assigned to a variable, a run-time error results:

```
var   Digit : 0..9;
          .
          .
   writeln (' Enter a digit between 0 and 9');
   readln (Digit)
```

```
=>
 Enter a digit between 0 and 9
 □22


 *** error ***           value out of range
 program terminated at offset 000076 in main program
```

The exact error message will depend on your computer system, but in any event the program will halt execution. Clearly, subranges cannot be used when there is any chance that an out-of-range value might be entered by a user working interactively. For interactive programs, data checking with **repeat-until** or **while-do** loops can be used to trap bad data before it is used in the program, as was shown in Chapter 8.

11.3 Character arrays

From the discussion of enumerated types, it is clear that being able to use words and phrases is an advantage in many programs. Words and phrases are frequently referred to as *strings*. Although **string** is not a data type in standard Pascal, many versions of Pascal allow **string** to be used as a standard data type. Because the extension for each version is different, this chapter will cover only the use of strings of characters in standard Pascal. To find out about the extensions for your version of Pascal, you will have to read the documentation for your compiler.

The data type that allows words and phrases to be treated as a single entity rather than as individual characters is the **packed array.** In a **packed array,** the data is stored so that more than one character occupies a single storage location in memory. You can visualize a packed array versus an unpacked array as follows:

Packed	Unpacked (normal)
coal	c
	o
	a
	l

Although an array of any data type can be packed, it does not make sense to pack real or integer data because a number usually requires an entire memory location. There are standard Pascal procedures called **pack** and **unpack** that will compress a numerical array and then return it to its original state, but they tend to make the program run more slowly and lead to problems with parameter lists. Unless the available memory on your computer is severely constrained, do not pack numerical arrays.

Declaring packed character arrays not only saves memory, more importantly, it allows arrays of characters to be treated as phrases—that is, as strings. The rest of this section will discuss the special characteristics of **packed arrays of char**acters, which are referred to as *string-type* variables. **String** is not a reserved word in standard Pascal; however, it is a standard word in many versions of Pascal including UCSD Pascal and Turbo Pascal. To avoid confusion, the user-defined identifier **CharString** will be used in the examples.

In standard Pascal, to use a *string-type* variable you must declare a data type that is a **packed array of char** which has a lower bound of 1 and which has more than one element. The following statement declares a data type called **CharString**

```
const MaxLetter = 10;

type  CharString = packed array [1..MaxLetter] of char;
```

CharString is a string-type. Once you have declared a string-type, you do not have to worry about using the procedures **pack** and **unpack**.

After defining your data type, you can declare variables to be of type **CharString**:

```
var  Fuel, AuxFuel : CharString;
```

This statement declares two variables, **Fuel** and **AuxFuel**, that are of the user-defined data type **CharString**.

For some constructions, there is no difference between a string-type and a regular **array of char**. For example, the variable **Fuel** declared as a **packed array of char** can be referenced element by element just like a normal array:

```
Fuel := '#6 Oil     ';
writeln;
writeln (' The fourth letter in the fuel type is ', Fuel [4])
```

=>

```
The fourth letter in the fuel type is O
```

The value of one string-type variable can be assigned to another string-type variable if they are both of the same length. For example, if

```
var   Fuel, AuxFuel : CharString;
```

the following statement will set every element in **AuxFuel** equal to the corresponding element in **Fuel**:

```
AuxFuel := Fuel
```

Reading and writing string-type variables

A string-type variable has advantages in other constructions because it can be referenced as if it were a single variable even though it is an array. In standard Pascal, you can print a string-type variable with a single **write** statement to any file of type **text**, including the standard output file. For example

```
writeln (' The fuel type is ', Fuel)
```
```
=>
The fuel type is #6 Oil
```

Reading string-type variables is not so simple in standard Pascal. In standard Pascal, the following statement:

```
readln (Fuel)
```

results in an error message. A common extension to standard Pascal is to allow the preceding statement if the read parameter is defined to be a **packed array of char.** Some versions of Pascal will allow the preceding statement only if the array is defined to be a **string**.

To allow for all the variations, in the remainder of the book, the procedure **ReadString** will be used whenever it is necessary to read a character string from input. The procedure **ReadString** reads characters from the terminal into a **packed array of char.**

```
procedure ReadString (NumChar : integer; var Word : CharString);
   (*     Procedure ReadString reads NumChar characters from the terminal   *)
   (*     into the character string Word.  The read pointer remains         *)
   (*     positioned after the last character that was read.                *)

var CharNum : integer;         (* for statement control variable. *)

begin
   for CharNum := 1 to NumChar do
      read (Word [CharNum])
end; (* procedure ReadString *)
```

Although the procedure **ReadString** is written in standard Pascal, it will not execute correctly on some compilers with special extensions for strings. If necessary, you can either modify **ReadString** for your compiler or replace it with the **readln** command.

Comparison of string-type variables

A string-type variable can be set equal to a string of characters enclosed in single quotation marks. The string must have exactly the same number of characters as the dimension of the array.

```
Fuel := 'tenletters'
```

If the number of characters between the quotation marks is not the same as the declared dimension of the array, the compiler will give an error message saying that the types are not compatible. In order to put a shorter word into the array, you must pad the words with blanks. It makes a difference whether the blanks are added on the left or on the right as shown in the following statements:

```
Fuel := '#6 Oil     ';

AuxFuel := '     #6 Oil'
```

Fuel and **AuxFuel** do not have the same value.

Another advantage of using string-type variables is that unlike any other type of array they can be compared using the relational operators. = (equal) means that each element of each array is identical. <> (not equal) means that one or more of the elements in the arrays are different. For example, in the following fragment, **Fuel** and **AuxFuel** are not equal:

```
Fuel := '#6 Oil     ';
AuxFuel := '     #6 Oil';

if Fuel <> AuxFuel then
   writeln (' The main fuel and the auxiliary fuel are different.')
else
   writeln (' The main fuel and the auxiliary fuel are the same.')
=>
 The main fuel and the auxiliary fuel are different.
```

The other relational operators, <, <=, >, and >=, can also be used to compare string-type variables. The strings are compared character by character from left to right until there is a difference between the strings. Then the relational operator is applied. Whether a value of true or false is returned depends on the order of the characters in the character set used by your computer. As you know, the two most common codes are the ASCII and EBCDIC codes, listed in Appendix A. In either character set, the following boolean expressions are all true:

$$' oil' < 'oil\ '$$

$$'gas\ ' < 'oil\ '$$

$$'coal' < 'coat'$$

But the expression

$$'Oil' < 'oil'$$

will be false in EBCDIC and true in ASCII.

Instead of using the tables in the appendix, you can use the standard functions **pred, succ,** and **ord** to determine the order in which characters are stored. **Pred, succ,** and **ord** all depend on the character code that is used on your computer system. For example, in ASCII the predecessor of 'j' is 'i' as you would expect, but in EBCDIC the predecessor of 'j' is an unprintable character. Before using the boolean operators or the standard functions with character strings, become familiar with the code that your system uses.

The function **ord** returns the decimal equivalent of the binary code that is used to store a character within the computer. For example, if you look in the ASCII table in Appendix A, you will see that the character '0' has the code 48 in decimal or 110000 in binary. The binary code is the way that the character '0', as opposed to the number 0, is stored.

Arrays of string-types

You can also create an array composed of string-type elements. If each packed array represents a word, the array of words might represent a sentence, or the array of words might contain the names of entities stored in an array, as in

```
const  MaxLetter = 10;           (* maximum number of letters in a name. *)
       MaxClass = 300;           (* maximum number of classes of fuel plant. *)

type   CharString = packed array [1..MaxLetter] of char;
       FuelArray = array [1..MaxClass] of CharString;

var    FuelName : FuelArray;     (* fuel name for each class of fuel plant. *)
       ClassNum : integer;       (* index for FuelName array. *)
```

FuelName is equivalent to a two-dimensional array that is 300 by 10. However, if **FuelName** is declared to be an **array** of a **packed array of char,** it can be visualized as a one-dimensional array:

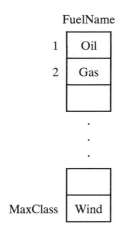

It is easier to manipulate the array as a one-dimensional array because only one subscript is required to identify the fuel type for a plant. For example, the name of the fuel can be specified in a single statement:

```
FuelName [ClassNum] := 'hydro     ';
```

Note that the number of characters between the apostrophes equals the dimension of **CharString.**

It is possible, although not usually necessary, to use two subscripts to address an element in a variable that is an array of an array. For example, in the following fragment, the third character is picked out of the name of a fuel class:

```
var    ThirdChar : char;
         .
         .
       FuelName [ClassNum] := 'hydro     ';
       ThirdChar := FuelName [ClassNum, 3];
```

ThirdChar has a value of 'd', given the assignment statements above.

Solution to Case Study 18

The steps outlined at the beginning of the chapter will be covered in order of the ease of their implementation. Step 3, replacing the numbers with the enumerated data type, can be carried out with a text editor. The integers used in the internal scheme can be automatically changed to the equivalent enumerated identifiers. Because enumerated types are equivalent to integers, no programming changes are required other than defining the enumerated type. The only difference is that the program is easier to read because the fuels are identified by name rather than by number.

The solution to Step 4, converting the enumerated identifiers to character strings for output, can be implemented using a case statement, as illustrated in the procedure **WriteName** developed in Section 11.1. One problem with any such output procedure is that if a new fuel type is added, the procedure must be modified to include the new fuel type. An alternative solution is to store the standard fuel names in an array of names that is indexed over the enumerated type:

```
const MaxLetter = 10;              (* maximum number of letters in a name. *)

type  CharString = packed array [1..MaxLetter] of char;
      FuelType = (Oil, Gas, Nuclear, Coal, Hydro, Sun, Wind );
      NameType = array [Oil..Wind] of CharString;

var   StandardName : NameType;     (* array of standard fuel names. *)
      Fuel : FuelType;             (* for loop control variable. *)
```

The standard names can be set in a procedure:

```
procedure InitializeNames (var Standards : NameType);
(*      This procedure initializes the names in the array Standards.     *)
(*      Any time a new fuel type is added, its name must be added to     *)
(*      this procedure.  The name of the new fuel must also be added to  *)
(*      the data type FuelType.                                          *)
begin
   Standards [Oil]     := 'Oil       ';
   Standards [Gas]     := 'Gas       ';
   Standards [Nuclear] := 'Nuclear   ';
   Standards [Coal]    := 'Coal      ';
   Standards [Hydro]   := 'Hydro     ';
   Standards [Sun]     := 'Sun       ';
   Standards [Wind]    := 'Wind      '
end; (* procedure InitializeNames *)
```

Once the names have been set at the beginning of the program by invoking the procedure,

> InitializeNames (StandardName)

the standard names can be treated like constants within the program, since they are never reset once they are initialized. The procedure to write the fuel name can be modified to

```
procedure PrintName (Standards : NameType; Fuel : FuelType);
(*      PrintName prints the name of the Fuel, which is an enumerated   *)
(*      type.  InitializeNames must be executed prior to this           *)
(*      procedure's invocation.                                         *)
begin
  write (Standards [Fuel])
end; (* procedure PrintName *)
```

With the data structure for character strings, **PrintName** is no longer necessary. Once the standard names have been defined, the variable **StandardName[Fuel]** can be used anywhere in the program where the name of the fuel is needed as a character string. A data structure has replaced the executable statement.

The solution for Step 2, converting a fuel name from a string to an enumerated type, can be implemented using a function that tests the value of a string and returns the equivalent enumerated type. The function could consist of a long sequence of **if** statements that compare the string to every possible fuel name. However, the function would be very long and would have to be modified every time a new fuel type was added. An alternative solution is to compare the fuel name to the standard fuel names stored in the character array:

```
function ConvertToScalar (Standards : NameType;
                          FuelString : CharString) : FuelType;
(*      ConvertToScalar takes the fuel name passed to the procedure in  *)
(*      FuelString and converts it to the enumerated type FuelType.     *)

var   FuelIndex : FuelType;     (* for statement control variable. *)

begin
  ConvertToScalar := Oil;       (* assign a value to ConvertToScalar in *)
                                (* case FuelString does not have the    *)
                                (* value of a standard string.          *)
  for FuelIndex := Oil to Wind do
    if Standards [FuelIndex] = FuelString then
      ConvertToScalar := FuelIndex
end; (* function ConvertToScalar *)
```

If the function is invoked with a correct argument, there is no need for the first executable statement that gives the function **ConvertToScalar** a value in case no match is found in the **StandardName** array. However, in any program there is always the possibility that something will go wrong, so a little protection does not hurt. How would you rewrite the function so that the user was warned if **ConvertToScalar** was essentially undefined?

Reading fuel names from input is the most difficult step. Because enumerated types cannot be read, it is necessary to have a proxy for the fuel type in the input file. If character strings are used, the spelling, spacing, and capitalization in the input string must be the same as in the standard string so that the input string can be compared to the standard string. The spacing and capitalization can be taken care of by procedures and functions that manipulate strings to convert capital to small letters and to remove spaces. However, spelling and alternative phrasing remain a problem. Another alternative is to require that the fuel types be entered as numbers. For example, a plant with oil for fuel would be listed in the data file as having fuel class 0. If the data on the power plants has to be entered only once into a large file by someone who knows the program, this solution may be acceptable.

A final, and increasingly common, solution is to write a program that allows the entire database to be constructed interactively. The fuel types can be presented in a menu from which the user must choose. Using an interactive program to construct the database can eliminate the problem of recognizing character strings and allows data to be checked before the program is run. For example, the following procedure could be used to construct part of the database for the power plants:

```
procedure EnterFuelType (Standards: NameType; NewName : CharString;
                         var PlantType : FuelType);
(*      This procedure asks the user to enter the fuel type of the plant   *)
(*      that has just been entered and passed into NewName.                *)

var    Fuel : FuelType;        (* for statement control variable. *)
       FuelNum : integer;      (* temporary variable for the fuel type. *)
begin
  writeln;
  writeln (' What type of plant is ', NewName, '?');
  writeln;
  for Fuel := Oil to Wind do
    begin
      write (ord (Fuel) + 1:3, '  ');
      PrintName (Standards, Fuel);
      writeln
    end;
  writeln;
  writeln (' Enter the number of the fuel for this plant');
  readln (FuelNum);

  for Fuel := Oil to Wind do
    if (FuelNum - 1) = ord (Fuel) then
      PlantType := Fuel;
```

```
        writeln;
        write (NewName, ' is a ');
        PrintName (Standards, PlantType);
        writeln (' plant')
     end; (* procedure EnterFuelType *)
```

Because a complete program already exists, all that is needed for the case study is to create a test program that will allow the new procedure to be debugged before it is included in the main program. Program 11.2 **TestFuels** is set up to allow the programmer to test the procedure **EnterFuelType** under different conditions.

PROGRAM 11.2   TestFuels

```
program TestFuels (input, output);
(*      Program TestFuels is a program designed to test the procedure    *)
(*      EnterFuelType that will be used as part of a larger program.     *)

const MaxLetter = 10;              (* maximum number of letters in a fuel *)
                                   (* type name. *)

type  CharString = packed array [1..MaxLetter] of char;*)
      FuelType = (Oil, Gas, Nuclear, Coal, Hydro, Sun, Wind );
      NameType = array [Oil..Wind] of CharString;

var   StandardName : NameType;     (* array of standard fuel names. *)
      PlantName : CharString;      (* name of the plant for which the fuel *)
                                   (* type will be entered. *)
      PlantFuel : FuelType;        (* name of the fuel for PlantName. *)
      FuelNum : integer;           (* fuel number of PlantName. *)

(*$i readstring *)
(*$i initializenames *)
(*$i printname *)
(*$i enterfueltype *)
(*$i converttoscalar *)


(*---------------------------------------------------------------------*)
begin (* main program *)
  InitializeNames (StandardName);

  writeln (' What is the name of the next power plant in the data base?');
  ReadString (MaxLetter, PlantName);
  writeln;
  writeln;

  EnterFuelType (StandardName, PlantName, PlantFuel);
  PlantFuel := ConvertToScalar (StandardName, StandardName [PlantFuel]);
```

```
     FuelNum := ord (PlantFuel) + 1;
     writeln (PlantName, ' belongs to class number ', FuelNum:1)
end. (* program TestFuels *)
```

```
=>
 What is the name of the next plant in the data base?
□Yankee 1

 What type of plant is Yankee 1  ?

    1  Oil
    2  Gas
    3  Nuclear
    4  Coal
    5  Hydro
    6  Sun
    7  Wind

 Enter the number of the correct fuel
□3

    Yankee 1 is a  Nuclear    plant
    Yankee 1 belongs to class number 3
```

APPLICATION
Searching

Searching and sorting are two tasks that are easily performed by people. Given the numbers

$$10 \quad 750 \quad -1 \quad 5670 \quad 2 \quad 32 \quad 45 \quad 60 \quad 209$$

you could easily pick out the largest number, the smallest number, any duplicate numbers, all the even numbers, etc. You could also quickly arrange the numbers in ascending or descending order. Writing a Pascal program that accomplishes these same tasks requires some work in order to translate an unconscious thought process into a computer algorithm. If the list is short and only has to be searched or sorted once, it may be easier to have a person do the searching and sorting. If, however, you are preparing a phone book or rearranging names in a file, writing a computer program will be easier. This section will discuss two algorithms for searching and four algorithms for sorting.

CASE STUDY 19
Searching for a power plant

A user of the program that was developed for Case Study 18 in the beginning of the chapter now would like to be able to search through the plants by name to find information on a particular power plant. Write a procedure that will search the database for the name of a power plant entered by the user.

Algorithm development

The problem for the case study is to create an algorithm that will allow a user to enter the name of a power plant and then receive the information that has been stored on that plant. Because the program to create the database already exists, the only unsolved problem is the method for searching through the database to find the power plant name. This section will develop the algorithms for two different searching techniques: the sequential search and the binary search.

Sequential search

To find the name of a particular power plant in a list of names, you can search through the list, comparing each plant's name to the given name. Once a match is found, the index of the plant is known, and once the index is known, the rest of the information on the plant can be retrieved. The algorithm can be outlined as follows:

ALGORITHM 11.1 First iteration of an algorithm for a sequential search

Loop through the plants
   For each plant, compare its name to the given name
   If the names match, then save the plant index
end loop

A **for** loop would be simple to implement because the number of plants is known. However, you know that the loop can be stopped as soon as a match is made, and therefore a **repeat-until** or **while-do** loop could also be used. The algorithm can be modified to

ALGORITHM 11.2 Alternative algorithm for a sequential search

repeat
   get the next plant
until the name matches the given name

Both Algorithms 11.1 and 11.2 are called *sequential searches* because the list of plants is searched in the order in which the plants are stored.

You must now decide which type of loop to use. Compare the two algorithms: You can see that if there are n names in the list, you will always make n comparisons using the **for** loop. If the name you are looking for is the last name in the list, the second algorithm also makes n comparisons. But if the name you are searching for is equally likely to be anywhere in the list, you can expect on the average to make $n/2$ comparisons using the second algorithm. So, in the worst case the two algorithms make the same number of comparisons, but on average the second algorithm makes half as many comparisons as the first algorithm. The second algorithm is more efficient.

To code the search algorithm in Pascal, you must know how the names are stored. Assume that the plant names are stored in an array called **PlantName**, which is of type **NameArray**:

```
const  MaxLetter = 10;              (* maximum number of letters in a *)
                                    (* character string. *)
       MaxPlants = 9;               (* maximum number of plants in the *)
                                    (* data base. *)

type   CharString = packed array [1..MaxLetter] of char;
       NameArray = array [1..MaxPlants] of CharString;

var    PlantName : NameArray;       (* names of the fuel plants in the *)
                                    (* data base. *)
       NumberOfPlants : integer;    (* number of plants in the data base. *)
       SearchName : CharString;     (* name of plant to be searched for. *)
```

The name to be searched for is stored in the variable **SearchName,** which is of type **CharString.**

The following program fragment implements the algorithm:

```
PlantIndex := 0;
repeat
  PlantIndex := PlantIndex + 1
until PlantName [PlantIndex] = SearchName
```

What happens if no plant has the name **SearchName**? When the name of the last plant is compared to **SearchName** and no match is found, the counter **PlantIndex** is incremented. Either **PlantIndex** will have a value outside the bounds of the array index or the value of **PlantName[PlantIndex]** will be undefined, so a run-time error will occur. The fragment can be rewritten so the search stops either when a match is made or when the end of the array is reached:

```
PlantIndex := 0;
repeat
  PlantIndex := PlantIndex + 1
until (PlantName [PlantIndex] = SearchName) or
      (PlantIndex = NumberOfPlants);

if PlantName [PlantIndex] <> SearchName then
   writeln (' Plant ', SearchName, ' is not in the data base')
```

Notice that you cannot test the value of **PlantIndex** to see whether the search was successful. If the **SearchName** is the last name in the list, both conditions of the **until** statement will be true when the loop is exited. If the algorithm is implemented as a procedure, there must be a signal to the main program when the name is not found. A boolean variable can be used for this:

```
procedure FindName (Names : NameArray; SearchFor : CharString;
                    var PlantIndex : integer; var Found : boolean);
(*      FindName searches sequentially through NameArray for the name   *)
(*      SearchFor. If the name is found, PlantIndex is set to the       *)
(*      array index where the name was found and Found is set to true.  *)
(*      If the name is not found, Found remains false.                  *)
begin
  Found := false;         (* initialize the flag and the plant index. *)
  PlantIndex := 0;

  repeat
    PlantIndex := PlantIndex + 1
  until (Names [PlantIndex] = SearchFor) or (PlantIndex = NumberOfPlants);

  if Names [PlantIndex] = SearchFor then Found := true
end; (* procedure FindName *)
```

Binary search

In the problem for Case Study 19, suppose that the plant names are entered in alphabetical order. The sequential search could not take advantage of this fact since it starts at the beginning of the array and checks every name even if the name being searched for starts with a letter at the end of the alphabet.

You can take advantage of the fact that the names are in alphabetical order. Think about how you look up a name in a telephone directory: you open the phone book to approximately the right section, flip several pages at a time until you get to a page with the correct first letter, flip pages a few more times until you get to the page with the correct first and second letter, and so on until you have located the name.

A computer can use an algorithm similar to the method that you use to find a name in a phone book. Start by dividing the phone book in two. Is the name being searched for in the first half or the second half? Suppose it is in the second half. Divide the second half in two. Now the search is confined to the third and fourth quarters of the phone book. Is the name in the third or the fourth quarter? Suppose it is in the third quarter. Divide the third quarter in half. Continue dividing the sections in half until the name is found. This algorithm is a *binary search*.

The Pascal procedure **BinarySearch** performs a binary search for the name stored in **SearchName**. The section being searched is between the array indices **HighIndex** and **LowIndex**. If **SearchName** is greater than the name stored in **Names[Half]**, which is halfway down the current section, then **SearchName** must be between **Half + 1** and **HighIndex**. So the array index **LowIndex** is reset to **Half + 1**. Similarily, if **SearchName** is less than **Names[Half]**, then **SearchName** must be between **LowIndex** and **Half − 1**. So **HighIndex** is reset to **Half − 1**. What happens if **SearchName** is misspelled or not in the list? The algorithm must include the condition that **SearchName** is not found as one of its stopping conditions.

```
LowIndex := 1;                  (* array index at which search begins. *)
HighIndex := NumberOfPlants;    (* array index at which search ends if *)
                                (* plant has not been found. *)
PlantIndex := 0;                (* the array index of the plant being *)
                                (* searched for once it is found. *)
Found := false;                 (* if found is false at the end of the *)
                                (* search, the search was unsuccessful. *)

repeat
  Half := (HighIndex + LowIndex) div 2;
  if PlantName [Half] = SearchName then
    begin
      PlantIndex := Half;
      Found := true
    end
  else
    if SearchName > PlantName [Half] then
      LowIndex := Half + 1
    else
      HighIndex := Half - 1
until (PlantIndex <> 0) or (LowIndex > HighIndex)
```

This algorithm depends on the fact that the letters are stored internally in ascending order; that is, a < b. Suppose they were stored in descending order. How would the algorithm change?

The binary search takes longer to figure out and code than does a sequential search. However, if the data is ordered and there is a lot of data, a binary search will be faster than a sequential search except in very special cases. A binary search will not work for data that is not ordered. For unordered data, a sequential search must be used.

Solution to Case Study 19

Assuming that the power plant names are entered in the database in a random order, you must use a sequential search. The database for the program in Case Study 19 has already been created, so you need only a test program to call the procedure **SequentialSearch.** Once the test program is running satisfactorily, the procedure **SequentialSearch** can be incorporated into the larger power plant program. Program 11.3 **TestSearch** allows the user to test the search program with different input data.

PROGRAM 11.3   TestSearch

```
program TestSearch (input, output);
(*      Program TestSearch tests the sequential search procedure FindName.  *)
(*      The program is set up to allow the user to search for several       *)
(*      different names.                                                    *)

const MaxLetter = 10;              (* maximum number of letters in a *)
                                   (* character string. *)
      MaxPlants = 9;               (* maximum number of plants in the *)
                                   (* data base. *)

type  CharString = packed array [1..MaxLetter] of char; *)
      NameArray = array [1..MaxPlants] of CharString;
      IntArray = array [1..MaxPlants] of integer;

var   PlantName : NameArray;       (* names of the fuel plants in the *)
                                   (* data base. *)
      NumberOfPlants : integer;    (* number of plants in the data base. *)
      Capacity : IntArray;         (* capacity of the plants. *)
      PlantIndex : integer;        (* array index of the plant when it is *)
                                   (* found. *)
      Found : boolean;             (* if the plant is not in the data *)
                                   (* base, Found is set to false. *)
      SearchName : CharString;     (* name of plant being searched for. *)
      plantfile : text;            (* file in which the plant data is *)
                                   (* stored. *)

(*$i readplants *)
(* the procedure readplants reads the PlantName array and the Capacity *)
(* array from the file plantfile.  *)

(*$i findname *)
(*$i readstring *)
(*$i yesentered *)
```

```
(*------------------------------------------------------------------*)
begin (* main program *)
  ReadPlants (NumberOfPlants, Capacity, PlantName);

  repeat
    Found := false;
    writeln;
    writeln (' What is the name of the plant that you want to find?');
    ReadString (MaxLetter, SearchName);
    writeln;

    FindName (PlantName, SearchName, PlantIndex, Found);
    if Found then
      writeln (SearchName, ' is plant number ', PlantIndex:1)
    else
      writeln (SearchName, ' is not in the data base');
    writeln (' Do you want to continue ');
  until YesEntered
end. (* program TestSearch *)
```

```
=>
 What is the name of the plant that you want to find?
□SeaRaven 1

SeaRaven 1 is not in the data base

 Do you want to continue
 Enter yes or no (y/n)
□y
 What is the name of the plant that you want to find?
□SeaRaven#1

SeaRaven#1 is plant number 1

 Do you want to continue
 Enter yes or no (y/n)
□n
```

 In the sample output of the program **TestSearch,** the plant name SeaRaven 1 was not found because it was entered in the database as SeaRaven #1. How could this type of error be prevented? How would the strategy for preventing this error change if there were 25 power plants in the database or 5000 power plants?

APPLICATION
Sorting

This section describes three different types of sorting algorithms. The first one, the selection sort, is easy to understand but not very efficient. The second one, the bubble sort, is comprehensible, and it is usually more efficient than the selection sort. The third one, the insertion sort, is more difficult to follow, but is reasonably efficient.

CASE STUDY 20
Sorting power plants

Another user of the program that was developed for Case Studies 18 and 19 would like an output report of the names of the power plants arranged in descending order by their megawatt power rating. The plants must be sorted by power rating before the report is printed. Write a Pascal procedure that will allow the user to sort the power plants in descending order of their megawatt capacity.

Algorithm development

Again, the main program for the case study exists and the problem is to modify it to meet another user's needs. Given the plant names in random order, how can you sort them in order of decreasing capacity? In the solution to the case study, a test program will be written to compare the three algorithms to sort the data.

So that the algorithms will be understandable, the following numbers will be used for each of the examples. The numbers are stored in an array called **Capacity,** which is of type **IntArray.**

250	750	50	1500	100	500	850	1000	1225

The procedure **Exchange** which was developed in Chapter 6 and which exchanges the integer values stored in array locations **First** and **Second** is used in two of the sorting solutions:

```
procedure Exchange (var First, Second : integer);
(*      Exchange puts the variable that is passed in First into Second   *)
(*      and the variable that is passed in Second into First, using a    *)
(*      temporary variable.                                              *)

var   Temp : integer;       (* temporary variable to hold value of First *)
                            (* during the exchange. *)
begin
  Temp := First;
  First := Second;
  Second := Temp
end;   (* procedure Exchange *)
```

For example,

Exchange (Capacity[8], Capacity[9])

would result in the array

| 250 | 750 | 50 | 1500 | 100 | 500 | 850 | 1225 | 1000 |

Selection sort

Looking at the initial array

| 250 | 750 | 50 | 1500 | 100 | 500 | 850 | 1000 | 1225 |

you can easily see that the largest number, 1500, must go at the beginning of the array; however, you need an algorithm that the computer can use to find the largest number in the array. The simplest algorithm for finding the largest number in an array is to search through the array, comparing the numbers and keeping track of the largest number so far:

```
IndexOfLargest := 1;
for Index := 2 to NumberOfPlants do
  if Capacity [IndexOfLargest] < Capacity [Index] then
    IndexOfLargest := Index
```

After this code has been executed, the index, that is, the *position* in the array, of the largest number is stored in **IndexOfLargest**. In this example, **IndexOfLargest** is 4. You can then execute **Exchange**:

```
Exchange (Capacity [IndexOfLargest], Capacity [1])
```

The execution of **Exchange** will cause the number in position 4 in the array to be exchanged with the number in position 1, resulting in

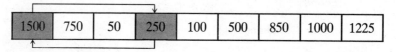

Since the largest number is now in the first position, you can ignore it and look for the largest number in the remainder of the array. So the search for the largest number is repeated, starting at position 2 in the array:

```
IndexOfLargest := 2;
for Index := 3 to NumberOfPlants do
  if Capacity [IndexOfLargest] < Capacity [Index] then
    IndexOfLargest := Index
```

When you find the next largest number, it will be exchanged with the second element of the array. Since 1225 is the next largest number, the exchange results in

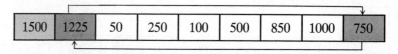

The algorithm repeats $n - 1$ times until the array is in descending order. The complete algorithm for the selection sort is

```
for Pass := 1 to (NumberOfPlants - 1) do
  begin
    IndexOfLargest := Pass;

    for Index := (Pass + 1) to NumberOfPlants do
      if Capacity [IndexOfLargest] < Capacity [Index] then
        IndexOfLargest := Index;

    Exchange (Capacity [IndexOfLargest], Capacity [Pass])
  end    (* for Pass *)
```

Application: Sorting

The procedure **SelectSort** uses the selection sort algorithm in a generalized procedure that sorts an array of integers.

```
procedure SelectSort (ArrayLimit : integer; var SortArray : IntArray);
(*      Procedure SelectSort uses a selection sort to sort the values   *)
(*      in SortArray from largest to smallest and returns the sorted    *)
(*      array in SortArray.  The procedures requires that IntArray has  *)
(*      been declared as an array type with the dimension ArrayLimit.   *)
(*      SelectSort uses the procedure Exchange.                         *)

var   Pass : integer;                (* for loop control variable -- counts *)
                                     (* the number of passes through the    *)
                                     (* array so far. *)
      Index : integer;               (* for statement control variable. *)
      IndexOfLargest : integer;      (* index of the largest element so far. *)

begin
  for Pass := 1 to ArrayLimit - 1 do
    begin
      IndexOfLargest := Pass;

      for Index := Pass to ArrayLimit do
        if SortArray [Index] > SortArray [IndexOfLargest] then
          IndexOfLargest := Index;

      Exchange (SortArray [IndexOfLargest], SortArray [Pass])
    end   (* for Pass *)
end; (* procedure SelectSort *)
```

Notice that the selection algorithm ignores the order of the original array. For example, 750 obviously does not belong at the end of the array. It was closer to its proper position before the second exchange was made. No matter what the order of the original data, the selection sort will always make $n(n + 1)/2$ comparisons in the process of finding the largest number remaining. It always makes $n - 1$ exchanges.

The selection sort executes the procedure **Exchange** even if an element is in the correct location. How could you modify **SelectSort** to make it more efficient?

Bubble sort

The bubble sort is so named because during the sort the large numbers seem to bubble slowly to the top and the small numbers seem to bubble quickly to the bottom. To see how the bubble sort works, start at the beginning of the original array:

250	750	50	1500	100	500	850	1000	1225

The first value is 250, the second is 750. Since the numbers are to be sorted in descending order, 750 should come before 250, so exchange the two values. In Pascal, the code is

```
if Capacity[1] < Capacity[2] then
    Exchange (Capacity[1], Capacity[2])
```

Now the array looks like

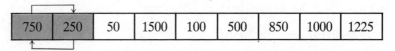

Compare the new value of **Capacity[2]** to the next value in the array. **Capacity[3]**, 50, is less than 250, so the values are not exchanged. Compare **Capacity[3]** to **Capacity[4]**; 1500 is bigger than 50, so exchange them:

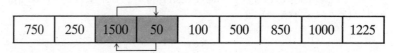

Now compare **Capacity[4]** to **Capacity[5]**; since 100 is bigger than 50, exchange them:

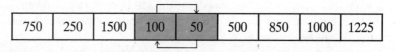

Once the smallest number is found, it will continue to bubble to the bottom, leaving the order of the rest of the array unchanged. Thus the number 50 will continue to bubble to the bottom. After the first pass, the array **Capacity** looks like

750	250	1500	100	500	850	1000	1225	50

Application: Sorting

In Pascal, the code for the first pass through the array is

```
for Index := 1 to (NumberOfPlants - 1) do
    if Capacity [Index] < Capacity [Index + 1] then
        Exchange (Capacity [Index], Capacity [Index + 1])
```

After the first pass through the array, the bubble sort always has the smallest number at the end of the array. Now that the smallest number, 50, is where it belongs, the rest of the array has $n - 1$ elements that remain to be sorted. Using the same algorithm, starting again at the beginning, compare 750 to 250. They are in order, so move on to 250 and 1500. They are out of order, so exchange them:

| 750 | 1500 | 250 | 100 | 500 | 850 | 1000 | 1225 | 50 |

Now compare 250 to 100. They are in order, so go on to compare 100 to 500. They are out of order, so exchange them:

| 750 | 1500 | 250 | 500 | 100 | 850 | 1000 | 1225 | 50 |

Now that 100, the smallest of the $n - 1$ remaining numbers, has been found, it will continue to be exchanged with each entry in the array until it is in position $n - 1$ in the array:

| 750 | 1500 | 250 | 500 | 850 | 1000 | 1225 | 100 | 50 |

Now that the two smallest numbers are at the bottom of the list, the $n - 2$ remaining numbers can be sorted. Start back at the top of the list, going from 1 to $n - 2$, moving the smallest remaining number to position $n - 2$. This algorithm is repeated $n - 1$ times until all the numbers are in order. In Pascal, the bubble sort algorithm can be coded as follows:

```
for Pass := 1 to (NumberOfPlants - 1) do
    for Index := 1 to (NumberOfPlants - Pass) do
        if Capacity [Index] < Capacity [Index + 1] then
            Exchange (Capacity [Index], Capacity [Index + 1])
```

Notice that as the small numbers bubble to the bottom, the large numbers bubble to the top. In each pass through the array, 1500 moves one step closer to the top where it belongs. So the array might possibly be in order before pass $n - 1$. How can you tell if it is in order? If no exchanges are made during any pass through the array, the array is in order. Therefore, you can add a boolean variable that is set to false if **Exchange** is called. If the variable is set to true at the beginning of a pass and is still true at the end of a pass, the array is in order.

```
Pass := 1;
Done := false;

while (not Done) and (Pass < NumberOfPlants) do
  begin
    Done := true;
    for Index := 1 to (NumberOfPlants - Pass) do
      if Capacity [Index] < Capacity [Index + 1] then
        begin
          Exchange (Capacity [Index], Capacity [Index + 1]);
          Done := false
        end;
    Pass := Pass + 1
  end (* while *)
```

The procedure **BubbleSort** is a generalized procedure for sorting an integer array.

```
procedure BubbleSort (ArrayLimit : integer; var SortArray : IntArray);
(*      This procedure sorts the values in SortArray from largest to    *)
(*      smallest using a bubble sort.  The sorted array is returned in  *)
(*      SortArray.  IntArray must have been previously declared as an   *)
(*      array type with dimension ArrayLimit.  The bubble sort skips    *)
(*      parts of the array that are already in order and thus           *)
(*      unnecessary comparisons are not made.  The procedure Exchange   *)
(*      is invoked from within this procedure.                          *)

var     Index : integer;           (* for loop control variable. *)
        LastIndex : integer;       (* index of the last array element that is *)
                                   (* not yet in order. *)
        Pass : integer;            (* counter to keep track of the number of *)
                                   (* passes through the array. *)
        SortDone : boolean;        (* flag that is initialized to false and *)
                                   (* set to true when no exchanges are made *)
                                   (* during a pass through the array. *)
```

```
begin
  SortDone := false;            (* initialize the flag and the number of *)
  Pass := 1;                    (* passes. *)

  while (not (SortDone)) and (Pass < ArrayLimit) do
    begin
      SortDone := true;
      LastIndex := ArrayLimit - Pass;
      for Index := 1 to LastIndex do
        begin
          if SortArray [Index] < SortArray [Index + 1] then
            begin
              Exchange (SortArray [Index], SortArray [Index + 1]);
              SortDone := false;
            end (* if *)
        end; (* for *)
      Pass := Pass + 1
    end     (* while *)
end; (* procedure BubbleSort *)
```

In the best case, if an array is already in order, the bubble sort will make $n - 1$ comparisons and no exchanges. In the worst case, if the array is in reverse order, the bubble sort will make $n(n - 1)/2$ comparisons and $n(n - 1)/2$ exchanges. In the worst case, the bubble sort is very inefficient, so it is best to use the bubble sort on an array that is already partially ordered.

Insertion sort

In the bubble sort, many unproductive exchanges are made as a value bubbles to the top or to the bottom; for example, 1500 is exchanged with every element until it reaches the top. Why not just put it at the top? That is what an insertion sort does. An insertion sort works by finding elements that are out of order and putting them in order. The insertion sort creates successively longer strings of ordered numbers until the whole array is ordered. Scanning the array from the beginning, you can see that the first element that is out of order is 750:

| 250 | 750 | 50 | 1500 | 100 | 500 | 850 | 1000 | 1225 |

Remove 750 from the array:

| 250 | | 50 | 1500 | 100 | 500 | 850 | 1000 | 1225 |

Move 250 down into the spot that was left empty when 750 was removed:

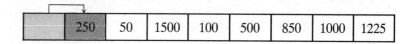

| | 250 | 50 | 1500 | 100 | 500 | 850 | 1000 | 1225 |

Then put 750 into the place left empty when 250 was removed:

| 750 | 250 | 50 | 1500 | 100 | 500 | 850 | 1000 | 1225 |

This process is an exchange stated in terms of the insertion sort algorithm.

Scanning from the beginning again, find the next element that is out of order: 1500. Rather than exchanging 1500 with another number, remove it from the array and save it:

| 750 | 250 | 50 | | 100 | 500 | 850 | 1000 | 1225 |

Comparing 1500 to the sequence of numbers 750, 250, and 50, you can see that 1500 is greater than all of them, and so it should go before all of them in the array. Move the three numbers keeping them in order, to fill the hole left by the removal of 1500, and insert 1500 in the remaining space:

| 1500 | 750 | 250 | 50 | 100 | 500 | 850 | 1000 | 1225 |

In Pascal, the logic of the preceding paragraph can be coded as follows:

```
SaveCapacity := Capacity[4];
IndexOfHole := 3;
```

Application: Sorting

```
while (IndexOfHole >= 1) and (SaveCapacity > Capacity [IndexOfHole]) do
  begin
    Capacity [IndexOfHole + 1] := Capacity [IndexOfHole];
    IndexOfHole := IndexOfHole - 1
  end;
Capacity [IndexOfHole + 1] := SaveCapacity
```

Now, search for the next number that is out of order. That number is 100, the fifth element of the array. Remove 100 from the array and save it:

| 1500 | 750 | 250 | 50 | | 500 | 850 | 1000 | 1225 |

Compare 100 to 50; 100 should come before 50. Compare 100 to 250; 100 should come after 250. So move 50 into the hole left when 100 was removed:

| 1500 | 750 | 250 | | 50 | 500 | 850 | 1000 | 1225 |

Put 100 in the empty space:

| 1500 | 750 | 250 | 100 | 50 | 500 | 850 | 1000 | 1225 |

Now there are five numbers that are in order. The insertion algorithm continues until all the numbers are in order. To complete the Pascal code, all you need is an outer loop to go through the array:

```
for Pass := 2 to NumberOfPlants do
  begin
    SaveCapacity := Capacity [Pass];
    IndexOfHole := Pass - 1;
    while (IndexOfHole >= 1) and (SaveCapacity > Capacity [IndexOfHole]) do
      begin
        Capacity [IndexOfHole + 1] := Capacity [IndexOfHole];
        IndexOfHole := IndexOfHole - 1
      end;
    Capacity [IndexOfHole + 1] := SaveCapacity
  end (* for loop *)
```

Procedure **InsertionSort** is a generalized algorithm for sorting an integer array. Note that the array being sorted has been redimensioned to start at 0. This modification is necessary because, on many compilers, both expressions in the compound boolean expression

```
(IndexOfHole >= 1) and (SaveCapacity > SortArray [IndexOfHole])
```

will be evaluted even if the first expression is false. Thus **SortArray[0]** must have a value to prevent an error from arising because of an undefined variable or an array index that is out of range when **IndexOfHole** has the value of 0 the last time the loop is executed. Can you think of other ways around this problem?

```
procedure InsertionSort (ArrayLimit : integer; var SortArray : IntArray);
(*     This procedure sorts the values in SortArray from largest to     *)
(*     smallest using an insertion sort.  The sorted array is returned  *)
(*     in SortArray.  IntArray must have previously been declared as an *)
(*     array type.  The array must be dimensioned from 0.               *)

var    Pass : integer;              (* counter for the number of times *)
                                    (* through the array. *)
       IndexOfHole : integer;       (* index for the array element that is *)
                                    (* currently empty. *)
       Save : integer;              (* value of the array element with the *)
                                    (* array index that corresponds to Pass *)
                                    (* -- this value is being saved for *)
                                    (* insertion at the final value of *)
                                    (* IndexOfHole. *)

begin
  SortArray[0] := 0;          (* prevents an undefined value in the while-do. *)

  for Pass := 2 to ArrayLimit do
    begin
      Save := SortArray [Pass];
      IndexOfHole := Pass - 1;

(* The following while-do loop will start execution when a value greater *)
(* than Save is found.  As long as the current value is greater than the *)
(* value stored at location IndexOfHole, move each value down one position *)
(* in the array. *)
```

```
    while (IndexOfHole >= 1) and (Save > SortArray [IndexOfHole]) do
      begin
        SortArray [IndexOfHole + 1] := SortArray [IndexOfHole];
        IndexOfHole := IndexOfHole - 1
      end; (* while *)

    SortArray [IndexOfHole + 1] := Save
  end    (* for *)
end; (* procedure InsertionSort *)
```

The insertion sort is similar to the bubble sort, except that there are fewer exchanges and the exchanges are more efficient. Like the bubble sort, the insertion sort will make $n - 1$ comparisons if the data is already sorted and $n(n - 1)/2$ if the data is in reverse order.

Solution to Case Study 20

Three algorithms have been developed for sorting the power plants. Any one of the sorting procedures can be added to the program and will allow the user to sort the plant names by megawatt power rating.

In order to show you how the different algorithms compare in efficiency, we modified the procedures **SelectSort, BubbleSort,** and **InsertionSort** so that the numbers of comparisons, exchanges, and assignment statements were counted. The output below is from a test program that executed each of the procedures. Each sort algorithm was executed with the original unordered data used throughout this section.

```
The initial order of the array for each sort is:
   250    750     50  1500    100    500    850   1000   1225

For the selection sort,
the number of exchanges was 8,
the number of comparisons was 44, and
the number of assignments was 22.

The final order of the array is:
  1500   1225   1000    850    750    500    250    100     50

For the bubble sort,
the number of exchanges was 26,
the number of comparisons was 52, and
the number of assignments was 52.
```

```
The final order of the array is:
 1500  1225  1000   850   750   500   250   100    50

For the insertion sort,
the number of exchanges was 0,
the number of comparisons was 52, and
the number of assignments was 77.

The final order of the array is:
 1500  1225  1000   850   750   500   250   100    50
```

Summary

Enumerated types make programs easier to read because they allow you to use names instead of numbers as **for** control variables, **case** control expressions, and array indices. The major disadvantage of enumerated types is that their values cannot be read or written.

Subranges are used to limit the values that can be assigned to a variable. A subrange can be declared for any variable that has an ordinal type.

Using packed arrays of characters to represent words or phrases makes input and output easier. In standard Pascal, packed arrays of characters must be of fixed length. Many versions of Pascal have extensions that allow the use of strings of characters of variable length.

Exercises

PASCAL

1. Identify and correct the syntax error(s) in each fragment below:

    (a) ```
        type Gases = (Helium, Neon, Hydrogen, Oxygen, Argon);
             Metals = (Lead, Tin, Aluminum, Zinc, Gold, Silver);
             GroupI = (Hydrogen, Lithium, Sodium, Potassium, Rubidium,
                       Cesium, Francium);
        ```

    (b) ```
        type MachineTools = (Lathe, MillingMachine, Grinder, DrillPress);

        var  Tool : MachineTools;
             Speeds : 20.0..500.0;   (* feet/minute *)
        ```

(c) var Language : (FORTRAN, LISP, C, Pascal, COBOL, BASIC);
.
.
.
writeln (LISP, ' is a very different sort of language');

(d) type CharString : array[1..20] of char;

var Word : CharString;
.
.
.
readln (Word);
writeln (' The word is ', Word)

2. What is the output from the following program fragments?

(a) Word := 'help!';
write (' The word is ');
for Index := 1 to 5 do
   write (Word [6 - Index]:1);
writeln

(b) Element := 'helium   ';
write (' The element is ');

Count := 1;
while Element [Count] <> ' ' do
  begin
     write (Element [Count]:1);
     Count := Count + 1
  end;
writeln;
writeln (' There are ', Count - 1:1, ' letters in the word ', Element)

Problems

PASCAL

1. Write a program that reads a word from the terminal no matter where the word is entered on the line. The program should generate an error message if the word contains any characters that are not letters. The output of the program should be a report of the length of the word and the number of vowels in the word.

2. Using the data on the heights and weights of students from Problem 13 in Chapter 7, write a program that reads the name of each student and prints out

two lists: one of the names of students who are taller than average and one of the names of students who are lighter than average.

3. Rewrite the fuel-consumption program from Problem 18 in Chapter 5 using character arrays. Store the input data, including the names of the aircraft, in arrays. Use a function to compute the fuel consumption. The program should write a well-formatted report giving the name of the aircraft, the passenger-miles per gallon, the seating capacity, and the gallons per mile.

4. The experiment from Problem 3 in Chapter 9 has been expanded to include blocks of four types of metal: lead, brass, aluminum, and iron. The mass and density of each block are measured five times. Write a program to find the average of the measurements. Use an enumerated data type as a dimension in an array and as a control variable in a loop. Use a case statement to label each mass and density by the type of metal.

Lead		Brass		Aluminum		Iron	
mass in kgs	density in gm/cm^3	mass in kgs	density in gm/cm^3	mass in kgs	density in gm/cm^3	mass in kgs	density in gm/cm^3
24.341	11.312	100.214	8.601	24.97	2.678	88.321	7.825
24.024	11.291	99.884	8.599	24.89	2.702	89.434	7.841
25.006	11.298	103.790	8.585	25.00	2.715	88.899	7.826
24.837	11.302	99.521	8.610	24.83	2.683	87.993	7.836
24.422	11.305	101.250	8.593	24.92	2.704	88.675	7.840

5. A palindrome is a word, such as radar, that is the same forward as it is backward. Write a program that reads a word from the terminal, writes the word forward and backward, and states whether it is a palindrome. Assume that the word will have no more than ten letters and no blanks.

6. A sentence can be a palindrome as defined in Problem 5. The sentence "Able was I ere I saw Elba" (attributed to Napoleon) is a palindrome. Write a program that reads a sentence of no more than 80 characters from the terminal and determines whether it is a palindrome. Remember that there will be blanks and capitals. To get fancier, you may want to include punctuation, as in the palindrome attributed to Adam on meeting Eve: "Madam I'm Adam."

7. You have an unproofread list of student names and IDs. The names should be composed only of letters and the IDs only of numbers. Write a program that reads the names and IDs as character arrays and determines which are valid names and which are valid IDs. The typist is erratic, but the keyboard has only letters and numbers so there are no symbols such as #. The program should list the names and IDs as they were read in and indicate which are valid and which are not. Note: It is easier to ask whether a character is or is not a digit than whether it is or is not a letter. Test your program on the following data:

Thomas A. Edison Jr	111223333	Sherlock Holmes	2K6801357
Blaise Pascal Jr	333112222	Na8cy Drew	135792468

Madame Curi3	222113333	Louisa May Alcott	012345678
Leonardo DaVinci	a11234567	Earl Grey	876543210
Albert Einstein Jr	123456789	Rosalyn Yalow	98765432H

8. Write a program to decode a sequence of secret messages. The sender enters a code on the first line of a **textfile** followed by the messages. Each encoded message is a sequence of integers with a 0 marking the end of a message. The number −1 marks the ends of the entire transmission. The code contains the 26 letters of the alphabet in capital letters with no embedded blanks. Set up a 1:1 correspondence between a number and a letter of the code; for example, 2 corresponds to the second letter. Any number larger than 26 should be replaced by a blank (a gap in the message). No message will be longer than 75 encoded integers. The program should print all of the decoded messages. Some sample input is:

ZYWXVUTSRQPONMLKJIHGFEDCBA
21 26 18   9 56 18   8 56 21 12   6 15 56 26 13 23 56 21 12   6 15 56 18   8
56 21 26 18   9   0 19 12   5 22   9 56   7 19   9 12   6 20 19   56   7 19 22 56
21 12 20 56 26 13 23 56 21 18 15   7 19   2 56 26 18   9   0 −1

The decoded message is:

FAIR IS FOUL AND FOUL IS FAIR
HOVER THROUGH THE FOG AND FILTHY AIR

Test your program on the entire transmission:

  3 19 22 13 56   8 19 26 15 15 56   3 22 56   7 19   9 22 22 56 14 22 22   7
56 26 20 26 18 13 56   0 18 13 56   7 19   6 13 23 22   9 56 15 18   20 19   7
13 18 13 20 56 12   9 56 18 13 56   9 26 18 13 56   0   3 19 22 13   56   7 19
22 56 19   6   9 15   2 56 25   6   9 15   2   8 56 23 12 13 22 56   3 19 22 13
56   7 19 22 56 25 26   7   7 15 22   8 56 15 12   8   7 56 26 13 23   56   3 12
13   0 21 26 18   9 56 18   8 56 21 12   6 15 56 26 13 23 56 21 12   6 15 56
18   8 56 21 26 18   9   0 19 12   5 22   9 56   7 19   9 12   6 20 19 56   7 19
22 56 21 12 20 56 26 13 23 56 21 18 15   7 19   2 56 26 18   9   0 −1

9. In Problem 8, after a message is decoded, it must be checked for the key word AIR. If the key word is found, it should be deleted from the message and the user should be informed of the censoring. The word should not be deleted if it is part of another word; that is, AIR should not be deleted from the word fair, nor should the user be notified. The decoded message from Problem 8 would be reported as:

This transmission contained a censored word
FAIR IS FOUL AND FOUL IS FAIR
HOVER THROUGH THE FOG AND FILTHY

10. Write a program that plays tic-tac-toe. The program should tell the user how to enter plays, check the input for invalid entries, display the board after each round of play, devise a reasonable strategy for the computer to follow, and let the user know who won.

11. Write a progam to be used by Trans-City Airlines in assigning seats on airplanes with a capacity of 42 passengers. The seats are in seven rows marked a to g, with six seats per row:

Row	Seat					
a	wl	ml	al	ar	mr	wr
b	wl	ml	al	ar	mr	wr
.	.	.	.	.	.	.
.	.	.	.	.	.	.
.	.	.	.	.	.	.
g	wl	ml	al	ar	mr	wr

"wr" stands for window right, "ml" for middle left, "al" for aisle left, etc.

Seats are assigned on a first come–first served basis. If the seat a passenger wants is taken, seat him or her in the same seat and row on the opposite side; that is, if a passenger requests Seat f,ar and it is taken but Seat f,al is free, assign Seat f,al. If that seat is also taken, start with Row a and find the next empty aisle seat. If all the aisle seats are taken, use the following priorities based on the passenger's first choice:

window: aisle   middle
aisle: window   middle
middle: aisle   window

That is, if a passenger requested a window seat and none are available, assign an aisle seat. If no aisle seats are available, find a middle seat. If a passenger requests a nonexistent seat such as q 9, you may either bump the passenger or assign any available seat.

The output of your program will be a table that shows the name of each person and the proper seat. You may truncate names to make your chart look better. Use enumerated types and subranges, for example, you might use:

```
type Location = (windowl, windowr, aislel, aisler, middlel, middler);
     Condition = (empty, full);
     Row = 'a'..'g';
```

The names and seat choices of the passengers for the next Trans-City flight are stored in a text file in the following format: name row seat.

Newton I	f ar	Byron, A	e wr	Marconi	g mr
Euclid	c ar	Alcott L M	d wr	Thoreau H	e wl
Franklin B	g wr	Bohr N	a al	Cicero	d wl

Pasteur L	a ml	Laplace P S	b ml	Pascal B		d wl	
Boole G	e ar	Euler L	f ar	Einstein A		g wl	
Hilbert D	a rm	Pythagoras	c wl	Planck M		g wr	
Curie P	f al	Curie M	f ml	Shaw G B		g ml	
Mahler G	d wr	Orwell G	a ml	Hardy T		b ar	
Pavlova A	b wr	Stein G	d wl	Hugo V		e ml	
Poe E A	a wl	Mead M	b mr	Astaire F		b wr	
Sartre J P	c rw	Babbage C	e mr	Tolstoy L		h ml	
Archimedes	c wl	Edison T A	f al	Bacon F		d ar	
Bell A G	f ml	Joyce J	e wl	Bessemer H		f al	
Whitney E	e wl	Eliot G	e wl	Boetheus		e wl	
Ethelred	c al	Maxwell J C	e wr	daVinci L		d wr	
Jefferson T	h wr	Darwin C	f wl	Copernicus		f ar	
Tesla N	d wr	Austin J	b mr	Howe E		g ar	
Descartes R	l al	Racine J	f wl	Gauss K		a wr	

APPLICATION

12. Exchange the largest and the smallest elements in a matrix of 20 by 20 real numbers. The output should include (a) the input matrix, (b) the values and original locations of the largest and smallest elements, and (c) the matrix with the elements exchanged.

13. Write a game in which the computer tries to guess the number between 1 and 100 that you are thinking of. Have the computer guess it within seven turns. Help the computer by telling it whether its guesses are too high or too low. (Don't cheat if you want your program to work.) Why is this program harder to write than the program from Problem 1 in Chapter 8?

14. You are given a 12 by 4 table of names that you are to search for a particular name that may or may not be there. Write a program that prints the original table, the name being searched for, and whether the name was found. Search for both a name that is in the table and one that is not. Test your program on the following data:

Marconi	Ethelred	Euclid	Alcott, L. M.
Babbage, C.	Franklin, B.	Bohr, N.	Cicero
Pasteur, L.	Laplace, P. S.	Pascal, Blaise	Descartes, Rene
Euler, L.	Einstein, A.	Gauss, K.	Hilbert, D.
Planck, M.	Maxwell, J. C.	Curie, Pierre	Curie, Marie
Darwin, C.	Mahler, G.	Orwell, G.	Hardy, T.
Pavlova, A.	Stein, G.	Hugo, Victor	Racine, J.
Mead, M.	Astaire, F.	Boetheus	Sartre, J. P.
Tolstoy, L.	daVinci, L.	Archimedes	Edison, T. A.

Copernicus	Bell, A. G.	Joyce, James	Bessemer, H.
Byron, Ada	Boole, G.	Austen, Jane	Bacon, Francis
Thoreau, H.	Pythagoras	Poe, E. A.	Tesla, N.

15. Write an interactive program that searches, by name or by ID, through a Pascal **textfile** containing 20 students' names and IDs. Each name is a string of 20 characters, and each ID is a string of 9 digits. The program should write a message to the terminal asking the user whether a name or an ID will be entered. If a name is entered, the program should search for the name and print the ID; if an ID is entered, it should search for the ID and print the name. If the student is not found, a message should be written. Test your program on the following data:

Byron, Ada	123456789	Zeppelin, Ferdinand	987654321
Edison, Thomas A.	234567890	Diesel, Rudolph	098765432
Euclid	345678901	Bohr, Niels	109876543
Pascal, Blaise	456789012	Ampere, Andre	210987654
Boole, George	567890123	Volta, Count	321098765
Einstein, Albert	678901234	Faraday, Michael	432109876
Copernicus, Nicholas	789012345	Fahrenheit, Gabriel	543210987
Kepler, Johannes	890123456	Maxwell, James C	654321098
Curie, Marie	901234567	Newton, Sir Isaac	765432109
LaGrange, Pierre S	012345678	daVinci, Leonardo	876543210

16. A binary search can be used to find roots of equations. Instead of searching through an array, you search through a function. For example, given the function $f(x) = 3x + 2$, start with end points $x = -3$ and $x = +3$. Since $f(-3) = -7$ and $f(3) = 11$, we know that $f(x)$ must cross the origin at least once between -3 and $+3$:

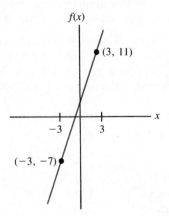

Dividing the interval in half gives $x = 0$. The function $f(x)$ evaluated at $x = 0$ is $+2$, so $f(x)$ must cross the origin between $x = -3$ and $x = 0$. Dividing the interval in half gives $x = -1.5$. The function $f(x)$ evaluated at $x = -1.5$ is -2.5, so $f(x)$ must cross the origin between $x = -1.5$ and $x = 0$. Divide that interval in half, continuing until $f(x)$ is equal to 0.

Write a program that uses a binary search to find the cube root of an equation. Choose your starting points carefully.

17. Write a program that sorts the students from Problem 15 into alphabetical order and then sorts them into numerical order by ID. For practice, use two different types of sorting algorithms.

18. Using the heights and weights of students from Problem 13 in Chapter 7, write a program to print three lists of students: one in alphabetical order, one from lightest to heaviest, and one from tallest to shortest.

19. Write a program that merges the lists of names from Problems 14 and 15. The program should read each list of names into a separate array (use only the first 15 characters for each name), and then create a single array that has all the names in alphabetical order with no duplicate names. However, the merged list will contain different names that refer to the same person, for example, Edison, T. A. and Edison, Thomas A. will both appear in the merged list. Before you begin to write your algorithm, think about ways to eliminate duplicate names.

20. The following table, which is stored in a **textfile**, represents typical power output in megawatts from a power plant over a period of eight weeks.

	Mon	Tue	Wed	Thur	Fri	Sat	Sun
Week 1	207	301	222	302	22	167	125
Week 2	367	60	120	111	301	400	434
Week 3	211	72	441	102	21	203	317
Week 4	401	340	161	297	441	117	206
Week 5	448	111	370	220	264	444	207
Week 6	21	313	204	222	446	401	337
Week 7	213	208	444	321	320	335	313
Week 8	162	137	265	44	370	315	322

Write a program using enumerated data types that prints (a) average daily power output—average power output on Monday for the eight-week period, average power output on Tuesday, etc.; (b) number of days in the eight-week period with greater than average power output; and (c) a list of days sorted by power output. The program should identify the days and list the associated power output. For example, the first item on the list after sorting is Monday week 5, output of 448.

CHAPTER 12

Sets and Program Design

The Pascal data type **set** corresponds closely to the mathematical definition of a set. This chapter is divided into three sections: Section 12.1 on the mathematical definitions and operations for sets, Section 12.2 on the implementation of sets in Pascal, and Section 12.3 on program design and validation.

The material in Section 12.1 is covered in many mathematics courses. If you are familiar with set theory, you may want to skim or skip this section. Some of the material in Section 12.3 was covered earlier in Chapter 4, but it is covered in greater detail in this chapter. Because you can now write more complicated programs and solve more complex problems, you have many more tools at your disposal than you had before; thus the need for program design and validation is greater.

CASE STUDY 21
The chemical elements

A chemistry teacher has requested a program that will help beginning chemistry students learn the periodic table (see Figure 12.1). The teacher would like the program to name an element and then ask the student questions about the element, such as

What group does it belong to?

What is its period?

Is it gas, liquid, or solid?

What is its atomic number?

The program should correct wrong answers and keep track of the number of correct and incorrect answers.

Group	I	II										III	IV	V	VI	VII	O
Period																	
1	H 1																He 2
2	Li 3	Be 4										B 5	C 6	N 7	O 8	F 9	Ne 10
3	Na 11	Mg 12				Transition elements						Al 13	Si 14	P 15	S 16	Cl 17	Ar 18
4	K 19	Ca 20	Sc 21	Ti 22	V 23	Cr 24	Mn 25	Fe 26	Co 27	Ni 28	Cu 29	Zn 30	Ga 31	Ge 32	As 33	Se 34	Br 35	Kr 36
5	Rb 37	Sr 38	Y 39	Zr 40	Nb 41	Mo 42	Tc 43	Ru 44	Rh 45	Pd 46	Ag 47	Cd 48	In 49	Sn 50	Sb 51	Te 52	I 53	Xe 54
6	Cs 55	Ba 56	* 57–71	Hf 72	Ta 73	W 74	Re 75	Os 76	Ir 77	Pt 78	Au 79	Hg 80	Tl 81	Pb 82	Bi 83	Po 84	At 85	Rn 86
7	Fr 87	Ra 88	† 89–103															
* **Lanthanide series**				La 57	Ce 58	Pr 59	Nd 60	Pm 61	Sm 62	Eu 63	Gd 64	Tb 65	Dy 66	Ho 67	Er 68	Tm 69	Yb 70	Lu 71
† **Actinide series**				Ac 89	Th 90	Pa 91	U 92	Np 93	Pu 94	Am 95	Cm 96	Bk 97	Cf 98	Es 99	Fm 100	Md 101	No 102	Lr 103

FIGURE 12.1
The Periodic Table.

Algorithm development

In Case Study 21, unlike most of the previous case studies, the problem does not involve the implementation of an algorithm. Instead it involves the overall design of a program. Although some of the data structures necessary for the solution to the case study will be covered in Section 12.2 on sets in Pascal, most of the solution to the case study, including the algorithm, will be developed in Section 12.3 on program design and validation.

12.1 Mathematical sets

In mathematics, a *set* is a collection of objects. For example, the collection of the chemical elements is a set. From this simple definition, *set theory,* or the algebra of sets, has been developed. Set theory is used in almost every field of mathematics, science, and engineering.

Definitions

Before you begin to study sets, there are several definitions and concepts that you must understand. To make the definitions clearer, we will discuss each definition in terms of the set of chemical elements. By coincidence, the objects that make up a set are also called *elements,* so each chemical element is an element in the set of all chemical elements.

The set of all the elements that are under consideration is the *space,* also called the *universal set* or *universal space.* All the chemical elements constitute the space in the example. The set that contains no elements is called the *null set,* or the *empty set.*

One set is said to be a *subset* of another set if every element in the subset is also an element of the other set. Gases are a subset of the set of chemical elements. The metals are another subset of the set of chemical elements. Subsets can also have subsets. For example, the inert gases are a subset of the set of gases, which is in turn a subset of the set of chemical elements. A set is a *proper subset* of another if at least one element of the set is not an element of the subset. The set of inert gases is a proper subset of the set of gases because many gases are not inert gases.

The *complement* of a set is the subset that contains all the elements in the universal set that are not in the set under consideration. The complement of the set of gases is the set of all chemical elements that are not gases. The complement of the universal set is the empty set, and the complement of the empty set is the universal set.

Mathematical notation

The definitions given above can be expressed in mathematical notation. A set is indicated by a capital or script letter, and the elements of the set are written between braces (curly brackets). The expression

$$A = \{a_1, a_2, a_3\}$$

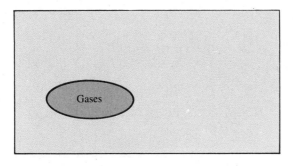

FIGURE 12.2
The set of gases and its complement, the set of all elements that are not gases.

is read as "The set A contains the elements a_1, a_2, and a_3." To indicate that a particular element is a member of a set, one says that the element is *in* the set. The expression

$$a_2 \in A$$

is read as "The element a_2 is in the set A," or "a_2 is an element in the set A."

The complement of a set is indicated by drawing a bar over the name of the set. The set $\overline{A}$ is the complement of the set A. The symbol $\overline{A}$ is usually read as "not A." The universal set is indicated by a capital S or U, and the empty set is indicated by the symbol $\emptyset$. By definition, $\overline{S} = \emptyset$ and $\overline{\emptyset} = S$.

A subset is indicated by a curved less-than sign, $\subset$. The expression $A \subset B$ is read as "A is a subset of B," or "A is contained in B."

Venn diagrams

A common way to illustrate sets is to use *Venn diagrams*. In a Venn diagram, a box or outline is understood to represent the universal set. In the example, the universal set is all the chemical elements. Within the universal set, various subsets can also be delineated with outlines. In the Venn diagram in Figure 12.2, the universal set is the set of all elements. The subset of gases is illustrated, along with its complement.

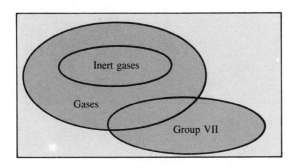

FIGURE 12.3
The sets of gases, inert gases, Group VII elements, and all remaining elements.

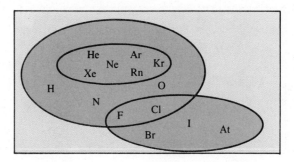

FIGURE 12.4
The elements of the sets of inert gases, gases, and Group VII elements.

The same universal set could also be divided into the four subsets gases, inert gases, Group VII elements, and all elements that are not gases, inert gases, or Group VII elements, as illustrated in Figure 12.3. A Venn diagram can also list the elements in each set. Figure 12.4 lists the elements of the sets of gases, inert gases, and Group VII elements.

There are two basic operations for sets: union and intersection. The *union* of two sets creates a new set that contains every element of the two sets. The union of the set of gases with the set of Group VII elements is a set that contains all the gases and all of Group VII. The union of two sets is indicated in a Venn diagram by shading the elements that are members of the new set, as in Figure 12.5. The new set contains hydrogen, helium, nitrogen, oxygen, fluorine, neon, chlorine, argon, bromine, krypton, iodine, xenon, astatine, and radon.

You could also take the union of the set of inert gases with the set of

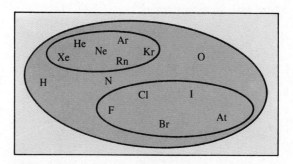

FIGURE 12.5
The union of the sets of gases and the elements of Group VII.

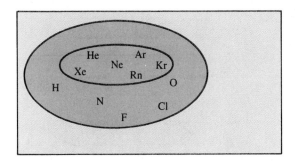

FIGURE 12.6
The union of the set of inert gases with the set of gases.

gases. The resulting set, shown in Figure 12.6, is the set of gases, because every inert gas is also a gas.

In mathematics, the notation for the union operation is ∪. The algebraic expression $A \cup B$ means the set of all elements that are in set A and/or set B and is read as "The set of A union B." A general version of the Venn diagram for union is given in Figure 12.7.

Although the union operation may look like the addition operation in arithmetic, it is fundamentally different. Because a set is a collection of elements that have no order, the elements of a set cannot be added in the same way that numbers are added. When the union of the set of inert gases and the set of gases is formed, each inert gas is included only once in the new set. The resulting set contains all the gases, including those that are inert gases. An element can never appear more than once in a set.

The second operation for sets is *intersection*. The intersection of two sets creates a new set that contains only the elements that belong to both sets. The intersection of the set of gases with the set of Group VII elements is the set that contains fluorine and chlorine. The intersection of sets is indicated on a

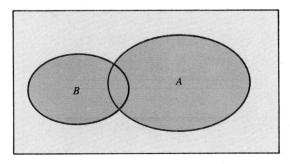

FIGURE 12.7
The set of A union B.

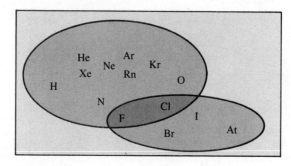

FIGURE 12.8
The intersection of the set of gases with the set of Group VII elements.

Venn diagram by shading in the elements that are common to both sets, as in Figure 12.8.

In mathematics, the intersection operation is indicated by an upside-down U, $\cap$. The algebraic expression $A \cap B$ means the set that contains the elements that are in both A and B and is read as "A intersection B," or "A intersect B." A Venn diagram for intersection is given in Figure 12.9.

The intersection of the set of inert gases with the set of elements in Group VII is the empty set because no elements in Group VII are also inert gases. Two sets are said to be *mutually exclusive,* or *disjoint,* when their intersection is the empty set. The inert gases and Group VII are mutually exclusive sets. Mutually exclusive sets have the property

$$A \cap C = \emptyset$$

The Venn diagram for mutually exclusive sets is shown in Figure 12.10.

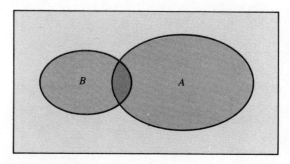

FIGURE 12.9
The intersection of sets A and B.

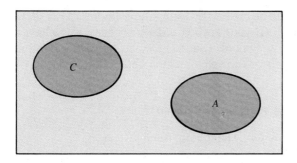

FIGURE 12.10
Mutually exclusive sets.

12.2 Sets in Pascal

Pascal has a structured data type called the **set** that has the same properties as a mathematical set. The syntax diagram for declaring a **set** data type is

→ **type** → *type-identifier* → = → **set** → **of** → { *ordinal-type-identifier* / *new-ordinal-type* } →

In Pascal, the data type of a **set** must be an ordinal data type, and all **sets** must have a data type. **Set**s can only contain members that are of the same data type, and this data type is called the *base type* of the **set.**

Formally, the elements of a Pascal **set** are called *members;* however, you will not be misunderstood if you refer to them either as elements or as members. Since the examples below involve the chemical elements, for clarity the elements of a set will be referred to as members of the set.

To create a **set** in Pascal, you can use the reserved word **set** to define a new data type. For example, if an enumerated type like **ElementType** has been declared, you can declare a data type **ElementSet** that is a **set** of the enumerated type **ElementType,** and you can declare **set** variables of the **ElementSet** data type.

```
type   ElementType = (Hydrogen, Helium, Nitrogen, Oxygen, Fluorine, Neon,
                     Chlorine, Argon, Bromine, Krypton, Iodine, Xenon,
                     Astatine, Radon );
       ElementSet = set of ElementType;

var    Gases : ElementSet;
```

The declaration of **Gases** creates a structured variable. The variable is uninitialized; that is, the members of the set **Gases** are undefined. To initialize the **set** variable, you need a method for assigning elements to the set. Within a program, a **set** can be created by listing the elements of the set between square brackets, [and]. (If your keyboard does not have these characters, you can use the alternative symbols (. and .) instead of [and].) To initialize the **set**, you can use either of the assignment statements below:

```
Gases := [Hydrogen, Helium, Nitrogen, Oxygen, Fluorine, Neon, Chlorine,
         Argon, Krypton, Xenon, Radon];
```

```
Gases := [Hydrogen..Argon, Krypton, Xenon, Radon]
```

Note that the second assignment statement uses a subrange of the data type **ElementType**. The two assignment statements are equivalent, since all the constants in **ElementType** between **Hydrogen** and **Argon** are gases. However, you should be careful when using subranges to declare **sets**. The type definition

```
type SomeElements = set of Radon..Hydrogen;
```

creates a **set** that can have no members because there are no elements between **Radon** and **Hydrogen**. Some, but not all, compilers will write an error message if the upper bound is less than the lower bound.

You can also use an implicit type when declaring a **set** variable. That is, you do not need to define a data type identifier for the **set**. Instead, you can use the type declaration syntax immediately after the variable name where the data type identifier would normally go:

```
var Elements : set of Hydrogen..Radon;
```

This statement appears to initialize the **set** to contain the members between **Hydrogen** and **Radon**. In fact, the declaration only states that these constants are legal members of the **set** variable **Elements**. In some implementations of Pascal, the statement above does initialize the **set**. You should not, however, depend on such a statement for initialization. You should execute an assignment statement such as

```
Elements := [Hydrogen..Radon]
```

to initialize the **set** variable **Elements**.

Set *operations in Pascal*

In the following example, after **GroupVII, Gases,** and **Elements** have been declared to be **sets** of **ElementType, GroupVII, Gases,** and **Elements** are initialized by listing their members between square brackets:

```
var    Gases, GroupVII, Elements : ElementSet;
begin
  GroupVII := [Fluorine, Chlorine, Bromine, Iodine, Astatine];
  Gases    := [Hydrogen..Argon, Krypton, Xenon, Radon];
  Elements := [Hydrogen..Radon]
```

> **REMINDER**
>
> Do not confuse sets with enumerated types. A **set** is delimited with square brackets; an enumerated type is delimited with parentheses. It is easy to confuse the two data types, both when you are writing a program and when you are reading someone else's program. In the example using the enumerated type, **ElementType** is declared by listing the constant identifier names between parentheses. The set variable **GroupVII** is declared to be an **ElementSet**, which is a **set** of **ElementType**. The members of the set **GroupVII** are listed between square brackets when the set is initialized within the program.

Union of sets

In Pascal, the operator that designates the union of **sets** is the plus sign (+). After the assignment statement

```
Elements := GroupVII + Gases
```

is executed, the **set** variable **Elements** has the members **Fluorine, Chlorine, Bromine, Iodine, Astatine, Hydrogen..Argon, Krypton, Xenon,** and **Radon.** The result of the union operation is a set containing every element that is in either or both sets. The union operator is not limited to declared **sets**. For example, the union of a **set** with a single member or list of members can be taken by using an implicit **set,** as in

```
Elements := GroupVII + [Krypton, Argon, Chlorine];

for Chemical := Hydrogen to Radon do
    Elements := Elements + [Chemical]
```

where **Chemical** is declared to be a variable of type **ElementType.**

The union operation is only valid for **set**s, and the **set**s must have the same base type. You cannot take the union of a **set** of characters with a **set** of chemical elements. The result of the union operation is always another **set** of the same base type as the original **set**s.

You must also be careful to ensure that you are taking the union of two **set**s. If **Elements** and **Gases** are sets of **ElementType** and **Hydrogen** is a constant in **ElementType,** the statements below will result in syntax errors:

```
                          Elements := Elements + Hydrogen;
***error***                                             $134
                          Gases := Gases + [Elements]
***error***                                             $136

error messages :
***************

 134: illegal type of operand(s).
 136: set element type must be ordinal or subrange.
```

The correct assignment statements are

```
         Elements := Elements + [Hydrogen];

         Gases := Gases + Elements
```

WARNING

You cannot tell whether an assignment statement involving **set**s is correct without knowing the data types of the identifiers.

REMINDER

When using sets, you must always distinguish between the container (the set) and the things contained (the members of the set).

In the **for** statement

```
         for Chemical := Hydrogen to Radon do
            Elements := Elements + [Chemical]
```

the set **Elements** must be initialized to the null set to ensure that **Elements** has no members before the **for** statement is executed. In Pascal, the null set (the set with no members) is indicated by a pair of square brackets with no members listed between them. The complete **for** statement becomes

```
Elements := [ ];
for Chemical := Hydrogen to Radon do
    Elements := Elements + [Chemical]
```

Although the statements above add the members from **Hydrogen** to **Radon** to the set **Elements,** you should be aware that the assignment statement

```
GroupVII := GroupVII + [Chlorine]
```

has no effect if **Chlorine** is already a member of the set **GroupVII.** Remember that a member can appear only once in a set and that a set has no order.

Intersection of sets

In Pascal, the intersection of two sets is indicated by the asterisk (∗). The result of the intersection operation is a set containing all the elements common to both sets. All of the rules about set compatibility are the same for the intersection operation as for the union operation. After the assignment statement

```
Elements := Gases * GroupVII
```

is executed, the set **Elements** contains the members **Fluorine** and **Chlorine.** After the statement

```
Elements := Gases * [Helium, Neon, Argon, Xenon, Radon, Krypton]
```

is executed, the set **Elements** contains the members that are inert gases—that is, the members listed between brackets. After the statement

```
Elements := GroupVII * [Helium, Neon, Argon, Xenon, Radon, Krypton]
```

is executed, the set **Elements** has no members; it is equal to the empty set.

Difference of sets

Pascal has a set operation called *difference* that is not a standard mathematical operation. The difference of two sets is the set containing members that are in the first set but not in the second set. In a Venn diagram, the difference of sets *A* and *B* is indicated by the shaded area, as illustrated in Figure 12.11.

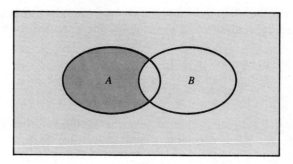

FIGURE 12.11
The difference of two sets.

The difference operator is the minus sign (−). The assignment statement

Elements := Gases - GroupVII

assigns the members **Helium, Neon, Argon, Xenon, Radon, Krypton, Hydrogen, Neon,** and **Oxygen** to the set **Elements.** The shaded area in Figure 12.12 is the set **Elements** after the difference operation has been performed.

The difference operator, unlike the union and intersection operators, is

not commutative; that is, the order in which the sets are listed affects the outcome. The statement

Elements := GroupVII -.Gases

assigns to **Elements** the members **Bromine, Iodine,** and **Astatine,** as shown in Figure 12.13.

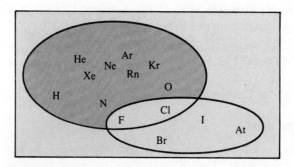

FIGURE 12.12
The difference of the set of gases and the set of Group VII elements.

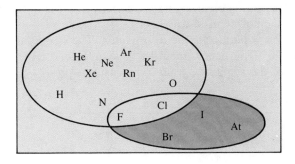

FIGURE 12.13
The difference of the set of Group VII elements and the set of gases.

With the difference operator, you can also use an implicit **set** just as you did for the union operation, as in

```
Elements := Gases - [Fluorine, Chlorine]
```

You must still be careful to distinguish between the set and the members of the set.

REMINDERS

The members of a set do not have any order. The set [Argon, Neon] is identical to the set [Neon, Argon].

A member can never appear more than once in a set. The set [Neon] + [Neon] is equal to the set [Neon].

Boolean operators for sets

The boolean operators can be used to compare sets. When used with sets, the boolean operators have the following meanings:

A = B If every member of set A is a member of set B and every member of set B is a member of set A, then **A = B** is true. Otherwise, **A = B** is false.

A <> B If **A <> B** is true, then **A = B** is false. That is, **A <> B** is true if there is some member of A that is not a member of B or some member of B that is not a member of A.

A <= B If every member of set A is also in set B, then set **A<=B** is true. If **A<=B** is true, then A is a subset of set B.

A >= B If every member of set B is also a member of set A, then **A >= B** is true. If **A >= B**, then B is a subset of A.

In standard Pascal, the boolean operators strictly less than, <, and strictly greater than, >, cannot be used with sets.

The last boolean operator is the **in** operator. The **in** operator is used to determine whether a particular element is a member of the set. The **in** operator is part of the syntax diagram for an expression:

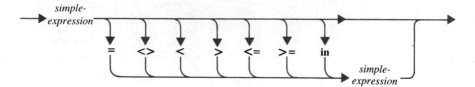

WARNING

The simple expression that precedes **in** must be an element of the base type of the set. The simple expression after the **in** must be a set.

Given the definitions of the sets **GroupVII** and **Gases,** after the statements

```
FirstBoolean := Xenon in Gases;

SecondBoolean := Xenon in GroupVII
```

are executed, **FirstBoolean** will have a value of true and **SecondBoolean** will have a value of false.

According to the syntax for the **in** operator, the following statement is illegal:

```
        if [Nitrogen] in Gases then AskGasQuestions (Nitrogen)
***error***                  $129

error messages :
***************

  129: type conflict of operands.
```

The correct statement is

```
        if Nitrogen in Gases then AskGasQuestions (Nitrogen)
```

With the **in** operator, you can use an implicit **set.** For example, if a character is read from the terminal, you can test to see if it is a vowel using the statement

```
if CharVar in ['a', 'e', 'i', 'o', 'u'] then writeln (' vowel')
```

Using the implicit **set,** you never have to declare a data type or a variable that is a set of characters. All the members listed in the implicit **set** must be of the same data type, and the member being tested must be of the same data type as the members of the **set.**

Sets *of characters and integers*

The sets used in the previous examples have been sets of an enumerated type. For many applications, it is useful to have sets of characters or sets of integers. Although standard Pascal does not have a specified maximum for the number of members a set can have, every implementation of Pascal imposes its own limits on the maximum number of members. In some implementations, the maximum set size is as small as 64 members. With such small sets, the declaration below, which most compilers will accept, will result in a syntax error:

```
type   CharSet = set of char;
```

In almost every Pascal implementation, the declaration

```
type  Numbers = set of integer;
***error***                     $169

error messages :
***************
  169: error in base type.
```

will result in a syntax error. You must use a subrange to declare a set that has the integers as its base type:

```
type   Numbers = set of 1..10;
```

On systems that have a small limit on the set size, it is necessary to use a subrange to declare a set of characters:

```
type  LetterSet =  set of 'a'..'z';
```

The definition for **LetterSet** creates a set that can contain all the letters between 'a' and 'z'.

Problems arise with sets of characters because of the differences between ASCII and EBCDIC. On a computer using EBCDIC, the set **LetterSet** would contain not only the small letters between 'a' and 'z', but also the nonalphabetic characters that are stored between 'a' and 'z' in EBCDIC.

Another declaration that would result in different sets in ASCII and EBCDIC is the declaration

```
type  Alphabet = set of 'a'..'Z';
```

In EBCDIC, this set would contain all the letters between small 'a' and capital 'Z', plus the unprintable characters that are also stored in that range. In ASCII, the set would be empty, because the capital letters are stored before the small letters. To create the set of capital and small letters in ASCII, you can use the declaration

```
type  Alphabet = set of 'A'..'z';
```

Sets of characters and integers are useful for checking data that is entered from the terminal. You can test whether or not a value read from the terminal is a member of a set, as in Program 12.1, **TestACharacter**.

PROGRAM 12.1 TestACharacter

```
program TestACharacter (input, output);
(*     TestACharacter reads a character from the terminal and writes a     *)
(*     message stating whether the character is a vowel, consonant,        *)
(*     digit, or something else.                                           *)

type  Alphabet = set of char;

var   Vowels : Alphabet;              (* set of vowels. *)
      Consonants : Alphabet;          (* set of consonants. *)
      Digits : set of '0'..'9';       (* set of characters corresponding to *)
                                      (* the digits 0 through 9. *)
      Answer : char;                  (* character read from the terminal. *)

begin
  Vowels := ['a', 'e', 'i', 'o', 'u',
             'A', 'E', 'I', 'O', 'U'];

  Consonants := ['A'..'z'] - Vowels;  (* The set of consonants consists *)
                                      (* of all the letters minus the set *)
                                      (* of vowels. *)
  Digits := ['0'..'9'];               (* the set Digits must be initial- *)
                                      (* ized. The declaration gives *)
                                      (* only the valid range. *)
```

```
    writeln (' Enter a character from the terminal.');
    readln (Answer);

    if Answer in Vowels then
      writeln (' That was a vowel.')
    else
      if Answer in Consonants then
        writeln (' That was a consonant.')
      else
        if Answer in Digits then
          writeln (' That was a number between 0 and 9.')
        else
          writeln (' That was not a vowel, consonant, or a digit.')
end.   (* program TestACharacter *)
```

```
=>
 Enter a character from the terminal.
□4
 That was a number between 0 and 9.
```

```
=>
 Enter a character from the terminal.
□&
 That was not a vowel, consonant, or a digit.
```

```
=>
 Enter a character from the terminal.
□E
 That was a vowel
```

```
=>
 Enter a character from the terminal.
□v
 That was a consonant
```

Without sets, Program 12.1 would be difficult to write because you would have to use **case** statements or complicated **if** statements to test whether a value was a consonant or a vowel.

12.3 Program design and validation

The goal of programming is to produce useful programs that are correct, error free, reliable, and efficient, but there is no algorithm for producing an efficient, reliable program or even a correct program. There are guidelines, recommendations, and standards that may help you to produce a good program, but you will encounter many problems whose solution will depend on your creativity and ingenuity.

When programs are developed by groups of people, accepted standards and systematic techniques must be followed. Otherwise the programs could not be coordinated and verified. Many of the techniques and suggestions given in this section were discovered and refined by people working on large programs. (For an entertaining discussion of the development of large programs, see Reference 1 at the end of the chapter.)

The steps that a program goes through as it is developed are often given as follows: statement of functional requirements, statement of program specifications, implementation, debugging, verification, testing, documentation, and maintenance. Each of these steps will be discussed in greater detail in the development of the program **PeriodicTable**.

Functional requirements

Most programs are written because someone has a task, problem, or need to which the computer's ability to store and manipulate data can be applied. For example, people in business have kept books and done financial analysis for thousands of years. With the advent of the computer, many of the functions that were previously done by hand are now being done on the computer, and programs have been developed to meet the needs of businesspeople.

When computer solutions to problems are first envisioned, the statement of the purpose of the program is usually not specific. The statement, in fact, can be quite general: write a program that performs financial analysis; write a program that plays chess; write a program that performs the mundane tasks in drafting; write a program that performs the job of receiving and dispensing money in a bank; write a program that edits text; or write a program to teach the periodic table to chemistry students.

After the general purpose of the program has been stated, many clarifications and limitations must be specified before the program is written. For any program, there must be a list of *functional requirements* that state what the program will and will not do when it is completed. The functional requirements do not state how the program is to be written, but rather state in some detail what the final program should do. For example, the functional requirements would state whether a chess program should play only beginner-level chess, only grand master-level chess, or any level between beginner and grand master at the discretion of the user. For a large program being written by a group of programmers, the decisions about what functions the program should perform will usually be made by many people.

You might organize the program to help beginning chemistry students learn the periodic table by deciding, with the chemistry teacher, that the program should perform the following tasks:

1. Name an element drawn randomly from the list of elements.

2. Ask four out of the eight following questions:

    a. What group does the element belong to?

    b. What is the element's period?

    c. Is the element gas, liquid, or solid?

    d. What is the element's atomic number?

    e. What is the element's symbol?

    f. What is the element's atomic weight?

    g. How many electrons are in the element's outer ring?

    h. How many neutrons does the element have?

3. If the student gets all four questions right, remove the element from the list of elements to be asked about; otherwise, the element remains in the list.

4. Keep running totals of the number of questions answered correctly and the number answered incorrectly. For the teacher, keep a running total of the elements whose properties are known, and a running total of the number of right and wrong answers for each question, so that the teacher can tell whether, for example, the student knows the symbols but not the atomic numbers.

5. Allow the student to stop the program after at least five elements have been covered.

Program specifications

Once the general functional requirements for the program have been decided upon, a list of specifications can be drawn up. The specifications usually define what the input and the output of the program will be and may specify the algorithm to be used. Specifications are almost always incomplete because they are written at a preliminary stage of program development. You must ask questions to be sure that you understand the specifications and to be sure that you are aware of the assumptions that are implicit in the specifications. For example, you may have to ask how the end of data is specified: Can you assume beforehand that the number of entries is known, or must the program read until the end of data is encountered?

Some companies have formal specifications sheets that must be completed before any code is written. These sheets usually cover the input/output requirements and the algorithm to be used. The specifications sheets serve as preliminary documentation and are used by other members of the group who must use the procedures defined by the specifications sheets before the procedures have been implemented on the computer.

The first major decision to be made for the case study is what types of data structures to use. An enumerated list can be used to keep track of the element names within the program. If you enter the elements in order into the enumerated list, you can use the standard function **ord** to return the element's atomic number. To store the elements' atomic weights, you will need a real array, and to store the elements' symbols, you will need an array of character strings. To write the name of the element for the student, you will also need to store the names of the elements in an array of character strings. To make the program easier to read and write, you can use the enumerated type as the index for each array so that, for example, the atomic weight of argon is stored in the element **AtomicWeight[Argon]** and the symbol for helium is stored in **ElementSymbol[Helium]**.

```
const NumberOfElements = 103;

type  ElementType = (Hydrogen, Helium, Lithium, Beryllium, Boron,
                            .
                            .
                            .
                     Fermium, Mendelevium, Nobelium, Lawrencium);

      ElementSet = set of ElementType;
      CharArray = packed array [1..15] of char;

(* The array data types below will be used for the variables that store *)
(* the information about the symbol, name, and atomic weight of each *)
(* element from Hydrogen..Lawrencium. *)

      SymbolArray = array [Hydrogen..Lawrencium] of CharArray;
      NameArray   = array [Hydrogen..Lawrencium] of CharArray;
      RealArray   = array [Hydrogen..Lawrencium] of real;

var   ElementSymbol : SymbolArray;     (* array of symbols for the *)
                                       (* chemical elements. *)
      ElementName   : NameArray;       (* array of names of the chemical *)
                                       (* elements. *)
      AtomicWeight  : RealArray;       (* array of the atomic weights of *)
                                       (* the chemical elements. *)
```

The elements can be organized into eight groups and seven periods. The sets that contain the groups and the periods can be organized into arrays for which the index is the group number or period number.

```
const NumGroups = 8;           (* number of groups in the periodic table. *)
      NumPeriods = 7;          (* number of periods in the periodic *)
                               (* table. *)

type  GroupSetArray = array [1..8] of ElementSet;
      PeriodSetArray = array [1..7] of ElementSet;

var   Group : GroupSetArray;   (* array of sets to store the elements of *)
                               (* each group. *)
      Period : PeriodSetArray; (* array of sets to store the elements of *)
                               (* each period. *)
```

In this data structure, **Group[1]** is the set of elements that are in Group I, and **Period[5]** is the set of elements that are in the fifth period.

Implementation

Having decided on the major data structures that will be used in the program, you can start working on developing the algorithm for solving the overall problem and then on solving the individual tasks. It will probably be necessary to add more variables as the program develops, but you need to have the general structure before beginning to work on the tasks.

Many different techniques exist for developing programs from problem statements. One of the most successful has been *structured programming*. As you know, the basic idea of structured programming, or top-down design, is to divide a problem into several smaller problems, then to divide each of those problems into still smaller problems until each problem can be solved. The structure of Pascal makes top-down design easier to implement than does the structure of many other programming languages. Also, using the top-down design strategy makes programs easier to test and verify.

The algorithm that solves the case study is not immediately apparent, so the problem must be broken down further. Having settled on the basic data structure, you can write an algorithm for the main program from the five tasks listed previously.

ALGORITHM 12.1 First iteration of algorithm for Case Study 21

begin algorithm
  1. Initialize data structures and introduce program to user
  2. Repeat until **NumElements** covered >= 5 and **RequestToStop**

        a. Select an **Element**
          for 4 questions about **Element**
          begin
            Select a new question from the list of 8 questions
            Write the question and read the response
            Determine if the response is correct
            Add 1 to the proper running totals
          end for
       b. **NumElements** ← **NumElements** + 1
       c. If **NumCorrect** = 4 then remove element from list
    end repeat
  3. Write progress reports for student and teacher
end algorithm

This outline is still quite general, and many decisions remain. For example, **RequestToStop**, which appears as a stopping criterion for the repeat loop, might be a boolean function that asks whether the user wants to stop, or it might be a boolean variable that is set during some other interaction with the user.

At this point, you can decide which tasks will be organized into procedures and begin to work on the individual procedures. The **for** statement might use procedures called **SelectAQuestion, AskTheQuestion, RecordStudentGrade, RecordForTeacher.** Algorithm 12.2 is a second version of the algorithm for the program.

ALGORITHM 12.2   Second iteration of Algorithm 12.1

begin algorithm
  1. Initialize data structures and introduce program to user
  2. Repeat
      SelectAnElement
      for **QuestionsAsked** := 1 to 4 do
        begin
          SelectAQuestion
          AskTheQuestion
          AskIfDone
          RecordStudentGrade
          RecordForTeacher
        end for
      NumElements ← **NumElements** + 1
      if **NumCorrect** = 4 then remove element from list
    until **NumElements** >= 5 and **RequestToStop**
  3. Write progress reports for student and teacher
end algorithm

Because the entire program is rather large, the initial version of the program will be restricted to a single question: asking the student to what group an element belongs. Also, some of the straightforward procedures like the initialization procedure will not be developed. Algorithm 12.3 outlines a solution to the restricted version of the original problem. Argument lists have been added to the procedures to make it clearer how the information flows within the program. In Algorithm 12.3, **Element** is the current element and **CorrectResponse** is a boolean variable that is set to true if the student answers the question correctly.

ALGORITHM 12.3 Restricted version of Algorithm 12.2

begin algorithm
1. Initialize data structures and introduce program to user
2. Repeat
    SelectAnElement(Element)
    QuestionNum ←1
    AskTheQuestion(Element, QuestionNum, CorrectResponse)
    AskIfDone(RequestToStop)
    RecordStudentGrade(Element, CorrectResponse)
    RecordForTeacher(Element, CorrectResponse)
    NumElements ← **NumElements** + 1
    until **NumElements** >= 5 and **RequestToStop**
3. Write progress reports for student and teacher
end algorithm

The first subprogram in Algorithm 12.3 is **SelectAnElement.** So that the program will be interesting for the student, the elements should be selected randomly. In Chapter 14, random number generators will be developed that will enable you to select elements in a random order. For now, while the program is under development, a reasonable decision is to work through the elements in the order in which they are given. Later, the modifications for random selection can be added.

ALGORITHM 12.4 Algorithm for the initial version of SelectAnElement

begin algorithm
1. Find the next element in the list after the current **Element**
    NextElement ← successor(**Element**)
2. Return the **NextElement** as the current **Element**
    Element ← **NextElement**
end algorithm

Chapter 12 □ Sets and Program Design

Using Algorithm 12.4, you could write **SelectAnElement** as a Pascal function that has a data type of the enumerated type **ElementType,** as shown below.

```
function SelectAnElement (LastElement : ElementType) : ElementType;
(*     This temporary version of SelectAnElement returns the element    *)
(*     that comes after LastElement in the list ElementType.  In its    *)
(*     final version the function will return a randomly selected       *)
(*     element.                                                         *)

begin
  SelectAnElement := succ (LastElement)
end; (* function SelectAnElement *)
```

The next step in Algorithm 12.3 is **AskTheQuestion.** In the restricted problem statement, only the question about what group the element is a member of will be asked. To keep the structure of the original program, you can list all the questions, but the program will always ask the question about the group. Algorithm 12.5 gives a strategy for achieving this objective.

ALGORITHM 12.5   Algorithm for AskTheQuestion

begin algorithm
   case **QuestionNum** of
     1 : **WhatGroup**
     2 : **WhatPeriod**
     3 : **GasLiquidOrSolid**
     4 : **WhatNumber**
     5 : **WhatSymbol**
     6 : **WhatWeight**
     7 : **HowManyElectrons**
     8 : **HowManyNeutrons**
   end case
end algorithm

Procedure **AskTheQuestion** is a direct translation of Algorithm 12.5.

```
procedure AskTheQuestion (Element : ElementType; QuestionNum : integer;
                          var Answer : boolean );
(*     AskTheQuestion selects a question based on the value of          *)
(*     QuestionNum.  In this temporary version, the question number is  *)
(*     always set to 1.                                                 *)

begin
  QuestionNum := 1;          (* This statement must be deleted in the final *)
                             (* version. *)
```

```
  case QuestionNum of
    1 : WhatGroup (Element, Answer);
    2 : WhatPeriod (Element, Answer);
    3 : GasLiquidOrSolid (Element, Answer);
    4 : AtomicNumber (Element, Answer);
    5 : WhatSymbol (Element, Answer);
    6 : WhatWeight (Element, Answer);
    7 : HowManyElectrons (Element, Answer);
    8 : HowManyNeutrons (Element, Answer)
  end (* case *)
end; (* procedure AskTheQuestion *)
```

The next step is to refine the procedure **WhatGroup.** This procedure must ask the question, read the response, and check to see if the response is correct. Algorithm 12.6 is a refinement of the task of asking what group an element is in.

ALGORITHM 12.6 Algorithm for WhatGroup

begin algorithm

1. Write a message asking to what group **Element** belongs; include a list of valid responses
2. Read the **Group** from the user
3. If **Element** in **Group** then
      CorrectResponse ← true
      write a message stating that the answer is correct
   else
      CorrectResponse ← false
      write a message stating the correct group

end algorithm

In Algorithm 12.6 sets are used to test whether **Element** is in the group entered by the student. When the algorithm is translated into Pascal, the correct **Group** can be referenced by using the group number as an array index. The only step requiring further refinement is writing the message stating the correct group if the student entered an incorrect group. To find the correct group, all the groups must be searched until the correct group is found. The procedure **WhatGroup** uses this strategy to test whether the student has entered the correct response and to find the correct group if the student enters the incorrect group.

```
procedure WhatGroup (Element : ElementType; var Answer : boolean);
(*      WhatGroup asks the user what group the chemical element is in.     *)
(*      It also determines whether the answer is correct.                  *)
```

```
    var    GroupNum : integer;         (* group number entered by the student from *)
                                       (* the terminal. *)
           Num : integer;              (* for loop control variable to search for *)
                                       (* the correct group number. *)
begin
  writeln (' What group is ', ElementName [Element], ' in?');
  writeln (' Enter an integer between 1 and 8 (Group0 = 8)');
  readln (GroupNum);

  if Element in Group [GroupNum] then
    begin
      Answer := true;
      writeln (' That is correct.', ElementName [Element], ' is in',
               ' Group ', GroupNum:1)
    end
  else
    begin
      Answer := false;
      write (' That is not correct.', ElementName [Element], ' is in');
      for Num := 1 to NumGroups do
        if Element in Group [Num] then
          writeln (' Group ', Num:1)
    end (* else *)
end; (* procedure WhatGroup *)
```

Program 12.2, **PeriodicTable,** combines the procedures and functions that have been developed so far. As it stands, the program is incomplete. At this stage, several students can be asked to try the program. Then their suggestions can be incorporated into the program as its development is completed.

Program 12.2 is a program *stub;* that is, the procedures that have been referred to but that have not been developed are included as just **begin-end** statements. A program stub can be compiled and tested; then as each procedure is developed, it can be added and tested until the program is complete. Program 12.2 is rather long; when you read it, remember to look for the main program and begin reading there.

PROGRAM 12.2 PeriodicTable

```
program PeriodicTable (input, output);
(*     PeriodicTable is a preliminary version of a program designed    *)
(*     to help chemistry students learn the periodic table of elements. *)
(*     This version is only capable of asking the question: What group *)
(*     does the element belong to?  The elements are selected sequent- *)
(*     ially instead of randomly as they would be in the final version *)
(*     of the program, and not all elements are included in the type   *)
(*     definition.                                                      *)
```

```
const NumberOfElements = 103;      (* number of elements about which the *)
                                   (* program will store information. *)

      NumGroups = 8;               (* number of groups in the periodic table. *)
      NumPeriods = 7;              (* number of periods in the periodic *)
                                   (* table. *)
      QPerElement = 1;             (* number of questions that are asked per *)
                                   (* element. *)
      MustGetRight = 1;            (* number of questions that must be *)
                                   (* answered correctly before an element is *)
                                   (* removed from the list. *)

type  ElementType = (Hydrogen, Helium, Lithium, Beryllium, Boron, Carbon,
                          .
                          .
                     Fermium, Mendelevium, Nobelium, Lawrencium);

      StateType = (Solid, Liquid, Gas);
      ElementSet = set of ElementType;
      CharArray = packed array [1..15] of char;

(* The array data types below will be used for the variables that store *)
(* the information about the symbol, name, and atomic weight of each *)
(* element from Hydrogen..Lawrencium. *)

      SymbolArray = array [Hydrogen..Lawrencium] of CharArray;
      NameArray = array [Hydrogen..Lawrencium] of CharArray;
      RealArray = array [Hydrogen..Lawrencium] of real;
      IntegerArray = array [Hydrogen..Lawrencium] of integer;
      StateArray = array [Hydrogen..Lawrencium] of StateType;
      GroupSetArray = array [1..NumGroups] of ElementSet;
      PeriodSetArray = array [1..NumPeriods] of ElementSet;

var   ElementSymbol : SymbolArray;        (* array of symbols for the chem-*)
                                          (* ical elements. *)
      ElementName : NameArray;            (* array of the element names. *)
      NumOuterElectrons : IntegerArray;   (* array for the number of elec- *)
                                          (* trons in the outer ring of *)
                                          (* each element. *)
      NumNeutrons : IntegerArray;         (* array for the number of neu- *)
                                          (* trons in each element. *)
      AtomicWeight : RealArray;           (* array of the atomic weights.*)
      Group : GroupSetArray;              (* array of sets to store the el- *)
                                          (* ements of each group. *)
      Period : PeriodSetArray;            (* array of sets to store the el- *)
                                          (* ements of each period. *)
      State : StateArray;                 (* array for the element state. *)
      Element : ElementType;              (* element about which questions *)
                                          (* are currently being asked. *)
```

```
        NextElement : ElementType;       (* next element about which ques- *)
                                         (* tions will be asked. *)
        NumElements : integer;           (* number of elements that have *)
                                         (* been covered so far. *)
        NumCorrect : integer;            (* number of correct answers for *)
                                         (* current element. *)
        QuestionsAsked : integer;        (* loop counter for the number of *)
                                         (* questions asked on current el- *)
                                         (* ement. *)
        RequestToStop : boolean;         (* allows student to request that *)
                                         (* the program stop. *)
        CorrectResponse : boolean;       (* set to true if the answer to *)
                                         (* current question is correct. *)
        QuestionNum : 1..QPerElement;    (* the number between 1 and *)
                                         (* QPerElement of the current *)
                                         (* question. *)
        infile : text;                   (* textfile in which the data on *)
                                         (* the elements is stored. *)

(*$i initializetheelementdata *)
(*   procedure InitializeTheElementData;                                    *)
(*       InitializeTheElementData sets up the name, symbol, and atomic      *)
(*       weight arrays and initializes the period and group sets.  All the  *)
(*       the data is read from a textfile.  The values that are input are   *)
(*       treated as if they were constants in the rest of the program.      *)

  procedure SetTheCounters (var FirstElement : ElementType);
  begin
    FirstElement := Hydrogen
  end;  (* procedure SetTheCounters *)

(*$i selectanelement *)
(*$i whatgroup *)

  procedure WhatPeriod (Element : ElementType; var Answer : boolean);
  begin end;

  procedure GasLiquidOrSolid (Element : ElementType; var Answer: boolean);
  begin end;

  procedure AtomicNumber (Element : ElementType; var Answer: boolean);
  begin end;

  procedure WhatSymbol (Element : ElementType; var Answer: boolean);
  begin end;

  procedure WhatWeight (Element : ElementType; var Answer: boolean);
  begin end;
```

```
  procedure HowManyElectrons (Element : ElementType; var Answer: boolean);
  begin end;

  procedure HowManyNeutrons (Element : ElementType; var Answer: boolean);
  begin end;

(*$i askthequestion *)

  procedure AskIfDone (var Stop : boolean);
  begin
    Stop := true       (* stop set so the program executes only once. *)
  end;   (* procedure AskIfDone *)

  procedure WriteProgressReport;
  begin end;

  procedure RecordStudentGrade (Element : ElementType; Answer : boolean);
  begin end;

  procedure RecordForTeacher (Element : ElementType; QuestionNum : integer;
                              Answer : boolean);
  begin end;

(*$i welcomestudent *)
(* procedure WelcomeStudent;                                                  *)
(*         Procedure WelcomeStudent writes the initial message to the         *)
(*         student.                                                           *)


(*--------------------------------------------------------------------------*)
begin (* main *)
  reset (infile);

  WelcomeStudent;
  SetTheCounters (NextElement);
  InitializeTheElementData;

  repeat
    Element := NextElement;
    for QuestionsAsked := 1 to QPerElement do
      begin
        AskTheQuestion (Element, QuestionNum, CorrectResponse);
        AskIfDone (RequestToStop);
        RecordStudentGrade (Element, CorrectResponse);
        RecordForTeacher (Element, QuestionNum, CorrectResponse)
      end;
    NumElements := NumElements + 1
  until(NumElements >= MustGetRight) and RequestToStop;
```

```
    WriteProgressReport
end. (* program PeriodicTable *)


=>
                    Periodic Table Program

This program is designed to help you learn the periodic table.
You will be asked 1 questions about an element.
You can request to stop after you have answered all 1 questions
correctly for at least 1 elements.

This program currently only asks 1 question about Hydrogen.

What group is         Hydrogen in?
Enter an integer between 1 and 8
□4
That is not correct.         Hydrogen is in Group 1
```

Debugging and testing

In order to debug and test the procedures developed in the previous section, you can write a test program whose sole purpose is to invoke the procedures. Occasionally, a test program requires the inclusion of subprograms that have not been developed yet. For example, in the program for the case study, procedures consisting only of **begin-end** pairs are used so that the overall structure of the program can be tested without worrying about every detail.

If the procedures that are not yet complete are part of a larger program and need to be used by other programmers, it may be necessary to create temporary procedures. Other programmers can use the temporary procedures in their programs. Later, when the correct versions become available, the programmers will not have to modify any of their code.

The initial phases of debugging a program or subprogram involve removing syntax and run-time errors. Once the program compiles and runs without errors, the next step is to test the program for a range of input values. When you are testing a procedure, you should test both typical values and values at the boundaries. For example, when the atomic number is requested from the user, you should begin the testing by entering several numbers between 1 and 103.

You should then try to think of possible responses that a user might enter when data is requested. Part of debugging is being able to pick good test data values. This may sound simple to do, but it is not. In debugging, programmers sometimes unconsciously do not pick values that they know will not work. For example, you might not enter a negative number or a real number when testing values for the atomic number. But the student who has not yet learned the atomic numbers may not know that a negative number or a real number is an inappropriate response. Some data values are so bizzare that it would never occur to you that someone would enter them either on purpose or by accident. If you ever are involved in program support, you will be amazed at the logical reasons program users have for choosing bizarre values.

Documentation

Documentation of programs is often neglected by programmers who write programs that will be run only by themselves or a few other people; however, documentation is essential if anyone except the programmer will be running the program. What is obvious to the program author is usually completely obscure to everyone else. Internal comments, when used correctly, can help someone who is trying to understand the program, but there is no substitute for documentation that can be read by anyone, programmer or not.

The documentation should answer some basic questions: What does the program do? What are its limitations? What does the user do to make it work? Documentation should also provide the user with information about how to get additional help, who wrote the program, when the program was written, and when the program was last updated.

Defining what a program does and the program's limitations is extremely important. No one wants to invest time and money figuring out how a program works, only to discover that it does not perform some crucial task. If your program finds real roots of equations but cannot find imaginary roots, the documentation should say so.

For large programs, the documentation should explain, from the user's point of view, how to run the program. In fact, good documentation is usually written by someone other than the person who wrote the program. Programmers tend to get caught up in the details of the program and the clever tricks that they discovered. Someone using the program will be more concerned with input requirements or storage-space requirements. Anyone running the program will need sample program runs with both the input and the output. They will also need a list of error messages and probable causes for the errors. Whenever you write documentation, you should try to keep in mind the type of person who will be running the program.

The user documentation for the case study might be written by the chemistry teacher with whom you are working. The teacher would have a better understanding of the students' experience with using computers and the level of instruction that they need to use the program. Although the chemistry teacher may not have an understanding of the programming details, he was involved in developing the functional requirements and understands what the program is supposed to do and what the students need to know to use it.

As part of the documentation, you should include a section on system dependence: Is the program written so that it can be compiled using any standard Pascal compiler? What parts of the program are system dependent? What parts of the program are hardware dependent? If the program uses extensions to standard Pascal, such as character strings, the extensions should be pointed out in the documentation and isolated within the program. Then, if someone wants to use a compiler that does not have these extensions, the program can be modified without being torn apart. Similarly, if you want to clear the screen in a program, you should write a procedure that clears the screen and give it a name such as **ClearScreen.** If you call **ClearScreen** every time the screen is cleared, a user who has a terminal with a different clear-screen control sequence can modify just the **ClearScreen** procedure and the program will work.

If you have looked at any of the documentation for your computer system, you have probably seen both good and bad documentation. Bad documentation is written in computer jargon without definitions, does not answer the questions that users have, and often gives technically correct, but misleading, information. It is remarkably easy to write bad documentation.

Maintenance

When a programming project is first completed, it is rarely finished. As soon as the program has been distributed to the people who will use it, two phenomena occur: users find bugs that were not detected during the testing phase, and users start making requests for modifications or extensions to the program. The process of eliminating undetected bugs and making modifications is called *maintenance*. Maintenance has been estimated to account for as much as two-thirds of the total cost of a program. Several factors contribute to this high cost. Maintenance is a continuous process that does not stop until the program becomes obsolete. As the program is modified over time, it tends to lose its modular structure, so fixing one bug creates another bug. New versions of the program and documentation must be released periodically to all the users, increasing the overall cost of the program.

Suppose that the program for the case study has been completed and is now being used by the chemistry students. You get a phone call from the chemistry teacher, who tells you that the program is working well. The only problem is that students become very frustrated when they enter data of the wrong data type. The program stops running, and they have to start over again. He asks if you can solve this problem. This problem was discussed in Chapter 8, but now that you know how to use sets you can develop a better solution than the one previously developed.

The solution to the problem of performing data checking is simple enough in scope that you can make a list of the functional requirements. Implicit in the problem statement is the requirement that the strategy should be transparent to the user. That is, when a real number is requested, the user should be able to enter the real number in any legal format, and the program will accept it without the user being aware of the strategy that is used to make the number acceptable to the computer. Also implicit is the requirement that all values entered as input data should be passed to the rest of the program in the appropriate format. These statements may seem obvious, but many strategies for preventing the user from entering invalid data do not meet these requirements.

The functional requirements for the solution to the problem can be stated as follows:

1. Accept integer, real, and character data when entered correctly.

2. Intercept character data when entered where numerical data is required.

Remember from Chapter 8 that the way to prevent the user from terminating the program with invalid data is to read all the data as characters. This solution, although it solves the major problem, creates several new problems. If real and integer numbers are read as characters, they must be translated from character format into equivalent numerical format. After the decision to read all data as characters has been made, the problem becomes

1. Read character variables the standard way.

2. To read an integer or real variable:

   a. Read the input data as characters.

   b. If there are any illegal characters, write a message to the user and request new data.

   c. Else convert the characters to their equivalent numerical value.

Looking at the second part of the problem, you should realize that the illegal characters for integer numbers are different from the illegal characters for real numbers. Therefore, the problem can be broken down into these major steps:

1. Use the standard procedure **read** to read character variables.

2. Write a procedure to replace the standard **read** procedure for integer variables.

3. Write another procedure to replace the standard **read** procedure for real variables.

This list immediately suggests a modular solution. That is, you can work on writing the procedure for integer numbers without having to worry about the procedure for real numbers.

When you are breaking down a problem, it is important for you to recognize problems that have been solved before. You do not, for example, need to write a procedure to read character data, because the standard **read** procedure can be used. Many algorithms involve tasks such as finding the average of an array, sorting a list, searching for a value, generating a random number, or finding the dot product. Once you or someone else has written subprograms to perform these tasks, you do not have to rewrite them each time you need them. As you write more programs, you will accumulate your own collection of useful, and correct, subprograms that you can incorporate into any program you write. This collection is sometimes referred to as a *toolkit*. The analogy is clear: You do not make a hammer every time you need one; you take one out of your toolkit and use it. Every once in a while you may have to modify a tool for a special job, but you rarely have to create a tool from scratch.

Refinement of the problem

One decision that should be made explicit is that the procedures should read *any* integer or real number that conforms to the following Pascal syntax diagrams.

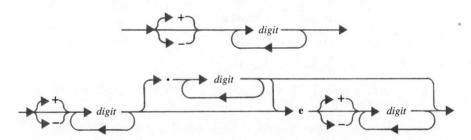

Since integers make up part of the syntax of real numbers, the procedure that you develop for reading integer numbers can be used within the procedure for reading real numbers.

Going back to the list of steps to solve the problem, you can check off Step 1 because it requires no work on your part. You can work on Step 2 and hold off working on Step 3 because the solution to Step 2 will help to solve Step 3.

Done	1.	Use the standard procedure **read** to read character variables.
Working	2.	Write a procedure to replace the standard procedure **read** for integer variables.
On hold	3.	Write a procedure to replace the standard procedure **read** for real variables.

The problem statement for Step 2, reading an integer, can be refined further by breaking it into smaller steps, as is done in Algorithm 12.7.

12.3 Program design and validation

ALGORITHM 12.7 Algorithm to read an integer

begin algorithm
    2. Read an integer
        a. Repeat
            read the input data as characters
           until a number with no illegal characters is entered
        b. Convert the characters to their equivalent integer values
end algorithm

For Algorithm 12.7, you can start by refining Step 2a, reading the input data as characters. Because a procedure cannot be written with a variable number of parameters, you cannot write a procedure that looks like the standard **read** procedure, which can have any number of variables in the list. But you should write a procedure that allows the user to enter more than one variable on an input line. Therefore, your procedure should emulate the standard **read** procedure, which advances until a nonblank character is encountered, then reads the integer that ends at the next blank.

The task of reading the input data as characters is broken into smaller steps in Algorithm 12.8.

ALGORITHM 12.8 Refinement of Step 2a of Algorithm 12.7

begin algorithm
    2a. Read the input data as characters
        (i) Read characters until you get to the first nonblank character
            read **Character**
            while **Character** = blank do
                read **Character**
        (ii) Read and save the nonblank characters
            save **Character**
            repeat
                read **Character**
                save **Character**
            until **Character** = blank
end algorithm

At this point, the algorithm for Step 2a is almost ready to be turned into a Pascal procedure. One feature that must be added to the algorithm is a check to be sure that the read pointer has not moved beyond the end of the line or beyond the end of the file.

Another decision that remains to be made is how to store the characters that make up the integer. The first step of the algorithm involves testing the characters and the second step involves converting them; storing the characters in an array as they are read will make them available to both steps. It will also be useful to know the number of characters in the integer. Given that a data type **CharString** has been declared, the procedure for Step 2a can be written as follows:

```
procedure ReadBetweenBlanks (var Word : CharString; var Length : integer);
(*     ReadBetweenBlanks reads the characters between the current read   *)
(*     pointer and the next blank.  At the end of the procedure, the     *)
(*     read pointer is positioned after the next blank.                  *)

const Blank = ' ';

var   Character : char;     (* temporary variable for the character *)
                            (* currently being read. *)
begin
  Length := 0;

  read (Character);
  while (Character = Blank) and (not eoln) do
    read (Character);

  repeat
    Length := Length + 1;
    Word [Length] := Character;
    read (Character)
  until (Character = Blank) or (eoln)
end; (* procedure ReadBetweenBlanks *)
```

Notice that the procedure **ReadBetweenBlanks** does no error checking. Its sole purpose is to read the characters between blanks from the input device; thus it can be used for reading real numbers as well as integer numbers. If any data checking were included in the procedure, the procedure could not be used to read real numbers because the legal characters for reals are different from the legal characters for integers. Also, the way the procedure has been written, it can be used for the general purpose of reading words from input.

The next step in reading integers is to check for illegal characters in the input string. The syntax diagram for integer numbers shows that the only legal characters are '+', '−', and the digits '0' to '9'. To check for illegal characters, you can make a set of the legal characters and then test whether every element in the input string is in the set. There is, however, a minor wrinkle in that '+' and '−' are only legal as the first character in the string, but they are not required to appear as the first character. So it is better to have two sets: one that contains the characters that correspond to digits and another that contains the characters '+' and '−'.

```
type   CharSet = set of char;

var    DigitSet, FirstSet : CharSet;
```

12.3 Program design and validation

```
begin
  DigitSet := ['0'..'9'];
  FirstSet := ['+', '-']
```

Because the exit condition for the **repeat-until** in Step 2a only tests whether the string is a valid integer, you can write a boolean function that returns the value true if the integer is valid and false if it is not.

```
function ValidInteger (CharDigits : CharString; Length : integer) : boolean;
(*      ValidInteger returns a value of true if the characters stored in   *)
(*      CharDigits between 1 and Length represent a valid integer. The     *)
(*      value of the function is false otherwise.                          *)

var   Position : integer;          (* for statement control variable. *)
      DigitSet : set of char;      (* set of characters that represent *)
                                   (* integers. *)
      FirstSet : set of char;      (* set of other characters that may be *)
                                   (* the first element of a valid integer. *)

begin
  DigitSet := ['0'..'9'];
  FirstSet := ['+', '-'];

  ValidInteger := true;            (* initialize the function to true. *)

  (* If the first character is not a plus sign, a minus sign, or a digit, *)
  (* the string is not a valid integer. *)

  if not (CharDigits [1] in (FirstSet + DigitSet)) then
    ValidInteger := false;

  (* If any of the remaining characters is not a digit, the string is not *)
  (* a valid integer. *)

  for Position := 2 to Length do
    if not (CharDigits [Position] in DigitSet) then
      ValidInteger := false
end; (* function ValidInteger *)
```

The function **ValidInteger** is not as efficient as it could be, because it continues to check all the elements in the string even after an invalid element is found; however, when you are developing an algorithm, it is best to use a clear algorithm that works. Later, if efficiency becomes an issue, the function can be streamlined so that it stops as soon as an invalid character is encountered.

The next step, Step 2b, is to convert the string to its equivalent integer value. Since you know that the string is legal, all you need is the algorithm to convert from the characters to the numerical value. The first subtask is to convert each digit from its character form to its numerical form. The algorithm for this task was developed in Chapter 8. Briefly, to find the numerical value of a character that represents an integer, you can take the ordinal value of the character and subtract the ordinal value of '0':

```
function ConvertToDigit (CharDigit : char) : integer;
(*      ConvertToDigit converts a character between '0' and '9' to its      *)
(*      equivalent numerical value using the standard function ord.          *)

begin
  ConvertToDigit := ord (CharDigit) - ord('0')
end; (* function ConvertToDigit *)
```

You must now figure out how to convert a sequence of separate digits to their integer value. Suppose the sequence of digits is 5, 6, 7, 8. Within the computer, each digit is stored in a separate storage location. The digits must be combined to form the integer 5678. An integer can be expressed as the sum of powers of ten:

$$5678 = 5 \times 10^3 + 6 \times 10^2 + 7 \times 10^1 + 8 \times 10^0$$

So the value of a string of digits representing an integer can be found by starting from the right and multiplying each digit by the appropriate power of 10. The algorithm is given in Algorithm 12.9.

ALGORITHM 12.9 Algorithm to convert a character string to an integer

begin algorithm

    2b. Convert the characters to their equivalent integer value
        (i) Work from right to left and do not convert the leftmost character (**CharDigits[1]**)
            Value ← 0.0
            for **Position** := **Length** downto 2 do
             begin
                convert **CharDigits[Position]** to its equivalent **Digit**
                Place ← **Length** − **Position**
                PlaceValue ← **Digit** ∗ **ZToPower**(10, **Place**)
                Value ← **Value** + **PlaceValue**
            end for
        (ii) Test if the first character is a + or − sign, or a digit
            if **CharDigits[1]** = '+' then ignore it
            if **CharDigits[1]** = '−' then
                Value ← **Value** ∗ (−1)
            if **CharDigits[1]** in **DigitSet** then
                repeat Step i for **Position** = 1
    end algorithm

Algorithm 12.9 can be written in Pascal as follows:

```
function IntegerConvert (CharDigits : CharString;
                         Length : integer) : integer;
(*      This function converts an array of character digits into their    *)
(*      equivalent integer value.  Length is the number of digits in      *)
(*      the number.                                                       *)

const Ten = 10;             (* used in computing the power of ten by *)
                            (* which each digit is multiplied. *)

var   Value : integer;      (* running total for the numerical value of *)
                            (* the character array. *)
      Place : integer;      (* intermediate value for the current tens *)
                            (* place. *)
      Digit : integer;      (* integer value of a character. *)
      Position : integer;   (* for loop control variable. *)

begin
  Value := 0;

  (* Starting from the right, go backward through the number converting *)
  (* each character digit to an integer, multiplying it by the appropriate *)
  (* power of 10, and adding it to the running total. *)

  for Position := Length downto 2 do
    begin
      Digit := ConvertToDigit (CharDigits [Position]);
      Place := Length - Position;
      Value := Value + Digit * ZToPower (Ten, Place)
    end;

  (* Test the first character in the number -- if it is a minus sign, *)
  (* multiply the number by -1.0; if it is a plus sign, leave the value *)
  (* of the number unchanged; if it is a digit, convert the character to *)
  (* a digit and include it in the value. *)

  if CharDigits[1] = '-' then
    Value := -1 * Value
  else
    if CharDigits[1] in ['0'..'9'] then
      begin
        Digit := ConvertToDigit (CharDigits[1]);
        Place := Length - 1;
        Value := Value + Digit * ZToPower(Ten, Place)
      end;

  IntegerConvert := Value
end; (* function IntegerConvert *)
```

You have now completed all the subtasks for writing a procedure that replaces the **read** procedure for integers. The only remaining step is to integrate the procedures and functions:

```
procedure ReadInt (var IntVar : integer);
(*      This procedure reads a single integer value from the terminal.  *)
(*      The procedure checks to be sure that the integer is in a valid  *)
(*      format.  If the format is invalid, the procedure prompts for a  *)
(*      new value to be entered.                                        *)

const LineLength = 80;

type  CharString = array [1..LineLength] of char;

var   HowLong : integer;              (* the number of digits in the number -- *)
                                      (* returned by ReadBetweenBlanks. *)
      IntString : CharString;         (* character string in which the number *)
                                      (* stored. *)
      OK : boolean;                   (* OK is set to true if the character *)
                                      (* string is a valid integer. *)

(*$i ztopower *)
(*$i readbetweenblanks *)
(*$i validinteger *)
(*$i converttodigit *)
(*$i integerconvert *)


begin
  OK := false;                        (* the integer is assumed to be invalid *)
                                      (* until it has been tested. *)

  repeat
    ReadBetweenBlanks (IntString, HowLong);

    (* If the number is a valid integer, convert it to its numerical value, *)
    (* otherwise, ask the user to enter the number again. *)

    if ValidInteger (IntString, HowLong) then
      begin
        OK := true;
        IntVar := IntegerConvert (IntString, HowLong)
      end
    else
      begin
        writeln;
        writeln (' That is not a valid integer.  Please re-enter')
      end
  until OK
end; (* procedure ReadInt *)
```

The procedure **ReadInt** has been implemented as a modular unit: it has its own data structures and its own procedures and functions. If other programmers want to use the procedure, they do not have to declare data types or variables for **ReadInt**. All they need to do is include the procedure, and all the definitions and declarations come with it.

Another advantage of this structure is that the identifiers used within **ReadInt** are local to it. Another programmer working on another part of the program could use a procedure called **ValidInteger,** with a completely different definition, and not interfere with the working of **ReadInt.** Thus the scope of identifier names in Pascal is a major advantage when modular procedures are being developed, since the limited scope reduces the possibility of side effects.

Once the procedure for reading an integer number has been developed, the task of writing a procedure to read real numbers can be started. This process is similar to the procedure for reading integers, but it is more complicated because of decimal places and exponents. In the chemistry program, the only real number that is read from the terminal is the atomic weight. Since it is unlikely that a student would enter this number in exponential format, the procedure to read real numbers will assume that the numbers are entered in the standard format using the decimal point. In Problem 6 at the end of the chapter, you are asked to write a procedure that will read a real number in any legal format.

The algorithm for reading a real number in decimal format involves searching for the position of the decimal point. Once the position of the decimal point has been found, the characters to its left can be converted to their numerical value using the same algorithm that was developed for converting to integers. The characters to the right of the decimal point can be converted to their numerical value using a modification of the integer conversion algorithm in which the digits are multiplied by 10 raised to the −**Place.** If no decimal point is found, the number can be converted using the integer conversion algorithm.

ALGORITHM 12.10 Algorithm to read a real number in decimal format

begin algorithm
   3. Write a procedure to replace **read** for real variables
      a. Repeat
         read the input data as characters
        until a number with no illegal characters is entered
      b. Convert the characters to a real number
         (i) Search for the **Location** of the decimal point
        (ii) If **Location** = 0 then
           use integer conversion
        (iii) If **Location** <> 0 then
           for the string stored in 1 to (**Location** − 1)
              use integer conversion including the test of the sign

>> for the string stored in (**Location + 1**) to **EndOfString**
>> use integer conversion, replacing 10****Place** by 10**−**Place** where **Place** is the number of places to the right of the decimal point
> end algorithm

Each step in Algorithm 12.10 is slightly more complicated than it was in the algorithm for reading integers. The function **ValidReal** is an extension of the function **ValidInteger** that tests whether the characters entered constitute a valid real number.

```
function ValidReal (CharDigits : CharString; Length : integer) : boolean;
(*     ValidReal tests whether the character string is a valid real     *)
(*     number.  ValidReal returns a value of true if the characters     *)
(*     stored in CharDigits between 1 and Length are a valid real       *)
(*     number.  This version of ValidReal will return a value of        *)
(*     false for a real number entered in exponential notation.         *)
const DecimalPoint = '.';

var    Position : integer;           (* for statement control variable. *)
       ValidSet : set of char;       (* set of characters that represent *)
                                     (* components of real numbers. *)
       FirstSet : set of char;       (* set of characters (+, -) that may *)
                                     (* precede a real number. *)

begin
   ValidSet := ['0'..'9', '.' ];     (* the set of characters that are valid *)
                                     (* in a real number. *)
   FirstSet := ['+', '-'];           (* the set of characters that are valid *)
                                     (* as the first character. *)

   ValidReal := true;                (* initialize the function to true. *)

   (* If the first character is not a digit, a sign, or the decimal point *)
   (* the real number is invalid. *)

   if not (CharDigits[1] in (FirstSet + ValidSet)) then
      ValidReal := false;

   (* If the first character is the decimal point, it is removed from the *)
   (* set of valid characters because it can occur only once. *)

   if CharDigits[1] = DecimalPoint then
      ValidSet := ValidSet - [DecimalPoint];

   (* Loop through the number testing to be sure that each character is a *)
   (* valid character for a real number.  Once the decimal point has been *)
   (* located, remove it from the set of valid chars. *)
```

```
    for Position := 2 to Length do
      begin
        if not (CharDigits [Position] in ValidSet) then
          ValidReal := false;
        if CharDigits [Position] = DecimalPoint then
          ValidSet := ValidSet - [DecimalPoint]
      end (* for *)
end; (* function ValidReal *)
```

The function **RealConvert** is an extension of the function **IntegerConvert** using Step 3b from Algorithm 12.10.

```
function RealConvert (CharDigits : CharString; Length : integer) : real;
(*      This function converts an array of characters that represent a    *)
(*      valid real number into the equivalent real number. Length is      *)
(*      the number of characters in the real number. The function can-    *)
(*      not convert a number in exponential notation.                     *)
const DecimalPoint = '.';
      Ten = 10;

var   Value : real;           (* running total for the numerical value of *)
                              (* the character array. *)
      Place : integer;        (* intermediate value for the current tens *)
                              (* place. *)
      Digit : integer;        (* integer value of a character. *)
      Position : integer;     (* for loop control variable. *)
      Location : integer;     (* location of the decimal point. *)
      FirstChar : char;       (* first character in the string. *)
begin
  Value := 0;
  Location := Length + 1;     (* if a decimal point is not found, it is *)
                              (* assumed to come after the last digit. *)

  (* The following for statement finds the location of the decimal point. *)

  for Position := 1 to Length do
    if CharDigits [Position] = DecimalPoint then
      Location := Position;

  (* For the numbers to the right of the decimal point, convert the digits *)
  (* to numbers and multiply them by 10**-Place. *)

  for Position := (Location + 1) to Length do
    begin
      Digit := ConvertToDigit (CharDigits [Position]);
      Place := Position - Location;
      Value := Value + Digit / ZToPower (Ten, Place)
    end;
```

```pascal
(* For the numbers to the left of the decimal point, convert the digits *)
(* to numbers and multiply them by 10**Place. *)

    for Position := (Location - 1) downto 2 do
      begin
        Digit := ConvertToDigit (CharDigits [Position]);
        Place := Location - Position - 1;
        Value := Value + Digit * ZToPower (Ten, Place)
      end;

(* Test the first character in the number -- if it is a minus sign, *)
(* multiply the number by -1.0; if it is a plus sign or the decimal *)
(* point, leave the value of the number unchanged; if it is a digit, *)
(* convert the character to a digit and include it in the value. *)

    FirstChar := CharDigits [1];
    if FirstChar = '-' then
      Value := -1 * Value
    else
      if (FirstChar <> '+') and (FirstChar <> DecimalPoint) then
        begin
          Digit := ConvertToDigit (FirstChar);
          Place := Location - 2;
          Value := Value + Digit * ZToPower (Ten, Place)
        end;

    RealConvert := Value
  end; (* function RealConvert *)
```

Both **ValidReal** and **RealConvert** are used in the procedure **ReadReal**.

```pascal
procedure ReadReal (var RealVar : real);
(*      The current version of ReadReal reads a real variable in char-  *)
(*      acter format and tests whether it is a valid real number entered *)
(*      in decimal format (xx.xx). The character string is then con-    *)
(*      verted to its numerical value. In its final version, ReadReal   *)
(*      will read any valid format for a real number.                   *)

const LineLength = 80;

type  CharString = array [1..LineLength] of char;

var   HowLong : integer;          (* number of digits in the real number *)
                                  (* -- returned by ReadBetweenBlanks. *)
      RealString : CharString;    (* character string in which the real *)
                                  (* number is stored. *)
      OK : boolean;               (* OK is set to true if the string is *)
                                  (* a valid real number. *)
```

```
(*$i ztopower *)
(*$i readbetweenblanks *)
(*$i converttodigit *)
(*$i validreal *)
(*$i realconvert *)


begin
  OK := false;                          (* the number is assumed to be in- *)
                                        (* valid until it has been tested. *)

  repeat
    ReadBetweenBlanks (RealString, HowLong);

    (* Test if the string is a valid real.  If it is, convert it to its *)
    (* numerical value.  Otherwise, request a new number. *)

    if ValidReal (RealString, HowLong) then
      begin
        OK := true;
        RealVar := RealConvert (RealString, HowLong)
      end
    else
      begin
        writeln;
        writeln (' That is not a valid real number.  Please re-enter')
      end
  until OK
end; (* procedure ReadReal *)
```

Program 12.3, **TestConvert**, is a test program for the procedures **ReadInt** and **ReadReal**. The output of **TestConvert** shows that it will reject input strings that contain letters and that it will correctly convert negative numbers.

PROGRAM 12.3   TestConvert

```
program TestConvert (input, output);
(*     Program TestConvert tests the procedures ReadReal and ReadInt.     *)

var    OneInt : integer;     (* integer to be read and converted..*)
       OneReal : real;       (* real number to be read and converted. *)


(*$i readint *)
(*$i readreal *)
(*$i yesentered *)
```

```
begin (* main program *)
  repeat
    writeln;
    writeln (' Enter an integer ');
    ReadInt (OneInt);
    writeln;
    writeln (' You entered the integer ', OneInt);

    writeln;
    writeln (' Enter a real number');
    ReadReal (OneReal);
    writeln;
    writeln (' You entered the real number ', OneReal);
    writeln (' Do you want to continue?')
  until not YesEntered
end.   (* program TestConvert *)
```

=>

 Enter an integer number
□ab
 That is not a valid integer. Please re-enter
□1
 You entered the integer   1

 Enter a real number
□5x
 That is not a valid real number. Please re-enter
□5.78
 You entered the real number   5.7800000000000e+00
 Do you want to continue?
 Enter yes or no (y/n)
□y

The next two sets of data, however, demonstrate error conditions that were not taken into account when the procedure was written. In the first case, the user entered a number that was larger than **maxint**, and in the second case, the user entered just a plus sign.

=>
 Enter an integer number
□1234873048927408923741823540293740297102983

                                        pascal termination log
                                        ------ ----------- ---
 *** error ***   program interrupt - fixed point overflow exception
                                  .
                                  .
                                  .
interrupt at offset 00009a in segment integerc

```
local vars: digit    = 8                   i       = 70                length
           = 79              runningt= 176673134
                    tenspowe= 1000000000

====> called from offset 00007e in segment readint

local vars: howlong = 79                   intvar  = <undefined>       ok
           = <true>

====> called from offset 0000ac in main program

local vars: oneint  = <undefined>          onereal = <undefined>

=>
 Enter an integer number
□+
 You entered the integer        0

 Enter a real number
□+
 You entered the real number    0.0000000000000e+00
 Do you want to continue?
 Enter yes or no (y/n)
□n
```

In order to avoid these errors, error checks must be added to the procedures.

Once the procedures **ReadInt** and **ReadReal** are working correctly, the **read** statements in **PeriodicTable** can be replaced by either **ReadInt** or **ReadReal.** One advantage of top-down design is that each procedure can be tested individually. Because each procedure is developed separately and performs only one task, each procedure can be debugged and tested before it is put into general use. For example, if **ReadReal** were being developed by one member of a team working on a large program, all the other programmers could be using a temporary version of **ReadReal** while the final version was being written and debugged. Once **ReadReal** is working correctly, it can be released to the group. If other programmers find problems with the procedure, the problems can be corrected and a new release of the procedure can be issued. If each procedure is tested individually, you can have much more confidence that the entire program works. If the procedures are put together all at once without testing, it becomes necessary to test the program for all possible combinations of input values, which is close to impossible for large programs. There is a branch of mathematics that deals with formal proofs that an algorithm has been implemented correctly. This process, called *program verification,* is beyond the scope of this text.

Summary

Mathematical sets are composed of collections of elements. The basic operations for sets are union and intersection. Sets and the operations performed on them are often represented by Venn diagrams.

In Pascal, a set can be declared as a structured data type. The Pascal operator for the union of sets is +, and the Pascal operator for the intersection of sets is *. Pascal also uses the difference operator, −, with sets.

The boolean operators <>, =, <=, and >= can be used to test whether sets are equal—that is, whether all the elements are the same or whether one set is a subset of the other. Pascal also has another boolean operator, **in,** that can be used to test whether a particular element is in a given set.

Large programs are often written by groups of people, and their design can become quite complicated. The steps in developing a program include the following: statement of functional requirements, statement of program specifications, implementation, debugging, verification, testing, documentation, and maintenance. Once the functional requirements of a program have been determined, top-down design can be used to break the problems down into smaller tasks and to refine each task in a step-by-step manner. Debugging, verification, and testing are done for each part of the program as it is developed. Documentation must be provided for both users and other programmers. Maintenance of programs consists of changes, extensions, and corrections of a program once it is in use.

References

1. F. P. Brooks, *The Mythical Man-Month,* Reading, Mass.: Addison-Wesley, 1972.

2. C. L. McGowan and J. R. Kelly, *Top-Down Structured Programming Techniques,* New York: Petrocelli/Charter, 1975.

3. P. Grogono and S. H. Nelson, *Problem Solving and Computer Programming*, 2nd ed., Reading, Mass.: Addison-Wesley, 1984.

4. E. W. Dijkstra, *A Discipline of Programming,* Englewood Cliffs, N.J.: Prentice-Hall, 1976.

Exercises

1. List the elements of the set that results from the set operations, given the following sets:

$$A = [1, 3, 5, 7, 9]$$
$$B = [2, 4, 6, 8, 10]$$
$$C = [0, 5, 6]$$

(a) $A \cup B$ (b) $A \cap B$ (c) $A \cap C$ (d) $B \cup C$
(e) $A - B$ (f) $C - B$ (g) $A \cup B \cup C$ (h) $A \cap (B \cup C)$

2. What is the value of each boolean expression, given the following sets?

$$A = ['c', 'a', 't'],$$
$$B = ['c', 'o', 'a', 't']$$
$$C = ['a', 'e', 'i', 'o', 'u', 'y']$$

(a) $A \subset B$ (b) $B \subset C$ (c) $'c' \in B$ (d) $['t'] \subset A$

3. Write the boolean expressions from Exercise 2 as Pascal expressions.

4. Given the declarations

```
type  DigitSet = set of '1'..'9';
      NumberSet = set of 1..9;
      CharSet = set of char;

var   Digits : DigitSet;
      Numbers : NumberSet;
      Characters : CharSet;
```

correct the syntax errors in the expressions and fragments below:

(a)   `[a] in DigitSet`

(b)   `Numbers <= Digits`

(c)   `Digits in Characters`

5. Given the declarations and initial values

```
type  FuelType = (Oil, Gas, Nuclear, Solar, Coal, Wind);
      FuelSet = set of FuelType;

var   Thermal, NonConventional, Air, Intersect, Union : FuelSet;
      Fuel : FuelType;

begin
  Thermal := [Oil, Coal, Gas, Nuclear];
  Air := [Oil, Gas, Coal];
  NonConventional := [Solar, Wind];
  Intersect := [];
  Union := [];
end
```

what is the output from each of the following fragments?

(a)
```
        for Fuel := Oil to Solar do
          begin
            if Fuel in Thermal then
              writeln (' A thermal plant');
                if Fuel in NonConventional then
                  writeln (' Not a thermal plant')
          end (* for Fuel *)
```

(b)
```
        if Air <= Thermal then
          writeln (' It''s a subset')
        else
          writeln (' It''s not a subset');

        Union := Air + Thermal;
        if Union >= Thermal then
          writeln (' Thermal is a subset')
```

6. Rewrite the function **ValidInteger** so that it stops checking the character string as soon as the first invalid character is encountered.

7. When many programmers are working on a large project, the modular structure of Pascal allows them to create subprograms using the same names that other programmers have used. If each subprogram is stored in a separate computer file, as long as each programmer works on a separate computer account, there is no confusion about which file contains which subprogram. When the subprograms are integrated into one large program, there may be problems with duplicate file names. There are several general solutions to this problem and several solutions that depend on the file structure of a particular computer operating system. Propose and discuss two general solutions and one solution particular to your operating system.

Problems

1. Rewrite the proofreading program from Problem 7 in Chapter 11 using sets. Write the program so that only letters are valid for the names and only numbers are valid for the IDs.

2. Write a program that counts the number of digits, letters, and punctuation marks entered in a line from the terminal.

3. In Program 4.11, **AnalyzeThreeCircuits2**, the user was required to enter the data for three circuits and to analyze each circuit in both a series and a parallel connection. Modify the program so that the user can request to see the analysis for a parallel connection, a series connection, or both connections. Allow the user to analyze as many circuits as he or she wants. Using a set of valid answers, have the program test to be sure that the response from the user is in the valid set. If it is not, the prompt should be repeated until a valid answer is entered. The program should tell the user how to stop the program and should stop on request without performing any more calculations.

4. Use sets to write a program that reads two five-letter words from the terminal and prints out the letters that appear in both words, the letters that appear in only one of the words, and all of the letters that appear in either word. It should also report whether the words have more letters that are the same or more that are different. Test your program on the following two sets of words: would, could; exist, digit.

5. Program 11.2, **TestFuels,** uses enumerated data types for the names of the fuels used by the electric power plants. Typically, oil, gas, coal, and nuclear power plants are labeled conventional, and hydro-electric, solar, and wind power plants are labeled nonconventional. Use sets to modify Program 11.2 so that it informs the user whether the power plant that has been entered is conventional or nonconventional. Also, use sets to prevent the user from entering a fuel number that is not one of the options.

6. Complete the procedure **ReadReal** so that it returns the value of a real number that is entered in any valid format. Test your program on the following real numbers:

    $+0 + 1e1, 4.5E-03, -1.0, -2.5e+10, -8.1e-2.0$

7. Using sets, write the procedures that ask the student what period an element is in. Integrate the procedures into the program **PeriodicTable.**

CHAPTER 13

Records

In Pascal, the data type **record** can be used to store and reference information that goes together but that does not have the same data type. The **record** is a structured data type that allows you to organize data in a manner quite different from the structured data types **array** and **file** that you have previously learned. Data can be organized in a logical manner with **record**s. This reduces the complexity of algorithms that require the use of related pieces of information that otherwise could not be stored in a single data type.

CASE STUDY 22
Tagging birds

You are working with a group of ecologists performing research on a species of migratory birds. These birds are known to use six breeding locations, and you want to determine whether each bird returns to the same breeding ground every year. For this project, you need to keep a record of the tag number, the sex of the bird, the date the bird was tagged, and the location where the bird was found. You also want to record the name and affiliation of the ecologist who found each bird. The program you write should allow you to enter the data interactively and store it in a Pascal **textfile**.

Algorithm development

Because the information you need to store for Case Study 22 is of different data types, you cannot put it into a multidimensional array. A **record** can be used so that the tag number, the sex of the bird, the location, and the date the bird was tagged can all be stored and referenced together. The following sections will introduce **record**s, and the program to store the information necessary for the case study in a Pascal **textfile** will be given in the solution to the case study.

13.1 Declaring a record

The data type for a **record** is declared using the following syntax diagram:

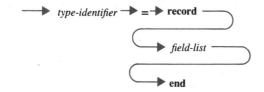

A simplified syntax diagram for the *field list* is

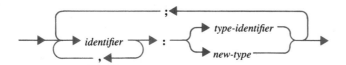

The *type-identifier* is the name that you give to the new **record** data type, and the *identifiers* in the field list are the names that you give to the components of the record. For example, you could declare a **record** to store the information on a single bird:

```
type  BirdRecord = record
        ID : integer;
        Site : packed array[1..20] of char;
        Sex : (male, female );
        Date : packed array[1..20] of char
      end; (* BirdRecord *)
```

In this **record,** the tag number is stored as an integer, the location is stored in a packed array of 20 characters, the sex of the bird is stored as an enumerated type, and the date is stored as a packed array of characters in mm/dd/yy format.

The tag number, location, sex, and date are the *fields* of the **record BirdRecord.** Notice that the syntax for declaring fields in a **record** is the same as the syntax for declaring variables. The type specified for an identifier in the field list can be any standard data type or user-defined data type. Another way of declaring the **BirdRecord** would be to declare explicit data types for the fields:

```
const MaxLetter = 20;              (* maximum letters in CharString. *)

type  Gender = (male, female);
      CharString = packed array [1..MaxLetter] of char;

      BirdRecordType = record
        ID   : integer;            (* tag id for the bird. *)
        Site : CharString;         (* site where bird was tagged. *)
        Sex  : Gender;             (* sex of the bird. *)
        Date : CharString          (* date the bird was tagged, stored in *)
                                   (* mm/dd/yy format. *)
      end; (* BirdRecordType *)
```

Notice that the **record** syntax has an **end** but no **begin.** The reserved word **record** takes the place of the **begin** in the definition of the fields of the record.

Using record variables

A **record** type definition is like any other user-defined type definition. In order to use the type definition, you must declare a variable to have the **record** data type. It is a good idea to give your data types descriptive names, like **BirdRecordType,** so that you do not forget that they are **record** data types and not variable names.

If the **record** type **BirdRecordType** is defined as given above, variables can be assigned to have the data type **BirdRecordType**:

```
var    Bird : BirdRecordType;
```

The variable **Bird** has the structure of the **record** type **BirdRecordType**; that is, it is composed of an integer, a character array, an enumerated type, and another character array.

The field-selector notation

The variable **Bird** is a **record** containing several different kinds of information. In order to access any of this information, you must specify which field in the **record** you want to reference. The information stored in the fields of **record** variables is referred to by giving the identifier of the **record** variable, followed by a period and the field identifier. For example:

Bird.ID is an integer for the bird's tag number,

Bird.Site is a packed array of characters for the bird's location.

13.1 Declaring a record

Formally, this notation is called *field-selector notation,* but it is more commonly referred to as *period notation* or *dot notation.*

Once you have specified a field of a **record** variable completely by giving the **record** variable name and the field name, the field of a **record** variable is like any other variable. You can assign values to fields, read them, and write them as long as you follow all the rules for the data type of the field. To assign a value to the **ID** of a bird that you have just found and tagged, you could write

```
Bird.ID := 1234
```

You could use a **read** statement to assign a value to a field within a record:

```
writeln (' What is the tag number for the bird?  Enter an integer.');
readln (Bird.ID)
```

And in standard Pascal, you could use the statement

```
writeln (' The bird was tagged on ', Bird.Date)
```

to print the date on which the bird was tagged. The fragment below reads and reports the information on one bird:

```
writeln (' What is the tag number for the bird?  Enter an integer.');
readln (Bird.ID);

writeln (' Where was the bird tagged?');
ReadString (MaxLetter, Bird.Site);
writeln;

writeln (' On what date was the bird tagged?  Enter in MM/DD/YY format');
ReadString (MaxLetter, Bird.Date);
writeln;

writeln;
writeln (' Bird ', Bird.ID:1, ' was tagged at location ', Bird.Site,
         ' on ', Bird.Date)
```

```
=>
 What is the tag number for the bird?  Enter an integer.
□12
 Where was the bird tagged?
□Dickens Farm R.I.
 On what date was the bird tagged?  Enter in MM/DD/YY format
□07/21/85

 Bird 12 was tagged at location Dickens Farm R.I.    on 07/21/85
```

In each of the Pascal statements in the fragment, **Bird.Date** and **Bird.Site** are treated as packed arrays of characters and **Bird.ID** is treated as an integer variable.

> *REMINDER*
>
> The **record** variable name (the first name) is given in the **var** declaration. The field names (the second names) are given in the **type** declaration.

The with *structure*

The preceding fragment that reads and reports the information about when and where a bird was tagged uses the field-selector notation. In this fragment, the record variable name **Bird.** is repeated six times. Because it is common to refer to many different fields of a record variable within a small section of a program, the **with** statement is available as a shorthand alternative to the field-selector notation. The syntax for the **with** statement is

What this syntax diagram does not tell you is that any record variable fields that appear in the *statement* of a **with** structure do not have to be preceded by the record variable name. For example, the following fragment and the preceding fragment execute exactly the same way:

```
with Bird do
  begin
    writeln (' What is the tag number for the bird?  Enter an integer.');
    readln (ID);

    writeln (' Where was the bird tagged?');
    ReadString (MaxLetter, Site);

    writeln (' On what date was the bird tagged?  Enter in MM/DD/YY',
             ' format');
    ReadString (MaxLetter, Date);

    writeln (' Bird ', ID:1, ' was tagged at location ', Site, ' on ',
             Date)
  end (* with Bird do *)
```

13.1 Declaring a record

The effect of the **with** statement is to put the record identifier, **Bird.**, in front of every field that is used within the statement following the **do.**

Notice that the **with** statement uses a **do**, which, like any other **do**, requires a compound statement using a **begin-end** pair if more than one statement is to be executed in the **with** statement.

In the following fragment, the temporary character variable **Answer** is used inside the **with** statement, but **Answer** is not one of the fields of **BirdRecord**:

```
with Bird do
  begin
    writeln (' What is the sex of the bird?  Enter m or f');
    readln (Answer);

    if Answer = 'm' then
      Sex := male
    else
      Sex := female
  end (* with Bird do *)
```

Will the compiler try to create a variable **Bird.Answer?** No, this statement will compile and run correctly because the record identifier **Bird** is placed only in front of the identifiers that are in the field list of the record. In a **with** statement, you can use both regular variables and fields of a record variable.

Scope of field identifiers

The fields of a record variable exist only as part of the record. The identifiers **Bird.ID, Bird.Site, Bird.Sex,** and **Bird.Date** are declared when the variable **Bird** is declared to be of type **BirdRecordType.** If it is not preceded by the **record** variable name, **ID** is an undeclared identifier, unless it is declared separately as follows:

```
type  BirdRecordType = record
        ID : integer;           (* tag id for the bird. *)
        Site : CharString;      (* site where bird was tagged. *)
        Sex : Gender;           (* sex of the bird. *)
        Date : CharString       (* date the bird was tagged, stored in *)
                                (* mm/dd/yy format *)
      end; (* BirdRecordType *)

var   Bird : BirdRecordType;
      ID : integer;
```

There are now two distinct integer variables: **Bird.ID** and **ID**. This situation can lead to confusion of the record field and the variable used outside the record. If you use the **with** construction

```
                with Bird do
                  readln (ID)
```

you will read a value into the variable **Bird.ID**. If you want to read a value into both variables, you can use the construction

```
                readln (Bird.ID, ID)
```

It is possible, although sometimes confusing, to use the same identifier in the field lists of different records, as in

```
type  BirdRecordType = record
        ID   : integer;         (* tag id for the bird. *)
        Site : CharString;      (* site where bird was tagged. *)
        Sex  : Gender;          (* sex of the bird. *)
        Date : CharString       (* date the bird was tagged, stored in *)
                                (* mm/dd/yy format *)
      end;  (* BirdRecordType *)

      MammalRecordType = record
        ID   : integer;         (* tag id for the mammal. *)
        Site : CharString;      (* site where mammal was tagged. *)
        Sex  : Gender;          (* sex of the mammal. *)
        Date : CharString       (* date the mammal was tagged, stored in *)
                                (* in mm/dd/yy format *)
      end;  (* MammalRecordType *)

var   Bird   : BirdRecordType;
      Mammal : MammalRecordType;
```

After the variables **Bird** and **Mammal** have been declared, the identifiers **Bird.ID** and **Mammal.ID** both exist as distinct integer variables. Your programs will be easier to read and write if the same field identifier is not used in more than one record, unless the structures of the records are parallel as in **Bird** and **Mammal**.

Type compatibility

All of the values of a **record**-structured variable can be assigned to another **record**-structured variable of the same type in a single assignment statement. For example, if you declare the variables

```
        var   NewBird, OldBird : BirdRecordType;
```

then the following assignment statement is valid:

```
                OldBird := NewBird
```

On execution of the statement, all the fields in **OldBird** will be assigned the values of the corresponding fields in **NewBird**.

You can set two **record** variables equal to each other only if they are declared to be of the same **type**. In the example below, **EnglishSparrow** and **HouseSparrow** are not of the same type:

```
type  BirdRecordType = record
        ID : integer;            (* tag id for the bird. *)
        Site : CharString;       (* site where bird was tagged. *)
        Sex : Gender;            (* sex of the bird. *)
        Date : CharString        (* date the bird was tagged, stored in *)
                                 (* mm/dd/yy format *)
      end;  (* BirdRecordType *)
var   EnglishSparrow : BirdRecordType;
      HouseSparrow : record
        ID : integer;
        Site : CharString;
        Sex : Gender;
        Date : CharString;
      end;  (* variable HouseSparrow *)
```

The declaration of the variable **HouseSparrow** may look strange at first. If you examine the syntax, you will see that it is the same as the short-hand syntax used previously for **array**s and **file**s. In the short-hand syntax, a type-identifier is not defined. Instead, the type definition is given on the right side of the **var** declaration.

Given the declarations above, the statement below results in a type conflict of operands error message:

```
            EnglishSparrow := HouseSparrow;
            ***error***                    $129

            error messages :
            ***************

            129: type conflict of operands.
```

Remember that only variables with the same named data type are considered to be of the same type.

13.2 Arrays of records

In the examples above, the **record** variable is for a single bird. Suppose you want to store data on all of the birds that you have tagged during the current breeding season; you can declare a new data type: an **array** that has a

record as its data type. Recall from Chapter 9 that the syntax diagram for declaring an array is

So to create an array of **records,** all you need is a data type that corresponds to a **record**—for example,

```
const  MaxBirds = 10;            (* maximum number of birds in data base. *)
       MaxLetter = 20;           (* maximum letters in CharString. *)

type   Gender = (male, female);
       CharString = packed array [1..MaxLetter] of char;

       BirdRecordType = record
         ID : integer;           (* tag id for the bird. *)
         Site : CharString;      (* site where bird was tagged. *)
         Sex : Gender;           (* sex of the bird. *)
         Date : CharString       (* date the bird was tagged, stored in *)
                                 (* mm/dd/yy format *)
       end; (* BirdRecordType *)

       BirdArray = array [1..MaxBirds] of BirdRecordType;

var    Sparrow : BirdArray;      (* data stored for sparrows. *)
```

Sparrow is now an array of **BirdRecordType**. To assign a value to the **ID** field of the first sparrow, you would write

```
Sparrow[1].ID := 1234
```

or to set the date when the third sparrow was sighted, you would write

```
Sparrow[3].Date := '03/25/85                          '
```

Notice that the array subscript follows the **record** name, not the field identifier. This notation makes sense because in the experiment you have many sparrows, each with an **ID,** not one sparrow with several **IDs**. If, however, you want to print out the month in which the third sparrow is found, the construction is

```
writeln (' The third sparrow was found in month ',
         Sparrow[3].Date[1], Sparrow[3].Date[2])
```

or, equivalently,

```
with Sparrow[3] do
   writeln (' The third sparrow was found in month ', Date[1], Date[2])
```

=>
```
The third sparrow was found in month 03
```

To assign two sparrows identical values, use the statement

```
Sparrow[2] := Sparrow[1]
```

which sets every field in the second **record** variable equal to the corresponding field in the first **record** variable.

13.3 Records of records

Since a field identifier can be of any valid Pascal data type and since a **record** is a valid Pascal data type, you can have a field identifier whose type is another **record.**

You can define a **record** type to store information about which ecologist found and tagged each bird:

```
const MaxLetter = 20;                     (* maximum letters in CharString. *)

type  Gender = (Male, Female);
      CharString = packed array[1..MaxLetter] of char;

      ResearchRecType = record
        Name : CharString;                (* researcher's name. *)
        Affiliation : CharString          (* researcher's affiliation. *)
      end;  (* ResearchRecType *)

      BirdEcologistType = record
        ID : integer;                     (* tag id for the bird. *)
        Site : CharString;                (* site where bird was tagged. *)
        Sex : Gender;                     (* sex of the bird. *)
        Date : CharString;                (* date the bird was tagged, *)
                                          (* stored in mm/dd/yy format *)
        Ecologist : ResearchRecType       (* person who found the bird. *)
      end;  (* BirdEcologistRecord *)

var   Bird : BirdEcologistType;
```

One of the fields in the **BirdEcologistType** is a **record** variable. That variable, **Ecologist,** has two components in its field list: **Name** and **Affiliation.** For

records of **record**s, the **record** variables are strung together with periods. To assign values to the fields of **Bird.Ecologist** using the field-selector notation, you can write

```
Bird.Ecologist.Name := 'Dr. Jane J. Smith    ';
Bird.Ecologist.Affiliation := 'State University   '
```

For the fields of **Bird** that are simple variables, you can assign values as you did before:

```
Bird.ID := 1234;
Bird.Sex := Female
```

The same statements can be written using the **with** construction:

```
with Bird do
  begin
    ID := 1234;
    Sex := Female
  end;   (* with Bird *)

with Bird.Ecologist do
  begin
    Name := 'Dr. Jane J. Smith    ';
    Affiliation := 'State University   '
  end   (* with Bird.Ecologist *)
```

or

```
with Bird, Ecologist do
  begin
    ID := 1234;
    Sex := Female;
    Name := 'Dr. Jane J. Smith    ';
    Affiliation := 'State University   '
  end (* with Bird, Ecologist *)
```

or

```
with Bird do
  begin
    ID := 1234;
    Sex := Female;
    with Ecologist do
      begin
        Name := 'Dr. Jane J. Smith    ';
        Affiliation := 'State University   '
      end (* with Ecologist *)
  end   (* with Bird *)
```

13.4 Variant records

The **record** structure in Pascal has an additional feature that allows records of the same data type to have different, or variant, field identifiers. This structure can be useful because data is seldom so uniform that all the entries can fit into an identical structure. For example, data on habitats would be much more extensive for migratory birds than for nonmigratory birds. With variant **records**, the data for both types of birds can be stored in the same data structure.

The complete syntax diagram for the field list of a **record** is

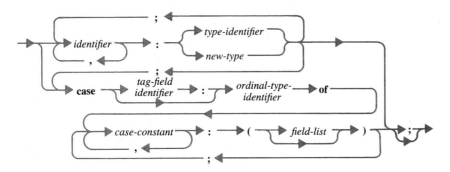

Notice that the syntax diagram for the field list uses the reserved word **case**. Indeed, the structure is analogous to that of a **case** statement except that here the **case** structure is used to control the fields that are declared instead of the statements that are executed. To declare a **record** for birds that are either migratory or nonmigratory, you could use the construction

```
type  BirdType = (Migratory, NonMigratory);

      NewBirdRecordType = record
        ID : integer;
        Sex : Gender;
        case FlightPattern : BirdType of
              Migratory : (LeavingDate : CharString;
                           ArrivalDate : CharString;
                           WinterHabitat : HabitatRecord;
                           SummerHabitat : HabitatRecord);
              NonMigratory : (Habitat : HabitatRecord)
      end; (* variant record *)
```

where **HabitatRecord** is another **record** that defines the data necessary to describe the habitat of the bird.

The syntax for declaring a variant **record** requires a little definition and explanation. First, the fields that are declared before the **case** are called the *fixed part* of the **record** and are just like the fields of any **record**. The part of the **record** defined by the **case** is called the *variant part* of the **record.** The variable that follows the reserved word **case** is called the *tag-field identifier,* and its data type is declared directly after it. In the preceding example, **FlightPattern** is the tag-field identifier, and its data type is the enumerated type **BirdType.** All possible values of the tag-field identifier must appear in the list. The fields of the variant part of the **record** are declared just like the fields of any other **record,** except that they are enclosed in parentheses. Finally, the **case** in the declaration of a variant record does not have an **end**; there is only one **end** at the end of the **record.**

Once a variant **record** has been declared, using it is not difficult. Suppose that two variables **Robin** and **Penguin** have been declared:

```
var   Robin, Penguin : NewBirdRecordType;
```

The variables in the fixed part of each **record** variable can be referenced exactly as they were before:

```
Robin.ID := 4321;

Penguin.Sex := Male
```

To reference the variable part of the record, you must assign a value to the tag-field identifier:

```
Robin.FlightPattern := Migratory;

Penguin.FlightPattern := NonMigratory
```

Once the tag field has been assigned a value, the rest of the record is defined. For the **record** variable **Robin,** the fields **LeavingDate, ArrivalDate, WinterHabitat** and **SummerHabitat** are defined. For the **record** variable **Penguin,** only the field **Habitat** is defined.

To conserve storage space, the fields of variant records are overlaid in storage. The storage locations for the variant part of the **NewBirdRecordType** can be visualized as

| LeavingDate | ArrivalDate | WinterHabitat | SummerHabitat |

| Habitat | | | |

If **BirdType** is assigned a value of **Migratory,** the top scheme is used. If **Birdtype** is assigned a value of **NonMigratory,** the bottom scheme is used.

Thus, the statement

```
Robin.LeavingDate := '10/10/85
```

assigns a value to the **LeavingDate** field of the migratory **Robin.** The

```
Penguin.LeavingDate := '10/10/85
```

statement will overwrite the **Habitat** field of the nonmigratory **Penguin.**

The data structure created by the variant **record NewBirdRecordType** could also have been created by declaring two separate **record**s, one for migratory birds and one for nonmigratory birds, or by declaring one **record** that had fields that were not used for some birds. The advantage of variant records is that they use less storage space during program execution.

Solution to Case Study 22

Because the problem for the case study was a problem of data organization, the solution follows almost directly from the preceding sections. To complete the program, all that is required is a procedure that writes the information to a Pascal **textfile** so that it can be retrieved at a later date. Program 13.1, **SaveTheBirds,** is the completed program that reads the data interactively and stores the data in a **textfile.**

PROGRAM 13.1   SaveTheBirds

```
program SaveTheBirds (input, output, birdfile, newfile);
(*      SaveTheBirds will create and update data about birds that have   *)
(*      have been tagged for an ecological study.  Each bird is identi-  *)
(*      fied by a tag number and its sex.  The date and site of the latest *)
(*      tagging along with the name and affiliation of the person who    *)
(*      tagged the bird are entered interactively from the terminal.    *)
(*      The data is stored in a textfile which is updated each time a   *)
(*      new bird is entered.                                             *)

const MaxBirds = 10;                    (* maximum number of birds in the *)
                                        (* data base. *)
      MaxLetter = 20;                   (* maximum letters in CharString. *)

type  Gender = (male, female);
      CharString = packed array [1..MaxLetter] of char;

      ResearchRecType = record
        Name : CharString;              (* researcher's name. *)
        Affiliation : CharString        (* researcher's affiliation. *)
      end; (* ResearchRecType *)
```

```
            BirdEcologistType = record
              ID : integer;                       (* tag id for the bird. *)
              Site : CharString;                  (* site where bird was tagged. *)
              Sex : Gender;                       (* sex of the bird. *)
              Date : CharString;                  (* date the bird was tagged, *)
                                                  (* stored in mm/dd/yy format *)
              Ecologist : ResearchRecType         (* person who found the bird. *)
            end;  (* BirdEcologistType *)

            BirdArray = array[1..MaxBirds] of BirdEcologistType;

    var     Birds : BirdArray;                    (* array of records in which the data *)
                                                  (* on the birds is stored. *)
            NumberOfBirds : integer;              (* number of birds in data base. *)
            Stop : boolean;                       (* set to true when the user indicates *)
                                                  (* that no more birds will be entered. *)
            birdfile : text;                      (* textfile in which the data on birds *)
                                                  (* entered previously is stored. *)
            newfile : text;                       (* textfile to which the data stored in *)
                                                  (* birdfile plus the birds that are *)
                                                  (* entered are written. *)

(*$i yesentered *)

(*$i readstring2 *)
(* procedure ReadString2 (var infile : text; NumChar : integer;      *)
(*                        var Word : CharString);                    *)
(*    Procedure ReadString2 reads a character string of length NumChar *)
(*    from the texfile infile.  With the file as a parameter, the    *)
(*    procedure can read from any file including input.              *)

  procedure ReadOldBirds (var OldBirds : BirdArray; var NumBirds : integer);
    (*    ReadOldBirds reads the data on the birds that have already been *)
    (*    stored in the data base.  NumBirds is the number of birds in    *)
    (*    the file birdfile.  This procedure uses ReadString2 that reads  *)
    (*    a character string from a file.                                 *)

  var   Answer : char;      (* character that indicates whether the bird is *)
                            (* male or female. *)
  begin
    reset (birdfile);
    NumBirds := 0;

    while not eof(birdfile) do
      begin
        NumBirds := NumBirds + 1;
        with OldBirds [NumBirds] do
          begin
            read (birdfile, ID);
            ReadString2 (birdfile, MaxLetter, Site);
            read (birdfile, Answer);
```

```
            if Answer = 'm' then
              Sex := male
            else
              Sex := female;

          ReadString2 (birdfile, MaxLetter, Date);
          with Ecologist do
            begin
              ReadString2 (birdfile, MaxLetter, Name);
              ReadString2 (birdfile, MaxLetter, Affiliation)
            end
        end (* with OldBirds *)
    end (* while *)
end; (* procedure ReadOldBirds *)


procedure ReadFinder (var Finder : ResearchRecType);
(*      procedure ReadFinder reads the name and affiliation of the       *)
(*      ecologist who found the bird.  This procedure uses ReadString2   *)
(*      to read a character string from the file input.                  *)

begin
  with Finder do
    begin
      writeln (' What is the name of the ecologist who found this bird?');
      ReadString2 (input, MaxLetter, Name);
      writeln;
      writeln (' What is the affiliation of the ecologist?');
      ReadString2 (input, MaxLetter, Affiliation);
      writeln;
    end   (* with Finder *)
end; (* procedure ReadFinder *)


procedure ReadBird (var NewBird : BirdEcologistType);
(*      procedure ReadBird reads the tag number and sex of a bird.  It   *)
(*      also reads where and when the bird was last tagged.              *)

var Answer : char;       (* temporary variable to store the character *)
                         (* entered for the sex of the bird. *)
begin

  with NewBird do
    begin
      writeln (' What is the tag number for the bird?  Enter an integer.');
      readln (ID);

      repeat
        writeln (' What is the sex of the bird?  Enter m or f');
        readln (Answer);
      until Answer in ['m', 'M', 'f', 'F'];
```

```pascal
                if (Answer = 'm') or (Answer = 'M') then
                  Sex := male
                else
                  Sex := female;

                writeln (' On what date was the bird tagged?  Enter in MM/DD/YY',
                         ' format.');
                ReadString2 (input, MaxLetter, Date);
                writeln;
                writeln (' Where was the bird tagged?');
                ReadString2 (input, MaxLetter, Site);
                writeln;
                ReadFinder (NewBird.Ecologist)
      end    (* with Bird do *)
end; (* procedure ReadBird *)

procedure WriteBirds (AllBirds : BirdArray; NumBirds : integer);
(*      WriteBirds writes all the birds, the birds that have just been   *)
(*      entered plus the birds that were originally in the file to a     *)
(*      new textfile.                                                    *)

var BirdNum : integer;        (* for statement control variable. *)
    Answer : char;            (* character that indicates whether the bird *)
                              (* is male or female. *)

begin
  rewrite (newfile);

  for BirdNum := 1 to NumBirds do
    begin
      with AllBirds [BirdNum] do
        begin
          if Sex = male then
            Answer := 'm'
          else
            Answer := 'f';

          writeln (newfile, ID:5, Site:20, Answer:1, Date:20,
                   Ecologist.Name:20, Ecologist.Affiliation:20)
        end  (* with AllBirds *)
    end  (* for *)
end; (* procedure WriteBirds *)

(*------------------------------------------------------------------------*)
begin (* main *)
  ReadOldBirds (Birds, NumberOfBirds);

  Stop := false;

  writeln (' You will be prompted for data on the birds until you request',
           ' to stop.');
  writeln;
```

```
   repeat
      NumberOfBirds := NumberOfBirds + 1;    (* there is now one more bird *)
                                             (* in the database. *)
      ReadBird(Birds[NumberOfBirds]);        (* read the data on the next *)
                                             (* bird. *)
      writeln;
      writeln (' Do you want to enter data on another bird?');
      if not YesEntered then
         Stop := true
   until Stop or (NumberOfBirds >= MaxBirds));

   WriteBirds(Birds, NumberOfBirds);
   writeln (' The new data base is stored in the file newbirds.dat')
end. (* program SaveTheBirds *)
```

=>
 You will be prompted for data on the birds until you request to stop.

 What is the tag number for the bird? Enter an integer.
□1004
 What is the sex of the bird? Enter m or f
□M
 On what date was the bird tagged? Enter in MM/DD/YY format.
□07/04/76
 Where was the bird tagged?
□Philadelphia
 What is the name of the ecologist who found this bird?
□Paul John Jones
 What is the affiliation of the ecologist?
□U. of Penna.

 Do you want to enter data on another bird?
 Enter yes or no (y/n)
□n
 The new data base is stored in the file newbirds.dat

Listing of oldbirds.dat

```
    1234            Bostonm          12/25/84         Jane Smith
State U.
      12            Bostonf          12/25/84         John Smith
Audubon Society
```

Listing of newbirds.dat

```
    1234            Bostonm          12/25/84         Jane Smith
State U.
      12            Bostonf          12/25/84         John Smith
Audubon Society
    1004            Philadelphiam    07/04/76         Paul John Jones
U. of Penna.
```

APPLICATION
Complex Numbers

CASE STUDY 23
Complex number package

Complex numbers are commonly used in science and engineering, and it will be useful for you to have a package of procedures to use whenever you need to write programs that use complex numbers. Develop a package that includes procedures to add, subtract, multiply, and divide complex numbers.

Many physical phenomena cannot be described using only the real number system to which you are accustomed. For example, alternating current and harmonic oscillation are described using the complex-number system. This section will not discuss the theory of complex numbers, but it will discuss how to represent and perform the basic arithmetic operations for complex numbers. The problems at the end of the chapter give several examples of the use of complex numbers in describing physical systems.

Representation of complex numbers

A complex number z can be written in the form

$$z = x + iy \qquad (13.1)$$

where i is the square root of -1. The first part, x, is referred to as the *real part*, and the second part, y, is referred to as the *imaginary part* of z. (Engineers frequently use the letter j for the square root of -1 because the letter i is used to represent electrical current.)

Because the square root of -1.0 is undefined, using the statement

```
Z := X + sqrt(-1.0) * Y
```

to represent a complex number in a program will result in a run-time error. You can avoid this obstacle by using a convention in which complex numbers are stored in parts. Complex numbers are usually stored in a computer in either rectangular or polar coordinates. In either coordinate system, it is understood that some part of the number is to be multiplied by the square root of -1, but this operation cannot actually be performed inside a computer.

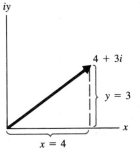

FIGURE 13.1
Complex number in rectangular coordinates.

Complex numbers are frequently represented geometrically by plotting the value of the real part on the x-axis and the value of the imaginary part on the y-axis. The plane described by the real x-axis and the imaginary y-axis is called the *complex plane*. The complex number z can be thought of either as the point (x, y) or as the vector that goes from the origin to the point in the complex plane. Representation and manipulation of complex numbers are similar to representation and manipulation of vectors. It may be helpful to review the application section from Chapter 9 before continuing with complex numbers.

A complex number expressed as the point (x, y) is said to be in *rectangular coordinates*. In rectangular coordinates, a complex number such as $4 + 3i$ would be represented as shown in Figure 13.1.

Recall that, for a vector in polar coordinates, r is the magnitude of the vector and θ is the angle between the x-axis and the vector. Thus the complex number $4 + 3i$ is represented by a vector that has a magnitude of 5 and makes an angle of 36.9°, or 0.64 radian, with the x-axis, as shown in Figure 13.2. From Figures 13.1 and 13.2, it should be clear that whether you use rectangular coordinates, (x, y), or polar coordinates, (r, θ), you will be describing the same point in the complex plane.

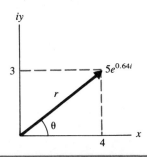

FIGURE 13.2 Complex number in polar coordinates.

Just as with vectors, some operations with complex numbers are easier to perform in rectangular coordinates, and some are easier to perform in polar coordinates. It will be useful to store both the (x, y) representation and the (r, θ) representation of a complex number so that any operation can be performed. A complex number can be defined using the **record** data type as

```
type  Complex = record
         X, Y : real;
         R, Theta : real
      end; (* Complex record *)

var   Z : Complex;
```

After the variable **Z** has been declared as a **record** variable of type **Complex**, **Z.X** can store the real part of the complex number, **Z.Y** can store the imaginary part, **Z.R** can store the magnitude, and **Z.Theta** can store the angle.

Conversion of coordinates

Converting from polar coordinates to rectangular coordinates requires simple geometry. Figure 13.3 shows the relationship between the (r, θ) coordinates and the (x, y) coordinates. Algebraically, the relationships are

$$x = r \cos \theta \quad \text{and} \quad y = r \sin \theta \qquad (13.3)$$

So, given the magnitude r and angle θ of a complex number, you can compute the value of the x- and y-coordinates. If the record variable **Z** is defined as previously and if **Z.R** and **Z.Theta** have been assigned values, **Z.X** and **Z.Y** can be computed using the **with** statement and Equations 13.3.

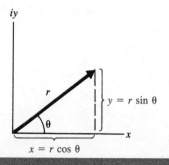

FIGURE 13.3
Geometric relationships among x, y, r, and θ.

```
           with Z do
             begin
               X := R * cos (Theta);
               Y := R * sin (Theta)
             end (* with Z do *)
```

These statements can be used in procedure **ComplexPoleToRec,** which converts a complex number from polar to rectangular coordinates:

```
procedure ComplexPoleToRec (var Z : Complex);
(*     ComplexPoleToRec converts a complex number in polar coordinates   *)
(*     to rectangular coordinates using the definitions of the sine      *)
(*     and cosine.                                                       *)
begin
  with Z do
    begin
      X :=  R * cos (Theta);
      Y :=  R * sin (Theta)
    end (* with Z *)
end; (* procedure ComplexPoleToRec *)
```

To see the similarities between notation for vectors and the notation for complex numbers, you can compare the procedure **ComplexPoleToRec** with the procedure **PoleToRec** in the application section of Chapter 9. Think about why **ComplexPoleToRec** cannot be a function that returns the complex number **Z**.

The formulas to convert a complex number from rectangular to polar coordinates can be found by referring to Figure 13.3. The magnitude of a complex number is the length r as shown in Figure 13.3. Given the values of x and y, the value of r can be found using the Pythagorean theorem:

$$r = \sqrt{x^2 + y^2} \qquad (13.4)$$

The value of θ is given by the geometric relation

$$\theta = \arctan \frac{y}{x} \qquad (13.5)$$

Equations 13.4 and 13.5 can be used to convert from rectangular to polar coordinates. So, if **Z.X** and **Z.Y** have been assigned values, **Z.R** and **Z.Theta** are given by

```
            with Z do
              begin
                R := sqrt (sqr(X) + sqr(Y));
                Theta := arctan(Y/X)
              end (* with Z do *)
```

The assignment statement that computes **Z.Theta** divides **Z.Y** by **Z.X**. Any time you perform the division operation, ask yourself if the divisor could ever be zero. For the complex numbers, **Z.X** could be zero. In this case, **Y / X** will be undefined, but from Figure 13.3 you can see that the angle θ will be 90° ($\pi/2$ radians) if y is greater than zero. If y is less than zero, θ will be $-90°$ ($-\pi/2$ radians). If y is equal to zero, θ will be undefined. Thus, when x is equal to zero, the value of the angle can be set depending on the sign of y. If both x and y are zero, the angle can be arbitrarily set to zero. Procedure **ComplexRecToPole** includes a test so that division by zero never occurs. **ComplexRecToPole** also includes a test to determine the quadrant in which the point lies.

```
procedure ComplexRecToPole (var Z : Complex);
(*      ComplexRecToPole converts rectangular coordinates to polar       *)
(*      using the Pythagorean theorem and the definition of the          *)
(*      arctangent.                                                      *)

const Small = 1.0e-6;       (* a real value close to small is considered *)
                            (* to be equal to zero. *)
      Pi = 3.1415926;       (* used to compute 90 degrees. *)

begin
  with Z do
    begin
      R := sqrt (sqr(X) + sqr(Y));        (* the magnitude is the length of *)
                                          (* the hypotenuse. *)
      Theta := 0.0;                       (* initialize Theta to zero; if Y *)
                                          (* is less than Small, Theta will *)
                                          (* equal 0.0. *)
      if abs (Y) > Small then
        begin
          if abs (X) > Small then         (* if X is not close to zero, Theta *)
            Theta := arctan (Y/X)         (* is the angle with tangent Y/X. *)
          else
            if Y > Small then             (* otherwise Theta is plus or minus *)
              Theta := Pi/2.0             (* 90 degrees, depending on the *)
            else                          (* quadrant of the point. *)
              Theta := -Pi/2.0;
```

```
            if X < -Small then            (* if X is negative, then Theta is *)
                Theta := Theta + Pi       (* in the 2nd or 3rd quadrant. *)
            end (* else Theta defined. *)
        end (* with Z do *)
end; (* procedure ComplexRecToPole *)
```

Equivalence of coordinate systems

In the discussion above, the complex number z was not written in its polar form,

$$z = re^{i\theta} \tag{13.6}$$

This form is similar to the polar form for vectors, except that the imaginary number i appears in the exponent. To show that the polar form in Equation 13.6 is equivalent to the rectangular form,

$$z = x + iy$$

you can use Euler's formula, which states that

$$\cos\theta + i\sin\theta = e^{i\theta} \tag{13.7}$$

Given the complex number in the form

$$z = x + iy$$

you can substitute Equation 13.3 for x and y:

$$z = r\cos\theta + ir\sin\theta$$

or
$$z = r(\cos\theta + i\sin\theta) \tag{13.8}$$

Combining Equations 13.7 and 13.8 gives

$$z = r(\cos\theta + i\sin\theta) = re^{i\theta}$$

or
$$z = re^{i\theta} \tag{13.9}$$

So, $z = x + iy$ is equivalent to $z = re^{i\theta}$.

Addition and subtraction of complex numbers

Addition of two complex numbers is similar to addition of two vectors. Complex addition is illustrated graphically in Figure 13.4. To add two complex

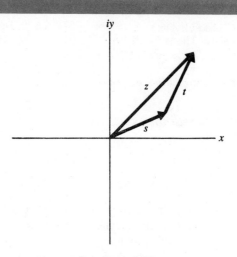

FIGURE 13.4 Complex addition.

numbers, you add the real part of the first complex number to the real part of the second complex number and the imaginary part of the first number to the imaginary part of the second number. For two complex numbers s and t where

$$s = a + ib \quad \text{and} \quad t = c + id$$

the sum $z = s + t$ is given by

$$z = (a + c) + i(b + d) \tag{13.10}$$

For example, the sum of the complex numbers

$$6 + 5i \quad \text{and} \quad 2 - 3i$$

is the complex number

$$(6 + 2) + [5 + (-3)]i \quad \text{or} \quad 8 + 2i.$$

Procedure **AddComplex** employs Equation 13.10 to add two complex numbers using the data structure **Complex**.

```
procedure AddComplex (Num1, Num2 : Complex; var Z : Complex);
(*     AddComplex finds the sum of two complex numbers, Num1 and Num2,   *)
(*     in rectangular coordinates.  The sum is returned in Z; all of     *)
(*     the fields of the new complex number, Z, are defined.             *)
```

```
begin
  Z.X := Num1.X + Num2.X;
  Z.Y := Num1.Y + Num2.Y;

  ComplexRecToPole(Z)      (* compute the polar coordinates of new complex *)
                           (* number. *)
end; (* procedure AddComplex *)
```

Complex subtraction is illustrated in Figure 13.5. To subtract two complex numbers, you subtract the real part of the second complex number from the real part of the first complex number and the imaginary part of the second number from the imaginary part of the first number. For the previous complex numbers s and t, the difference $z = s - t$ is given by

$$z = (a - c) + i(b - d) \tag{13.11}$$

For example, the difference of the complex numbers

$$6 + 5i \quad \text{and} \quad 2 - 3i$$

is the complex number

$$(6 - 2) + [5 - (-3)]i \quad \text{or} \quad 4 + 8i.$$

Procedure **SubtractComplex** employs Equation 13.11 to subtract two

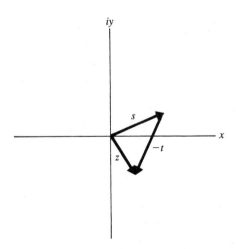

FIGURE 13.5 Complex subtraction.

complex numbers using the data structure **Complex**.

```
procedure SubtractComplex (Num1, Num2 : Complex; var Z : Complex);
(*      This procedure finds the difference between two complex       *)
(*      numbers, Num1 and Num2, using rectangular coordinates.  The   *)
(*      difference is returned in Z.  All of the fields of the new    *)
(*      complex number, Z, are defined.                               *)
begin
  Z.X := Num1.X - Num2.X;
  Z.Y := Num1.Y - Num2.Y;

  ComplexRecToPole(Z)     (* compute the polar coordinates of new complex *)
                          (* number. *)
end; (* procedure SubtractComplex *)
```

Multiplication and division of complex numbers

Complex numbers can be multiplied in rectangular form by expanding the products:

$$z = st = (a + ib)(c + id)$$

It is much easier, however, to multiply complex numbers in polar form. Converting the complex numbers s and t to polar form yields

$$s = r_s e^{i\theta_s}$$

and

$$t = r_t e^{i\theta_t}$$

For the complex number s, the magnitude and the angle are found by substituting into the formulas

$$r_s = \sqrt{a^2 + b^2}$$

and

$$\theta_s = \arctan\left(\frac{b}{a}\right)$$

respectively. The magnitude and angle for the complex number t are found in a similar manner. Given the polar coordinates for s and t, the product z is given by

$$z = st = r_s r_t e^{i(\theta_s + \theta_t)} \tag{13.12}$$

Notice that the complex number z is in polar form

$$z = r_z e^{i\theta_z}$$

with the magnitude r_z given by

$$r_z = r_s r_t \qquad (13.13)$$

and the angle given by

$$\theta_z = \theta_s + \theta_t$$

So the algorithm for multiplying complex numbers is to multiply their magnitudes and add their angles. The procedure **MultiplyComplex** uses Equations 13.13 to multiply two complex numbers.

```
procedure MultiplyComplex (Num1, Num2 : Complex; var Z : Complex);
(*      MultiplyComplex finds the product of two complex numbers, Num1   *)
(*      and Num2, using polar coordinates.  The product is returned in Z. *)
(*   ·  All of the fields of the new complex number, Z,  are defined.    *)
begin
  Z.R := Num1.R * Num2.R;
  Z.Theta := Num1.Theta + Num2.Theta;

  ComplexPoleToRec(Z)     (* compute the rectangular coordinates of Z. *)
end; (* procedure MultiplyComplex *)
```

The quotient z of the complex numbers s and t in polar coordinates is given by

$$z = \frac{s}{t} = \frac{r_s e^{i\theta_s}}{r_t e^{i\theta_t}} \qquad (13.14)$$

or

$$z = \frac{r_s}{r_t} e^{i(\theta_s - \theta_t)} \qquad (13.15)$$

The complex number z is now in polar form,

$$z = r_z e^{i\theta_z}$$

with the magnitude r_z given by

$$r_z = \frac{r_s}{r_t}$$

and the angle θ_z given by

$$\theta_z = \theta_s - \theta_t \qquad (13.16)$$

The algorithm for dividing complex numbers is to divide their magnitudes and subtract their angles. The procedure **DivideComplex** uses Equations 13.16 to divide complex numbers.

```
procedure DivideComplex (Num1, Num2 : Complex; var Z : Complex);
(*    DivideComplex finds the quotient of two complex numbers, Num1    *)
(*    and Num2, using polar coordinates.  The quotient is returned in  *)
(*    Z.  All of the fields of the new complex number, Z, are defined. *)
begin
  Z.R := Num1.R / Num2.R;
  Z.Theta :=  Num1.Theta - Num2.Theta;

  ComplexPoleToRec(Z)     (* compute the rectangular coordinates of Z. *)
end; (* procedure DivideComplex *)
```

Solution to Case Study 23

The procedures for performing operations with complex numbers have been developed. The only remaining task is to write a test program invoking all the procedures to ensure that they operate correctly. Program 13.2, **TestComplex**, is an interactive program that allows the user to perform operations on complex numbers.

PROGRAM 13.2   TestComplex

```
program TestComplex (input, output);
(*    This program converts complex numbers from rectangular to polar  *)
(*    coordinates and vice versa.  It also computes sums, differences, *)
(*    products, and quotients of complex numbers.                      *)

type   Complex = record
         X, Y : real;
         R, Theta : real
       end; (* record Complex *)
```

```
var   Number1, Number2 : Complex;       (* two complex numbers read from *)
                                        (* the terminal. *)
      Operation : char;                 (* operation to be performed on the *)
                                        (* complex numbers: +, -, /, or *. *)
      NewNum : Complex;                 (* complex number resulting from *)
                                        (* performing the operation on *)
                                        (* complex Number1 and Number2. *)

(*$i complexrectopole *)
(*$i complexpoletorec *)
(*$i addcomplex *)
(*$i subtractcomplex *)
(*$i multiplycomplex *)
(*$i dividecomplex *)


(*$i readinrectangular *)
(*  procedure ReadInRectangular (var Num1 : Complex);                       *)
(*      This procedure reads a complex number in rectangular coord-         *)
(*      inates and calls ComplexRecToPole to compute the polar coord-       *)
(*      inates of the new complex number.                                   *)


  procedure ReadInPolar (var Num1 : Complex);
     (*     Procedure ReadInPolar reads a complex number in polar coord-    *)
     (*     inates.                                                         *)

  const Pi = 3.14159;              (* used to convert degrees to radians. *)

  var   Answer : char;             (* indicates whether the angle will be *)
                                   (* entered in degrees or radians. *)
        AngleInDegrees : real;     (* if the angle is entered in degrees, *)
                                   (* it must be converted to radians. *)
begin
  writeln (' Enter the magnitude of the complex number');
  readln (Num1.R)

repeat
  writeln (' Will you enter the angle in degrees or radians? ');
  writeln (' Enter d or r');
  readln (Answer)
until Answer in ['r', 'R', 'd', 'D'];
```

```
      if Answer in ['d', 'D'] then
        begin
          writeln (' Enter the angle in degrees for the complex number');
          readln (AngleInDegrees);
          Num1.Theta := AngleInDegrees * Pi / 180.0
        end
      else
        begin
          writeln (' Enter the angle in radians for the complex number');
          readln (Num1.Theta)
        end;

      ComplexPoleToRec (Num1)     (* compute the rectangular coordinates. *)
end; (* procedure ReadInPolar *)


procedure ReadComplexNumber (var Num1 : Complex);
(*     This procedure reads a complex number in polar or rectangular     *)
(*     coordinates from the user.                                        *)

var    PorR : char;       (* indicates whether the number will be entered *)
                          (* in polar or rectangular coordinates. *)

begin
  repeat
    writeln (' Do you want to enter the number in polar or rectangular',
             ' coordinates?');
    writeln (' Enter p or r');
    readln (PorR)
  until PorR in ['p', 'r', 'P', 'R'];

  if PorR in ['r', 'R'] then
    ReadInRectangular (Num1)
  else
    ReadInPolar (Num1)
end; (* procedure ReadComplexNumber *)


procedure ReadOper (var Num1, Num2 : Complex; var Oper : char);
(*     ReadOper asks the user which operation is to be performed on      *)
(*     the complex numbers.  The user can stop the program by entering   *)
(*     "s".  The user is asked whether the numbers are in polar ('p')    *)
(*     or rectangular ('r') coordinates.  Finally, the x and y coord-    *)
(*     inates or r and theta for the two complex numbers are read.       *)
```

```pascal
    begin
      writeln;
      writeln (' This program will add, subtract, multiply, or divide two',
               ' complex numbers');
      repeat
        writeln (' Do you want to add, subtract, multiply, or divide?');
        writeln (' Enter +, -, *, /, or s to stop.');
        readln (Oper)
      until Oper in ['+', '-', '*', '/', 's', 'S'];

      if not (Oper in ['s', 'S']) then
        begin
          writeln (' Enter the first complex number');
          ReadComplexNumber (Num1);
          writeln (' Enter the second complex number');
          ReadComplexNumber (Num2)
        end (* if not stop *)
    end; (* procedure ReadOper *)

(*$i writecomplex *)
(*   procedure WriteComplex (Num1, Num2, Z : Complex; Oper : char);         *)
(*         WriteComplex  writes out the result of the complex operation     *)
(*         requested by the user.                                           *)


(*--------------------------------------------------------------------------*)
begin (* main program *)
  repeat
    ReadOper (Number1, Number2, Operation);

    if not (Operation in ['s', 'S'] ) then
      begin
        case Operation of
          '+' : AddComplex (Number1, Number2, NewNum);
          '-' : SubtractComplex (Number1, Number2, NewNum);
          '*' : MultiplyComplex (Number1, Number2, NewNum);
          '/' : DivideComplex (Number1, Number2, NewNum)
        end; (* case statement *)

        WriteComplex (Number1, Number2, NewNum, Operation)
      end (* if not stop *)
  until Operation in ['s', 'S']
end.   (* program TestComplex *)
```

```
=>
 This program will add, subtract, multiply, or divide two complex numbers
 Do you want to add, subtract, multiply, or divide?
 Enter +, -, *, /, or s to stop.
□-
 Enter the first complex number
 Do you want to enter the number in polar or rectangular coordinates?
 Enter p or r
□p
 Enter the magnitude of the complex number
□1.0
 Will you enter the angle in degrees or radians?
 Enter d or r
□d
 Enter the angle in degrees for the complex number
□90.0
 Enter the second complex number
 Do you want to enter the number in polar or rectangular coordinates?
 Enter p or r
□r
 Enter the real part of the complex number
□1.0
 Enter the imaginary part of the complex number
□1.0

  First number:     0.00 + i  1.00
 Second number:     1.00 + i  1.00

 Operation:  -

 Answer in polar coordinates:     1.00 exp(i  0.00)

    0.00 radians =    0.00 degrees

 Answer in rectangular coordinates:    -1.00 + i  0.00


 This program will add, subtract, multiply, or divide two complex numbers
 Do you want to add, subtract, multiply, or divide?
 Enter +, -, *, /, or s to stop.
□s
```

Program 13.2 is a useful program to have when you are working with complex numbers. However, as a general utility program, it has several shortcomings: only one operation can be performed on a set of numbers, the operations of conversion between coordinate systems cannot be requested by the user, and only two complex numbers can be operated on at one time. In Problem 8, you are asked to remedy some of these shortcomings.

Summary

Information of different types can be stored together using the structured data type **record.** The pieces of information can be stored as the fields of a record. A **record** is defined by declaring the variable identifiers that make up the fields of the record. Each variable can be of a different data type, simple or structured. Within the body of the program, field-selector notation, or dot notation, is used to specify which field of the record is being referenced. With the dot notation, the name of the **record** variable is given, followed by a period and the field identifier. The **with** statement can be used to place the name of the **record** variable in front of the field identifiers for all the statements between a **begin-end** pair so that the **record** identifier does not have to be typed for every field identifier.

Variant **record**s allow **record**s with the same data type identifier to have different field identifiers. This construction is useful if a **record** is used to store information about similar categories that have only a few fields of information that are different. A variant **record** is created using syntax similar to that of a **case** statement.

Exercises

PASCAL

1. Given the type definitions and variable declarations

```
type  DayType = (Sun, Mon, Tues, Weds, Thurs, Fri, Sat);

      WeekRecord = record
        Dates : array [1..7] of integer;
        Day : DayType
      end; (* record *)

var   Week : WeekRecord;
```

correct the syntax errors in each of the following fragments:

(a) with WeekRecord do
```
        writeln (' The day of the week is currently', ord(Day):5)
```

(b) with Week[5] do
```
        writeln(' This is day', ord(Day), ' of the week')
```

(c) type NewRecord = record
```
        Dates : array [1..7] of integer;
        case Day : DayType of
              Sat, Sun : ( );
            Mon, Weds, Fri : (Classes : WordArray;)
                             (Tests : WordArray);
              Tues, Thurs : SportsSchedule : WordArray)
        end (* case *)
    end; (* record *)
```

2. Create a data structure to store all the information in an address book.
   (a) Use records and arrays.
   (b) Use variant records and arrays.

APPLICATION

3. Convert the following complex numbers from rectangular to polar coordinates:
   (a) $1.2 - 5.1i$ (b) $-6.3 + 50.2i$ (c) $-9.3 - 9.3i$ (d) $3 + 4i$

4. Convert the following complex numbers from polar to rectangular coordinates (remember that the angles are given in radians):
   (a) $1.0e^{i\pi}$ (b) $-4.5e^{-1.2i}$ (c) $-1.4e^{3.0i}$ (d) $6.3e^{0.3i}$

5. If $z = a + ib$, its *complex conjugate* is the number $a - ib$. The complex conjugate is written as $\bar{z}$. The product of a complex number with its conjugate is
$$z\bar{z} = (a + ib)(a - ib) = a^2 + b^2$$
Write a subprogam that returns the complex conjugate of a complex number. Use the preceding result to test your subprogram.

6. Simplify each of the following complex expressions until it is in the form $a + ib$. You may need to use the complex conjugate from Exercise 5.
   (a) $(-4.3 + 2.1i) + (1.2 - 6.3i)$ (b) $(-3.2 - 6.1i) - (-1.2 - 4.6i)$
   (c) $(-2 - 3i)(2 + 4i)$ (d) $3.0 + 4i + 2.2e^{0.3i}$
   (e) $(3.1 + 5.9i)(6.5e^{3.6i})$ (f) $\dfrac{1}{3 + 4i}$
   (g) $\dfrac{R}{R + i\omega L}$ (h) $\dfrac{3.2e^i}{5 - i}$

Problems

PASCAL

1. Using the (poor) approximate formula for the density of a person,
$$\text{density} = \text{weight}/\text{height}^3$$
   write a program to find the student with the lowest density. Read the name, height, and weight of the five students from Problem 13 in Chapter 7. Store the data on each student in a **record** variable. For the student with the lowest density, print out the name, height in centimeters, mass in kilograms, and density in kilograms per cubic centimeter.

2. The widget factory from Problem 5 in Chapter 8 has expanded and now produces 10 different types of widgets. Each widget has a part number (an integer between 1 and 10). The robot reads the part number and the diameter of each widget and then counts how many widgets of each type were produced that hour. One hundred widgets are produced each hour. Write a program that reads the widget data into an array of records. Print a chart with the part number and the number of parts of that type that were produced during the hour.

3. Rewrite the aircraft program from Problem 18 in Chapter 5 to store the data on each aircraft in a **record** variable. The input and the output should be the same as they were for the original problem statement.

4. Write a program to do fractional arithmetic. Have the user enter a fraction, an operation (**/** or *), and another fraction—for example,
$$1/15 \ / \ 1/3$$
   Report the result as the fraction 3/15 (not 0.2). Store each fraction in a record variable.

5. Expand the program from Problem 4 so that it reduces the fraction to its simplest form. In the example above, you would report the original fraction, 3/15, the factor 3, and the final form, 1/5.

6. Expand the program from Problem 4 so that the user can also request addition or subtraction of fractions.

7. Modify Progrm 13.1, **SaveTheBirds,** so that when data on a bird is entered, the program checks to see if the bird has been sighted before. The program should list the previous data on the bird and save both the new and the old information.

APPLICATION

8. Using the subprograms from Program 13.2, **TestComplex,** write an interactive program that allows the user to perform more than one operation on a set of numbers, and to enter more than two complex numbers at a time.

9. In polar coordinates, to raise a complex number to an integer power

$$\text{if } z = re^{i\theta} \quad \text{then} \quad z^n = r^n e^{ni\theta}$$

Write an interactive program that raises a complex number to an integer power. Request that the user enter a complex number in rectangular coordinates and then request the power to which the number is to be raised. Reject noninteger values of n. Write the result to the terminal in rectangular coordinates. Store the complex number in a **record** variable.

10. Pascal has the standard functions **sqr, sqrt,** and **exp** for real numbers. Write your own subprograms to compute the same functions for complex numbers. Write a program to test your subprograms. Test your program using some numbers for which you are certain of the answer—for example, $0.0 + 0.0i$, $1.0 + 0.0i$, and $0.0 + 1.0i$. Then test your program on some other numbers for which you do not know the exact answers.

11. From Euler's formula, Equation 13.7, the cosine can be written as the sum of two complex numbers:

$$\cos \theta = \frac{e^{i\theta} + e^{-i\theta}}{2.0}$$

for any angle θ in radians.

Using this formula, write your own subprogram that returns the cosine. Test your subprogram by comparing your values to the values returned by the standard Pascal function **cos**. **cos** is a real-valued function, and your value is a complex number. If x is the real part and y is the imaginary part of your value, the formulas will be identical if

$$|x - 2.0 \cos \theta| = 0 \quad \text{and} \quad |y| = 0$$

Compute the difference between Euler's formula and the standard function for the following values of θ: $-\pi, -3\pi/4, -\pi/2, \ldots, 3\pi/4, \pi$.

12. If wire is wrapped into a coil and current flows through the coil, a magnetic field is created. If the current varies, the magnetic field varies and a magnetic flux is induced. The coil itself is called an *inductor,* and the constant that relates the change in current to the induced flux is called the *inductance* of the coil. Inductance is measured in henrys (a volt-second per ampere) and is given the abbreviation L. In an *RL* series circuit—that is, a circuit with a resistor and an inductor in series—the voltage is given by

$$V = (R + i\omega L)I$$

where V = voltage in volts
I = current in amperes
R = resistance in ohms
L = inductance in henrys
ω = angular velocity in radians per second

For a circuit with $R = 1.2 \, \Omega$, $L = 0.5$ mH, and $\omega = 372.0$ rad/s, write an interactive program that computes the complex voltage given the complex current and computes the complex current given the complex voltage.

13. Write a program that finds the real and imaginary roots of a quadratic equation.

14. A square wave with a period of T seconds can be represented by the Fourier series

$$f(t) = \sum_{n=-\infty}^{\infty} C_n e^{in\omega t}$$

where $C_n = \begin{cases} 1/(in\pi) & \text{if } n \text{ is even} \\ 0 & \text{if } n \text{ is odd} \end{cases}$

$\omega = 2\pi/T$

The square wave looks like

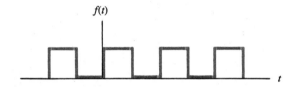

Write a program that generates $f(t)$ at twenty equally spaced values of t between 0 and T for a square wave with a period of 1 second. Compute the Fourier series for $n = -50$ to $n = +50$.

CHAPTER 14

Files

The **textfile** that was covered in Chapter 10 is a specific instance of a more general data type known in Pascal as a **file**. A **textfile** is restricted to the data type **char,** but a **file** can be constructed of any data type, such as **real, integer, record,** or **array.** In Pascal, a **file** is a sequential data type of components that all have the same data type.

CASE STUDY 24
Constructing a stellar data base

A data base of information on stars has been constructed for use with several different astronomical programs. Each star is identified by its catalogue number, name, constellation, right ascension, and declination. In addition, the following information is stored for each star: spectral class, proper motion in seconds per year, radial velocity in kilometers per second, distance from the sun in parsecs, temperature in degrees Kelvin, luminosity relative to the sun, whether the star is a binary star, whether the star is a variable star, and whether the star belongs to a special class such as pulsar, quasar, nebula, or X-ray star.

The data base contains thousands of stars and is continually being expanded. An interactive program to add the new entries to the data base and store the updated data base is required.

Algorithm development

The problem for the case study can be restated as follows: For each star to be added to the data base, read the information on the star and then add the new information to the existing computer file of information on stars.

From what you know about **textfile**s and computer files, you may realize that new information cannot be added to a file in a single step. To add information to an existing computer file, it is necessary to read the original information from the existing file and then write both the original information and the new information to a new computer file. These steps are listed in Algorithm 14.1, which is constructed so that the new information plus all of the original data base must be stored in memory before the new file is written.

ALGORITHM 14.1   First iteration of an algorithm for Case Study 24

begin algorithm
1. Read the new information from the user on each star to be added to the data base
2. Read the original information from the computer file
3. Write the original information plus the new information to a computer file

end algorithm

It would be more efficient to refine the algorithm so that the information about a star is read from either the computer file or the user and then written to the new file immediately. With this refinement, it is necessary to store only one star at a time in the computer's memory.

The problem statement does not say where in the data base the new information should be added. In Algorithm 14.2 it is assumed that the new information is to be added after the original data.

ALGORITHM 14.2   Second iteration of an algorithm for Case Study 24

begin algorithm
1. Read each star from the old file and write it to the new file
2. Read each new star from the user and write it to the new file
end algorithm

Algorithm 14.2 is incomplete; several questions must still be answered. For example, how can you read each star from the old file unless you know how many stars there are? In Chapter 10 you learned how to use the boolean function **eof** to test whether you have reached the end of a file. You have also written subprograms to determine whether the user wants to add more information. You can refine the steps in Algorithm 14.2 to include these tasks, as shown in Algorithm 14.3.

ALGORITHM 14.3 Algorithm for the case study

begin algorithm
1. while not eof (original file)
   a. Read a star from the original file
   b. Write the star to the new file
2. repeat
   a. Read a new star from the user
   b. Write the star to the new file
   until the user is done
end algorithm

At this point, you could write the program using **record** variables and **textfiles**, but implementation of the algorithm using **record** variables and the more general **file** structure will result in a program that is easier both to write and to understand.

14.1 File structure

The data type of a **file** can be any Pascal data type including user-defined data types. Like the components of a **textfile**, the components of a **file** must all have the same data type.

The syntax diagram for declaring a **file** data type is

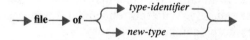

The components of a file may be any data type except another **file** or a structured data type that includes a **file** data type. The following statements declare two **files**. The first file, **realfile**, is declared to be of type **FileType** which is a user-defined type-identifier that defines a **file** of real numbers. The second file, **integerfile**, a **file** of integer numbers, is declared using the **var** statement in which the data type is not given an identifier. As with **textfile** identifiers, **file**-type identifiers will be written in all lowercase letters to make the input/output statements easier to read.

```
type   FileType = file of real;

var    realfile : FileType;
       integerfile : file of integer;
```

For the astronomy problems in the case study, you could declare a **record** called **StarRec** to organize the information on each star:

14.1 File structure

```
const MaxLetter = 20;

type  CharString = packed array [1..MaxLetter] of char;

      StarRec = record
        CatalogueNumber : integer;
        Name : CharString;
        Constellation : CharString;
        SpectralClass : CharString;

        RightAscension : real;      (* hours. *)
        Declination : real;         (* degrees. *)
        ProperMotion : real;        (* seconds of arc. *)
        RadialVelocity : real;      (* km/sec. *)
        Distance : real;            (* from earth in parsecs. *)
        Temperature : real;         (* temperature in degrees Kelvin. *)
        RelativeLuminosity : real;  (* luminosity relative to the sun. *)

        BinaryStar : boolean;
        VariableStar : boolean;
        SpecialClass : ( None, Pulsar, Quasar, Nebula, XRayStar )
      end;   (* StarRec *)
```

After the data type **StarRec** has been declared, a Pascal **file starfile** of the data type **StarRec** can be defined:

```
var    starfile : file of StarRec;
       Star : StarRec;
```

You can visualize the file created by this type declaration as a sequence of boxes, each potentially containing the information about one star. When you declare the file, you create the skeletal structure

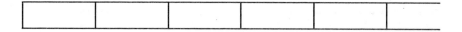

This structure is similar to the structure created when you declared an **array.** There are two major differences between a **file** and an **array**: a **file** does not have a fixed length, and the components of the **file** cannot be accessed randomly. For an **array** the number of components must be given in the declaration, but for a **file** the number of components is indeterminate. A **file** is a sequential data structure; the components that are stored in the file can be accessed only one at a time and only in the order in which they were stored. The sequential nature of **files** has several ramifications that will be covered later in the discussion.

14.2 Internal files

Like arrays, files can be used to organize, store, and transfer data within a program. A Pascal **file** that is used only within a program is referred to as an *internal file*. An internal file is never stored in long-term memory; it exists only during the execution of the program. A Pascal **file** that is associated with a computer file stored in long-term memory is referred to as an *external file*. Because this latter use of files is system dependent, internal files will be covered first.

Writing files

Because the components of the **file** are undefined when the **file** is declared, it is first necessary to write values to the **file**. As with a **textfile**, you must use the standard procedure **rewrite** to go to the beginning of a **file** before you can write to it. The **rewrite** command has the same effect as erasing and rewinding a tape. After the statement

```
rewrite (starfile)
```

is executed, the end-of-file marker, ⟨eof⟩, and the write pointer, ↑, are positioned at the beginning of the **file starfile:**

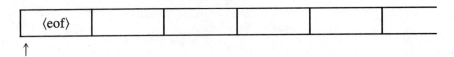

The syntax for writing to a **file** is the same as that for writing to a **textfile**. In the **write** statement, you must include the name of the **file**, followed by a comma and the name or names of the variables to be written to the **file**. The following statement, when executed, assigns the values stored in the **record** variable **Star,** which has the data type **StarRec,** to the **file starfile:**

```
write (starfile, Star)
```

To see the usefulness of **files,** suppose that for the case study you have written an interactive procedure **EnterStarData** that reads and checks the data for one star. The data could be stored in an array of **StarRec,** but the number of stars that a user will want to enter is not known and may vary each time the procedure is executed. Rather than arbitrarily limit the number of stars, you can use a **file** to store the data. Each time the procedure **EnterStarData** is invoked, the data that is entered can be written to a **file** for storage.

14.2 Internal files

Assume that the heading for the procedure **EnterStarData** is

```
procedure EnterStarData (var Star : StarRec; var Stop : boolean);
```

where the **record** variable, **Star,** stores the data that was entered about the star and **Stop** returns with a value of true if the user indicates that the last star has been entered. If **NextStar** is declared to be of data type **StarRec,**

```
var    NextStar : StarRec;
```

the following fragment will read the data from the terminal and store it in the **file starfile** until the user enters a request to stop:

```
rewrite (starfile);              (* move the write pointer and eof *)
                                 (* to the beginning of the file. *)
repeat
  EnterStarData (NextStar, RequestToStop);

  write (starfile, NextStar)     (* write NextStar to starfile. *)
until RequestToStop
```

The values stored in **NextStar** are written at the end of the **file,** which is indicated with the end-of-file marker. After the statement

```
        write (starfile, NextStar)
```

has executed one time, **starfile** can be visualized as

NextStar	⟨eof⟩				
↑					

where **NextStar** represents all the information stored in the **record** variable. Each time the **write** statement is executed, the data on the next star is added at the end of the **file,** and the end-of-file marker and write pointer are advanced. After the information on two stars has been written to the **file, starfile** can be visualized as

NextStar$_1$	NextStar$_2$	⟨eof⟩			
	↑				

Subscripts have been added to the variable names to indicate that the values stored in the **record** variable **NextStar** are different each time the **write** statement is executed.

It must be emphasized that only variables whose data type is the same as that of the **file** can be written to the **file.** Given the variable declarations

```
type   StarArray = array [1..MaxStars] of StarRec;

var    starfile : file of StarRec;      (* a file of star records. *)
       Stars : StarArray;               (* an array of star records. *)
       Sirius : StarRec;                (* a single star record. *)
       Velocity : real;                 (* a single real variable. *)
```

where the data type **StarRec** has the previous definition, the following **write** statements are correct:

```
write( starfile, Stars[1]);      (* write a single element from the *)
                                 (* array to starfile. *)
write (starfile, Sirius)         (* write a single record to starfile. *)
```

Each **write** statement puts a variable that has the data type **StarRec** into the **file.** The following **write** statements will not compile, given the previous declarations:

```
              write (starfile, Velocity);
***error***                         $116

              write (starfile, Stars);
***error***                         $116

error messages :
***************

116: error in type of standard procedure parameter.
```

The first **write** statement does not compile because a variable of type **real** is written to a **file** with the basic data type of **StarRec**; the second statement does not compile because the variable **Star** is an **array** of **record**s rather than a single **record.** For the **file** and the variable to be compatible, a declaration such as

```
type   StarArray = array [1..MaxStars] of StarRec;

       FileArray = file of StarArray;
```

```
var   newfile : FileArray;
      Stars : StarArray;
            .
            .
            .
   write (newfile, Stars)        (* write the entire array of stars to the *)
                                 (* file newfile. *)
```

would have to be used.

The **files** that you write do not have to be **files** of **record**s. They can be **files** of any valid data type (**real, integer, boolean, char,** or a user-defined data type). But you cannot write data of different data types, such as **char** and **integer,** to the same **file** unless they are part of a more complex data type such as a record.

REMINDER

Every variable that is written to a **file** must have a data type that is identical to the data type of the components of the **file.**

The statements to write information to a Pascal **file** are more compact than the statements to write information to a **textfile.** With a **textfile,** each component of the record must be written and formatted according to its data type. With a **file,** the entire record can be written as a whole.

Once the data on the stars has been assigned to the **file** variable, the entire set of data can be referred to by the **file** variable name. The **file** variable, like any other variable, can then be passed to other subprograms, with one restriction: the value of a **file** variable cannot be assigned to any other variable. Technically, the data type **file** is never *assignment compatible*. The following assignment statement will always result in an error message even though **infile** and **outfile** have the same data type:

```
         var    infile, outfile : text;

         begin
           outfile := infile;
         ***error***        $146

         error messages :
         ***************

           146: assignment of files not allowed.
```

Because **files** are not assignment compatible, a **file** *cannot* be passed as a value parameter to a subprogram (see Chapter 6). A **file** *must* be passed as a variable parameter.

Reading files

After data has been assigned to the **file** variable using the **write** statement, another part of the program can inspect the data using the **read** statement. A **file** must be either in the **write** state or in the **read** state. The file is shifted to the **read** state by using the **reset** statement. Execution of the **reset** statement causes the read pointer to move to the beginning of the file. This action is similar to rewinding a tape. The following statement moves the read pointer to the beginning of **starfile**:

```
reset (starfile)
```

Assuming that information on five stars has previously been written to **starfile**, after execution of the **reset** statement the file can be visualized as

NextStar$_1$	NextStar$_2$	NextStar$_3$	NextStar$_4$	NextStar$_5$	⟨eof⟩

↑

The **reset** statement moves only the read pointer. The end-of-file marker stays at the end of the **file**. If you mistakenly use **rewrite,** thus moving the end-of-file to the beginning of the **file,** all the information in the **file** will be erased.

After the statement

```
read (starfile, NextStar)
```

is executed, the read pointer advances so that it is positioned immediately after the **file** component that has just been read. After the information on the first star has been read, **starfile** can be visualized as

NextStar$_1$	NextStar$_2$	NextStar$_3$	NextStar$_4$	NextStar$_5$	⟨eof⟩

       ↑

The only way to move the read pointer backward is to move it all the way back to the beginning using the **reset** statement.

Suppose you are interested only in the information on one of the stars. Unlike the data in an array, the data in a **file** is sequential and cannot be accessed randomly. If you want the fifth component in the file, you must read the first four components to advance the **read** pointer to the fifth component:

```
reset (starfile);

for StarNum := 1 to 5 do
   read (starfile, FifthStar)
```

After the fragment above is executed, the fields of the **record** variable **FifthStar,** which has a data type of **StarRec,** will be equal to the fields of the fifth star that was written to the file.

REMINDER

A variable with the data type **file** must be in either the inspection (**read**) state or the modification (**write**) state. After a value has been written to a **file,** the only way to inspect the value is to **reset** the **file** and read until the value is encountered.

14.3 External files

For many applications, an internal Pascal **file** may be associated with a computer file in the long-term memory. For example, entering the data on all the stars for the case study each time the program is run is not feasible; for the program to be practical, the data must be stored in a computer file. The Pascal **file** can be associated with a computer file containing the information on the stars. The syntax for external **files** used within a Pascal program is the same as that for internal **files**. Like internal **files**, external **files** can be passed as parameters, rewritten, and reset. As with **textfiles**, it is easy to confuse a Pascal **file** with a computer file stored in long-term memory. Where confusion could arise, the terms Pascal **file** and computer file will be used to distinguish between them.

Because a **file** can be composed of any data type, there is a major difference between **textfiles** of characters and **files**. A **file** *cannot* be created using the computer's editor nor can a computer file that has been created from a Pascal **file** be printed or listed using a computer system's standard programs. Because a **file** has a special internal structure and is not stored in character format, it can be written only by using a Pascal program and can be read only by using a Pascal program. (To get an idea of what the internal representation looks like, you can try listing a computer file that has been created with a Pascal program on your computer system.) These restrictions may appear to limit the usefulness of **files**, but, as you have already seen, **files** can streamline programs in which large amounts of data must be read and/or written.

Assigning disk files to Pascal files

As with **textfiles**, it is necessary to use a system command that is specific to your computer's operating system in order to associate a Pascal **file** with a computer file stored in the computer's memory. The syntax for the command should be the same as that for **textfiles**. The system command assigning the Pascal **file** to the computer file must be issued at the beginning of the program. For most compilers, all external **files** must be declared in the program heading along with **input** and **output,** which are also external **files**.

Within a program, writing to an external **file** is the same as writing to an internal **file.** The difference is that after the program has executed, all the information that was written to the external Pascal **file** is stored in a computer file, but the data that was written to the internal **file** no longer exists in computer memory.

The steps for writing a Pascal **file** that is to be saved in long-term memory are as follows:

1. Use the system command for your computer that associates the Pascal **file** with the computer file.

2. Declare the Pascal **file** in the program heading.

3. Declare the data type that the **file** will use.

4. Declare the **file** in the **var** statement.

5. **Rewrite** the **file** before the first variable is written to the **file.**

6. **Write** the variables to the **file.**

7. If necessary on your system, close the **file** at the end of a program to save the computer file in long-term memory.

The following fragment includes the heading, the declarative part of the program, and the statements necessary to write the first record of the data on the stars to the external **file.**

```
program WriteAstro (input, output, starfile);
(*      This program writes the data stored in the variable NextStar    *)
(*      to the external file starfile.                                  *)
type   StarRec = record
                   .
                   .
                   .
         end; (* StarRec *)

  var  starfile : file of StarRec;
       NextStar : StarRec;

(*$i getstardata *)
```

```
begin
  rewrite (starfile);
  GetStarData (NextStar);
  write (starfile, NextStar);
    .
    .
    .
```

Reading an external **file** is similar to reading an internal **file**. The major difference is that the Pascal **file** into which data is read is associated with a computer file that has been created previously by a Pascal program.

The Pascal **file** into which data is read must have a data type identical in structure to the data type declared for the Pascal **file** used to create the computer file. The basic component of the **starfile** is a record with an integer, two packed arrays of twenty characters, two reals, etc. In order to read from the computer file in which the data from the Pascal **file starfile** is stored, a data type with the same structure must be declared:

```
const MaxLetter = 20;

type  CharString = packed array [1..MaxLetter] of char;

      AstroRec = record
        CatalogueNumber : integer;
        StarName : CharString;
        Constellation : CharString;
        SpectralClass : CharString;

        RightAscension : real;       (* hours. *)
        Declination : real;          (* degrees. *)
        ProperMotion : real;         (* seconds of arc. *)
        RadialVelocity : real;       (* km/sec. *)
        ParSecs : real;              (* from earth in parsecs. *)
        Temperature : real;          (* temperature in degrees Kelvin. *)
        RelativeLuminosity : real;   (* luminosity relative to the sun. *)

        BinaryStar : boolean;
        VariableStar : boolean;
        SpecialClass : ( None, Pulsar, Quasar, Nebula, XRayStar )
      end; (* AstroRec *)
var   astrofile : file of AstroRec;
      Star : AstroRec;
```

Even though some of the fields have different names, the structure of the **record AstroRec** is the same as that of the **record StarRec**. The structure is like a template that the program places over the data in the **file**. If the data format matches the structure of the template, the data can be read and placed in the fields of the **record** variable in the **read** statement. If the data does not match the template, an error message about mismatched data types will occur.

The steps for reading data from a computer file created with a Pascal **file** are as follows:

1. Use the system command for your computer that associates the Pascal **file** with the computer file.

2. Declare the **file** in the program heading.

3. Declare a data type that is compatible with the data type of the **file** that is being read.

4. Declare the **file** in the **var** statements.

5. Before the first **read** statement, **reset** the **file** to move the **read** pointer to the beginning of the **file.**

6. Read the components of the **file** sequentially.

The following fragment includes the heading, the declarative part of the program, and the statements necessary to read the first **record** of the data on the stars from the computer file that will be created on execution of the Pascal fragment that was given for writing the data to the computer file.

```
program ReadStar (input, output, astrofile);
(*       This program reads stars from the external file that was        *)
(*       created using program WriteAstro.                                *)
type   AstroRec = record
                .
                .
                .
       end; (* AstroRec *)

var    astrofile : file of AstroRec;
       Aster : AstroRec;

begin
  reset (astrofile);
  read (astrofile, Aster)
```

Sequential versus random-access data structures

As stated previously, a **file** is a sequential data structure, whereas an **array** is a random-access data structure. In many programs the fact that a **file** is a sequential data type is not limiting. Programs that use **arrays** frequently use **for** statements that process the **arrays** sequentially. You can look back through the algorithms you have written to see which algorithms use the random-access property of **arrays** and which algorithms use **arrays** as a

14.4 Standard file procedures and functions

convenient method for storing data without using the random-access property. If random access is not required, a **file** can be used instead of an **array**.

Textfile*s versus* file*s*

In some situations **textfiles** are superior to **files**. In many output files that contain data of different types, the data cannot be combined into a single data structure. Because variables of any data type can be written to a **textfile**, **textfiles** must be used for these files. For a report containing both character strings and numerical data, the output file must be a **textfile.** Input and output files that are to be listed or printed must be **textfiles**.

VOCABULARY WARNING

In standard computer vocabulary, "a logical record" is the basic unit of a computer file. Out of context, "a file of records" can mean either a Pascal **file** whose basic data type is a **record** or a computer file made up of logical records. A logical record is the record as used and interpreted by a person, as opposed to the physical record that the computer stores. You may have heard the term lrecl (pronounced el-rek'-el), which stands for logical record length. For a computer file created by a text editor, the lrecl is usually a line of text. This term has nothing to do with Pascal **files** of **record**s. Do not become confused or think that you have missed something.

14.4 Standard file procedures and functions

There are several standard procedures and functions that can be used in Pascal with both internal and external **files**. You are already familiar with many of them. A complete list is provided at the end of this section of the standard functions and procedures for use with **files**.

The buffer variable

Because the implementation of Pascal **files**, and input/output in general, is strongly system dependent, Pascal uses a *buffer variable* to store data being put into or retrieved from **files**. The buffer variable is automatically declared when the **file** is declared, and its data type is the data type of a single component of the **file.** When the **file starfile** is declared:

```
var    starfile : file of StarRec;
```

the buffer variable **starfile**↑ is created with the same data type as the components of the **file**. In this case, the buffer variable has the data type **StarRec,** so all the information on a single star can be stored in the buffer variable **starfile**↑.

Depending on the compiler, the symbol for the buffer variable is one of the following symbols: the up arrow (↑), a circumflex or a caret (^), or an at sign (@). You must consult the documentation for your compiler to determine which symbol you should use for the buffer variable.

When you are using a **file** only one component of the **file** is accessible to you at a time. You can think of the buffer variable as a window into the **file** that allows you to see and manipulate the current component. If the **file** is in the **write** state, that is, once the **file** has been rewritten

```
rewrite (starfile)
```

the buffer variable can be used to modify the current component. When a variable is written to the **file,** the information stored in the variable is assigned to the buffer variable, and then the information in the buffer variable is put into the **file** and the end-of-file marker is advanced.

If the **file** is in the **read** state, the buffer variable allows you to inspect the current component of the **file**. When a **file** is reset,

```
reset (starfile)
```

the buffer variable **starfile**↑ is set to the first component in the **file**. When information is read from the **file,** the information stored in the buffer variable is assigned to the variable named in the read-argument list and the buffer variable is advanced to the next component in the **file.**

The put *and* get *procedures*

Two standard procedures, **put** and **get,*** can be used for input and output at a lower level than **read** and **write.** In most cases the use of **read** or **write** is preferable, but an understanding of **put** and **get** clarifies the use of the buffer variable and the implementation of **files.**

The procedure **put** writes the information stored in the buffer variable to the **file.** So, the statements

```
starfile↑ := Star;      (* assign the value of the variable Star to the *)
                        (* buffer variable starfile. *)
put (starfile)          (* write the value of the buffer variable after *)
                        (* the last entry in file starfile. *)
```

---

*The Turbo Pascal compiler does not support **put** and **get.**

14.4 Standard file procedures and functions

are equivalent to

```
write (starfile, Star)
```

For the **put** procedure to execute correctly, the file must be in the **write** state, that is, a **rewrite** must have been issued for the **file.**

The procedure **get** retrieves information from the **file** and stores it in the buffer variable. When the file is first **reset,** the buffer variable is assigned the value of the first component in the file. So, the statements

```
get (starfile);
Star := starfile↑
```

assign the values stored in the first component of the **file** to the variable **Star,** then the **get** statement advances the buffer variable to the next component in the **file.** Executing the **get** statement again,

```
get (starfile);
Star := starfile↑
```

will result in **Star** being assigned the value of the second record in the **file.**

The following fragment illustrates the use of **put** and **get** to transfer the data on one of the stars from the original **file, astrofile,** to the new **file, starfile,** without using **read** or **write.**

```
reset (starfile);                (* reset assigns the first element in the *)
                                 (* file to the buffer variable. *)
rewrite (astrofile);
while not eof (starfile) do
  begin
    astrofile↑ := starfile↑;
    put (astrofile);
    get (starfile)
  end
```

The eof *function*

As the buffer variable is moved through the file, the boolean function **eof** remains false as long as the buffer variable is an element of the **file.** As soon as the buffer variable is moved beyond the last element in the **file,** the buffer variable contains the eof character and the function **eof** returns a value of true.

Upon execution of the following fragment each component of the **file** will be retrieved, the values will be assigned to the variable **Star,** and the procedure **ProcessNextStar** will be invoked until the end-of-file marker is encountered:

```
reset (starfile);                    (* reset assigns the first element in *)
                                     (* the file to the buffer variable. *)
while not eof (starfile) do
  begin
    Star := starfile↑;
    ProcessNextStar (Star);
    get (starfile)
  end
```

When a **file** is empty, the buffer variable contains the eof character, so the function **eof** is true. Therefore, whenever you read from **files**, you should invoke **eof** before the buffer variable is referenced or before a **read** is executed to be sure that the file contains data.

The standard function **eoln** which was used with **textfiles** is not defined for **files**. The function **eoln** returns a value of true if the read pointer is at the end of a line, but a **file** does not have lines the way a **textfile** does. If you write a loop that tests the **eoln** of a **file** as the terminating condition, you may have an infinite loop. **Readln** and **writeln** are also only defined for **textfiles.**

The following list summarizes the standard procedures and functions used in input/output in Pascal. The ellipses (. . .) in the last four procedures indicate that a parameter list can be used.

eof(file)	returns a value of true if the buffer variable **file**↑ contains the eof character.
eoln(file)	returns a value of true if **file**↑ currently points to the end of the line. **eoln** can be used only for **textfiles**.
get(file)	advances the buffer variable **file**↑ to the next component in the **file**. If the **file** does not have a next component, **file**↑ contains the eof character and **eof** is true.
put(file)	appends the contents of the buffer variable **file**↑ to the end of **file. eof** remains true. **file**↑ contains the eof character after it has been **put**.
reset(file)	initializes **file** for input by setting the buffer variable **file**↑ to the first component of **file** and by setting **eof** to false if the **file** is not empty. All **files** except **input** must be **reset** before a **read** or **get** is issued.
rewrite(file)	initializes a **file** for output by setting **eof** to true. All **files** except **output** must be rewritten before a **write** or **put** is issued.
read(file, . . .)	reads data from **file** until the argument list data requirements are satisfied.
readln(file, . . .)	reads data from **file,** until the argument list requirements are satisfied, then the buffer variable advanc-

	es to the first character of the next line. **readln** can be used only with **textfile**s.
write(file, . . .)	writes the data in the argument list to **file.**
writeln(file, . . .)	writes the data in the argument list to **file** and also writes an end-of-line indicator after the last argument in the list. **writeln** can be used only with **textfile**s.

Solution to Case Study 24

The solution to the case study can be written almost directly from the original outline, assuming that the interactive procedure **EnterStarData,** which elicits the data from the user, has been written. Program 14.1, **StarCatalogue,** reads the data on the stars already in the data base from the computer file, then reads the data on the stars to be added from the terminal, and finally writes all the data to a new computer file. While it is not necessary to have two computer files (the old version and the new version) this strategy is safer. If the execution of the program is interrupted for any reason, a Pascal **file** that is in write-access mode will be lost on most computer systems. Therefore, the original data base should be maintained as a separate computer file until the program **StarCatalogue** has completed execution and the results have been verified.

PROGRAM 14.1 StarCatalogue

```
program StarCatalogue (input, output, starfile, astrofile);

begin (* main *)
  reset (starfile);
  rewrite (astrofile);

  (* The while-do statement copies the data from starfile to astrofile by *)
  (* getting data from starfile while eof has not been reached. *)

  while not eof (starfile) do
    begin
      astrofile := starfile↑;
      put (astrofile);          (* put the star data in astrofile. *)
      get (starfile)            (* get the next component of starfile. *)
    end; (* while *)

  (* The repeat-until statement reads data on the next star until the user *)
  (* requests to stop.  Each time data is entered it is added to the end *)
  (* of astrofile using the write statement. *)
```

```
   repeat
     EnterStarData (NextStar, RequestToStop);
      write (astrofile, NextStar)
   until RequestToStop
end. (* program StarCatalogue *)
```

Many operating systems allow you to designate a computer file as a *read-only* file, that is, a computer file that cannot be modified without special permission. If your system supports read-only computer files, you should use them whenever you have a data base or a program which should not be modified.

APPLICATION
Random Number Generators

CASE STUDY 25
Satellite transmission

The transmission link to a communications satellite is subject to failures. If the satellite can transmit at time t, the probability that it will still be able to transmit at time $t + 1$ is 0.90. If the satellite cannot transmit at time t, the probability that it will be unable to transmit at time $t + 1$ is 0.60. When the satellite can transmit, its state is called *up*. When it cannot transmit, its state is called *down*.

Write a program that simulates the operation of the satellite over T time periods. In other words, write a program that prints the state of the satellite for T time periods. As part of your program, write a function that uses a random number generator to return the state (up or down) of the satellite at time $t + 1$, given its state at time t. The output from this program is to be stored in a computer file so that it can be used as the input to another program.

Algorithm development

The basic problem for the case study is to write a program that updates and prints the state of the satellite at the end of each time period. In Algorithm 14.4 it is assumed that a function called **NextState** can be written to return the new state of the satellite at time $t + 1$ given the state at time t.

ALGORITHM 14.4 First iteration of the algorithm for Case Study 25

begin algorithm
    1. Write a message to the user and initialize the program
        a. read **CurrentState, UpGivenUp, DownGivenDown**
        b. create the file for the output
    2. for **Time** := 1 to **T** do
        a. **OldState** ← **CurrentState**
        b. **CurrentState** ← **NextState(OldState, UpGivenUp, DownGivenDown)**
        c. write **CurrentState** to the file
end algorithm

The program itself is not complicated. The major problem is to write the function **NextState**. The algorithm for **NextState** will be developed in the following sections on random number generators.

Generating random numbers using a computer

Random variables were introduced in the application section of Chapter 10 on probability. The material on probability is not necessary for this chapter, although it will be used in some of the problems at the end of the chapter.

Briefly, a random variable can be used to represent the unknown outcome of an *experiment* such as tossing a coin, rolling a die, or throwing a dart. Each performance of the experiment is called a *trial,* and the outcome of a trial can be given a numerical value. For example, in the coin toss experiment, a head can be assigned the value of 1 and a tail the value of 0. When an experiment is performed, the random variable is *realized;* that is, it takes on a numerical value. The resulting number is referred to as a *random number*. For example, if a coin is tossed four times and the result is the sequence head-head-tail-head, the sequence of random numbers associated with the experiment is 1-1-0-1.

If the outcome of one trial does not affect the outcome of another trial, the trials are said to be *independent.* Coin tosses are independent because the outcome of one trial has no effect on the outcome of the next trial—knowing that a head was just tossed does not help you guess what the next toss will be. Unless told otherwise, you can assume that a random number sequence is a sequence of digits in which each digit is independent of every other digit and each digit is as likely to occur as any other digit.

Many algorithms require random numbers. Since it is impractical to toss a coin each time a random number is needed within a program, computer programs use *pseudorandom numbers.* Pseudorandom numbers are numbers indistinguishable from true random numbers, but pseudorandom numbers are generated by an algorithm rather than by the outcomes of experimental trials. Sometimes pseudorandom numbers are stored in tables rather than being computed within a program. Tables such as those in Reference 3 at the end of this chapter, which contain a million random digits, can be used to set up a computer file which can be associated with a Pascal **file** to retrieve random numbers as they are needed.

There are many algorithms for generating pseudorandom numbers. One of the simplest and best algorithms for generating random numbers is the *linear congruent algorithm,*

$$r_k = (xr_{k-1} + y) \bmod \max \tag{14.1}$$

where r_k is the random number generated and x, y, and max are constants. For a given number r_{k-1}, Equation 14.1 will generate an integer random number r_k between 0 and max $- 1$. This random number then serves as the *seed* for the next random number. So each random number serves two purposes: it can be used as a random number; and it can be used as the seed to create a new random number.

The values of x and y are arbitrary, but their values do affect the *goodness* of the random numbers that are generated. The goodness of a pseudorandom number generator is a measure of how hard it is to tell the difference between a pseudorandom sequence and a truly random sequence.

To see how Equation 14.1 works, suppose you need a sequence of random numbers between 0 and 9. You could set r_1 equal to 3, x equal to 4, and y equal to 2; max will be equal to 10. With these initial values, the sequence of random numbers in Table 14.1 is generated.

The values for r_k, the random numbers, are not very random in this sequence. As soon as r_k equals 4, the sequence alternates between 4 and 8. This random number generator has a *period* of 2, meaning that the pattern repeats every two numbers. The values that were chosen for x and y were particularly bad ones. Suppose that instead of $x = 4$ and $y = 2$, you had chosen $x = 11$ and $y = 7$. With $r_1 = 3$ as before, the sequence in Table 14.2 is generated.

With $x = 11$ and $y = 7$, the sequence has a period of 10. Every number between 0 and 9 is generated once, and then the cycle repeats. There are rules

TABLE 14.1
Pseudorandom Numbers Generated with $x = 4$, $y = 2$, max $= 10$

k	r_k	r_{k+1}
1	3	$r_2 = (4 * 3 + 2) \bmod 10 = 4$
2	4	$r_3 = (4 * 4 + 2) \bmod 10 = 8$
3	8	$r_4 = (4 * 8 + 2) \bmod 10 = 4$
4	4	$r_5 = (4 * 4 + 2) \bmod 10 = 8$
5	8	$r_6 = (4 * 8 + 2) \bmod 10 = 4$
6	4	$r_7 = (4 * 4 + 2) \bmod 10 = 8$

TABLE 14.2
Pseudorandom Numbers Generated with $x = 11$, $y = 7$, max $= 10$

k	r_k	$r_{k\times 1}$
1	3	$r_2 = (11 * 3 + 7) \bmod 10 = 4$
2	4	$r_3 = (11 * 4 + 7) \bmod 10 = 1$
3	1	$r_4 = (11 * 1 + 7) \bmod 10 = 8$
4	8	$r_5 = (11 * 8 + 7) \bmod 10 = 5$
5	5	$r_6 = (11 * 5 + 7) \bmod 10 = 2$
6	2	$r_7 = (11 * 2 + 7) \bmod 10 = 9$
7	9	$r_8 = (11 * 9 + 7) \bmod 10 = 6$
8	6	$r_9 = (11 * 6 + 7) \bmod 10 = 3$
9	5	$r_{10} = (11 * 3 + 7) \bmod 10 = 0$
10	0	$r_{11} = (11 * 0 + 7) \bmod 10 = 7$
11	7	$r_{12} = (11 * 7 + 7) \bmod 10 = 4$

that can be used to select x and y so that the sequence has the maximum period (see Reference 2).

Because some programs require hundreds of random numbers, a random sequence that repeats every 10 numbers is not satisfactory. Choosing a value for max that is much greater than the maximum number required circumvents this problem. For example, if the random numbers must be between 0 and 9, max can be chosen to be 1000. The sequence will generate numbers between 0 and 999. If the number generated is z, the random digit can be selected by a formula such as z **div** 100, or z **mod** 10, or (z **div** 10) **mod** 10, or any other formula that yields a number between 0 and 9.

In Table 14.3, r_k', the random digit between 0 and 9, is found by taking r_k **div** 100.

Because a pseudorandom sequence is generated by an algorithm, the sequence will always be the same if the initial conditions are the same. That is, if the program starts with the same data, it will always give the same result. Under normal conditions, the consistency of the computer is an advantage; however, when a random number is needed, this consistency is an obstacle. A sequence that is always the same is not usually satisfactory for a random number generator. You need a way to randomize the initial seed. One way is to ask the user to enter a number for the seed at the start of the program. Another way is to use a number, like the time, that may be supplied by your

TABLE 14.3
Pseudorandom Numbers Generated with $x = 11$, $y = 7$, max = 1000

k	r_k	$r_{k'}$	r_{k+1}
1	3	0	$r_2 = (11 * 3 + 7) \bmod 1000 = 40$
2	40	0	$r_3 = (11 * 40 + 7) \bmod 1000 = 449$
3	449	4	$r_4 = (11 * 449 + 7) \bmod 1000 = 946$
4	946	9	$r_5 = (11 * 946 + 7) \bmod 1000 = 413$
5	413	4	$r_6 = (11 * 413 + 7) \bmod 1000 = 550$
6	550	5	$r_7 = (11 * 550 + 7) \bmod 1000 = 57$
7	57	0	$r_8 = (11 * 57 + 7) \bmod 1000 = 634$
8	634	6	$r_9 = (11 * 634 + 7) \bmod 1000 = 981$
9	981	9	$r_{10} = (11 * 981 + 7) \bmod 1000 = 798$
10	798	7	$r_{11} = (11 * 798 + 7) \bmod 1000 = 785$
11	785	7	$r_{12} = (11 * 785 + 7) \bmod 1000 = 642$

computer system and that is different every time the program is executed. The function **FirstSeed** uses the system-dependent procedure **Time** to compute the seed for a random number generator. You can check the documentation for your computer to see if your system has a similar procedure.

```
function FirstSeed : integer;
(*      FirstSeed returns a value that is generated from a value    *)
(*      supplied by the procedure Time.  The first number in a random *)
(*      sequence is generated using the time that the function is   *)
(*      entered.  Time is a system supplied procedure that returns the *)
(*      current time in a packed array in hh/mm/ss format.  You will *)
(*      have to check the documentation for your system to see whether *)
(*      you have access to a procedure similar to Time.             *)


type    CharString = packed array [1..8] of char;

var     NewTime : CharString;       (* current time - returned by Time. *)
        TimeDigit : integer;        (* character converted to a digit. *)
        Index : integer;            (* for statement control variable. *)
        Sum : integer;              (* running total for the seed. *)
        Offset : integer;           (* location of '0' in the character set. *)
```

```
begin
  Time (NewTime);              (* invoke the system-dependent procedure *)
                               (* Time. *)
  Sum := 0;
  OffSet := ord ('0');

  (* Loop through the characters in NewTime, discarding the slashes stored *)
  (* in indices 3 and 6.  Convert each character to its equivalent *)
  (* numerical value and add it to the running total. *)

  for Index := 1 to 8 do
    if (Index <> 3) and (Index <> 6) then
      begin
        TimeDigit := ord (NewTime [Index]) - OffSet;
        Sum := Sum + TimeDigit * Index
      end; (* if *)
  FirstSeed := Sum
end; (* function FirstSeed *)
```

On this system, the time is returned as a character string, so **FirstSeed** uses the standard Pascal function **ord** to convert the characters in the time into numbers.

Since the function **FirstSeed** can produce random numbers, why not keep using **FirstSeed** to generate new random numbers if several are needed in a program? The seed produced by **FirstSeed** is based on the time at which your program is executed. Once your program starts executing, it executes in fractions of a second. Therefore, the time, as measured by the computer's clock, is constant while your program executes. If you keep calling **FirstSeed**, you will always get the same number, unless the seconds just happen to flip while your program is executing. Unless your program is very long, you will be unlikely to get more than two different numbers from **FirstSeed**.

The function **RanNum** uses Equation 14.1 to generate a random integer between 0 and 9 given a seed:

```
function RanNum1 (var NewSeed : integer) : integer;
(*      Function RanNum1 computes a random number using a linear    *)
(*      congruent algorithm.  The random number is between 0 and 9  *)
(*      given NewSeed between 0 and 999.  RanNum1 also creates the  *)
(*      NewSeed for the next call to RanNum1.                       *)

const X = 11;                  (* X, Y, and Max are constants for the *)
      Y = 7;                   (* linear congruent algorithm, *)
      Max = 1000;              (* Equation 14.1.*)
```

```
var    Argument : integer;          (* intermediate variable. *)
begin
  Argument := X * NewSeed + Y;
  NewSeed  := Argument mod Max;     (* compute the seed for the next call *)
                                    (* to the random number - Eq. 14.1. *)
  RanNum1 := NewSeed mod 10         (* the random number is between 0 and 9. *)
end; (* function RanNum1 *)
```

Notice that **NewSeed** is a **var** in the function parameter list. In general, **var** parameters should not be used for functions; however, in this case there are extenuating circumstances. If **NewSeed** were not a **var** parameter, it would be necessary to rewrite **RanNum** as a procedure instead of a function so that two values, the random number and the seed, could be returned. Since the generation of a random number is easier with a function and since the seed is used only by the function, it is permissible in this case to use a **var** parameter in a function.

Random numbers can be very useful for generating test data for programs. For example, in Problem 3 in Chapter 10, 100 pieces of data were required. Writing a program using a random number generator is much easier than making up 100 numbers and entering them into the computer. The program can generate 100 random numbers within the specified range and write them to a computer file so that they can be used as input for another program.

Solution to Case Study 25

Now that random number generators have been covered, Step 2b in Algorithm 14.4 can be completed. Algorithm 14.5 is an algorithm for the function **NextState**, which returns the state of the satellite at time $t + 1$, given its state at time t.

ALGORITHM 14.5   An algorithm for the function NextState

2b. **CurrentState** ← **NextState(OldState, UpGivenUp, DownGivenDown)**
      if **OldState** = up then
          NextState ← up with probability **UpGivenUp** and
          NextState ← down with probability 1 − **UpGivenUp**
      if **OldState** = down then
          NextState ← up with probability 1 − **DownGivenDown** and
          NextState ← down with probability **DownGivenDown**

The first problem in creating a function that returns the state of the satellite at time $t + 1$, given its state at time t, is to figure out how to implement the statements

NextState ← up with probability **UpGivenUp**
NextState ← down with probability $1 -$ **UpGivenUp**

In the problem statement, it was given that if the satellite is up at time t, the probability that it will be up at time $t + 1$ is 0.90. How can the random number generator be used to simulate this probability?

You can restate the probability 0.90 as 90 chances in 100. If a random number generator generates numbers between 1 and 100, there are 90 chances in 100 that a number will be between 1 and 90. To find out if the satellite is still up at $t + 1$, you can call the random number generator. If the number is less than or equal to 90, the satellite is still up. If the number is greater than 90, the satellite is down. Similarly, if the satellite is down, there is a probability of 0.60 that it will be down at $t + 1$. If the random number generator returns a number less than or equal to 60, the satellite is down at time $t + 1$. If it returns a number greater than 60, the satellite is up at time $t + 1$.

Within the program, it is useful to have the probability stated in terms of chances per hundred. In order to make the program easier to write, you might decide to request that the probabilities be entered as the number of chances out of 100. Another argument in favor of this method is that most people, except for mathematicians, are quite comfortable thinking of probabilities in this manner.

Implementing the strategy above requires a random number generator that returns numbers between 1 and 100. The function **RanNum** developed in the chapter returns numbers between 0 and 9. **RanNum** can be made more general by including the maximum number **Modulo** as one of the parameters. The random number can also be adjusted so that it is between 1 and **Modulo** instead of between 0 and **Modulo** $- 1$:

```
function RanNum (var NewSeed : integer; Modulo : integer) : integer;
   (*    Function RanNum computes a random number using the linear     *)
   (*    congruent algorithm.  The random number is between 1 and Modulo *)
   (*    given NewSeed between 0 and 999.  RanNum also creates the      *)
   (*    NewSeed for the next call to RanNum.                           *)

   const X = 11;                    (* X, Y, and Max are constants for the *)
         Y = 7;                     (* linear congruent algorithm, *)
         Max = 1000;                (* Equation 14.1.*)
```

```
    var   Argument : integer;          (* intermediate variable. *)
begin
  Argument := X * NewSeed + Y;
  NewSeed  := Argument mod Max;        (* compute the seed for the next *)
                                       (* call of RanNum - Eq. 14.1. *)
  RanNum   := (NewSeed mod Modulo) + 1 (* the random number is between *)
                                       (* 1 and Modulo. *)
end; (* function RanNum *)
```

If an enumerated type is declared for the state of the satellite,

```
type  State = (down, up);
```

a function that returns the state of the satellite at time $t + 1$, given its state at time t, can be written as

```
function NextState1 (CurrentState : State;
                     UpGivenUp, DownGivenDown : integer) : State;
(*     Function NextState returns the state of the satellite in the     *)
(*     next time period given its CurrentState.  UpGivenUp is the       *)
(*     number of times out of 100 trials that the satellite is up at    *)
(*     time t + 1 given that it was up at time t.  DownGivenDown is     *)
(*     the number of times out of 100 that the satellite is down at     *)
(*     time t + 1 given that it was down at time t.                     *)

const Modulo = 100;         (* largest random number. *)

var   RanDigit : integer;   (* random number returned by Rannum. *)
begin
  RanDigit := Rannum (Seed, Modulo);
  if CurrentState = down then
    if RanDigit <= DownGivenDown then
      NextState1 := down
    else
      NextState1 := up
  else
    if RanDigit <= UpGivenUp then
      NextState1 := up
    else
      NextState1 := down
end; (* function NextState1 *)
```

where **UpGivenUp** is the number of times in 100 trials that the satellite will be up at time $t + 1$, given that it was up at time t, and **DownGivenDown** is the number of times in 100 trials that the satellite will be down at time $t + 1$, given that it was down at time t.

The function **NextState** uses the variable **Seed,** which is not declared either in the parameter list or as a local variable. **Seed** cannot be a local variable because the function **FirstSeed** for the random number generator must be called only once in a program. The best way to keep the function modular is to add **Seed** to the parameter list of the function **NextState.**

Another modification can be made to **NextState** to make it more efficient. Since the satellite is usually **up,** the value of **NextState** can be set equal to **up** when the function is entered. The value of **NextState** is then changed only if the satellite state at $t + 1$ is **down.** The final version of the function **NextState** is as follows:

```
function NextState (CurrentState : State;
                    UpGivenUp, DownGivenDown : integer;
                    var Seed : integer) : State;
(*      Function NextState returns the state of the satellite in the    *)
(*      next time period given its CurrentState.  UpGivenUp is the      *)
(*      number of times out of 100 trials that the satellite is up at   *)
(*      time t + 1 given that it was up at time t.  DownGivenDown is    *)
(*      the number of times out of 100 that the satellite is down at    *)
(*      time t + 1 given that it was down at time t.  Seed is the       *)
(*      seed for the random number generator.                           *)

const Modulo = 100;            (* largest random number. *)

var    RanDigit : integer;     (* random number returned by Rannum. *)

begin
  RanDigit := Rannum (Seed, Modulo);
  NextState := up;
  if CurrentState = down then
    begin
      if RanDigit <= DownGivenDown then
        NextState := down
    end
  else
    if RanDigit > UpGivenUp then
      NextState := down
end; (* function NextState *)
```

Finally, you must write the main program that calls the function **NextState T** times and writes the program output to a **file**. For this program, a **file** of **StateType** can be created,

```
type   State = (down, up);

var    reportfile : file of State;
```

so that the state of the satellite can be written directly to the **file**.

PROGRAM 14.2 TestSatellite

```
program TestSatelliteState (input, output, reportfile);
(*      Program TestSatellite finds the state (up or down) of a       *)
(*      satellite for 100 time increments.  The states are written to *)
(*      a file of an enumerated type.                                 *)

type  State = (down, up);

var    reportfile : file of State;        (* output report file. *)
       TimeCount : integer;               (* for statement control variable. *)
       CurrentState : State;              (* current state of the satellite. *)
       OldState : State;                  (* previous state of the satellite. *)
       ChancesUp : integer;               (* number of times out of 100 the *)
                                          (* satellite is up at time t if it *)
                                          (* was up at t-1. *)
       ChancesDown : integer;             (* number of times out of 100 the *)
                                          (* satellite is down at time t if *)
                                          (* it was down at t-1. *)
       Seed : integer;                    (* seed for random number generator. *)
       TimeSteps : integer;               (* number of time steps for the *)
                                          (* simulation run. *)

(*$i firstseed *)
(*$i rannum *)
(*$i nextstate *)


(*$i readdata *)
(*   procedure ReadData (var UpGivenUp, DownGivenDown, Steps : integer;  *)
(*                      var FirstState : State);                         *)
(*      This procedure reads the data on the probability that the       *)
(*      satellite is up at time t given it was up at t-1 and the        *)
(*      probability that it is down at time t given it was down at      *)
(*      time t-1.  It also reads the number of time steps in the        *)
(*      simulation and the initial state of the satellite (up or down). *)
```

```
(*--------------------------------------------------------------------*)
begin (* main *)
  rewrite (reportfile);

  Seed := FirstSeed;
  ReadData (ChancesUp, ChancesDown, TimeSteps, OldState);

  for TimeCount := 1 to TimeSteps do
    begin
      CurrentState := NextState (OldState, ChancesUp, ChancesDown, Seed);
      write (reportfile, CurrentState);
      OldState := CurrentState
    end;
  writeln;
  writeln (' The output is in report.fil, a file of a structured data type');
end.  (* program TestSatelliteState *)
```

=>

This program simulates the states (up and down) of a
satellite given the probabilities that it moves from one
state to the other, and given its original state.

How many times out of 100 is the satellite up at time
t given that it was up at time t - 1. Enter a number
between 0 and 100
▫90

How many times out of 100 is the satellite down at time
t given that it was down at time t - 1. Enter a number
between 0 and 100
▫60

What is the initial state of the satellite?
Enter a 0 for down and a 1 for up.
▫1

How many time periods do you want to run the simulation?
▫100

The output is in report.fil, a file of a structured data type

Notice that the relationship between subsequent trials for the satellite is different from that for the coin. For the satellite, the state at time $t + 1$

is not independent of the state at time t. Knowing that the satellite is transmitting at time t, you know it is more likely to still be up at time $t + 1$. This is not true for the coin. The knowledge that a head was tossed at time t does not give you any new information on the likelihood of tossing a head at time $t + 1$.

The satellite is said to form a two-state Markov chain. The state diagram of the chain is illustrated in Figure 14.1. The arrows in Figure 14.1 represent transitions that the satellite can make from one state to another. The numbers next to each arrow represent the probability that the transition will take place. For an excellent discussion of Markov chains and their use in science and engineering, see Reference 1.

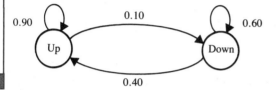

FIGURE 14.1
State diagram of the Markov chain for the satellite.

Summary

In Pascal, a **file** is a sequential data structure that can be used to streamline the transfer of data within and among programs. A **textfile** is a specific kind of file. The major difference between a **file** and a **textfile** is that the basic element of a **file** is not limited to characters. A **file** is a structured data type whose elements can be any valid data type except another **file.** Only variables that have the same data type as the elements of the **file** can be read from or written to the **file.**

A **file** may be an internal or an external **file.** An internal **file** is like any other variable in that it exists only during the execution of the program. An external **file** is treated just like any other variable while the program is executing; however, an external **file** is an unusual type of variable because it can be transferred to and saved in the long-term memory of the computer after a program has completed execution. The implementation of **files** is system dependent because when a program uses external **files,** the operating system must mesh the storage of information within a Pascal program with the physical storage of information in a computer file.

References

1. A. Drake, *Fundamentals of Applied Probability Theory,* New York: McGraw-Hill, 1967.

2. A. M. Law and W. D. Kelton, *Simulation Modeling and Analysis,* New York: McGraw-Hill, 1982.

3. *A Million Random Digits with 100,000 Normal Deviates,* New York: The Rand Corporation, The Free Press, 1955.

Exercises

PASCAL

1. Given the declarations

    ```
    type  RealArray = array [1..100] of real;

          OutRec = record
            Distances : RealArray;
            Numbers : array [1..30] of integer
          end; (* OutRec record *)

          RecArray = array [1..566] of OutRec;

    var   datafile : file of OutRec;
          allfile : file of RecArray;
          newfile : file of RealArray;
          DataPoint : OutRec;
    ```

    which of the following statements will result in a compile-time error?

    (a)   readln (datafile, DataPoint)

    (b)   write (allfile, DataPoint)

    (c)   write (allfile [3], DataPoint)

    (d)   write (newfile, datafile.RealArray)

2. Write a program that uses the procedure **get** to read 1 character, 3 integer numbers, and 3 real numbers that are stored in a computer file in the format below:

    ```
       x   25   984312
      10   75.02
     6.14   1.009
    ```

3. Write the even integers from -10 to 10 to a Pascal **file** of integers, and store the result in a computer file. Write another program that reads the computer file and prints the contents of the file at the terminal.

APPLICATION

4. Compute a table of values for r_k using Equation 14.1 with $x = 7$, $y = 3$, and max $= 10$.

5. Use the function **RanNum** to write a new function that returns a random number with the following distribution:

$$P(X = x) = \begin{cases} 0.11 & \text{for } x = 1, 3, 5, \text{ or } 7 \\ 0.10 & \text{for } x = 2, 4, 6, \text{ or } 8 \\ 0.16 & \text{for } x = 9 \\ 0.00 & \text{otherwise} \end{cases}$$

Call the new function 100 times and print the random numbers. Are exactly 16 out of the 100 numbers generated equal to 9? Should this be the case?

Problems

PASCAL

1. Write a program that reads the name and annual minimum and maximum temperatures for each of five cities from a Pascal **file** of **record**s. Find the city with the lowest minimum temperature and the city with the highest maximum temperature. Write your report on these cities to an external Pascal **textfile**. The data on the cities is written using the following Pascal statements:

```
const NumCities = 5;

type  CityRec = record
        Name : packed array [1..20] of char;
        MinTemp, MaxTemp : real
      end; (* CityRec record *)

      CityArray = array [1..NumCities] of CityRec;

var   cityfile : file of CityRec;
      Cities : CityArray;
      .
      .
      .
   for Num := 1 to NumCities do
      write (cityfile, Cities [Num])
```

2. A company stocks 100 different kinds of machine parts. An inventory of the parts is stored in a computer file that was created with a Pascal **file**. Write a program that reads from this computer file, writes another Pascal **file** that contains the parts that must be ordered again, and stores the new file in

computer memory. A part needs to be reordered if **ReOrder** is true in the original computer file. The following fragment creates the inventory file:

```
type  PartRec = record
         PartNumber : integer;
         PartName : packed array [1..20] of char;
         QuantityInStock : integer;
         Price : real;
         ReOrder : boolean
      end;

      PartArray = array [1..100] of PartRec;

var   Parts : PartArray;
      inventory : file of PartArray;
             .
             .
             .
      write (inventory, Parts);
```

3. A company has recently received a new mailing list that it wishes to merge with its master computer file. The computer files are both in alphabetical order by the first letter in the **Name** field. The number of names in each file is unknown. The company wants to produce a new master file that is in alphabetical order and has the same format as the original master file. The master file was created as a Pascal **file** with components of type **ClientRec:**

```
type  CharArray= packed array [1..10] of char;

      AddressRec = record
         StreetNumber : integer;
         StreetName : CharArray;
         City : CharArray;
         State : CharArray;
         ZipCode : CharArray
      end; (* Address record *)

      ClientRec = record
         Name : CharArray;
         Home : AddressRec
      end; (* ClientRec *)
```

If the same name appears in both of the original files, it should appear only once in the new master file. If the address for this person is different in the two original files, the address in the new file should be used because it contains more recent information.

(a) Write a Pascal program to create the new master file using the standard procedures **read** and **write.**

(b) Write the program using the standard procedures **get** and **put.**

4. The data for a nutritional experiment involving many mice is stored in a computer file created with a Pascal program. The data for each mouse is stored in a **record:**

```
type  MouseRec = record
         ID : integer;
         Strain : (C57B6, CD1, AJR, BalbC);
         GroupAssignment : integer;
         Weight : array [1..MaxDays] of real;
         DateOfBirth : integer;       (* days - measured relative *)
         DateOfDeath : integer        (* to start of experiment. *)
      end;
```

MaxDays is a constant representing the maximum number of days a mouse's weight is followed. The file is updated every day, and the newly recorded weight of each mouse is added to the **array** of weights for that mouse. The **array** index for **Weight** is the day number in the mouse's life, not the day of the experiment.

Write a Pascal program that will allow the technician to enter the weights each day. The program should request the current day number from the technician and check the date for consistency with the stored data. Then the program should display the **ID, Strain,** and **DateOfBirth** for a mouse and request the weight. The program must also allow the technician to indicate when a mouse has died. Once a mouse has died, the program should not request information on that mouse again. When the program is completed, the new Pascal **file** containing the current day's data should be stored in a computer file in the same format as the original file.

5. Modify the program from Problem 4 to keep track of both the file from Problem 4 and a separate computer file containing data on mice that have died. When a mouse dies, its data is written to this file, and the **record** is removed from the file of mice that are still alive.

APPLICATION

6. Generate 100 random numbers between 0 and 9, compute the average of the 100 random numbers, and print the numbers and their average at the terminal. What do you expect the average to be?

7. Using a random number generator, simulate the tossing of a coin. If the random number is even, a head was tossed. If the number is odd, a tail was tossed. Toss the coin 50 times and print out the sequence of tosses—for example, head–head–tail–head.

8. Rewrite Problem 7 using **sets.** Make two **sets,** one of the even digits and one of the odd digits. When you call the random number generator, test to see whether the result is in the **set** of odds or the **set** of evens. Report the number of heads and tails in a sequence of 100 coin tosses.

9. Generate a sequence of 100 random numbers between 0 and 9. Count how many random numbers there are between each pair of identical random numbers. Print a table of the number of times each count occurs. For example,

out of 10 pairs, there may be 1 pair with 3 calls between them, 3 pairs with 5 calls between them, and so on. Note that there is no maximum number of calls between pairs. You could have a sequence of 100 random numbers with no pairs. What you are computing and tabulating is the waiting time between pairs of numbers in a probability distribution.

10. In a *random walk,* a particle moves randomly up and down through space. At every time increment, the particle must move 1 step either upward or downward. Suppose that the particle moves up 1 step with a probability of p and down 1 step with a probability of $1 - p$. For each experimental trial, the particle starts at 0 and moves for 10 time increments.

    Write a program that reads the value of p from the terminal and computes the path of the particle for 20 different trials. For each trial, the particle starts again at 0.

    (a) For how many of the trials is the final position of the particle above 0? What is the average distance of the final position from 0?

    (b) Suppose there are reflecting barriers at $+5$ and -5. That is, if the particle reaches $+5$, it must move down at the next time increment, and if it reaches -5, it must move up at the next time increment. Compute the same values as in part (a).

    (c) Suppose there are absorbing barriers at $+5$ and -5. If the particle reaches either of these positions, it must stay there until the end of the trial. Compute the same values as in part (a).

11. Write a Pascal program that plays a game according to the following rules. The computer constructs a random set of digits (the integers 0 through 9). The player enters a set of digits using the terminal keyboard. The game ends when the player enters a set identical to the one chosen by the computer. If the set is not identical to the computer's set, the computer tells how many digits the two sets have in common.

    The number of elements in the set should be a random number between 1 and 8. (If the number of elements is 9, the game is too easy to be interesting.) The function **FirstSeed** can be used to generate a seed for the function **RanNum**. **RanNum** can then be used to generate the random numbers. The program should check that valid data is being typed in at the terminal and reject any entries that are not digits. Use sets to check the data.

12. Clock solitaire is played with a standard deck of 52 cards. The cards are dealt face down into 13 piles of 4 cards. Twelve piles are arranged in a clock-like pattern, and the thirteenth pile is placed in the center. A move consists of taking the top card from a pile and placing it face up on the bottom of the pile that corresponds to its face value. For example, a 1 goes under the 1 o'clock pile. A jack is 11, and a queen is 12. A king is 13, and the kings' pile goes at the center. The next card is taken from the top of the pile where the last card was placed. If all of the cards in a pile are face up, the next card is taken from the next pile (in the clockwise direction) that still has face-down cards. The first card can be drawn from any pile. The game terminates when the four kings have been placed on the center pile. The game is won if all the other cards are correctly placed when the last king is placed in the center.

Write a program that simulates clock solitaire using a random number generator to shuffle the deck before dealing the cards. Print out the board at the beginning of the game, after every 10 moves, and at the end of the game.

13. The game of Life, developed by the mathematician John Conway, models the life and death of a population living on an m by n board. There are three rules to the game of Life: (1) Every empty cell with three living neighbors will come to life in the next generation. (2) Any cell with 1 or 0 neighbors will die of loneliness, and any cell with 4 or more neighbors will die of overcrowding. (3) Any cell with 2 or 3 neighbors will live into the next generation.

   All births and deaths occur simultaneously. The population that is on the board at the end of a turn is considered to be a generation. Play your game on a small board, say 5 × 5, until you are sure that you have implemented the rules correctly. Write a procedure that uses a random number generator to decide which cells are occupied initially. Write a procedure to print out the board after each generation. (Use a * for a member and blanks for empty cells.)

   Write a program that plays the game of Life. Note that the game of Life is not a game that you can participate in. All you can do is watch the patterns as they develop and change. Be sure that you specify how the game stops.

14. The growth of the colony of rabbits from Case Study 11 in Chapter 8 can be described by the equation:

$$R_k = (1 + c)R_{k-1}$$

where R_k is the population of the end of time period k and c is the growth rate of the rabbits.

   Suppose that predators have been introduced into the same territory as the colony of rabbits. The predators feed on the rabbits, reducing their number, but the growth rate of predators depends on the number of rabbits. If the number of rabbits falls below 1000, the predators start to die off.

$$R_k = R_{k-1}(1 + c) - SP_{k-1}$$
$$P_k = P_{k-1} + \beta(R_{k-1} - 1000)$$

where R_k = number of rabbits at time k
   P_k = number of predators at time k
   c = birth rate of rabbits
   S = predation rate
   β = growth rate of the predators relative to the number of rabbits

Write a program that computes and reports the number of rabbits and the number of predators each time period for 20 time periods. Initially there are 3000 rabbits and 4 predators. The birth rate of the rabbits is 5 per pair per time period. The predation rate is 115 rabbits per predator per time period. The growth rate of the predators is 0.02 new predator per rabbit per time period.

15. You have been hired by a bank to study how long its customers wait in line. The bank opens at 9:00, with no customers. The bank closes after the last

customer who arrived before 3:00 has been served. The number of customers who arrive each minute is stored in a computer file that was created with a Pascal program as a **file** of integers. The number of customers who depart each minute is a random number between 0 and the number of customers in the line.

Compute the length of the line for each minute. At the end of the day, the president would like a summary stating the total number of minutes during which the line was of each length. For example, there were 10 minutes when there were no customers in line, 5 minutes when there was one customer in line, etc.

The length of the line at a given minute is the length of the line in the previous minute, minus those customers who were served during the minute, plus those who arrived during the minute. Since the function for the number of departures depends on the line length, the line is never negative.

CHAPTER 15

Recursive Subprograms and Subprograms as Parameters

This chapter covers two Pascal topics that extend the uses of subprograms: recursion and the use of subprograms as parameters. The topics can be covered any time after you have learned the **array** data structure in Chapter 9.

Recursion occurs when a subprogram invokes itself, causing its own action to be repeated. Recursion can be used in Pascal to create concise and powerful programs. The ability to use subprograms as parameters allows you to write general procedures and functions that can be included in many different programs.

The syntax for subprograms as parameters is also used in the applications numerical differentiation, numerical integration, and root finding in the *Advanced Applications* supplement. The applications on numerical differentiation, numerical integration, and root finding assume some familiarity with calculus. The material in the application on root finding uses the results from the application on numerical differentiation.

CASE STUDY 26
Computing the length of a river

You have been asked to write a program that computes the length of a river and its tributaries. You are told that the smallest scale that can be measured is s. The size of the scale affects the accuracy of the measurement because the smaller the scale, the more small curves in the river will be included in the measurement. The river has developed in a regular manner. The main branch has two tributaries that are smaller but completely similar to the main branch; each of these tributaries has two tributaries that are smaller but completely similar to the main branch; and so on. The distance between the point where each tributary enters its parent tributary and the parent's base is a fixed ratio of the distance between the point where that tributary entered its parent tributary and that parent's base. The length of each tributary is a fixed ratio of the length of its parent tributary.

For this particular river, each tributary enters at a distance d from the base of its parent tributary. The distance d is given by

$$d = h\frac{1}{2^n} \tag{15.1}$$

where h is the length of the parent tributary and n is a fixed integer. Each new tributary has a length h':

$$h' = h\left(1 - \frac{1}{2^n}\right) \tag{15.2}$$

Each length h' is added to the total length until the length of the next tributary is smaller than the scale s. In your program, you can vary the parameters s and n to see the effect that each one has on the measured length of the river.

Algorithm development

The total length of the river can be computed using either a looping structure or a recursive structure. For the original algorithm development, a solution using a looping structure will be developed. At the end of the section, after recursion has been covered, the algorithm will be restated using a recursive structure.

In computing the total length of the river, you must remember that each tributary has two tributaries of its own. One strategy for writing the algorithm to compute the length of the river would be to follow each tributary's tributary until you reach a tributary that is smaller than the scale, just as you would do if you were measuring the river by following a bank. This strategy will be used when the recursive algorithm is developed; however, it does not take advantage of the symmetry of the river, and if you think about how you would keep track of where you were on the river, you can see that this strategy will result in a complicated algorithm.

To take advantage of the symmetry, you must realize that the length of all the branches at each level of branching is the same. That is, the four tributaries off the two main tributaries are all the same length, the eight tributaries of the four secondary tributaries are all the same length, etc. In addition, you can compute that the number of tributaries at any level of branching is k^i, where i is the number of branchings that have occurred and k is the number of tributaries per branch.

The length of each new tributary can be computed by substituting the length of the last tributary into Equation 15.2. The number of tributaries of that length is k^i. Algorithm 15.1 uses a **while-do** statement to compute the length of the river.

ALGORITHM 15.1   Iterative algorithm for Case Study 26

begin algorithm
1. Initialize the program
   a. Read the parameters that describe the river
      read **Length,** the length of the main branch
      read **N**, the integer that determines where the branches occur
      read **TribsPerParent,** the number of tributaries off each parent
      read **Scale,** the length of the ruler
   b. Initialize the counters
      TotalSoFar ← Length
      LevelOfBranching ← 1
      LengthFactor ← 1.0 − (1.0 / ZToPower(2, N))
2. Compute the sum of the lengths of the tributaries
   while **Length >= Scale** do
      a. Compute the length of each new tributary using Equation 15.2
         NewLength ← Length ∗ LengthFactor
      b. Compute the number of tributaries of length **NewLength**
         LevelOfBranching ← LevelOfBranching + 1
         Num ← ZToPower(TribsPerParent, LevelOfBranching)
      c. Add the lengths to the running total
         TotalSoFar ← TotalSoFar + Num ∗ NewLength
      d. Set the length of the parent for the next level of branching
         Length ← NewLength
   end while
3. Report the **TotalSoFar,** the total length of the river
end algorithm

LengthFactor, from Equation 15.2, is computed at the beginning of the program because it is the same for all the branches. If the river had formed randomly rather than regularly, this factor would be different for every branch.

15.1 Recursive subprograms

In a *recursive* definition, something is defined in terms of itself. For example, you may look up a word that you do not know in a dictionary and find that the word is defined in terms of another word that you do not know. When you look the new word up, you find another word that you do not know. You keep going until you are referred to a word that you do know. All dictionaries are recursive because they can only define words in terms of other words. The algorithm for finding the definition of a word in a dictionary can be stated recursively as follows: Look up a word to find a new word until the definition of the new word is known. Notice that this algorithm may terminate after you look up one word or may not terminate at all. If you do not know the definitions of enough words, when you look up a word you may eventually be referred back to the original word that you do not know! When a recursive definition does not terminate, it is said to be *infinitely recursive*.

There are many mathematical functions that can be defined in terms of themselves. For example, in Chapter 7 the factorial sequence was defined recursively as

$$n! = n(n-1)! \quad \text{for } n > 0 \text{ where } 0! = 1 \qquad (15.3)$$

Another recursive sequence is the Fibonacci sequence:

$$a_{n+1} = a_n + a_{n-1} \quad \text{for } n > 2$$

where $\qquad(15.4)$

$$a_1 = 1 \quad \text{and} \quad a_2 = 1$$

Raising a number to an integer power can be expressed recursively as

$$x^n = x \cdot x^{n-1} \quad \text{for } n = 1, 2, \ldots$$

where $\qquad(15.5)$

$$x^0 = 1$$

Notice that each recursive definition includes a nonrecursive definition for at least one value of the function. If such a definition were not included, the functions would be infinitely recursive.

In Pascal, you can define functions and procedures that invoke themselves recursively. A recursive subprogram can be mathematically elegant, but it can also be difficult to write and debug. It is easy to inadvertently create an infinitely recursive subprogram. At the end of this section, the problems created by recursive subprograms will be covered.

Recursive functions

You may have noticed that the functions defined recursively in Equations 15.3 through 15.5 have been implemented previously using **for** statements.

15.1 Recursive subprograms

Solutions that involve **for, while-do,** or **repeat-until** statements are called *iterative solutions* because they repeat a sequence of steps to arrive at the solution. One way to write a function that computes $n!$ is to use a **for** statement that expands the formula for $n!$ into $(1)(2)(3)(4)\cdots(n)$. Each time through the loop, the running product is multiplied by the next term in the series.

```
function IterFact (N : integer) : integer;
(*    The function IterFact computes N! using a for statement that    *)
(*    computes the value of 1 * 2 * 3 * ...* N in a running product.  *)

var   Index : integer;        (* for statement control variable. *)
      Temp : integer;         (* temporary variable to store N!, the running *)
                              (* product. *)

begin
  Temp := 1;                  (* initialize running product. *)

  for Index := 1 to N do
    Temp := Temp * Index;     (* compute N factorial *)

  IterFact := Temp
end; (* function IterFact *)
```

The factorial function can also be implemented as a recursive function. The Pascal statement

$$\text{Fact} := \text{Fact}(N - 1) * N$$

looks like the definition of the factorial function given in Equation 15.3. In this Pascal statement, the function **Fact** invokes itself.

To see how this recursive function works, suppose it is invoked with **N** equal to 3. The first time the function is executed, the statement **Fact(N − 1) * N** is evaluated. When **N** is equal to 3, **Fact(2) * 3** must be computed. But the value of **Fact(2)** is unknown, so the function **Fact(2)** must be evaluated before the function **Fact(3)** can be assigned a value. Therefore, a copy of all the values of the variables in the first call to **Fact** must be kept so that once **Fact(2)** is known the value of **Fact(3)** can be computed. When **Fact** is invoked with **N** equal to 2, the value of **N** is known to be 2 but the value of **Fact(1)** is unknown, so **Fact(1)** must be evaluated. With **N** equal to 1, the expression **Fact(0) * 1** is evaluated.

If the expression

$$\text{Fact} := \text{Fact}(0) * 1$$

were evaluated using just the statement **Fact := Fact(N − 1) * N**, then **Fact(0)** would be evaluated as **Fact(−1) * 0.** Then **Fact(−1)** would be evaluated, then

Fact(−2), etc. However, from the definition of the factorial function, you know that **Fact(0)** should be evaluated as 1.

Clearly, a recursive program must have a stopping condition. Otherwise, a recursive subprogram will continue to invoke itself until the computer-time or storage-space limits are exceeded. The last call to a recursive program, for which the stopping condition is true, is sometimes referred to as the *limit call*. The following recursive function includes the condition for **N = 0**:

```
function Fact (N : integer) : integer;
(*     This function computes N factorial recursively from the defin-   *)
(*     ition N! = N (N - 1)! If N > 0, the function calls itself with   *)
(*     N - 1 as its argument, (N - 1)! is computed and multiplied by N. *)
(*     The result of this computation is assigned to function Fact. The *)
(*     recursion stops when N = 0 because 0! is defined to be 1.        *)
begin
   if N > 0 then
      Fact := Fact (N - 1) * N
   else
      Fact := 1
end; (* function Fact *)
```

Once the value of **Fact(0)** is known, **Fact(1)** can be computed; once the value of **Fact(1)** is known, **Fact(2)** can be computed; and once the value of **Fact(2)** is known, **Fact(3)** can be computed.

To understand how the computer keeps track of all the different values of the parameter **N**, you must understand the way value parameters are treated: Each time the function is invoked, a copy of the value passed through the parameter list is created as a local variable. You can visualize the process more easily if you give the local variables names—for example, N_1 for the local variable created in the first call to the function, N_2 for the local variable created in the second call, and so on. You can imagine a chain passing the values from one function call to the next, as in Figure 15.1.

You may wonder how an *unresolved* variable (that is, a partially computed value such as **Fact(2) * 3** where **Fact(2)** is unknown) is stored until the recursion has reached the bottom and **Fact(0), Fact(1),** and then **Fact(2)** can be computed. The answer to this question requires knowledge of the way the memory of the computer is organized. Most computers have a *stack* that is used to store temporary data. The stack works on much the same principle as the spring-loaded plate stackers used in cafeterias. In a plate stacker, plates can be loaded in only from the top, and plates can be removed only from the top. When a plate is loaded into the stacker, the rest of the plates are pushed down. When a plate is removed from the top, the rest of the plates pop up so that the next plate is on the top of the stack. The plate stacker has a property called *last in/first out;* that is, the last plate put on the stack is the first plate to be taken off. Although a computer stack stores data rather than plates, the

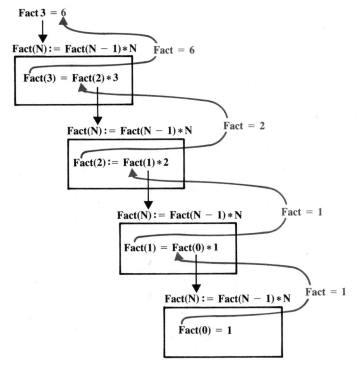

FIGURE 15.1
Passing values between function calls.

principle is the same: Data is *push*ed onto the top of the stack to be stored and *pop*ped off the top of the stack to be retrieved. Any data that is not on the top of the stack is temporarily inaccessible, and data is available on a last-in/first-out basis.

To see how the stack is used in recursion, imagine that there is a line of people and each one pushes a plate with his or her name on it onto the stack. When the last plate has been put on the stack, the line reverses. The last person to put a plate on takes the first plate off the top. Each person takes a plate until the last person has removed the last plate. When the line is finished, each person has his or her own plate back. In recursion, each procedure pushes the values of its variables onto the stack. When the limit call is reached, the order is reversed and each procedure in turn retrieves the values of its variables from the stack.

The contents of the stack as the factorial function is executed can be visualized as follows:

Stack after call to **Fact** with **N** = 3:

| **Fact** := **Fact(2)** * 3 | N = 3 |

Stack after call to **Fact** with **N** = 2:

| Fact := Fact(1) * 2 | N = 2 |
| Fact := Fact(2) * 3 | N = 3 |

Stack after call to **Fact** with **N** = 1:

Fact := Fact(0) * 1	N = 1
Fact := Fact(1) * 2	N = 2
Fact := Fact(2) * 3	N = 3

In the limit call, **Fact(0)** is evaluated and assigned a value of 1. When control returns to the function **Fact(1),** the top value on the stack is removed. **Fact(1)** is assigned a value of 1 * 1. The stack is now

| Fact := Fact(1) * 2 | N = 2 |
| Fact := Fact(2) * 3 | N = 3 |

When control returns to the function **Fact(2),** the top value is again removed from the top of the stack and **Fact(2)** is assigned a value of 1 * 2. The stack is now

| Fact := Fact(2) * 3 | N = 3 |

The top value is again removed from the top of the stack, and **Fact(3)** is assigned a value of 2 * 3 or 6.

Recursion versus iteration

All of the algorithms used in this section could be defined either recursively or iteratively (using a loop structure). Mathematically the recursive definitions are concise and compact, but their implementation as recursive Pascal functions often requires more computer time, more storage space, and more programming time than does implementation of iterative solutions. However, certain complex algorithms are more easily implemented using recursive subprograms.

15.1 Recursive subprograms

There are some algorithms that beginning programmers seem to implement naturally as recursive algorithms before they have learned recursion. For example, in Chapter 6 a function **YesEntered** was developed to ask the user to enter 'y' or 'n' from the terminal. Frequently the following algorithm is proposed: Request the user to enter 'y' or 'n'; if neither is entered, request the user to enter 'y' or 'n' until one or the other is entered. Algorithm 15.2 is a formal version of this algorithm.

ALGORITHM 15.2 Recursive algorithm to request yes or no from a user

begin **RequestYesOrNo**
  1. Write a message to the user to enter 'y' for yes or 'n' for no
  2. Read **Entry**
  3. If **Entry** <> 'y' and **Entry** <> 'Y' and **Entry** <> 'n' and **Entry** <> 'N' then
     RequestYesOrNo
end algorithm

Algorithm 15.2 is implemented as the recursive procedure **RequestYesOrNo.**

```
procedure RequestYesOrNo (var Entry : char);
(*      RequestYesOrNo is a recursive procedure that asks the user to   *)
(*      enter a yes or a no at the terminal.  The character returned in *)
(*      the var parameter Entry will be 'y', 'Y', 'n', or 'N'; if any   *)
(*      other character is entered, the procedure calls itself and     *)
(*      requests another character.  The procedure will continue to    *)
(*      invoke itself and request a y or n until one of the four pre-  *)
(*      vious characters is entered.                                    *)
begin
  writeln (' Enter yes or no (y/n)');
  readln (Entry);

  if (Entry <> 'y') and (Entry <> 'Y') and (Entry <> 'n') and (Entry <> 'N')
    then
      RequestYesOrNo (Entry)
end; (* procedure RequestYesOrNo *)

=>
 Enter yes or no (y/n)
□what
 Enter yes or no (y/n)
□7
 Enter yes or no (y/n)
□y
```

The **forward** *statement*

Another form of recursion occurs when one subprogram invokes another subprogram that in turn invokes the first subprogram. Although this may sound contorted, consider the following problem: You want to write a program that plays a two-person board game, such as checkers, where each player has a different strategy but each player's move depends on the other player's move. Let us call the procedures that determine the players' moves **YourMove** and **MyMove**. Each procedure must at some point invoke the other, to allow the appropriate player to enter a move.

If **YourMove** calls **MyMove** and **MyMove** calls **YourMove**, one or the other procedure must be invoked before it is defined, which is counter to the standard syntax of Pascal. The **forward** statement circumvents this problem. For the procedure **YourMove** to compile, only the header of **MyMove** is required. The **forward** statement allows you to declare just the header of a subprogram at the beginning of the declarative part of a program, and then declare the entire subprogram in a later part. For example, the header of **MyMove** could be declared with the statement

```
procedure MyMove (var CurrentBoard : BoardMatrix); forward;
```

Once the header of procedure **MyMove** has been declared using the **forward** statement, the procedure **YourMove**, which invokes **MyMove**, can be declared. Then the procedure **MyMove** can be declared. Since the heading of **MyMove** has already been declared, the parameter list does not have to be repeated and in fact cannot be repeated when the procedure body is declared. A good practice is to include the parameter list as a comment with the main body of the procedure for clarity, particularly in a long program. The declaration of the two procedures is

```
procedure MyMove (var CurrentBoard : BoardMatrix); forward;

procedure YourMove (var CurrentBoard : BoardMatrix);
   (*     Procedure YourMove finds your next move in a board game and      *)
   (*     tests if you have won the game.  If you have won, it writes a    *)
   (*     message and ends the game.  Otherwise, it invokes MyMove.        *)
   (*     The procedure YourStrategy, which is invoked within this pro-    *)
   (*     cedure, computes your move based on the current board and        *)
   (*     updates the board with your move.                                *)

   var   YourWin : boolean;      (* set to true if the status of the board *)
                                 (* indicates that you have won. *)
```

```
begin
  YourStrategy (CurrentBoard);
  YourWin := CheckForWin (CurrentBoard);

  if not (YourWin) then
    MyMove (CurrentBoard)
  else
    WriteCongratulations
end; (* procedure YourMove *)

procedure MyMove; (* var CurrentBoard : BoardMatrix *)
(*      Procedure MyMove finds my next move in a board game and tests    *)
(*      whether I have won the game.  If I have won, it writes a         *)
(*      message and ends the game.  Otherwise, it invokes YourMove.      *)
(*      The procedure MyStrategy, which is invoked within this pro-      *)
(*      cedure, computes my move based on the current board and          *)
(*      updates the board with mymove.                                   *)

var   MyWin : boolean;      (* set to true if the status of the board *)
                            (* indicates that I have won. *)
begin
  MyStrategy (CurrentBoard);
  MyWin := CheckForWin (CurrentBoard);

  if not (MyWin) then
    YourMove (CurrentBoard)
  else
    WriteCongratulations
end; (* procedure MyMove *)
```

Problems with recursion

The most common problems when writing recursive subprograms are an incorrect condition for the limit call or an incorrect value of a local variable due to a misunderstanding of the operation of the stack. When writing a recursive program, you should be absolutely certain that the boolean expression in the limit call is stated correctly so that the subprogram will terminate. In theory, checking the limit call in a recursive subprogram is no different from checking the terminating condition of a conditional loop; however, in practice, checking the limit call is more difficult because you must be certain of the values of the local variables at the time of the limit call.

Another problem that indirectly results from recursion is an inadvertently infinite recursive subprogram. This problem arises when a subprogram is defined in terms of itself by mistake. For example, if you create a function called **Next**, which has no arguments, and in the function include the statement

$$\text{Next} := \text{Next} + 1$$

Next is an infinite recursive function. It will continue to execute itself because there is no stopping condition.

Program 15.1 illustrates an infinitely recursive **read** procedure.

PROGRAM 15.1   RecursiveRead

```
program RecursiveRead (input, output);
(*     This program shows what can happen if you inadvertently redefine   *)
(*     the standard procedure read and then try to use the standard       *)
(*     procedure.  If the number of parameters in the heading of the      *)
(*     new procedure is the same as the number of variables that you try  *)
(*     to read with the now redefined standard procedure, you will create *)
(*     an infinitely recursive procedure that will execute until the      *)
(*     program runs out of storage space.  Some compilers treat stan-     *)
(*     dard words as reserved words, so this program would terminate      *)
(*     with a simple message that read cannot be redefined.               *)
var    A, B, C, D : real;

  procedure Read (var W, X, Y, Z : real);
  (*     This procedure reads 4 real variables from the terminal.  It     *)
  (*     also inadvertently redefines the standard procedure read.        *)

  begin
    read (W, X, Y, Z)
  end; (* procedure Read *)

begin (* main program *)
  Read (A, B, C, D);
  writeln (' The input values were : A =', A:10:2, ' B =', B:10:2, ' C =',
        C:10:2, ' D =', D:10:2)
end. (* program RecursiveRead *)

=>
                                                     pascal termination log
                                                     ------ ----------- ---
*** error ***           stack overflow - storage exhausted
program terminated at offset 000034 in segment read

local vars: w       = <undefined>        x       = <undefined>          y
      = <undefined>           z       = <undefined>
```

```
====> called from offset 000034 in segment read

local vars: w        = <undefined>        x       = <undefined>              y
            = <undefined>        z       = <undefined>


====> called from offset 000034 in segment read

local vars: w        = <undefined>        x       = <undefined>              y
            = <undefined>        z       = <undefined>

                    .
                    .
                    .
       2429 procedures not traced
                    .
                    .
                    .

====> called from offset 000084 in main program

local vars: a        = <undefined>        b       = <undefined>              c
            = <undefined>        d       = <undefined>
```

In Program 15.1, the standard procedure **read** is redefined in the first procedure of the program. (Remember that **read** is a standard procedure but not a reserved word, so it can be redefined.) Within the new **Read** procedure, the program author invokes what he thinks is the standard procedure **read**. However, the standard **read** procedure has been replaced by his procedure **Read** (the compiler does not distinguish between capital and lowercase letters), so the new procedure **Read** is invoked. Notice that the number and type of parameters in the procedure header and in the invocation agree. When the new procedure **Read** is invoked, it once again encounters a call to the procedure **Read,** which is invoked again. There is no way out of this loop. So the program continues to invoke itself until the computer runs out of space to store all the variables created at each level.

Solution to Case Study 26

The problem in the case study can be restated recursively as follows: Each new branch generates new branches until the length of each of the new branches is less than the measurement scale. Algorithm 15.3 is a recursive algorithm for creating new branches. Algorithm 15.3 is a simplified version of the case-study problem because the algorithm ignores the problem of keeping track of the total length and assumes that each branch has only one tributary.

ALGORITHM 15.3 Simplified algorithm for creating new branches

begin **CreateNewBranch(Length)**
  NewLength ← **Length** * **LengthFactor**
  if **NewLength** >= **Scale** then
    ComputeNewLength(NewLength)
end **ComputeNewLength**

In Algorithm 15.3, the **Length** of the main river is passed into the procedure and a branch is created of length **NewLength.** If the boolean expression **NewLength** >= **Scale** is true, the procedure **CreateNewBranch** calls itself with the **NewLength** that has just been computed as its argument. **NewLength** is now passed into the procedure, and its value is assigned to the parameter **Length.** The **NewLength** of the second branch off the main river is created from **Length,** which now has the value of the length of the first branch that was created, and if the boolean expression remains true, **CreateNewBranch** is again invoked with **NewLength** as its argument. The procedure continues to invoke itself and to create new branches based on the length of the previously created branch until the boolean expression becomes false. If the procedure is called recursively three times before the boolean expression becomes false, the river can be visualized as in Figure 15.2.

Each branch of the river has one tributary after the execution of this algorithm. Algorithm 15.4 takes into account the fact that each branch has not just one tributary but **TribsPerParent** (in the case study, 2) tributaries.

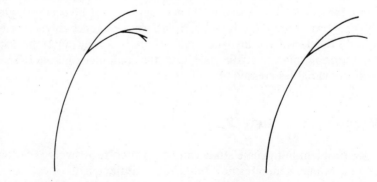

FIGURE 15.2
Main river with first three tributaries.

FIGURE 15.3
Main river with one tributary.

ALGORITHM 15.4 Recursive algorithm to create new branches

begin **CreateNewBranch(Length)**
  for **Branch** := 1 to **TribsPerParent** do
    begin for
      NewLength ← **Length** ∗ **LengthFactor**
      if **NewLength** >= **Scale** then
        CreateNewBranch(NewLength)
    end for
end **CreateNewBranch**

The execution of Algorithm 15.4 is more complicated than the execution of Algorithm 15.3. **Length** again is passed to the procedure, but this time all the branches of the river will be created by invoking the procedure **CreateNewBranch.** You can visualize the execution of the program by drawing the river and the branches each time **NewLength** is computed. On execution of the procedure, the control variable of the **for** statement is first set to 1 and the length of the first tributary of the river is computed. The river and its first tributary can be visualized as in Figure 15.3.

Assuming that the boolean expression is true, **CreateNewBranch** is again invoked, this time with **NewLength** as its argument. Notice that the **for** statement in the first call to **CreateNewBranch** has not yet finished execution when the procedure is called for the second time. **Branch** still has a value of 1 and has yet to be incremented to **TribsPerParent;** thus it is an unresolved local variable that will be stored in the computer's stack until the program execution returns to its position in the stack. The value of **Length** at this point in the program execution will also be stored in the stack and used later. For the second execution of **CreateNewBranch,** the **for** control variable will again take the value of 1, and the length of a new branch will be computed based on the **Length** that was passed into the

FIGURE 15.4
Main river with first two tributaries.

FIGURE 15.5
Main river with first three tributaries.

FIGURE 15.6
River with two tributaries on second branch.

FIGURE 15.7
Main river with half of its tributaries.

procedure. The river and its first two tributaries can be visualized as in Figure 15.4.

If the boolean expression is still true, the procedure will be invoked again. The second execution of the **for** statement is also incomplete, so the values of its control variable and the current value of **Length** will each be put onto the stack on top of the values from the first execution of the **for** statement. After the length of the third tributary has been computed, the river can be visualized as in Figure 15.5.

Assume that at this point in the procedure's execution, the boolean expression becomes false—that is, **NewLength** < **Scale**. The control variable of the **for** statement that is currently being executed will be incremented. The **NewLength** of the second tributary of the second river branch will be computed, and the river can be visualized as in Figure 15.6.

The execution will proceed to complete the next most recent **for** statement, taking the values of the control variable **Branch** and the variable **Length** from the top of the stack. When this **for** statement has completed execution, the lengths of all the tributaries on one side of the river have been computed and the river can be visualized as in Figure 15.7.

The procedure finally returns to the completion of the initial **for** statement. **TribsPerParent** is incremented to 2, and the **NewLength** of the first branch on the other side of the main river is computed. The procedure executes in the same way as before except that the **for** control variable from the first call to **CreateNewBranch** is now 2. The final river can be visualized as in Figure 15.8.

Procedure **CreateNewBranch** is a completed version of Algorithm 15.4. Statements have been added to compute the total length of the river.

FIGURE 15.8
Main river with all its tributaries.

The variables **Scale** and **LengthFactor** are included in the parameter list so that the procedure is modular.

```
procedure CreateNewBranch (Length, Scale, Factor : real;
                          TribsPerParent : integer; var TotalSoFar : real);
(*     This procedure is a recursive procedure that creates new branches  *)
(*     from a parent branch.  Each parent branch has TribsPerParent       *)
(*     tributaries branching from it.  The length of the original branch  *)
(*     is Length.  The length of each new branch is Factor times the      *)
(*     Length of its parent branch.  New branches are created until the   *)
(*     length of the new branch is less than Scale, the smallest measure- *)
(*     ment that can be made.                                             *)

var   NewLength : real;      (* length of the new branch. *)
      Branch : integer;      (* for statement control variable -- the *)
                             (* branch number of the current branch. *)

begin
  for Branch := 1 to TribsPerParent do
    begin
      NewLength := Length * Factor;

      if NewLength >= Scale then
        begin
          CreateNewBranch (NewLength, Scale, Factor, TribsPerParent,
                           TotalSoFar);
          TotalSoFar := TotalSoFar + NewLength;
        end (* if *)
    end (* for *)
end; (* procedure CreateNewBranch *)
```

Program 15.2, **RiverLength,** is a recursive solution to the case study that incorporates the procedure **CreateNewBranch.** The program is executed twice to show the effect of different values of *n*.

PROGRAM 15.2   RiverLength

```
program RiverLength (input, output);
(*     Program RiverLength computes the length of a river. The program   *)
(*     creates new tributaries of the river with lengths that are a      *)
(*     fixed fraction of the length of the parent branch.  Each parent   *)
(*     branch has TributariesPerParent tributaries that branch from it.  *)
(*     The length of the river is measured with a ruler of length        *)
(*     MeasurementScale.  Any branch whose length is less than Measure-  *)
(*     mentScale cannot be measured and is not included in the total     *)
(*     length. The new branches are created in a recursive procedure     *)
(*     called CreateNewBranch.                                           *)

var    MainLength : real;            (* length of the main river -- *)
                                     (* input. *)
       TotalLength : real;           (* computed length of the main *)
                                     (* river and its tributaries. *)
       MeasurementScale : real;      (* smallest length that can be *)
                                     (* measured by the ruler -- input. *)
       TributariesPerParent : integer;  (* number of tributaries off each *)
                                     (* parent branch -- input. *)
       N : integer;                  (* fractal dimension. *)
       LengthFactor : real;          (* computed ratio of length of the *)
                                     (* new branch to length of parent. *)


(*$i ztopower *)
(*$i createnewbranch *)

(*$i readinitialdata *)
(*     procedure ReadInitialData (var Length, Scale : real;              *)
(*                                var N, TribsPerParent : integer);      *)
(*          This procedure reads the data on the river from the terminal.*)


   procedure SetInitialVars (N : integer; Main : real;
                             var Factor, TotalSoFar : real);
      (*     Procedure SetInitialVars initializes the running total for the  *)
      (*     length of the river, TotalSoFar, to the length of the main      *)
      (*     trunk of the river and computes the ratio, Factor, of the       *)
      (*     length of each branch to its parent.                            *)

   begin
      TotalSoFar := Main;
      Factor := 1.0 - (1.0 / ZToPower(2, N));
   end; (* procedure SetInitialVars *)
```

```
(*------------------------------------------------------------------*)
begin (* main *)
  ReadInitialData (MainLength, MeasurementScale, N, TributariesPerParent);

  SetInitialVars (N, MainLength, LengthFactor, TotalLength);

  CreateNewBranch (MainLength, MeasurementScale, LengthFactor,
                   TributariesPerParent, TotalLength);
  writeln;
  writeln (' The total length of the river including the main branch is',
           TotalLength:10:2)
end. (* program  RiverLength *)
```

```
=>
  This program computes the length of a river and its tributaries

 How long is the main branch of the river?
□50
 How many tributaries does each branch have?
□2
 What is the scale on the ruler used to measure the river?
□10
 Each branch is (1 - 1/(2 ** N)) times the length of the previous branch.
 Enter an integer for N
□1

 The total length of the river including the main branch is     150.00
```

```
=>
  This program computes the length of a river and its tributaries

 How long is the main branch of the river?
□50
 How many tributaries does each branch have?
□2
 What is the scale on the ruler used to measure the river?
□10
 Each branch is (1 - 1/(2 ** N)) times the length of the previous branch.
 Enter an integer for N
□3

 The total length of the river including the main branch is   96183.92
```

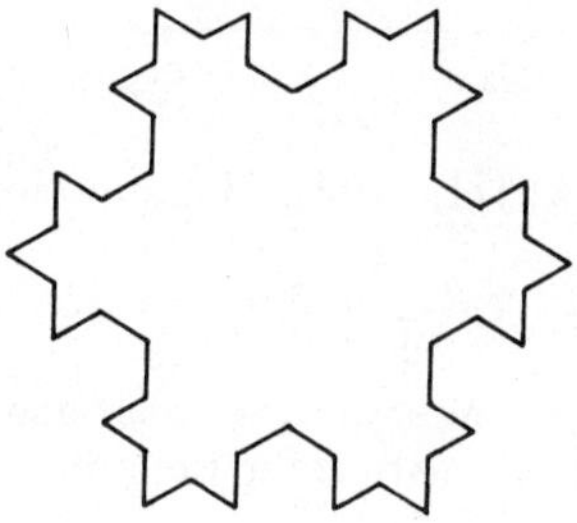

FIGURE 15.9
A fractal triangle.

Fractal geometry

The algorithm for creating the river in the case study uses *fractals,* which have been developed by the mathematician Benoit Mandelbrot (1924–). Fractals can be used to describe regular and irregular patterns in nature such as coast lines, the branching of trees and of blood vessels, mountains, and clouds.

Fractals are created by repeatedly subdividing a curve as was done with the river in the case study and as illustrated in Figure 15.9, a fractally subdivided triangle. The curves can be subdivided at fixed ratios or at random intervals. In other words, fractals can be either deterministic or random. In Figure 15.9, the triangle is subdivided and reproduced at a fixed interval.

One property of fractals is that they are infinitely self-similar. That is, the main branch of the river in Case Study 26 and one of its tributaries look the same if there is no reference scale. Another property of fractals is that the total length of the curve depends on the fineness of the ruler. If you measure the length of a river by following one bank, the length of the river bank will depend on how closely you follow the bends and tributaries. The finer the ruler, the longer the river.

In computer graphics, fractals are often used to create natural looking scenes. Standard Pascal does not have graphics capabilities, but if you have a graphics system, you may want to try drawing some pictures using fractals.

15.2 Subprograms as parameters

Standard Pascal allows functions and procedures to be passed into subprograms as parameters; however, many versions of Pascal including UCSD Pascal and Turbo Pascal do not support this feature. Before writing programs that pass subprograms as parameters, check your compiler documentation to be sure that this feature is included.

The format of this section is different from that of the earlier chapters. Using subprograms with parameters will be introduced first, and then a problem that can be solved by passing a generalized subprogram into another subprogram will be developed.

If your compiler does not allow you to pass a subprogram as a parameter, you can still use the functions and procedures developed in the application section, but you will not be able to pass subprograms to them. You will be able to use the generalized subprograms, but you will need to redefine the function or procedure on which the subprogram acts each time you use the subprogram.

Passing subprograms as arguments

In standard Pascal, one subprogram can be passed into another subprogram through the parameter list. This feature is useful for writing subprograms that performs tasks such as plotting curves. For such tasks the algorithm is the same for all functions of the form $y = f(x)$—that is, functions where the value of y is a function of (depends on) the value of x. For example, the algorithm for plotting the sine is the same as the algorithm for plotting the cosine, with the appropriate change of function.

The syntax diagram for a subprogram *parameter list* that may include a procedure or a function is

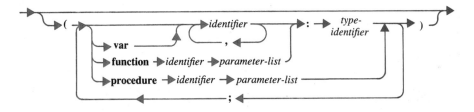

The parameter list can be used within the heading of either a **function** or a **procedure**. From the preceding syntax diagram, the following **procedure** heading can be created:

```
procedure UseFunction (function F (Z : real) : real);
```

This declaration allows a real-valued function with a single, real, value parameter to be passed to the procedure **UseFunction**. A subprogram in a parameter list must always be declared as a value parameter because the subprogram itself cannot be modified by another subprogram.

If a function, call it **FunctionX,** is declared as

```
function FunctionX (Number : real) : real;
```

that function matches the declaration in the parameter list of **UseFunction** and can be passed to the procedure **UseFunction** as

```
UseFunction (FunctionX)
```

Any real-valued function with a single, real-valued, value parameter can be passed into the function **UseFunction**. For example, the function

```
function Temperature (Pressure : real) : real;
```

could also be passed as an argument to **UseFunction** as follows:

```
UseFunction (Temperature)
```

Notice the difference between passing the function itself and passing the value of the function. Because a function returns a single value of a basic data type, you can always pass the *value* of a function into a parameter that is of the same type. For example, a procedure with the heading

```
procedure OneValue (X : real);
```

can be invoked by passing in any real value including the result of **FunctionX** declared above:

```
OneValue (FunctionX (RealVar))
```

In this statement, **FunctionX(RealVar)** is evaluated and the resulting real number is passed into the procedure **OneValue**. This is quite different from passing in the function.

From the syntax diagram for a parameter list, you can see that procedures can also be passed as parameters. Using subprograms as parameters, you can write general utility procedures that you can include in any of your programs. For example, for use in writing reports you might write a **NewPage** procedure that starts a new page, prints a title, and prints and increments the page number. A procedure that specifies the title to be printed can then be passed to **NewPage** by declaring the **WriteTitle** procedure in the parameter list of **NewPage**, as follows:

```
procedure NewPage (procedure WriteTitle; var PageNum : integer);
```

Various title-printing procedures can then be passed to **NewPage** so that the title can say anything you want, can have as many lines as you want, and can be printed on any part of the new page. You could call the procedure **NewPage** at the beginning of the program to write a title page:

```
NewPage (ProgramTitle, PageNumber);
```

You could call it again to start a new page while the output is being written:

```
if NumLines >= 55 then
   NewPage (SubTitle, PageNumber);
```

Notice that passing one subprogram into another is quite different from simply invoking one subprogram from within another. Just as including a variable in the parameter list allows a subprogram to operate on any variable that is passed as an argument, including a subprogram in the parameter list allows the subprogram to invoke any subprogram that is passed as an argument.

Because the standard Pascal functions and procedures do not have the same declarative syntax as user-defined subprograms, standard subprograms cannot be passed as arguments. This problem can be circumvented by declaring dummy user-defined subprograms that return the value of the standard subprograms, as illustrated in the first execution of the procedure for the solution to Case Study 27.

APPLICATION
Plotting a Function

CASE STUDY 27
Plotting the gravitational force

A weather balloon is to be released at sea level and allowed to rise to a predetermined height. One factor in determining how quickly it rises is the gravitational force between the weather balloon and the earth. The gravitational force between a spherical body and the earth is given by the equation

$$F = \frac{GmM_e}{r^2} \tag{15.6}$$

where m = the mass of the balloon
M_e = the mass of the earth
G = the gravitational constant
r = the distance between the center of the earth and the center of the balloon

Write a program that computes and plots the force between the balloon and the earth starting from sea level. The radius of the earth is

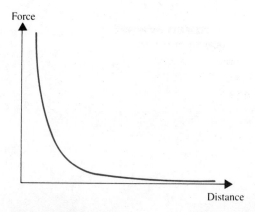

FIGURE 15.10
Force versus distance.

6.37×10^3 km. The mass of the balloon and its maximum height are to be interactively entered. The other values that you need can be found in the tables in Chapter 3. The plot should be in the standard orientation, with the *y*-axis vertical and the *x*-axis horizontal, and should resemble Figure 15.10. Use SI units for the variables in the graph.

Algorithm development

The first problem in Case Study 27 is to compute the values of the force at different distances. A Pascal function is clearly suited to this task. The second part of the problem is to plot the force at different distances. The solution to this second problem is not so clear; however, because the strategy for creating the plot depends strongly on the form of the function, we will take some time to set up the function.

There are several variables on the right-hand side of Equation 15.6. A change in the value of any of these variables will result in a different value for the force *F*. It is obvious that the gravitational constant and the mass of the earth do not change. Since the same balloon will be used for each calculation of the force, *m* will also be a fixed value. Therefore, the force will depend only on the distance between the center of the weather balloon and the center of the earth. The function for the case study can be written in the form

$$F = f(r) \qquad (15.7)$$

This equation is read as "The force is a function of the distance."

The basic steps in solving the case study are given in Algorithm 15.5. This algorithm will require major refinements before it can be implemented as a program. For example, Step 2 is written using a **for** statement that would cause an unnecessarily high number of values to be computed for the force.

ALGORITHM 15.5   Algorithm to plot the force versus the distance

begin algorithm
1. Initialize the program
2. Compute the force versus the distance over the given range
   for **Distance := SeaLevel** to **MaxDistance** do
      begin
         a. **Force ← F(Distance)**
         b. Plot **Force** versus **Distance** on a line printer
      end for
end algorithm

The only step of Algorithm 15.5 that you cannot readily solve is Step 2b, plotting the function. The rest of this section will concentrate on developing a procedure that will plot any given function. This plotting procedure then can be invoked with the function that returns the force given the distance as a parameter.

Plotting a function

When an equation is written in the form $y = f(x)$, the value of y depends on the given value of x. Also, for each given value of x, there is only one value of y. Thus, x is referred to as the *independent variable*, and y is referred to as the *dependent variable*. On a graph, the dependent variable always appears on the vertical y-axis and the independent variable on the horizontal x-axis. Thus in the case study, the force F is the dependent variable, corresponding to y, and is a function of the distance r, which is the independent variable, corresponding to x.

The first step in developing the algorithm is to decide on the basic elements of the printed plot. For a line graph, the basic elements are the x-axis, the y-axis, and the series of points (x, y). If you think about how a line graph is usually created—by drawing the axes and then plotting the points—and if you think about how a line printer works, you can see that a conflict exists: If the axes are printed first, there is no way to back up and print the points. Therefore, the printing of the axes and the printing of the points must be done in parallel.

Another problem with printing a plot on a line printer is that although the x and y variables may be continuous variables, the output page cannot be divided into an infinite number of lines and columns. The number of increments in the x-axis is limited to the number of print columns, and the number of increments in the y-axis is limited to the number of lines in the output page. Thus the point (x, y) must be rounded to the nearest point that can be printed.

Another consequence of the fact that the print position can only move forward is that the rows must be printed from top to bottom and the data points in each row must be printed from left to right. When the printer is positioned to print a line, the y value is known. The problem is to find the values of x that correspond to that value of y. One solution would be to find the inverse function, $x = f^{-1}(y)$; however, the inverse function may not be one-to-one and may not even exist. A better solution is to compute the value of the function $f(x)$ for all values of x. The value (or values) of x that corresponds to the current value of *y* can then be determined and the point (or points) plotted. This strategy requires that the function be evaluated for every value of *x* each time a new row is printed. The algorithm would be more efficient if the evaluation of the function were performed only once for each value of *x*.

This realization leads to an alternative strategy: Create a two-dimensional array corresponding to the rows and columns of the output page. Because an array is a random-access device, the points can be put into the array in any order. Once all the points have been evaluated and the array is complete, the array can be printed as a single entity. The array is an image of the final plot.

For the case study, an array of character data is most suitable because it is easy to print character data and to create the image of the graph in characters. Each element of the character array **Image** represents an element of the graph. If a point occurs at (column, line) of the graph, the value of **Image[Column, Line]** is set to a character such as '+'. Otherwise, it remains the blank character. Algorithm 15.6 which creates an image of the plot in the array **Image**, is a general algorithm that depends only on the function $f(x)$.

ALGORITHM 15.6 **An algorithm to create an array image of a plot**

begin algorithm
   1. Initialize the procedure
      Set **MinX,** the first value of *x* on the plot
      Set **MaxX,** the last value of *x* on the plot
      Set **Symbol,** the symbol to be printed at each plot point
   2. Compute the points on the graph
      for **X** := **MinX** to **MaxX** do
        begin
          $Y \leftarrow f(X)$
          NewY ← **Y** rounded to the nearest row
          Image[X, NewY] ← **Symbol**
        end for
   3. Complete the plot
      Add the *x*-axis and *y*-axis to **Image**
end algorithm

Algorithm 15.6 needs refinement because neither the distinction between the value of **X** and the column number nor the distinction between the value of **NewY** and the row number has been made clear.

To complete the procedure that plots any function that returns a single real value, several other parameters must be specified: the number of rows and columns in the plot, the first value of *x*, and the last value of *x*. Another parameter might also be the symbol to be used in creating the plot. A complete heading for the **Plotter** procedure is as follows:

```
procedure Plotter (function F (Z : real) : real; NumRows, NumCols : integer;
             MinY, MaxY : real; Symbol : char);
```

The main data structure of a two-dimensional array of characters has already been decided upon for the **Plotter** procedure. In the program development a few new variables will be required, but the major variables are as follows:

```
const MaxRows = 70;
      MaxCols = 80;

type  ImageArray = array [1..MaxCols, 1..MaxRows] of char;

var   Image : ImageArray;
```

Before Algorithm 15.6 can be translated into Pascal, it is necessary to know the minimum and maximum values of *y* so that the rows of the output page can be labeled. Given that *x* ranges from **MinX** to **MaxX,** the first step is to find the minimum and maximum values of *y*. The minimum and maximum values of *y* do not necessarily occur at the minimum and maximum values of *x*, so it is necessary to search over the range of *x* for the minimum and maximum values of *y*. An algorithm for finding the minimum and maximum values is given in Algorithm 15.7. In this algorithm, **MaxReal** is a constant that has been defined as a large real number.

ALGORITHM 15.7 Algorithm for finding the minimum and maximum values

begin procedure **MinMax**

1. Initialize the procedure
   MaxY ← −**MaxReal**
   MinY ← **MaxReal**

2. Search for **MinY** and **MaxY**
   for **X** from **MinX** to **MaxX** do

```
                begin
                    Temp ← f(X)
                    if Temp < MinY then
                        MinY ← Temp
                    if Temp > MaxY then
                        MaxY ← Temp
                end
        end procedure MinMax
```

In Step 2 of Algorithm 15.7, f is the function being plotted that will be passed into the procedure.

Given that the output page is **NumRows** by **NumCols,** and given the minimum and maximum values of x and y, the next step is to find the length of each increment on the x-axis and the y-axis. That is, if the values of x range from **MinX** to **MaxX** and there are **NumCols** in the plot, each column represents an increment of (**MaxX** − **MinX**) / **NumCols.** With the inclusion of Algorithm 15.7, the algorithm for initializing the variables for the **Plotter** procedure becomes Algorithm 15.8.

ALGORITHM 15.8 Refinement of Step 1 of Algorithm 15.6

begin algorithm

1. Initialize the procedure
    a. Set **MinX** and **MaxX** to the first and last values of x in the plot
    b. Compute **MinY** and **MaxY** using Algorithm 15.7
    c. Find the spacing for each x and y value
        SpaceX ← (**MaxX** − **MinX**) / **NumCols**
        SpaceY ← (**MaxY** − **MinY**) / **NumRows**
    d. Set the **Image** to blanks

end algorithm

Algorithm 15.8 is translated directly into Pascal as the procedure **Initialize,** which uses the procedure **MinMax.**

```
procedure MinMax (XMin, XSpace : real; var YMin, Ymax : real);
(*      This procedure finds the minimum and maximum values of Y given    *)
(*      the function F, the minimum value of X, and the spacing of the    *)
(*      values along the X axis.  MinMax uses the global constant         *)
(*      NumCols as an upper bound in the for statement to compute all     *)
(*      the values of X.                                                  *)
```

```
        const MaxReal = 1.0e30;        (* a large real number. *)

        var  Col : integer;            (* for statement control variable. *)
             X : real;                 (* value of X variable within the loop. *)
             Temp : real;              (* temporary variable for F(X). *)


        begin
          YMax := -MaxReal;                   (* set the maximum value of Y to a *)
                                              (* small value. *)
          YMin :=  MaxReal;                   (* set the minimum value of Y to a *)
                                              (* large value. *)


        (* The following statements compute the equivalent value of X for each   *)
        (* column in the plot.  The X value for the first column is XMin, and    *)
        (* the X value for the last column in XMax.  For each value of X, F(X)   *)
        (* is computed, and the smallest and largest values of F(X) are saved    *)
        (* as YMin and YMax. *)

          for Col := 1 to NumCols do
            begin
              X := XMin + (Col - 1) * XSpace;
              Temp := F(X);

              if Temp < YMin then
                YMin :=  Temp;
              if Temp > YMax then
                YMax := Temp
            end
        end; (* procedure MinMaX *)


        procedure Initialize (XMin, XMax: real; var YMin, XSpace, YSpace : real;
                              var PlotImage : ImageArray);
        (*      This procedure sets up the parameters for the rest of the pro-   *)
        (*      gram.  It finds the minimum and maximum values of F(X), the      *)
        (*      spacing that each row and column represents, and initializes     *)
        (*      the array image to blanks.                                       *)

        var Ymax : real;                       (* largest value of Y. *)
            Row, Col : integer;                (* for statement control *)
                                               (* variables. *)
```

```
begin
  XSpace := (XMax - XMin) / NumCols;      (* compute the X increment *)
                                          (* that each column *)
                                          (* represents. *)
  MinMax(F, XMin, XSpace, YMin, YMax);    (* compute the minimum and *)
                                          (* maximum values of Y. *)
  YSpace := (YMax - YMin) / NumRows;      (* compute the Y increment *)
                                          (* that each row represents. *)

  for Row := 1 to NumRows do              (* set the plot image to *)
    for Col := 1 to NumCols do            (* blank characters. *)
      PlotImage [Col, Row] := ' '
end; (* procedure Initialize *)
```

Finally, Algorithm 15.9 for computing and printing the points can be written. The major difference between Algorithm 15.9 and Algorithm 15.6 is that Algorithm 15.9 uses **SpaceX** and **SpaceY** to compute the indices of the array that correspond to the point (x, y).

ALGORITHM 15.9 Refinement of Algorithm 15.6

begin algorithm

1. Initialize the procedure (see Algorithm 15.8)

2. Compute the points on the plot
   for **Col** := 1 to **NumCols** do
     begin
       $X \leftarrow \text{MinX} + \text{Col} * \text{SpaceX}$
       $Y \leftarrow f(X)$
       $\text{Row} \leftarrow \text{round}(Y / \text{SpaceY})$
       Image[Col, Row] $\leftarrow$ Symbol
     end

3. Print the axes and (x, y) pairs
   for **Row** := **NumRows** downto 1 do
     begin
       print the value for the y-axis for this line
       for **Col** := 1 to **NumCols** do
         print **Image[Col, Row]**
     end for
   Print the x-axis

end algorithm

Procedure **Plotter** is a general procedure that plots a function using Algorithm 15.9. In the process of writing the procedure, two additions were made to the algorithm. The first was the logic required to print the values for the *x*- and *y*-axes. A decision was made to print the values at every fifth increment on the *x*-axis and every tenth increment on the *y*-axis. How could you change the procedure so that the values for the axes were automatically scaled and printed in multiples of 10?

```
procedure Plotter (function F (Z : real) : real;
                   NumRows, NumCols : integer; MinX, MaxX : real;
                   Symbol : char);
(*      Procedure Plotter plots the function F(Z) at the terminal.      *)
(*      The plot is NumRows by NumCols.  There are NumCols values of X  *)
(*      that range between MinX and MaxX.  The points on the curve are  *)
(*      marked by Symbol.                                               *)

const MaxRows = 70;             (* maximum number of rows. *)
      MaxCols = 80;             (* maximum number of columns. *)

type  ImageArray = array [1..MaxCols, 1..MaxRows] of char;

var   MinY, MaxY : real;        (* minimum and maximum values of Y that *)
                                (* occur in the plot. *)
      SpaceX, SpaceY : real;    (* X and Y increment represented by each *)
                                (* row/column. *)
      Image : ImageArray;       (* array to store the image of plot as *)
                                (* characters. *)


(*$i minmax *)
(*$i initialize *)


   procedure ComputePoints (XMin, YMin, XSpace, YSpace : real;
                            Character : char; var PlotImage : ImageArray);
   (*      This procedure computes the points on the curve F(Z) and puts  *)
   (*      them in the array, ImageArray.                                 *)

   var  Col : integer;          (* for control variable for the column number. *)
        X : real;               (* X value corresponding to Col.*)
        Y : real;               (* Y value computed from Y = F(X). *)
        Row : integer;          (* row corresponding to the Y value. *)
```

```
begin
  for Col := 1 to NumCols do
    begin
      X := XMin + (Col - 1) * XSpace;        (* compute the value of X *)
                                             (* for column Col. *)
      Y := F(X);
      Row := round ((Y - Ymin) / YSpace);    (* find the row that cor- *)
                                             (* responds to Y. *)
      PlotImage [Col + 1, Row] := Character  (* put the symbol in array *)
                                             (* element Col, Row to *)
                                             (* indicate that F(X) has a *)
                                             (* a value at that point. *)
    end
end; (* procedure ComputePoints *)

procedure PrintPoints (XMin, YMin, XSpace, YSpace : real;
                       PlotImage : ImageArray);
(*     This procedure prints the points stored in PlotImage at the    *)
(*     terminal.                                                      *)

var    Row, Col : integer;             (* for statement control *)
                                       (* variables. *)
       X, Y : real;                    (* X and Y values to be plotted. *)

begin
  writeln;
  for Row := NumRows downto 1 do       (* start at the top row. *)
    begin
      if ((Row - 1) mod 5) = 0 then    (* if this row divides evenly *)
        begin                          (* by 5, print the Y-axis value. *)
          Y := YMin + Row * YSpace;
          write (Y:8:1, ' |')
        end
      else                             (* also print the line for the *)
        write('|':10);                 (* Y-axis for this row .*)
      for Col := 1 to NumCols do       (* for each column in this row, *)
        write (Image [Col, Row]:1);    (* print the array that stores *)
      writeln                          (* the image of the plot. *)
    end; (* for Row *)

  write(' ':10);                       (* print -'s for the X-axis. *)
  for Col := 1 to NumCols do
    write('-');
  writeln;
```

```pascal
        for Col := 1 to NumCols + 1 do      (* print a tick mark every tenth *)
          if ((Col - 1) mod 10) = 0 then    (* value along the X-axis. *)
            write('|':10);
        writeln;

        for Col := 1 to NumCols + 1 do      (* print the values for the X- *)
          if ((Col - 1) mod 10) = 0 then    (* axis every tenth column. *)
            begin
              X := MinX + (Col - 1) * XSpace;
              write (' ', X:8:1)
            end;
        writeln
      end; (* procedure PrintPoints *)

  begin (* main body procedure Plotter *)
    Initialize (MinX, MaxX, MinY, SpaceX, SpaceY, Image);

    ComputePoints (MinX, MinY, SpaceX, SpaceY, Symbol, Image);

    PrintPoints(MinX, MinY, SpaceX, SpaceY, Image)
  end; (* procedure Plotter *)
```

Program 15.3, **PlotSine,** uses the procedure **Plotter** to plot the standard function **sin.** Program 15.3 shows how a user-defined function can be written using a standard function so that it can be passed as a parameter.

PROGRAM 15.3 PlotSine

```pascal
program PlotSine (input, output);
(*     This program plots the standard sine function using the proce-    *)
(*     dure Plotter. To pass the function sin as a parameter, it is      *)
(*     reassigned to a user-defined function called sine.                *)

const NumRows = 20;       (* number of rows in the plot. *)
      NumCols = 60;       (* number of columns in the plot. *)

var   MinX : real         (* minimum value of X to be plotted. *)
      MaxX : real;        (* maximum value of X to be plotted. *)


  function Sine (X : real) : real;
  (*     This function is a user-defined function that returns the       *)
  (*     value of the standard Pascal function sin(X). A user defined    *)
  (*     function is required so that the function can be passed as a    *)
  (*     parameter to Plotter.                                           *)
```

```
  begin
    Sine := sin(X)
  end;


(*$i plotter *)


(*----------------------------------------------------------------------*)
begin (* main program *)
  MinX := 0.0;            (* set the minimum value for X in the output plot. *)
  MaxX := 10.0;           (* set the maximum value for X in the output plot. *)

  Plotter (Sine, NumRows, NumCols, MinX, MaxX, 'x')
end. (* program PlotSine *)
```

Here is the output from program **PlotSine:**

=>

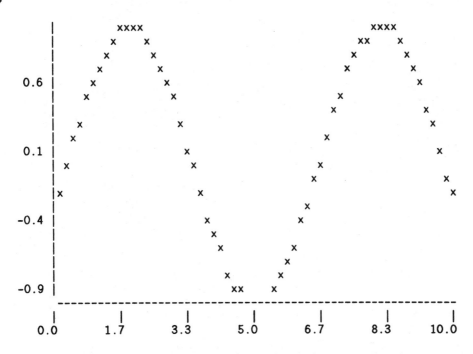

Solution to Case Study 27

Program 15.4, **PlotForce,** uses the procedure **Plotter** to print the force versus distance as required for Case Study 27.

PROGRAM 15.4   PlotForce

```
program PlotForce (input, output);
(*      This program computes and plots the force between a weather    *)
(*      balloon and the earth as it rises from the earth's surface.    *)
(*      This program uses the Plotter procedure and passes the function *)
(*      Force to it as an argument.                                    *)

const G = 6.67e-11;             (* gravitational constant in N-m**2 *)
                                (* kg**2. *)
      EarthMass = 5.98e24;      (* mass of the earth in kg. *)
      EarthRadius = 6.370e06;   (* radius of earth in m. *)
      NumRows = 24;             (* number of rows in the plot. *)
      NumCols = 60;             (* number of columns for the plot. *)

var   Distance : real;          (* distance between earth and balloon in *)
                                (* km. *)
      Mass : real;              (* mass of the balloon in kg. *)
      MaxDistance : real;       (* maximum height of the balloon in m. *)
      Symbol : char;            (* symbol for a point on the plot. *)


   function Force (R : real) : real;
   (*      This function computes the force between the earth and an-  *)
   (*      other body as a function of the distance between them.      *)


   begin
     Force := (G * EarthMass * Mass) / sqr(R)
   end; (* function Force *)

(*$i plotter *)
```

```
(*$i initializeplotter *)
(*   procedure InitializePlotter (var Character : char);              *)
(*        InitializePlotter writes a message to the user and sets the *)
(*        symbol for the points to be printed on the plot.            *)


(*$i readdata *)
(*   procedure ReadData (var Kilograms, Meters : real);               *)
(*        ReadData reads the mass of the weather balloon in kilograms *)
(*        and the distance to which it rises in kilometers.           *)


(*$i writeplotheading *)
(*   procedure WritePlotHeading (Kilograms : real);                   *)
(*        This procedure writes the heading for the plot of the force.*)


(*-------------------------------------------------------------------*)


begin (* main program *)
   InitializePlotter (Symbol);
   ReadData (Mass, MaxDistance);

   WritePlotHeading (Mass);
   Plotter (Force, NumRows, NumCols, EarthRadius,
            MaxDistance + EarthRadius, Symbol)
end. (* program PlotForce *)
```

Following is a listing of the output from the program **PlotForce:**

```
=>
 This program computes and plots the force between a weather
 balloon and the earth starting from the earth's surface.

 What is the mass of the weather balloon in kilograms?
□75.0
 How high does it rise in meters?
□500000.0

 The force between the earth and a      75.00 kg weather balloon is plotted
 versus the distance between the earth and the balloon in meters.
```

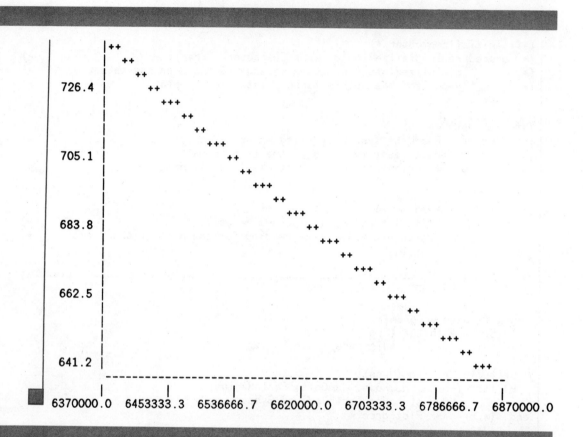

Summary

Topics that expand upon the ways that subprograms can be used within programs were covered in this chapter.

Recursive subprograms in which functions or procedures are defined in terms of themselves can be written in Pascal. When you write a recursive subprogram, you must be careful to include a terminating condition so that the recursion is not infinite. You must also be careful not to use recursion unintentionally; inadvertent use of recursion can result in a program that will never terminate. Unresolved variables are stored on the computer's stack in a last in/first out basis.

The **forward** statement can be used if one subprogram invokes another subprogram that in turn invokes the first subprogram. This statement indicates that the definition of a procedure or function will be included later in the program and allows the compilation of the subprogram that contains the as yet undefined identifier.

Functions and procedures can be passed as parameters into other subprograms. The subprogram is declared as a value parameter of the function or procedure into which it is passed. Passing a subprogram as a parameter allows you to write a general subprogram that performs a task such as plotting a curve or writing a title page. A subprogram that performs the specific task at hand can then be passed into the general subprogram.

Exercises

PASCAL

1. What is the output from the following fragment with a recursive function?

   ```
   function Mystery (X : real; N : integer) : real;
   begin
     if N = 0 then
       Mystery := 1.0
     else
       Mystery := X * Mystery (X, N - 1)
   end; (* function Mystery *)

   begin
     writeln (' The answer is', Mystery (2.0, 3):10:2)
   end
   ```

2. State the reason for and correct the syntax errors in each of the following fragments:

   (a) `procedure ComputeAndPrint (procedure Dummy (var Old, New : integer);`

   (b) `procedure NextFunction (function Z (RealVar : Real); X : real);`

   (c) `function LineIntegral (var function F (X : real) : real) : real;`

APPLICATION

3. Modify the procedure **Plotter** so that the scale for either the x-axis or the y-axis can be specified to be logarithmic. Print the force curve from Case Study 27 using a logarithmic scale for the x-axis.

4. Modify the procedure **Plotter** so that a heading can be printed at the top of the graph and so that the axes can be labeled. Store the output from the procedure in a file.

Problems

PASCAL

1. Write a recursive program that computes x^n where x is a real number and n is a positive integer.

2. Write a recursive function for the difference equation

$$y(k) - \frac{1}{2} y(k - 1) = 0$$

which has initial value $y(0) = 1$. Given a value of k between 0 and **maxint**, the function returns the value of $y(k)$. How can you test your function? Once your program is working, try a different value for $y(0)$.

3. Rewrite the binary search procedure from Chapter 11 as a recursive procedure.

APPLICATION

4. Using the random number generator developed in Chapter 14, write a program that generates a river in which the length of the tributaries is a random variable that is uniformly distributed between 0 and h' as given in Equation 15.2.

5. Write a general-purpose procedure to plot horizontal bar graphs in the form

```
│XXXXXXXX
│XXXXX
│X
│XXX
│XXXX
└─────────
```

Use this procedure in a program that performs the following tasks for the local weather bureau, given a list of the highest temperature for each day of a month:

(a) Computes the minimum and maximum high temperatures for the month.
(b) Computes the average high temperature for the month.
(c) Computes the standard deviation of the high temperatures for the month.
(d) Plots a horizontal bar graph showing the high temperatures in chronological order. The vertical axis is the day number and the horizontal axis is the temperature.
(e) Plots a horizontal bar graph of the number of days that have the same high temperature. The vertical axis is the temperature and the horizontal axis is the number of days on which that temperature occurs.

The data for the temperatures is stored in a **textfile**.

6. Experimental data that must be plotted is often stored in arrays. Modify the procedure **Plotter** so that it plots points stored in a two-dimensional array rather than plotting a function. Test the modified procedure using the data on the student heights and weights from Problem 13 in Chapter 7.

CHAPTER 16

Dynamic Variables

The variables that you are accustomed to using are *static* variables; the value of a static variable can change as a program executes, but a fixed storage location is always allocated when the variable is declared. In this chapter you will learn to use *dynamic* variables, which can be created and destroyed as the program executes. The ability to create variables and allocate storage as needed is a powerful tool; for example, if you create a static variable such as an array, the number of storage locations set aside to store the array is fixed by the initial declaration, but if you create a dynamic variable, the number of storage locations required can be increased or decreased as the program executes.

Dynamic variables can be created using *pointer*s. A pointer performs a simple function: it points to a location where information is stored. This ability to point to a storage location is the basis for a second level of data structures in Pascal. Pointers can be used to form data structures such as linked lists, queues, trees, and stacks, which are covered in advanced programming courses.

CASE STUDY 28
A flexible manufacturing system

In a flexible manufacturing system (FMS), many different parts can be manufactured on the same set of machines. Each machine in an FMS is controlled by a microprocessor and can be programmed to perform a

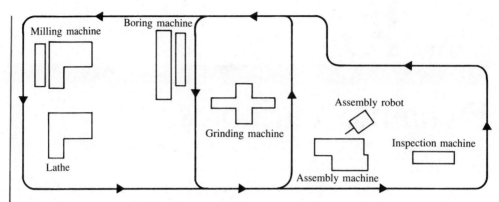

FIGURE 16.1 Floor plan for a flexible manufacturing system.

different sequence of operations for each part. In a typical FMS, the machines might be a 5-axis milling machine, a vertical turret lathe, a boring machine, a grinding machine, an assembly robot, an assembly machine, and an inspection machine. A materials transport system moves the parts from one machine to the next, as illustrated in Figure 16.1.

The routing sheet also identifies the sequence of instructions to be performed by a given machine. For example, the milling machine may be capable of drilling 50 different types of holes. Each type of hole requires a different *part program*—that is, a different sequence of instructions. The routing sheet must specify which part program is to be executed so that the correct hole is drilled in the part.

The routing sheet for a particular part may specify that the part visit the same machine more than once during its manufacture. The order in which the different parts visit the machines is not necessarily the same, and the number of machines visited by each part may differ. If one of the machines breaks down, the routing sheets must be changed so that production of the parts can continue while the machine is being repaired.

The entire system is controlled by a central computer that sends control signals to the machine microprocessors and keeps track of the status of the machines, transport system, and parts. Two important and difficult functions of the central computer are the coordination of the routing sheets for all the parts in the system and the modification of the routing sheets when a machine is out of service.

A program is being written so that the central computer can determine the best schedule for manufacturing the parts. Three procedures that this program will require are a procedure to create a routing sheet for each part from input data, a procedure to add machines to the routing sheet, and a procedure to delete machines from the routing sheet. Create a data structure to store the parts and their routing sheets and then write the three procedures.

Algorithm development

The solution to Case Study 28 will use a *linked list.* A linked list is an ordered list in which each element gives the necessary information to find the next element in the list. In Pascal, linked lists are usually implemented using the pointer data structure that is covered in this chapter. In this section of the chapter, we will develop an algorithm that solves the case study using a linked list, but that does not use the pointer data structure. At the end of the chapter, the algorithm will be modified to use pointers.

A linked list can be implemented in a number of ways; the simplest is through the use of two parallel arrays. The routing sheet could be made up of two arrays, one of which stores the name of the machine and the other the position of the machine on the routing list. For example, if the identifiers for the arrays are **Machine** and **Order,** the routing sheet for a part that visits Machine A, then Machine B, then Machine C, and finally Machine D can be visualized as in Figure 16.2.

Suppose the routing sheet is changed and the part must visit Machine E between Machines C and D. Insertion of a new machine into the list requires adding the new machine to the **Machine** array and rearranging the elements in the **Order** array so that the machines are visited in the proper order, as shown in Figure 16.3. If there were more machines in the list after Machine D, the elements of the **Order** array corresponding to these machines would need to be changed.

If the routing sheet is changed again so that the part does not visit Machine B, the **Order** of Machine B can be changed to an arbitrary number, such as 0, to indicate that the machine has been removed from the list. Again, the **Order** of the machines that follow Machine B must be modified so that they appear in the correct locations in the list.

The basic concept used in the linked lists for the routing sheet is that one variable can be used to indicate the position of another variable. For example, the 2 stored in the third element of the **Order** array tells you that the third element in the **Machine** array is the second machine in the routing sheet. This same basic concept will be used in the development of the pointer data structure.

Order	Machine
1	Machine A
2	Machine B
3	Machine C
4	Machine D

FIGURE 16.2
A linked list stored in two parallel arrays.

Order	Machine
1	Machine A
2	Machine B
3	Machine C
5	Machine D
4	Machine E

FIGURE 16.3
Insertion of a new machine into the linked list.

Order	Machine
1	Machine A
0	Machine B
2	Machine C
4	Machine D
3	Machine E

FIGURE 16.4
Elimination of Machine B from routing sheet.

16.1 Pointers

The **Order** array in the previous example serves the function of a pointer. The **Order** array indicates which element of the **Machine** array contains the next piece of information. A pointer stores the value of an address in computer memory where information is to be found. It is like a sign that points to a storage location. For the case study, a pointer can be used to indicate which storage location in computer memory stores the name of the next machine to be visited. Figure 16.5 illustrates a pointer to Machine D.

Pointer-type variables can be created in Pascal even though there is no data type called pointer. Instead, you declare a pointer variable by declaring the data type of the information to which it will point. The syntax for creating pointers is covered in the next section.

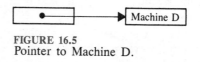

FIGURE 16.5
Pointer to Machine D.

16.2 Dynamic variables

When static variables are used, the memory space that your program or subprogram will use is allocated when the declaration statements are executed. If you declare a variable at the beginning of a program, a storage location is allocated even if you do not subsequently use the variable. In some languages, such as FORTRAN, storage locations are allocated for variables declared in a subprogram when the main program begins execution and remain allocated until the program terminates. In Pascal, however, storage locations for static variables declared within a subprogram are allocated each time the subprogram begins execution and are deallocated when the subprogram terminates. This process is invisible except for the fact that a value assigned to a variable during one execution of a subprogram is not accessible in the next execution. The advantage of this scheme is that, at any given time, the computer's stack has to be only large enough to contain the variables declared in the main program and those subprograms that are currently executing. Variables declared in Pascal subprograms are referred to as static variables even though they may be allocated and deallocated many times during program execution.

For large programs, static allocation can be wasteful, since declaration of all variables that might be needed ties up substantial memory space. When pointers are used, variables can be allocated dynamically in storage under the control of the programmer. With dynamic allocation, storage can be allocated

16.2 Dynamic variables

for a variable when it is needed rather than at the time of execution. When the information stored in a variable is no longer needed, it is possible to release the storage allocation and free it for later use.

When you create a dynamic variable in Pascal, you do not allocate a memory location in which to store the dynamic variable as you would for a static variable. Instead, you allocate a memory location in which to store a pointer to a memory location in which the dynamic variable will be stored.

For example, for the routing sheet used above, the information about the machine names could be stored in a variable with the following data type:

```
const MaxLetter = 9;

type  CharString = packed array [1..MaxLetter] of char;
```

The pointer variable will point to this data type. The syntax for creating a pointer type is

──▶ **type** ──▶ *type-identifier* ──▶ = ──▶ ↑ ──▶ *type-identifier* ──▶

Like the buffer variable, a pointer is indicated using one of the following symbols: ↑, @, or ^, depending on the system you are using.

Given the syntax diagram above, a pointer type can be defined as follows:

```
const MaxLetter = 9;

type  CharString = packed array [1..MaxLetter] of char;
      MachinePointerType = ↑CharString;
```

Then a variable of data type **MachinePointerType** can be declared to point to a machine name:

```
var   NextMachinePointer : MachinePointerType;
```

This variable declaration creates pointer **NextMachinePointer** that can point to a location of type **CharString**. However, the declaration creates only the pointer, which can be visualized as

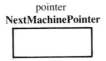

pointer
NextMachinePointer

Like a static variable declaration, the pointer declaration **NextMachinePointer**

Chapter 16 □ Dynamic Variables

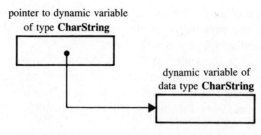

FIGURE 16.6
Visualization of a pointer and its dynamic variable.

assigns a specific location in memory for the pointer. Also, the value of the pointer, just like that of any other variable, is initially undefined.

The declaration of the pointer **NextMachinePointer** does *not* allocate a location for the dynamic variable. A machine name cannot be stored in the dynamic variable pointed to by **NextMachinePointer** until the standard procedure **new** has been executed with **NextMachinePointer** as its argument:

```
new (NextMachinePointer)
```

This statement creates a storage location for a dynamic variable of type **CharString** and assigns the address of this memory location as the value of the pointer **NextMachinePointer.** The pointer and its associated dynamic variable now can be visualized as shown in Figure 16.6. Thus, the standard procedure **new** creates a storage location for the dynamic variable **NextMachinePointer**↑ and assigns the pointer **NextMachinePointer** the value of the address of this storage location. The dynamic variable can be accessed only as the variable pointed to by the pointer **NextMachinePointer.** The name of the dynamic variable, **NextMachinePointer**↑, reflects this fact. The following statements allocate a storage location for the dynamic variable, assign the storage location's value to the pointer, and then assign the name of the first machine to the dynamic variable.

```
var   Machine1 : CharString;
      NextMachinePointer : MachinePointerType;

begin
  Machine1 := 'Machine A';

  new (NextMachinePointer);
  NextMachinePointer↑ := Machine1
```

These statements assign the value stored in static variable **Machine1** to the address that **NextMachinePointer** points to. Notice that **Machine1** has a data type **CharString** and that **NextMachinePointer** is a pointer to a variable of data type **CharString**. After the statements have been executed, the pointer variable and the dynamic variable can be visualized as shown in Figure 16.7.

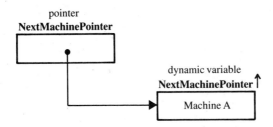

FIGURE 16.7
NextMachinePointer points to dynamic variable
NextMachinePointer ↑ .

Again, you must make the distinction between the address that the pointer references and the value that is stored at that address. In the example above, the address for the variable of type **CharString** is stored in **NextMachinePointer,** and the value of type **CharString** stored at that location is **NextMachinePointer**↑ . That is,

NextMachinePointer stores the address where a variable of type **CharString** is stored

NextMachinePointer↑ stores a value of data type **CharString** at address **NextMachinePointer**

Making the distinction between the pointer and the dynamic variable is particularly important in assignment statements. Suppose two pointers, **FirstMachinePointer** and **LastMachinePointer,** have been declared and their associated dynamic variables have been created:

```
var   FirstMachinePointer, LastMachinePointer : MachinePointerType;

begin
  new (FirstMachinePointer);
  new (LastMachinePointer)
```

In the following statements, first the character string 'Machine Z' is assigned to the dynamic variable **FirstMachinePointer**↑ and then is assigned to the dynamic variable **LastMachinePointer**↑ :

```
            FirstMachinePointer↑ := 'Machine Z';
            LastMachinePointer↑ := 'Machine Z'
```

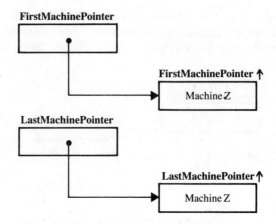

FIGURE 16.8
Two dynamic variables of the same value.

After these statements have been executed, the pointer and dynamic variables can be visualized as shown in Figure 16.8.

Continuing with the same example, suppose the following statements were executed next:

```
FirstMachinePointer↑ := 'Machine C';
LastMachinePointer := FirstMachinePointer
```

At first glance, these statements might appear to assign the value 'Machine C' to **LastMachinePointer**↑. However, closer examination shows that this is not

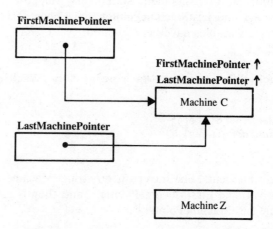

FIGURE 16.9
Two pointers to the same dynamic variable.

what happens. In fact, what happens is that the second statement assigns the address stored in the *pointer* **LastMachinePointer** to the address stored in the *pointer* variable **FirstMachinePointer,** so *both pointer variables now point to the same dynamic variable.* See Figure 16.9. The data that was stored in the dynamic variable **LastMachinePointer**↑ is no longer accessible, because no pointer points to it.

DEBUGGING HINT

Be sure to distinguish between the pointer and the dynamic variable. If you use a convention such as giving your pointers identifiers that include the word *pointer,* you will know that an identifier such as **NewPointer** is the pointer and **NewPointer**↑ is the dynamic variable.

The data structure created so far cannot store the routing sheet because the pointer points only to a single name. There is no way to point to the next machine in the list. After a discussion of the rest of the standard procedures and constants used for pointers, the data structure for a list will be covered.

The constant nil

When the standard procedure **new** is executed, the compiler assigns a value for the address of the object that is pointed to by the pointer. Assigning addresses to pointers is the prerogative of the compiler. Attempting to assign a numerical value to a pointer will result in an error:

```
          NextMachinePointer := 55
***error***                    $129

error messages :
***************

129: type conflict of operands.
```

There is one value, however, that the user can assign to a pointer, and that is the standard constant **nil**:

```
          NextMachinePointer := nil
```

With the data structure we have been discussing, the following fragment sets the values for the first and last machines in the line:

```
        var  FirstMachinePointer, LastMachinePointer : MachinePointerType;

    begin
      new (FirstMachinePointer);
      FirstMachinePointer↑ := 'Machine A';

      new (LastMachinePointer);
      LastMachinePointer := nil
```

After these statements have been executed, the computer memory can be visualized as shown in Figure 16.10. Notice that there is no dynamic variable associated with **LastMachinePointer.** The procedure **new** was not executed to create **LastMachinePointer**↑, a storage location of type **CharString,** so the dynamic variable does not exist.

In the example above, the pointer variable points to a character string. A pointer can also point to a more complicated data structure such as a record. For example, in the data structure

```
type  RouteRecord = record
         MachineName : CharString;        (* name of the machine. *)
         PartProgramID : integer          (* ID number for the program *)
                                          (* to be executed for this *)
                                          (* part on this machine. *)
      end;  (* RouteRecord *)

      MachinePointerType = ↑RouteRecord;  (* pointer-type to an entry in *)
                                          (* the routing sheet. *)

var   NextStopPointer : MachinePointerType;  (* pointer to the next entry *)
                                             (* in the routing sheet. *)
```

the pointer **NextStopPointer** points to a dynamic variable of the data type

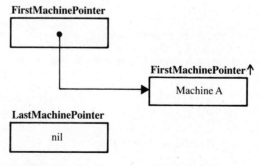

FIGURE 16.10
LastMachinePointer assigned the value **nil**.

16.2 Dynamic variables

RouteRecord. The pointer points to the entire record, not just a single variable. The statement

```
new (NextStopPointer)
```

will create a place in memory with the structure of **RouteRecord**. Values can now be assigned to the dynamic variable with the data type of **RouteRecord**. The fields of the record are accessed using the standard record syntax, but the dynamic record variable itself must be accessed through the pointer, by coding, for example,

```
writeln (' The name of the next machine that the part visits is ',
         NextStopPointer↑.MachineName)
```

or

```
NextStopPointer↑.PartProgramID := 109
```

Pointer addresses

The value of a pointer—that is, the address where the dynamic variable is stored—cannot be read or printed. The statement

```
            writeln (' The address of the first machine is ',
                    FirstMachinePointer)
***error***                                $116

    error messages :
    ****************

    116: error in type of standard procedure parameter.
```

will result in an error message. A pointer, however, can be compared to another pointer of the same data type or to the constant **nil** with the relational operators equal and not equal, as in

```
    if LastMachinePointer = nil then
      writeln (' LastMachinePointer does not point to any address');

    if LastMachinePointer <> FirstMachinePointer then
      writeln (' LastMachinePointer and FirstMachinePointer do not point',
               ' to the same address')
=>
LastMachinePointer does not point to any address
LastMachinePointer and FirstMachinePointer do not point to the same address
```

The relational operators >, >=, <, and <= cannot be used with pointers because it makes no sense to ask whether one address is greater than or less than another.

Dispose

There is one more standard procedure that can be used with pointers: **dispose**. The task performed by **dispose** is the opposite of that performed by procedure **new**. Execution of **dispose** releases the storage location pointed to by a pointer. The statement

```
dispose (LastMachinePointer)
```

frees the storage locations pointed to by **LastMachinePointer** so that the locations can be used again if required by the program. After **dispose** is executed, the specified pointer plus any other pointers to the same dynamic variable are undefined. Executing the **dispose** statement in effect purges the dynamic variable associated with the pointer.

Most computers have a large enough storage capacity that keeping variables no longer being used does not create problems. If you are working on a computer with an extremely limited memory, use of **dispose** may be necessary. You should be aware, however, that **dispose** should be used with great care because entire data structures can be lost with a single **dispose** statement.

16.3 Linked lists

Having a pointer that simply points to a dynamic variable is of limited use; however, pointers can be combined with other variables to create more complicated data structures. Imagine that you are on a treasure hunt, and you must find clues at each of several different locations. First, you need to know where to begin. You then need to know the first place you will find a clue, and once you have found the first clue, you need to know where the next one is. A sign that points to a place where you will find a clue will not be useful if you cannot find the sign and once you have found the first sign, the information that you find at the location where it sends you must include the next place you are to go.

By including pointers inside record variables, one element of the record can point to the next record in the list, thus creating a linked list. For example, you can make a linked list for the routing sheet of a part. Included in the information on each machine is a pointer to the next machine in the list. In this construction, both the record and the pointer are dynamic variables.

16.3 Linked lists

```
type  MachineLink = ↑RouteSheetRec;      (* pointer-type to a routing sheet *)
                                         (* entry. *)

      RouteSheetRec = record
         MachineName : CharString;       (* name of the machine. *)
         PartProgramID : integer;        (* ID number for the program to be *)
                                         (* executed for this part on this *)
                                         (* machine. *)
         NextPointer : MachineLink;      (* pointer to the next entry in *)
                                         (* the routing sheet. *)
      end; (* RouteSheetRec *)

var   CurrentPointer : MachineLink;      (* pointer to the current entry. *)
                                         (* in the routing sheet. *)
```

Note that the definition of a pointer type can precede the definition of the data type to which it points. In the example, **MachineLink** is defined in terms of **RouteSheetRec** prior to the definition of **RouteSheetRec.** This sole exception to the Pascal rule that an identifier must be defined before it can be used allows the record to include a pointer as one of its fields.

In this structure, the storage locations for the following variables contain the following data:

CurrentPointer	address where a variable of type **RouteSheetRec** is stored
CurrentPointer↑.MachineName	name of the machine stored in the record
CurrentPointer↑.PartProgramID	ID of the part program stored in the record
CurrentPointer↑.NextPointer	address where the next record is stored
CurrentPointer↑.NextPointer↑	record stored at address **CurrentPointer↑.NextPointer**
CurrentPointer↑.NextPointer↑.MachineName	name of the machine stored in the record stored at address **CurrentPointer↑.NextPointer**

The data structure created above can be visualized as shown in Figure 16.11.

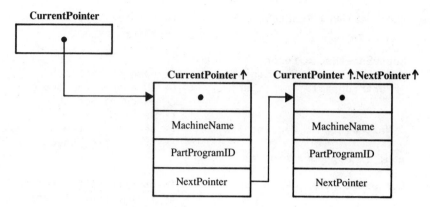

FIGURE 16.11
A linked list created using pointers.

Creating a linked list

So far, the linked list to store the routing sheet has no values assigned to the fields of the records. If the data for the fields is to be read interactively, an algorithm for reading the data from the terminal and linking each machine to the next machine can be created as shown in Algorithm 16.1

ALGORITHM 16.1   Creating an element in a linked list

begin algorithm

1. Declare a pointer **CurrentPointer** to point to an entry in the linked list
2. Create storage locations for the dynamic record variable that store the data on the entry in the routing sheet; put the address of this dynamic variable in the pointer **CurrentPointer**
    new (CurrentPointer)
3. Request the name of the machine that the part visits and request the ID number of the part program executed by the machine for this part
    read **CurrentPointer↑.MachineName**
    read **CurrentPointer↑.PartProgramID**
4. Create the next entry in the list to which **CurrentPointer↑.NextPointer** will point
    new (CurrentPointer↑.NextPointer)

end algorithm

The steps in Algorithm 16.1 create a single entry in the linked list. In Algorithm 16.2, Algorithm 16.1 is embedded in a conditional loop so that the entire linked list can be read. Notice that Steps 2 and 3 in Algorithm 16.2

create and save the pointer to the first entry in the list. If these steps were not taken, the list could not be accessed because the location of the first entry would not be known. If the first entry cannot be accessed, none of the list can be accessed.

ALGORITHM 16.2 Creating a linked list

begin algorithm
1. Declare a pointer **FirstPointer** to point to the first entry in the linked list
2. Create storage locations for the dynamic record variable that store the data on an entry in the routing sheet; put the address of this dynamic variable in the pointer **CurrentPointer**
    new (CurrentPointer)
3. Save the value of the first pointer
    FirstPointer ← CurrentPointer
4. Read the entries into the linked list
    repeat
        a. Request the name of the machine that the part visits and request the ID number of the part program that is executed by the machine for this part
            read **CurrentPointer↑.MachineName**
            read **CurrentPointer↑.PartProgramID**
        b. Create the entry in the list to which **CurrentPointer↑.NextPointer** will point
            new (CurrentPointer↑.NextPointer)
        c. Make the next entry the current entry
            CurrentPointer ← CurrentPointer↑.NextPointer
    until the user indicates that the last entry has been read
5. Set the last pointer to **nil**
    CurrentPointer↑.NextPointer ← nil
end algorithm

The final step, setting the last pointer to **nil,** is important because the **nil** pointer can be used to test for the end of the list.

Once the list has been created, the next task is to inspect the entries in the list. To inspect the values stored in the linked list, you can start from the first entry and work your way down the list through the pointers, as illustrated in Algorithm 16.3.

ALGORITHM 16.3 Inspecting a linked list

begin algorithm
1. Start at the beginning of the linked list
    CurrentPointer ← FirstPointer

2. Use the **NextPointer** of each element in the linked list to find the element following **CurrentPointer** until the pointer is the **nil** pointer indicating the end of the list
   while **CurrentPointer** <> **nil** do
      a. Write the name of the machine that the part visits and the ID number of the part program that is executed by the machine for the part
        write **CurrentPointer↑.MachineName**
        write **CurrentPointer↑.PartProgramID**
      b. Advance to the next entry in the list
        CurrentPointer ← **CurrentPointer↑.NextPointer**
   end while
end algorithm

Again, notice that Step 1 in Algorithm 16.3 requires that the pointer to the first dynamic variable in the list, **FirstPointer,** be defined so that the first element in the list can be accessed.

Doubly linked list

The linked list above can be accessed in only one direction. For any particular element in the list, you know the next element but you do not know the previous element. If a list is doubly linked, with pointers to both the next element and the previous element, you can go either forward or backward in the list. The following data type definition for an element in a doubly linked list can be used for the routing sheet:

```
type  MachineLink = ↑RouteSheetRec;       (* pointer to a routing sheet *)
                                          (* entry. *)

      RouteSheetRec = record
        MachineName : CharString;         (* name of the machine. *)
        PartProgramID : integer;          (* ID number for the program to be *)
                                          (* executed for this part on this *)
                                          (* machine. *)
        NextPointer : MachineLink;        (* pointer to the next entry in *)
                                          (* the routing sheet. *)
        PrevPointer : MachineLink         (* pointer to the previous entry *)
                                          (* in the routing sheet. *)
      end; (* RouteSheetRec *)

var   CurrentPointer : MachineLink;       (* pointer to the current entry in *)
                                          (* the routing sheet. *)
```

The list can be visualized as shown in Figure 16.12.

In this data structure, the field **NextPointer** stores the value of the pointer to the next entry and **PrevPointer** stores the pointer to the previous entry. The

16.3 Linked lists

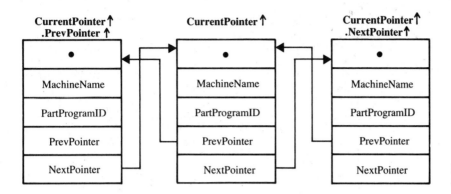

FIGURE 16.12
A doubly linked list.

procedure **CreateList** creates a doubly linked list based on Algorithm 16.2, in which the simple linked list was created. An extra step is required in creating the doubly linked list in order to keep track of the pointer to the previous record. Notice that the only parameters in the parameter list are **FirstPointer** and **LastPointer**. With the value of either of these pointers, the entire list can be accessed.

```
procedure CreateList (var FirstPointer, LastPointer : MachineLink);
(*      This procedure creates a doubly linked list for a routing sheet    *)
(*      for a part that is to be manufactured.  The data is requested      *)
(*      from the user at the terminal.                                     *)

var     CurrentEntry : MachineLink;     (* entry in the routing sheet cur- *)
                                        (* rently being read. *)
        Stop : boolean;                 (* flag that is set to true when the *)
                                        (* the user is done. *)

(*$i yesentered *)
(*$i readstring *)

begin
  Stop := false;
  new (FirstPointer);                   (* get the first pointer variable. *)
  FirstPointer↑.PrevPointer := nil;     (* there is no entry before the *)
                                        (* first entry, so set Prev to nil. *)
  CurrentEntry := FirstPointer;         (* CurrentEntry is the first entry. *)

  repeat
    writeln;
    writeln (' What is the name of the next machine that the part visits?');

    ReadString (MaxLetters, CurrentEntry↑.MachineName);
    readln;
```

```
    writeln (' Enter the number of the part program that is run on',
             ' this machine.');
    readln (CurrentEntry↑.PartProgramID);

    (* Ask if more machines are to be entered.  If yes, get a new pointer *)
    (* to the next entry, set NextPointer from the previous entry to *)
    (* point to the new entry, and make the new pointer the current *)
    (* entry.  If no more machines are to be entered, set Stop to true. *)

    writeln (' Are there more machines that this part visits?');
    if YesEntered then
      begin
        new (CurrentEntry↑.NextPointer);
        CurrentEntry↑.NextPointer↑.PrevPointer := CurrentEntry;
        CurrentEntry := CurrentEntry↑.NextPointer
      end (* if *)
    else
      Stop := true
  until Stop = true;

  (* Save the pointer to the last entry in the list, and since no entry *)
  (* follows the last entry, set NextPointer to nil. *)

  LastPointer := CurrentEntry;
  LastPointer↑.NextPointer := nil
end; (* procedure CreateList *)
```

Inserting an element into a doubly linked list

In order to insert an element into a doubly linked list, a storage location for the element must be created by executing the **new** procedure. Values can be assigned to the nonpointer fields:

```
new (NewMachinePointer);

with NewMachinePointer↑ do
  begin
    MachineName := 'Machine E';
    PartProgramID := 210
  end
```

Suppose that this element, representing Machine F, is to be inserted between Machine A and Machine D, which currently have the order shown in Figure 16.13. In order to insert the new element, **NextPointer** of Machine A is changed to point to Machine F, and **PrevPointer** of Machine D is changed to point to Machine F. Similarly, the pointers of Machine F are set to point back to Machine A and forward to Machine D. See Figure 16.14. Algorithm 16.4 states in algorithmic terms the steps for inserting an element into a doubly linked list, assuming that **InsertPointer,** the pointer to the element to be inserted, and **BeforePointer,** the pointer to the element that will precede the

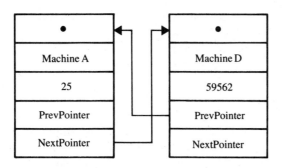

FIGURE 16.13
Two elements in a doubly linked list.

new element, are known. Again, both **InsertPointer** and **BeforePointer** point to record variables of data type **RouteSheetRec**.

ALGORITHM 16.4 Inserting an element in a doubly linked list

begin algorithm

1. Find the pointer to the element that will follow the new element in the list
   AfterPointer ← **BeforePointer↑.NextPointer**
2. Set **NextPointer** of the new element to point to the element that will come after it, and set its **PrevPointer** to point to the element before it
   InsertPointer↑.NextPointer ← **AfterPointer**
   InsertPointer↑.PrevPointer ← **BeforePointer**

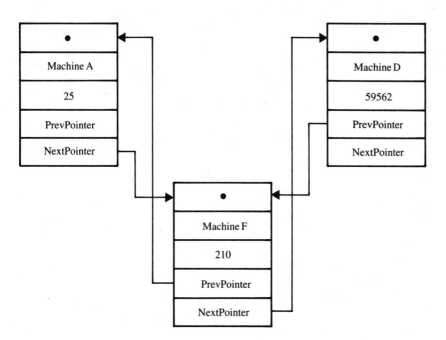

FIGURE 16.14
Insertion of a new element in a doubly linked list.

3. The element after **BeforePointer** is now **InsertPointer**↑, and the element before **AfterPointer** is now **InsertPointer**↑
   BeforePointer↑.**NextPointer** ← **InsertPointer**
   AfterPointer↑.**PrevPointer** ← **InsertPointer**
end algorithm

Algorithm 16.4 needs one refinement to account for the tail of the list. If the element is inserted after the last element, **LastPointer** must be reset to **InsertPointer.** Incorporating this refinement, the procedure **InsertAfter** inserts the element passed into the parameter **InsertPointer** after the element passed into the parameter **BeforePointer.**

```
procedure InsertAfter (var LastPointer, BeforePointer,
                                       InsertPointer: MachineLink);
(*      This procedure inserts the element InsertPointer after the     *)
(*      element BeforeElement in a doubly linked list.                  *)

var   AfterPointer : MachineLink;       (* AfterPointer is the element that *)
                                        (* that follows BeforePointer in the *)
                                        (* doubly linked list. *)

begin
  AfterPointer := BeforePointer↑.NextPointer;

  (* If the new element comes after the last element, set LastPointer to *)
  (* the new element. *)

  if BeforePointer = LastPointer then
    LastPointer := InsertPointer;

  (* Set the element that follows the new element to the element that used *)
  (* to follow BeforePointer, and set the element preceding the new ele-  *)
  (* ment to the one that used to precede BeforePointer*)

  InsertPointer↑.NextPointer := BeforePointer↑.NextPointer;
  InsertPointer↑.PrevPointer := BeforePointer;

  (* Set NextPointer and PrevPointer of the appropriate elements in the *)
  (* linked list to point to the new element. *)

  BeforePointer↑.NextPointer := InsertPointer;
  if AfterPointer <> nil then
    AfterPointer↑.PrevPointer := InsertPointer
end; (* procedure InsertAfter *)
```

Notice that procedure **InsertAfter** does not allow you to insert an element at the beginning of the list.

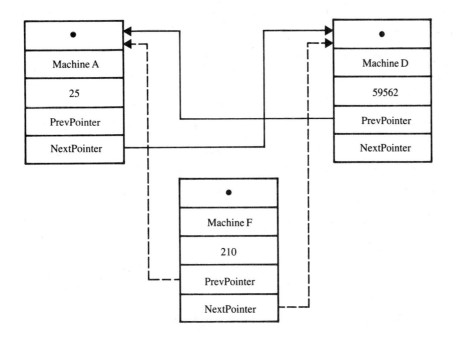

FIGURE 16.15
Machine F eliminated from doubly linked list.

Removing an element from a doubly linked list

Removing an element from a doubly linked list is straightforward. The only precaution that must be taken is to ensure that the necessary pointer values are not lost before the element is removed. When an element is removed, the pointers on either side of the removed element are redirected to point to each other. In Figure 16.15, Machine F is removed from the list. The procedure **RemoveElement** removes the element passed in the parameter **DeletedElement**. Again, the algorithm must account for the fact that the deleted element may be the head or the tail of the list.

```
procedure RemoveElement (var FirstPointer, LastPointer,
                                DeletedElement : MachineLink);
(*      This procedure removes the element, DeletedElement, from the   *)
(*      doubly linked list.  If the deleted element is the first or    *)
(*      last element in list, the value of FirstPointer or LastPointer *)
(*      must be reset.                                                 *)

begin
   (* If the deleted element is not the last element, set PrevPointer *)
   (* of the following element to point back to the element before the *)
   (* deleted element.  If the deleted element is the last element, reset *)
   (* LastPointer to the element before the deleted element. *)
```

```
      if DeletedElement <> LastPointer then
        with DeletedElement↑ do
          NextPointer↑.PrevPointer := PrevPointer
      else
        begin
          LastPointer := DeletedElement↑.PrevPointer;
          LastPointer↑.NextPointer := nil              (* nothing follows the *)
                                                       (* last element.      *)
        end;

      (* Set NextPointer of the element preceding the deleted element. *)

      if DeletedElement <> FirstPointer then
        with DeletedElement↑ do
          PrevPointer↑.NextPointer := NextPointer
      else
        begin
          FirstPointer := DeletedElement↑.NextPointer;
          FirstPointer↑.PrevPointer := nil             (* nothing precedes the *)
                                                       (* first element.       *)
        end
  end; (* procedure RemoveElement *)
```

Solution to Case Study 28

The data structure and procedures needed to create and modify the routing sheets in Case Study 28 have all been developed. Program 16.1, **TestFMS,** is a test program that creates a routing sheet from data that is read from the terminal. The program allows a user to select whether to add a machine, delete a machine, or print the current order of the routing sheet.

PROGRAM 16.1   TestFMS

```
program TestFMS (input, output);
(*     This program tests the procedures that were written to read the   *)
(*     routing sheet, to add machines to the list, and to delete         *)
(*     machines from the list.                                           *)

const MaxLetters = 9;

type   CharString = packed array [1..MaxLetters] of char;

       MachineLink = ↑RouteSheetRec;      (* pointer-type to an entry in the *)
                                          (* routing sheet. *)
```

```
            RouteSheetRec = record
              MachineName : CharString;          (* name of the machine. *)
              PartProgramID : integer;           (* ID number for the program to be *)
                                                 (* executed for this part on this *)
                                                 (* machine. *)
              NextPointer : MachineLink;         (* pointer to the next entry in *)
                                                 (* the routing sheet. *)
              PrevPointer : MachineLink          (* pointer to the previous entry *)
                                                 (* in the routing sheet. *)
            end; (* RouteSheetRec *)

   var   FirstMachine : MachineLink;             (* pointer to the first entry in *)
                                                 (* the routing sheet. *)
         LastMachine : MachineLink;              (* pointer to the last entry in the *)
                                                 (* routing sheet. *)
         Operation : char;                       (* operation selected by the user. *)

(*$i readstring *)
(*$i createlist *)
(*$i insertafter *)
(*$i removeelement *)

  procedure PrintList (FirstPointer, LastPointer : MachineLink);
  (*      This procedure prints the routing sheet in the current order,    *)
  (*      starting from the First entry and going to the Last.             *)

  var   CurrentEntry : MachineLink;     (* entry currently being printed. *)

  begin
    CurrentEntry := FirstPointer;
    writeln;
    writeln (' The current order of the list is:');
    writeln;
    writeln (' Machine Name':15, 'Part Program Number':25);

    while CurrentEntry <> nil do
      begin
        writeln (CurrentEntry↑.MachineName:15,
                 CurrentEntry↑.PartProgramID:20);
        CurrentEntry := CurrentEntry↑.NextPointer
      end
  end; (* procedure PrintList *)

  function FindPointer (FirstPointer, LastPointer : MachineLink;
                        Name : CharString) : MachineLink;
  (*      This function returns the value of the pointer-variable that     *)
  (*      points to the machine called Name.                               *)

  var   CurrentEntry : MachineLink;     (* current entry in the list. *)
        Temp : MachineLink;             (* temp stores the value of the *)
                                        (* function. *)
```

```
begin
  Temp := nil;                         (* assign the function a value in *)
                                       (* case Name is not found. *)
  CurrentEntry := FirstPointer;        (* start at the top of the list. *)

  (* The while-do statement executes until the name is found or until *)
  (* the end of the list is reached. *)

  while (Temp = nil) and (CurrentEntry <> LastPointer↑.NextPointer) do
    begin
      if CurrentEntry↑.MachineName = Name then
        Temp   := CurrentEntry;
      CurrentEntry := CurrentEntry↑.NextPointer
    end;
  FindPointer := Temp
end; (* function FindPointer *)


procedure ReadMachineName (var FirstPointer, LastPointer,
                                            Link : MachineLink);
(*     This procedure requests the user to enter a machine name. The   *)
(*     pointer to that machine name is returned in the parameter Link. *)

var  Name : CharString;     (* temporary variable for the machine name. *)

begin
  repeat
    writeln (' Enter the name of the machine');
    ReadString (MaxLetters, Name);
    readln;

    (* Use FindPointer to find the pointer to the name that was entered. *)

    Link := FindPointer (FirstPointer, LastPointer, Name);

    (* If Name is not found execute the loop again. *)

    if Link = nil then
      writeln (' That name is not in the list')
  until Link <> nil
end; (* procedure ReadMachineName *)


procedure AddMachine (var FirstPointer, LastPointer : MachineLink);
(*     This procedure adds a new machine to the list. The name, part  *)
(*     program ID, and name of the machine that this machine is to be *)
(*     inserted are requested from the user.                          *)

var  NewMachine : MachineLink;        (* machine to be added. *)
     PreviousMachine : MachineLink;   (* machine before the one to be *)
                                      (* added. *)
     Index : integer;                 (* for control variable. *)
```

```
    begin
      new (NewMachine);                     (* get a pointer for the new *)
                                            (* machine. *)
      with NewMachine↑ do
        begin
          writeln (' What is the name of the new machine?');
          ReadString (MaxLetters, MachineName);
          readln;

          writeln (' Enter the program number for the part on this machine.');
          readln (PartProgramID)
        end;

      writeln (' After which machine should the new machine be added?');
      ReadMachineName (FirstPointer, LastPointer, PreviousMachine);

      InsertAfter (LastPointer, PreviousMachine, NewMachine)
    end; (* procedure AddNewMachine *)


procedure DeleteMachine (var FirstPointer, LastPointer : MachineLink);
(*      This procedure deletes a machine from the doubly linked list,    *)
(*      using the procedure RemoveElement.                               *)

var     Link : MachineLink;     (* entry to be removed. *)

begin
  writeln (' What machine do you want to delete from the list?');
  ReadMachineName (FirstPointer, LastPointer, Link);

  RemoveElement (FirstPointer, LastPointer, Link)
end; (* procedure DeleteMachine *)


procedure ExecuteMenu (FirstPointer, LastPointer : MachineLink;
                       var Answer : char);
(*      Procedure ExecuteMenu presents the menu of options to the user,  *)
(*      reads the user's response, and executes the appropriate          *)
(*      procedure.                                                       *)

begin
  repeat
    writeln;
    writeln (' What would you like to do next?');
    writeln;
    writeln ('    L    list the machines in order');
    writeln ('    A    add a machine to the list');
    writeln ('    D    delete a machine from the list');
    writeln ('    S    stop the program');
    writeln;
    writeln (' Enter a single character');
    readln (Answer);
  until Answer in ['l', 'L', 'a', 'A', 'd', 'D', 's', 'S'];
```

```
      case Answer of
        'L', 'l' : PrintList (FirstPointer, LastPointer);
        'A', 'a' : AddMachine (FirstPointer, LastPointer);
        'D', 'd' : DeleteMachine (FirstPointer, LastPointer);
        'S', 's' :
      end (* case *)
  end; (* procedure ExecuteMenu *)


(*$i writeintro *)
(*  procedure WriteIntro;                                                  *)
(*       This procedure writes an introductory message to the user.        *)



(*-------------------------------------------------------------------------*)
begin (* main program *)
  WriteIntro;

  CreateList (FirstMachine, LastMachine);

  repeat
    ExecuteMenu (FirstMachine, LastMachine, Operation)
  until Operation in ['s', 'S']
end. (* program TestFMS *)

=>
This is a test program that creates a routing sheet for a
part in a flexible manufacturing system.

The name of each machine that the part visits and the ID
of the part program that is executed are entered from the terminal.
Machines can then be added, deleted, or exchanged in the routing sheet


What is the name of the next machine that the part visits?
▫Machine 1
Enter the number of the part program that is run on this machine.
▫23
Are there more machines that this part vists?
Enter yes or no (y/n)
▫y

What is the name of the next machine that the part visits?
▫Machine 2
Enter the number of the part program that is run on this machine.
▫786
Are there more machines that this part visits?
Enter yes or no (y/n)
▫y
```

```
What is the name of the next machine that this part visits?
▫Machine 3
Enter the number of the part program that is run on this machine.
▫8743
Are there more machines that this part visits?
Enter yes or no (y/n)
▫y

What is the name of the next machine that the part visits
▫Machine 4
Enter the number of the part program that is run on this machine
▫78
Are there more machines that this part visits?
Enter yes or no (y/n)
▫n

What would you like to do next?

    L    list the machines in order
    A    add a machine to the list
    D    delete a machine from the list
    S    stop the program

Enter a single character
▫l

The current order of the list is:

    Machine Name         Part Program Number

      Machine 1                  23
      Machine 2                 786
      Machine 3                8743
      Machine 4                  78

What would you like to do next?

    L    list the machines in order
    A    add a machine to the list
    D    delete a machine from the list
    S    stop the program

Enter a single character
▫a

What is the name of the new machine?
▫Machine 5
Enter the program number for the part on this machine?
▫34
After which machine should the new machine be added?
Enter the name of the machine
▫Machine 4
```

```
What would you like to do next?

    L    list the machines in order
    A    add a machine to the list
    D    delete a machine from the list
    S    stop the program

Enter a single character
□s
```

Summary

The storage locations for dynamic variables can be created and destroyed as a program executes. Creating a dynamic variable requires declaring a *pointer variable*. The pointer variable is used to store the address where the data for the dynamic variable is stored. Dynamic variables are powerful because they make possible a system of storing the address where information is stored, rather than storing the information directly. Storage is not allocated to a dynamic variable at the beginning of a program. Instead, when a new dynamic variable is required during program execution, the standard procedure **new** is executed to create a location in memory in which to store the new information. If the information stored in a dynamic variable is no longer needed, the standard procedure **dispose** can be used to dispose of a dynamic variable and free the storage locations for other information.

Dynamic variables can be used to create data structures such as linked lists in which the data is ordered and each element in the linked lists points to the elements that are adjacent to it.

Exercises

PASCAL

1. Given the pointer definitions and variable declarations

```
type   IntegerPointer = ↑integer;
       CharPointer = ↑char;

var    Point1 Point2 : IntegerPointer;
       Point3, Point4 : CharPointer;
```

what is the output from the following fragments?

(a)
```
new (Point1);
new (Point2);
Point1↑ := 6;
Point2↑ := Point1↑ + 20;
writeln (' Point1↑ equals', Point1↑:3, ' Point2↑ equals', Point2↑:3)
```

(b) ```
new (Point2);
Point2↑ := 2;
Point2↑ := sqr (Point2↑);
new (Point1);
Point1↑ := Point2↑ mod 3;
writeln (' Point1↑ equals', Point1↑:3, ' Point2↑ equals', Point2↑:3);
```

(c) ```
new (Point3);
new (Point4);
Point3↑ := 'Z';
Point4↑ := pred (Point3↑);
writeln (' Point3↑ equals', Point3↑:3, ' Point4↑ equals',
         Point4↑:3)
```

2. Given the same declarations as in Exercise 1, describe and correct the syntax error(s) in each of the following fragments:

(a) ```
new (Point1);
new (Point2);
new (Point3);

readln (Point3↑);
Point2↑ := Point1;
writeln (' Point3↑ equals', Point3↑:3,' Point4↑ equals',
         Point4↑:3)
```

(b) ```
new (Point1);

Point1 := Point2;
Point1 := 3.5 * Point1↑
```

(c) ```
new (Point1);
new (Point4);

Point1↑ := 48;
Point4↑ := chr (Point1);
Point1 := Point4
```

# Problems

## PASCAL

1. Rewrite Program 16.1, **TestFMS**, by implementing the linked list using parallel arrays. Write subprograms to insert and delete elements in the linked list.

2. Modify the procedure **AddMachine** in Program 16.1 so that machines can be added at the beginning of the routing sheet.

3. The data structure called a *queue* is based on the common phenomenon of waiting lines. A queue is formed, for example, when people arrive at a theater
```

to buy tickets. A person can enter a queue only at the end and must wait until the people standing in front have gotten tickets before buying a ticket. A queue is a first-in/first-out (FIFO) system.

The data structure of a queue can be created using a linked list in which each person points to the person behind. A queue, like any other waiting line, must have a beginning and an end, or, as they are commonly called, a *head* and a *tail*. A queue can be visualized as

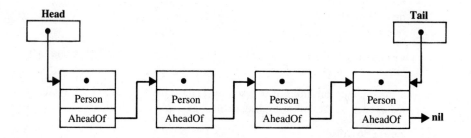

The data structure above looks like a linked list, but in a queue there are restrictions on the way information is added or removed from the list. With a queue, new information can be added only to the tail of the list and can be removed only from the head of the list.

Write a program that creates a queue of people waiting for tickets. Create a simple data base in which you are given the number of tickets that each person wants to buy. Once the queue has been created, start selling tickets at a rate of 3 per minute. Compute and report the length of time each person waits in line before reaching the head of the queue.

4. A *stack* is a last-in/first-out (LIFO) system. Stacks were discussed in Chapter 15 to explain the way in which the computer stores unresolved variables when recursion is used. A stack can be modeled using a linked list in which the pointer that allows entry to the list points to the value that has most recently been added to the list. Many inventory systems for nonperishable items are LIFO systems in which the last stock received is the first stock shipped.

A warehouse stores the following information on its stock: the date logged in, the serial number, and the aisle/shelf location. Write a set of procedures that will allow the warehouse to log in new stock, ship out stock, and compute the average length of time that an item is in stock. (All of the items in stock are identical.) Create a sample data base of items that come into the warehouse and demands for items leaving the warehouse. Write a test program for your procedures that logs in the items, ships them out, and computes the average time that an item is in the warehouse.

5. The warehouse in Problem 4 has considered using a FIFO inventory system. Write another set of procedures that processes the inventory as a queue instead of a stack. For the same sample data base, compute the average time that an item is in stock for the LIFO and the FIFO systems.

6. In a computer graphics program that stores three-dimensional data, it is necessary to be able to tell where the edges and surfaces of an object are. For example, the three objects below all have the same vertices, but one is a wire frame, one is an open box, and one is a closed cube:

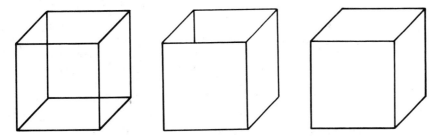

One method for storing the necessary data is to create a *tree*, with each level of the tree representing a different level in the geometry of the object. The lowest level of the tree is the vertices, the second level is the edges, and the highest level is the surfaces. Each element on a given level points to the element on the next level with which it is associated. Suppose the elements of the closed cube are labeled as follows:

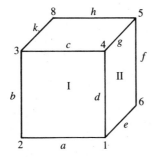

Vertices 1 and 2 are elements of edge *a*, and vertices 2 and 3 are elements of edge *b*, so the edges are represented as follows:

Edges *a*, *b*, *c*, and *d* are elements of suface I, and edges *d*, *e*, *f*, and *g* are elements of surface II, so the surfaces are represented as follows:

In the representation of the cube that is open on the top, the edges *c*, *g*, *h*, and *k* would not point to the surface level.

Create a data structure for storing three-dimensional objects. Write a test program that creates a pyramid, a cube with one open face, and a wire-frame octahedron.

7. Suppose you are creating a computer program to play the game Twenty Questions. In this game, one player thinks of an object and the other player is allowed to ask up to twenty questions that have yes or no answers. Based on the answers, the second player must decide what object the first player is thinking of. In the computerized version of Twenty Questions, the program user will think of the object, the computer will ask the questions, the user will answer yes or no, and the program will decide what the object is.

Create a data structure using pointers that stores appropriate questions given the sequence of yes or no responses from the user. The data structure should also include the name of the object when the responses are sufficient to identify it. For example, the computer should identify the object as a platypus if the user answers yes to the following questions: Is it a mammal? Does it lay eggs? Does it have webbed feet? (Hint: In a *binary tree,* each node has at most two branches. Each branch leads to a new node, which also has at most two branches. Any node with no branches is called a *leaf.*)

To create the program, you must limit the questions to a particular area—for example, mammals, presidents, or computer manufacturers. Also, you should start by creating a program to play Five Questions rather than Twenty Questions. (How many questions must be stored in the computer if all twenty questions are required to identify each object?)

Appendix A
Reserved and Standard Elements and Character Codes

A.1 Reserved and standard elements of Pascal syntax

Reserved elements of the Pascal syntax must be used as prescribed by the syntax rules and cannot be redefined. A compile-time error occurs if a reserved element is used other than as prescribed. *Standard* syntax elements are predefined by the Pascal syntax. The standard definition is assumed unless the identifier is redefined in a program. Often redefinition of standard identifiers is inadvertent.

Reserved words

and	end	mod	repeat
array	file	nil	set
begin	for	not	then
case	forward	of	to
const	function	or	type
div	goto	packed	until
do	if	procedure	var
downto	in	program	while
else	label	record	with

In some versions of Pascal, **otherwise, extern,** and **fortran** are also reserved words.

Reserved punctuation

:=	Assignment operator	The expression(s) to the right of the operator are evaluated and the result is put in the variable on the left of the assignment operator

Appendix A

=, <, >, <>, <=, >=	Relational operators	Compare variables and constants of the same data type, resulting in a boolean value of true or false
−, +, *, /	Arithmetic operators	For addition, subtraction, multiplication, and division
[ ] or (. .)	Square brackets	Enclose the indices of an array or elements of a set
,	Comma	Separates variables and expressions
.	Period	Signifies the end of a program or separates the name of a record from its field
..	Two periods	Define subranges
:	Colon	Used in declarative statements, case statements, and in output formatting
;	Semicolon	Separates executable statements
^ or ↑ or @	Carat	Used in pointers and buffer variables
( )	Parentheses	Clarify or change the precedence order and surround arguments of functions and procedures; also surround the names of files that a program will use
'	Single quote	Delineates character expressions in write or assignment statements
{ } or (* *)	Curly brackets	Surround comment statements

Standard constants

false   true   maxint

Standard data types

real   integer   char   boolean   text

In some versions of Pascal, **string** is also a standard data type.

Standard files

input   output

Standard trigonometric and algebraic functions

	data type of argument	*data type of result*	*description of value returned*
abs(X)	real or integer	same as **X**	absolute value of **X**

arctan(X)	real or integer	real	the angle whose tangent is **X**
cos(X)	real or integer	real	cosine of angle **X** (**X** must be in radians)
exp(X)	real or integer	real	e^X (exponential function)
ln(X)	real or integer	real	natural logarithm of **X**
odd(X)	integer	boolean	true if **X** is odd, otherwise false
round(X)	real	integer	**X** rounded to the nearest integer
sin(X)	real or integer	real	sine of angle **X** (**X** must be in radians)
sqr(X)	real or integer	same as **X**	square of **X**
sqrt(X)	real or integer	real	square root of **X**
trunc(X)	real	integer	**X** truncated to its smallest integer part

Standard functions used with ordinal data types

	data type of argument	data type of result	description of value returned
chr(X)	integer	char	character whose ordinal value is **X**, that is, the **X**th character in the set
ord(X)	any ordinal type	integer	position or ordinal value of **X** within its data type
pred(X)	any ordinal type	same as **X**	predecessor of **X** within its data type
succ(X)	any ordinal type	same as **X**	successor of **X** within its data type

Standard functions used with files

	data type of argument	data type of result	description of value returned
eof(f)	file	boolean	true if the next character in the file **f** is the end-of-file marker
eoln(f)	file	boolean	true if the next character in the file **f** is the end-of-line marker

Standard procedures used with files

read (f, variable list)	Reads data from file **f**, assigning values to variables in the listed order, until each variable has a value. If **f** is not specified, the default is the standard input file.
readln(f, variable list)	Same as **read**, except that **f** must be a textfile and after the variables have been assigned values, the read pointer is positioned at next line in **f**.

Appendix A

write(f, argument list)	Writes data in argument list to file **f**. If **f** is unspecified, the default is the standard output file.
writeln(f, argument list)	Same as **write**, except that **f** must be a textfile and after the argument list has been written, the write pointer is positioned at beginning of next line in **f**.
get(f)	Advances the buffer variable **f↑** and assigns it the value of the next component in the file, **f**.
page(f)	An implementation-dependent procedure that usually advances the write pointer to a new page in file **f**. Two pages in a row have the same effect as a single page.
put(f)	Appends the contents of the buffer variable **f↑** to the end of the file **f**.
reset(f)	Places the read pointer at beginning of file **f**.
rewrite(f)	Places the write pointer at beginning of file **f**. Previous contents of file **f** are lost.

Standard procedures used with arrays

pack(A, I, P)	Copies array **A**, beginning with the element whose subscript is **I**, into packed array **P**.
unpack(P, A, I)	Copies packed array **P**, beginning with its first element, into array **A** with the first element in **P** placed a subscript position **I** in array **A**.

Standard procedures used with pointers

new(P)	Allocates a storage location for pointer **P**.
dispose(P)	Releases the storage location of pointer **P**.

A.2 Operators and their order of precedence

Operators used in Pascal

( )	Parentheses can be used to override the standard operator precedence rules.
− and not	The negative sign when placed before a number and the logical **not** are called unary negators. The minus sign performs a different role when it is used to signify the operation of subtraction.
*	The star is the operator for multiplication of real or integer values and for intersection of sets.
/	The slash is used to perform division of real values.
div	**Div** can be used only for integer division. There must be one space before and one space after the word **div**.
mod	**Mod** determines the modulus or remainder for division of integers.

and	**And** is a boolean operator. **A and B** is evaluated as true if both boolean expressions, **A** and **B**, are true. It is false otherwise.
+	The plus sign is the operator for addition of real or integer values. It is also the operator for union of sets.
−	The minus sign performs subtraction and determines the difference of real or integer values and the difference of sets.
or	**Or** is a boolean operator. **A or B** is evaluated as true if either or both of the boolean expressions, **A** and **B**, are true. It is false only if both **A** and **B** are false.
in	**In** is a boolean operator that determines whether an element is in a set.

The relational operators compare two expressions that are of compatible data types.

=	equal
<>	not equal
<	less than
<=	less than or equal to
>	greater than
>=	greater than or equal to

Precedence order

highest	( )
.	+ − **not** (unary + and −)
.	*** / mod div and**
.	+ − **or**
lowest	= <> < <= > >= **in**

If two or more operators of the same order of precedence appear in the same expression, they are evaluated from left to right. For example,

$$X := A * B / C * D \text{ is evaluated as } X := ((A * B) / C) * D$$

If any other order is required, parentheses must be used.

A.3 ASCII and EBCDIC character codes

ASCII (American Standard Code for Information Interchange)

Table A.1 Printable character codes for ASCII

Left digit	\multicolumn{10}{c}{Right Digit}

Left digit	0	1	2	3	4	5	6	7	8	9
3			blank	!	"	#	$	%	&	'
4	(	)	*	+	,	−	.	/	0	1
5	2	3	4	5	6	7	8	9	:	;
6	<	=	>	?	@	A	B	C	D	E
7	F	G	H	I	J	K	L	M	N	O
8	P	Q	R	S	T	U	V	W	X	Y
9	Z	[	\	]	^	_	`	a	b	c
10	d	e	f	g	h	i	j	k	l	m
11	n	o	p	q	r	s	t	u	v	w
12	x	y	z	{			}	~	DEL		

Code 127 in ASCII is an unprintable control character that deletes or rubs out a character. Codes 00 to 031 are also control characters that are not printable. The codes for control characters 00 to 31 are given in Table A.2.

Table A.2 Nonprintable control codes for ASCII

Left digit	0	1	2	3	4	5	6	7	8	9
0	^@	^A	^B	^C	^D	^E	^F	^G	^H	^I
1	^J	^K	^L	^M	^N	^O	^P	^Q	^R	^S
2	^T	^U	^V	^W	^X	^Y	^Z	^[	^\	^]
3	^^	^_								

The control codes you are most likely to use are:

    ^G   Ring bell
    ^H   Backspace

	Tab
^I	Tab
^J	Line feed
^K	Vertical tab
^L	Form feed
^M	Carriage return
^[	Escape

EBCDIC (Extended Binary Coded Decimal Interchange Code)

Table A.3 Printable character codes for EBCDIC

Left digit	\multicolumn{10}{c}{Right Digit}

Left digit	0	1	2	3	4	5	6	7	8	9
6					blank					
7					¢	.	<	(	+	|
8	&									
9	!	$	*	)	;	¬	-	/		
10							^	,	%	_
11	>	?								
12			:	#	@	'	=	"		a
13	b	c	d	e	f	g	h	i		
14						j	k	l	m	n
15	o	p	q	r						
16			s	t	u	v	w	x	y	z
17								\	{	}
18	[	]								
19				A	B	C	D	E	F	G
20	H	I								J
21	K	L	M	N	O	P	Q	R		
22							S	T	U	V
23	W	X	Y	Z						
24	0	1	2	3	4	5	6	7	8	9

Codes 00 to 63 and 250 to 255 represent nonprintable control characters.

Appendix B
Syntax Diagrams

signed integer

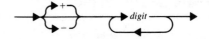

signed real

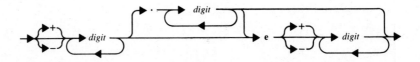

identifier

program

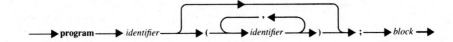

Syntax Diagrams A9

block

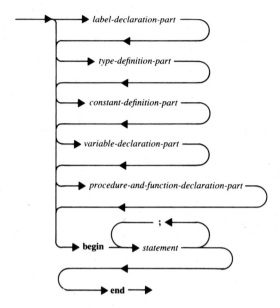

label-declaration part

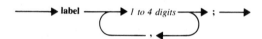

type-definition part

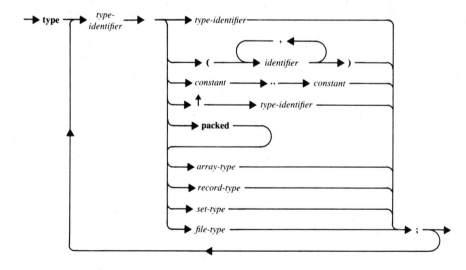

Appendix B

constant-declaration part

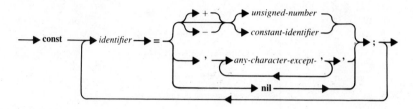

variable-declaration part

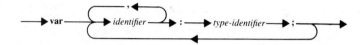

procedure-declaration part

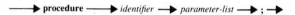

function-declaration part

parameter list

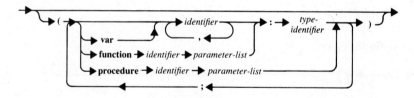

ordinal type

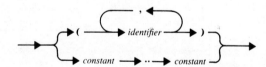

array type

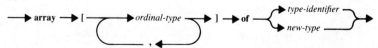

set type

file type

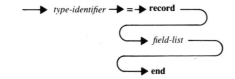

record type

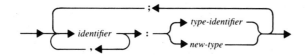

fixed-field list

variable-field list

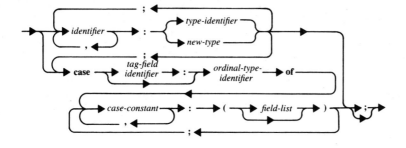

Appendix B

variable

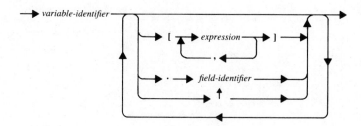

factor

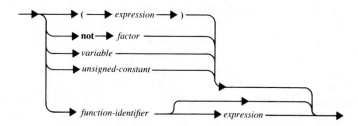

term

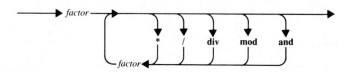

simple expression

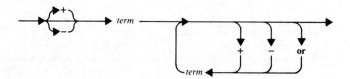

expression

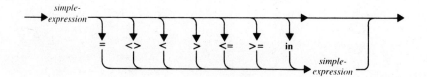

assignment statement

compound statement

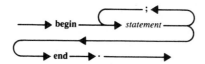

read statement

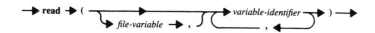

readln statement

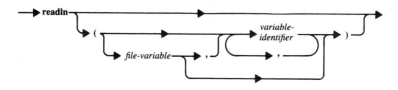

write statement

writeln statement

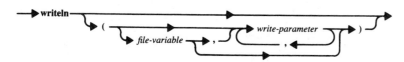

write parameter

Appendix B

if statement

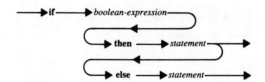

case statement

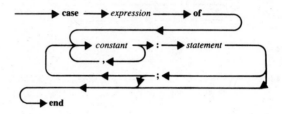

for statement

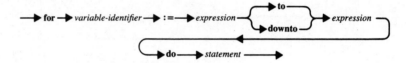

while statement

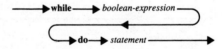

repeat statement

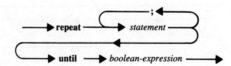

with statement

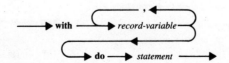

labeled statement

⟶ *1 to 4 digits* ⟶ ; ⟶ *statement* ⟶

goto statement

⟶ **goto** ⟶ *1 to 4 digits* ⟶

Appendix C
Binary, Octal, Decimal, and Hexadecimal Number Systems

When you run a program, most input to and output from the computer will be in decimal numbers, but the representation of the numbers within the computer is in binary. Normally, the internal representation is hidden from you, but there are some occasions when it is necessary to be able to understand other number systems. Table C.1 shows the number systems that you may encounter.

For example, when you install some large programs (such as word processing or data base management software) on microcomputers, the control codes that you must enter may be in base 16, or *hexadecimal* (often shortened to *hex*). In this case, you will enter codes in hex, and the computer will convert them to binary numbers.

On mainframe computers, you may occasionally see error messages with hexadecimal numbers. Such error messages occur when you have made what is known as a *fatal error*, after which the computer can no longer follow your instructions. The contents of the storage locations that you were using at the time of your error are *dumped* in hex—that is, the binary numbers within the storage locations are converted to hex and printed out. When you have gained more knowledge and experience with computers and computer programming these numbers may be useful to you in determining what went wrong in your program's execution. At this point, it is sufficient to know that if you see error messages containing hexadecimal numbers, you have a serious bug in your program. On minicomputers, a fatal dump may be in octal (base 8) numbers instead of hex.

Hexadecimal and octal numbers are used instead of decimal numbers in the previous situations because there is an exact and simple algorithm for converting from hexadecimal and octal to the internal representation in binary or vice versa. Conversion between the decimal number system and the binary number system was covered in Chapter 1.

Appendix C

Table C.1 Binary, octal, decimal, and hexadecimal numbers

Binary	Octal	Decimal	Hexadecimal
0	0	0	0
1	1	1	1
10	2	2	2
11	3	3	3
100	4	4	4
101	5	5	5
110	6	6	6
111	7	7	7
1000	10	8	8
1001	11	9	9
1010	12	10	A
1011	13	11	B
1100	14	12	C
1101	15	13	D
1110	16	14	E
1111	17	15	F
10000	20	16	10
10001	21	17	11
10010	22	18	12
10011	23	19	13
10100	24	20	14

Conversion between binary and hexadecimal numbers is not difficult. To convert a binary integer to a hexadecimal integer, the first step is to mark the binary number into groups of four digits, starting at the right. For example, the binary number 10100 can be divided into two groups of four digits by padding three 0s on the left. The first number 0001 is the hex number 1. The second number 0100 is the hex number 4.

$$0001\ 0100_2 = 14_{16}$$

To convert from a hexadecimal number to a binary number, replace each hex digit by its four digit binary equivalent. For example,

$$F1_{16} = 1111\ 0001_2 \quad \text{and} \quad 15_{16} = 0001\ 0101_2$$

Similarly, the binary number 10100 can be converted to octal by dividing it into groups of three starting from the right. So, the number is written as 010 100 and each binary group is replaced by its octal equivalent as follows:

$$010\ 100_2 = 24_8$$

The conversion from an octal number to a binary number involves replacing each octal digit by its three digit binary equivalent. For example,

$$13_8 = 001\ 011_2 \quad \text{and} \quad 20_8 = 010\ 000_2$$

Appendix D
Other Pascal Constructions

Four Pascal statements were not covered in the main text; these constructions are included here to make the discussion of Pascal complete. The **goto, pack, unpack,** and **page** statements were omitted because they are rarely useful, and, in fact, the use of a **goto** statement violates the principle of modular design and is usually considered a sign of a poorly designed program.

The **goto** *statement*

Flow of control is a phrase that is commonly used in discussions of programming, but that has been rarely used in this text. In many languages instead of writing a statement such as **repeat-until**, you must instruct the computer to execute a section of code and then return to the first statement in the code and again begin execution at that point. The flow of control is thus transferred from the statement that has just executed to a statement somewhere else in the program and the program execution continues from that statement. Pascal is designed in such a way that the flow of control is so clear that you do not have to think about it. The **goto** is a control structure that is available in Pascal to transfer the flow of control of a program directly to a statement, somewhere else in the program, which is identified by a **label**.

The syntax diagram for the **goto** statement is

⟶ **goto** ⟶ *1 to 4 digits* ⟶

Appendix D

For example, the statements

```
    if Error then
      goto 9999
    NextProcedure;
         .
         .
         .
    9999 : writeln (' Program terminating')
    end.
```

will transfer the control to the end of the program if **Error** is true. Otherwise, the procedure **NextProcedure** will be executed.

Any statement within a program can be labeled with an integer between 0 and 9999 followed by a colon. Two statements cannot have the same label. The labels that a program will use must be declared using the reserved word **label** in the declarative part of the program or subprogram. The **label** declaration is placed before the **const, var,** and **type** definitions. For the preceding statements, the following declaration would appear after the program heading:

```
    label 9999;     (* declare a label for the end of the program. *)
```

The scope of a label is the same as the scope of a variable. So, the **goto** in the preceding program transfers control to the end of the program, but if a **label** were declared within the procedure, a **goto** statement in the main program could not transfer control into the procedure. The **goto** can transfer control either forward or backward within a program, but only within the block where the label is declared or to an outer block.

The label on a statement is ignored unless a **goto** is used to transfer control to that statement. That is, the statement

```
    9999 : writeln (' Program terminating')
```

is executed even if **Error** false. In order to prevent the **writeln** statement from executing whether **Error** is true or false, another **goto** or an **if** statement is required.

```
            .
            .
            .
       ReportFinalResults;
    goto 1000;     (* normal termination - goto end - otherwise print *)
                   (* terminating message and continue to end. *)
    9999 : writeln (' Program terminating');
    1000: end. (* program DataProcessing *)
```

The **goto** statement is frowned on because it tends to obscure the logical flow of a program. For example, can you tell what the equivalent Pascal construction to the following statements is?

```
      X := 2;
  10: if (X > 50) then
        goto 20;
      X := X * X;
      goto 10;
  20: writeln (' The value of X is ', X:3)
```

The purpose of the following **while** statement:

```
X := 2;
while X <= 50 do
  X := X * X;

writeln (' The value of X is ', X:3)
```

is much clearer than the equivalent statements using **goto**s.

Occasionally, the **goto** statement may be preferred, as in the case when a program must stop executing due to a serious data error. The flow of control may be transferred to the end of the program using cascading **if** statements:

```
begin
  ReadData (ErrorFlag, DataArray);
  if not ErrorFlag then
    begin
      ProcessData (ErrorFlag, DataArray, ProcessedData);

      if not ErrorFlag then
        begin
          ProcessMoreData (ErrorFlag, DataArray, MoreProcessedData);

          if not ErrorFlag then
            ReportData (DataArray, ProcessedData, MoreProcessedData)
        end (* no errors in first process *)
    end; (* no errors in reading data *)

  if ErrorFlag then
    writeln (' Program terminating due to error condition.')
end. (* program DataProcessing *)
```

However, the program would be easier to follow if the programmer announced the convention that all fatal errors would result in control being transferred to the last statement, as illustrated in Program D.1.

PROGRAM D.1 DataProcessing

```
program DataProcessing (input, output);
(*    This program processes a lot of data.  Whenever a serious error   *)
(*    occurs, control is transferred to the end of the program which    *)
(*    is labeled as 9999.  Therefore whenever the statement:            *)
(*    goto 9999    is executed, a fatal error has occurred and the      *)
(*    program will halt execution.                                      *)

label 1000, 9999;

const MaxEntries = 8760;
```

Appendix D

```
        type  YearArray = array[1..MaxEntries] of real;

        var   ErrorFlag : boolean;
              DataArray, ProcessedData, MoreProcessedData : YearArray;

     (*$i readdata *)
     (*$i processdata *)
     (*$i processmoredata *)
     (*$i reportdata *)

     begin
       ReadData (ErrorFlag, DataArray);
       if ErrorFlag then goto 9999;

       ProcessData (ErrorFlag, DataArray, ProcessedData);
       if ErrorFlag then goto 9999;

       ProcessMoreData (ErrorFlag, DataArray, MoreProcessedData);
       if ErrorFlag then goto 9999;

       ReportData (DataArray, ProcessedData, MoreProcessedData);

       goto 1000;        (* normal termination - goto end - otherwise print *)
                         (* terminating message and continue to end. *)

     9999: writeln(' Program terminating due to error condition.');

     1000: end. (* program DataProcessing *)
```

The standard procedures pack *and* unpack

Character strings and packed arrays of characters were covered in Chapter 11. In that chapter, you were told that arrays of characters should be packed to take advantage of the special attributes of string-type variables, but that numerical arrays should not be packed. However, if you are working on a system with limited memory, you may need to pack a noncharacter array or some other large data structure. When a packed data type is declared, several elements are compressed so that they can share a location in memory, thus, in effect, increasing the storage space available to the program. But packing an array can also increase the execution time of the program because additional processing is required to access the elements of packed data structures.

You can declare any variable to be a packed data type by using the reserved word **packed** in front of the data type. For example:

```
        type  PackedType = packed array[1..10] of real;

        var   PackedRealArray : PackedType;
```

Information can be read into the variable **PackedRealArray** using the standard procedures **readln** and **read**, and for the most part, the elements of a packed structured variable can be treated like the elements of any other structured variable. The process of accessing the individual elements is handled invisibly by the computer.

However, there is one major difference between a packed data structure and a regular data structure: elements of a packed data structure cannot be passed as arguments into variable parameters of subprograms. If values are to be passed from a subprogram into elements of a packed data structure, the data must be passed through a nonpacked data structure and then transferred to the packed data structure using the standard procedure **pack**.

To use the standard procedure **pack**, you need two array variables: one that has been declared as a **packed** data type and another that has the same data type, but is not packed. The array that is not packed must have at least as many elements as the packed array.

```
type    PackedType = packed array[1..10] of real;
        RegularType = array[1..10] of real;

var     PackedRealArray : PackedType;
        NormalRealArray : RegularType;
```

The array variable **NormalRealArray** can be packed into the array variable **PackedRealArray** using the standard procedure **pack**

```
StartingSubscript = 1;
pack (NormalRealArray, StartingSubscript, PackedRealArray)
```

where **StartingSubscript** is the index of the element in **NormalRealArray** which is packed into the first element of **PackedRealArray**.

Similarly, the standard procedure **unpack** will unpack the array **PackedRealArray** into the array variable **NormalRealArray**

```
StartingSubscipt := 1;
unpack (PackedRealArray, NormalRealArray, StartingSubscript)
```

where **StartingSubscript** is the index of the element of **NormalRealArray** and where the first element of **PackedRealArray** is assigned. Because of their special attributes, character strings—that is, **packed arrays of char**, cannot be used as the arguments of either **pack** or **unpack**.

Pack and **unpack** are not implemented on either the UCSD Pascal compiler or on the Turbo Pascal compiler. Because these compilers are commonly used on microcomputers which have storage limitations, packing and unpacking are performed automatically by both compilers.

The standard procedure page

The procedure **page** is designed to give the programmer more control over the printing of output. Different line printers and terminals have differing numbers of lines per output page, and rather than requiring the programmer to use a variable number of **writeln** statements to advance to the top of a new page, the **page** procedure is supposed to allow the programmer to begin a new page at any time. However, in standard Pascal, the effect of **page** is implementation-dependent. So, executing the **page** procedure may have no effect, may cause the output device to advance to a new page, or may have some other effect on the output device. The default file for **page** is **output**. The statement

```
page (newtextfile)
```

would cause the **page** procedure to be executed for the **textfile**, **newtextfile**.

Appendix E
External Subprograms in Pascal

There are numerous documented and well-tested scientific and mathematical subprograms written in FORTRAN and assembly language. Commercially available libraries of subprograms perform tasks such as statistical analysis of variance, solving nonlinear and differential equations, and inverting sparse matrices. Although these libraries could be translated into Pascal, many Pascal compilers allow subprograms written in other languages to be used within a Pascal program. The ability to use the vast number of existing subprograms makes learning the idiosyncratic syntax for *external* subprograms worthwhile. Because of the compiler and system dependent nature of external subprograms, this discussion will not be detailed. Your compiler documentation will tell you whether external subprograms are supported, how they are implemented, and what languages are supported. You will also need to know which libraries are available on your computer system and how to access their subprograms.

The Turbo Pascal, Pascal VAX-11, and UCSD Pascal compilers, which are discussed individually in Appendices F through H, all allow the use of external subprograms. Because mainframe computers are likely to have extensive libraries available, our examples use the VAX-11 Pascal compiler, which is a compiler for use on mainframes; the programs in this section were run on a VAX-11/780 under the VMS operating system. This section uses features of the VAX-11 Pascal compiler that are discussed in more detail in Appendix H. The examples use FORTRAN subprograms because it is a common language for libraries and most compilers support external FORTRAN subprograms.

Suppose a Pascal program calls a FORTRAN subprogram, WRITEMAT, that prints a matrix. WRITEMAT is referred to as a *subroutine*, which is analagous to a Pascal procedure. The FORTRAN subroutine is as follows:

```
            SUBROUTINE WRITEMAT(N, A)
            DIMENSION A(N,N)
C
C           Subroutine WRITEMAT prints an N by N matrix at the terminal
C
            DO 10 I=1,N
               WRITE(6,800) (A(I,J),J=1,N)
10          CONTINUE
800         FORMAT(8F10.2)
            RETURN
            END
```

External Subprograms in Pascal   A23

WARNING
The word size of a real variable may be different in Pascal than it is in another language; therefore, you must check the documentation for each compiler to find the default size. It may be necessary, for example, to declare real variables in the FORTRAN program as DOUBLE PRECISION and to use the double precision versions of FORTRAN library subroutines.

WRITEMAT, which is stored in a VMS file **writemat.for**, can be compiled using the VMS command **fortran**. After execution of the following command, the compiled version will be stored in **writemat.obj**.

```
$ fortran writemat
```

After compilation, the external subprogram can be included as a procedure in the Pascal program. In the declarative part of the main Pascal program, you declare **writemat** using a Pascal procedure heading. Only the heading, which must be equivalent to the heading of the external subprogram, is included. The number and type of the arguments declared in the external subprogram and in the subprogram heading in the Pascal program must agree. (Note that in FORTRAN identifiers starting with I, J, K, L, or M denote integer variables unless declared otherwise.) The subprogram is declared to be external to the program by the word **extern** as shown in Program E.1, **TestExternal**, which uses the external subprogram WRITEMAT.

PROGRAM E.1   TestExternal

```
program TestExternal (input, output);
(*      This Pascal program includes the FORTRAN subroutine WRITEMAT    *)
(*      in a Pascal program.  TestExternal sends a matrix to WRITEMAT   *)
(*      to determine whether the versions of Pascal and FORTRAN store   *)
(*      matrices in  different orders (by rows or by columns).          *)

const NumDim = 2;

type  Matrix =  array[1..NumDim, 1..NumDim] of real;

var   AMatrix : Matrix;
      Order : integer;


   procedure WriteMat (var N : integer; var A : Matrix); extern;
   (*      The FORTRAN subroutine WRITEMAT prints a matrix at the       *)
   (*      terminal.                                                    *)


   procedure ReadMatrix (var InMatrix : Matrix);
   (*      Procedure ReadMatrix reads the matrix from the user at the   *)
   (*      terminal.                                                    *)
```

```
        var Row, Col : integer;          (* for-statement control variables for *)
                                         (* reading the matrix. *)

    begin
      writeln (' You will be asked to enter the rows of a ', NumDim:1,
               ' by ', NumDim:1,' real matrix');
      for Row := 1 to NumDim do
          begin
             writeln (' Enter the ', NumDim:1,' elements for row ', Row:1);
             for Col := 1 to NumDim do
                 read (InMatrix [Row, Col])
          end
    end; (* procedure ReadMatrix *)


(*---------------------------------------------------------------------------*)

begin (* main program*)

   ReadMatrix (AMatrix);

   Order := NumDim;
   writeln;
   writeln (' The matrix printed in the FORTRAN subroutine is:');
   WriteMat (Order, AMatrix)
end.   (* program TestExternal *)
```

Assuming that **TestExternal** is stored in the file **testextern.pas**, it can be compiled using the VMS command **pascal**, thus creating the file **testextern.obj**.

```
$ pascal testextern
```

The object files for **testextern** and **writemat** must then be linked in an executable module. The VMS **link** command creates the file **testextern.exe**.

```
$ link testextern,writemat
```

Finally, **testextern.exe** can be executed with the VMS **run** command as follows:

```
$ run testextern



=>
 You will be asked to enter the rows of a 2 by 2 real matrix

 Enter the 2 elements for row 1
□1.0 2.0
 Enter the 2 elements for row 2
□3.0 4.0

 The matrix printed in the FORTRAN subroutine is:
      1.00     3.00
      2.00     4.00
```

The output from **TestExternal** shows that the FORTRAN subroutine printed the transpose of the matrix read in the Pascal procedure **ReadMatrix**. Because the internal storage scheme for arrays differs among languages and compilers, an array may be stored either by columns or by rows. If the two compilers store arrays in the same fashion, **TestExternal** will execute smoothly, but if the compilers use opposite storage schemes, the output will be incorrect. You can run **TestExternal** on your system to determine whether the storage schemes for arrays are compatible; if they are not, you can use a procedure like **Transpose** to transpose matrices passed to or returned by an external subprogram. Note that this procedure is necessary only for matrices, it is not necessary for one-dimensional arrays.

```
procedure Transpose (N : integer; OldMat : Matrix; var NewMat : Matrix);
(*      Transpose transposes the N by N matrix, OldMat and returns the    *)
(*      transposed matrix in NewMat.                                      *)

var Row, Col : integer;       (* for statement control variables. *)

begin
  for Row := 1 to N do
    for Col := 1 to N do
      NewMat [Row, Col] := OldMat [Col, Row]
end; (* procedure Transpose *)
```

Program E.2, **TestExternal2**, uses the **Transpose** procedure to read the matrix in a Pascal procedure and print it in a FORTRAN subroutine.

PROGRAM E.2   TestExternal2

```
program TestExternal2 (input, output);
(*      This Pascal program includes the FORTRAN subroutine WRITEMAT      *)
(*      in a Pascal program.  This is the second version of TestExtern    *)
(*      which now uses the procedure Transpose to transpose matrices      *)
(*      before they are passed to a FORTRAN subroutine.                   *)

const NumDim = 2;

type  Matrix = array[1..NumDim, 1..NumDim] of real;

var   AMatrix : Matrix;
      Order : integer;


%include 'transpose.pas'
(* procedure Transpose (N : integer; OldMat : Matrix; var NewMat : Matrix);*)
(*      Transpose transposes the N by N matrix, OldMat and returns the    *)
(*      transposed matrix in NewMat.                                      *)


  procedure WriteMat (var N : integer; var A : Matrix); extern;
  (*      The FORTRAN subroutine WRITEMAT prints a matrix at the          *)
  (*      terminal.                                                       *)


  procedure ReadMatrix (var InMatrix : Matrix);
  (*      Procedure ReadMatrix reads the matrix from the user at the      *)
  (*      terminal.                                                       *)
```

A26   Appendix E

```
          var Row, Col : integer;      (* for-statement control variables for *)
                                       (* reading the matrix. *)

       begin
         writeln (' You will be asked to enter the rows of a ', NumDim:1,
                  ' by ', NumDim:1,' real matrix');
         for Row := 1 to NumDim do
            begin
               writeln (' Enter the ', NumDim:1,' elements for row ', Row:1);
               for Col := 1 to NumDim do
                  read (InMatrix [Row, Col])
            end
       end; (* procedure ReadMatrix *)


(*--------------------------------------------------------------------------*)

begin (* main program*)

  ReadMatrix (AMatrix);

  Transpose (NumDim, AMatrix);      (* transpose the matrix before it is *)
                                    (* passed to the FORTRAN subroutine. *)
  Order := NumDim;
  WriteMat (Order, AMatrix)
end.   (* program TestExternal2 *)


$ pascal testext2
$ link testext2,writemat
$ run testext2

=>
  You will be asked to enter the rows of a 2 by 2 real matrix

  Enter the 2 elements for row 1
1.0 2.0
  Enter the 2 elements for row 2
□3.0 4.0

  The matrix printed in the FORTRAN subroutine is:
       1.00      2.00
       3.00      4.00
```

The next example uses a FORTRAN subroutine LINV1F to invert a matrix. LINV1F is from the IMSL library, an extensive library of scientific and mathematical subroutines available for many different computers. The LINV1F subroutine has the following heading:

    SUBROUTINE LINV1F(A, N, IA, AINV, IDGT, WRKAREA, IER)

where A is the N by N matrix of real numbers that is to be inverted, N is the order of the matrix A, IA is the declared dimension of the matrix A, AINV is the N by N output matrix, IDGT is an accuracy flag that must be set by the user, and WRKAREA is a work vector of length N, and IER is an error flag that is returned by the subroutine. If IER is returned as 0, the inverse exists and has been computed to within the desired accuracy.

An equivalent Pascal procedure heading is declared as an external procedure in Program E.3, **TestInvert**.

PROGRAM E.3 TestInvert

```
program TestInvert (input, output, infile);
(*     This Pascal program uses the IMSL FORTRAN subroutine LINV1F that   *)
(*     inverts a matrix.                                                  *)

const NumDim = 2;          (* dimension of the matrix to be inverted. *)

type Matrix =  array[1..NumDim, 1..NumDim] of real;
     WorkVector = array[1..NumDim] of real;

var  OriginalMatrix : Matrix;      (* matrix to be inverted. *)
     TransMatrix : Matrix;         (* transpose of the original matrix -- *)
                                   (* to be passed to LINV1F. *)
     TransInverse : Matrix;        (* matrix inverse returned by LINV1F. *)
     Inverse : Matrix;             (* inverse of OriginalMatrix. *)
     Product : Matrix;             (* product of the original matrix and *)
                                   (* its inverse. *)
     Order : integer;              (* dimension of the matrix to be *)
                                   (* inverted. *)
     WorkArea : WorkVector;        (* work vector used by LINV1F. *)
     Ierror : integer;             (* error flag returned by LINV1F.  If *)
                                   (* Ierror = 129, the matrix is singular *)
                                   (* and if Ierror = 34, the accuracy *)
                                   (* test failed. *)
     Accuracy : integer;           (* accuracy flag required as input *)
                                   (* to LINV1F.  If Accuracy > 0, then *)
                                   (* Accuracy digits are assumed to be *)
                                   (* significant.  If Accuracy = 0, the *)
                                   (* accuracy check is bypassed. *)
     infile : text;                (* text file to store the input matrix. *)


%include 'transpose.pas'
     (*     procedure Transpose (N : integer; OldMat : Matrix;            *)
     (*                          var NewMat : Matrix);                    *)
     (*     Include the procedure that takes the transpose of the N by N  *)
     (*     matrix OldMat and returns it in NewMat.                       *)


     procedure LINV1F (var A : Matrix; var N, N2 : integer;
                       var AInverse : Matrix; var Idgt : integer;
                       var Work : WorkVector; var Err : integer) ; fortran;
     (*     The FORTRAN subroutine LINV1F takes the inverse of matrix A   *)
     (*     and returns it in AInverse.  N is the order of the matrix and *)
     (*     N2 is the declared dimension of the matrix.  Idgt is an       *)
     (*     accuracy flag.  If Idgt is equal to 0, the accuracy tests are *)
     (*     bypassed.  Work is a work vector used within the the FORTRAN  *)
     (*     subroutine.  The FORTRAN subroutine is stored in the IMSL     *)
     (*     library.  It is linked to TestInvert when the .exe module is  *)
     (*     created.                                                      *)
```

```
    procedure WriteMatrix (AMatrix : Matrix);
    (*      Procedure WriteMatrix prints AMatrix at the terminal in        *)
    (*      matrix format.                                                 *)

    var Row, Col : integer;       (* for-statement control variables for *)
                                  (* reading the matrix. *)

    begin
      writeln;
      for Row := 1 to NumDim do
        begin
          for Col := 1 to NumDim do
            write (' ', AMatrix [Row, Col]:8:2);
          writeln
        end;  (* for Row *)
      writeln
    end; (* procedure WriteMatrix *)

%include 'readmatrix.pas'
    (*      procedure ReadMatrix (var InMatrix : Matrix);                  *)
    (*      Procedure ReadMatrix reads the NumDim by NumDim matrix from the *)
    (*      textfile, infile.                                              *)


    procedure PrintReport (Save, AInverse, Multiplied : Matrix);
    (*      Procedure PrintReport prints three matrices: the original      *)
    (*      matrix, the inverse matrix, and the product of the original    *)
    (*      and the inverse.  The product should be equal to the           *)
    (*      identity matrix.                                               *)

    begin
      writeln;
      writeln (' The original matrix is:');
      WriteMatrix (Save);

      writeln (' The inverted matrix is:');
      WriteMatrix (AInverse);

      writeln (' The product of the matrix with its inverse is');
      WriteMatrix (Multiplied);
    end; (* procedure PrintReport *)


%include 'multiply.pas'
    (*      procedure  MultiplyMatrices (Num : integer; A, B : Matrix;     *)
    (*                                   var C : Matrix);                  *)
    (*      Procedure MultiplyMatrices from Chapter 9 multiplies two       *)
    (*      Num by Num matrices and returns the product.                   *)



(*----------------------------------------------------------------------*)

begin (* main program*)
  open (infile, 'testinv.dat', History := old);
  reset (infile);
```

```
ReadMatrix (OriginalMatrix);

(* Because matrices are stored in row order in VAX-11 Pascal and in *)
(* column order in VAX-11 FORTRAN, the transpose of the matrix must *)
(* be passed to the FORTRAN procedure.  The transpose of the inverse *)
(* matrix returned by LINV1F also must be computed so that the final *)
(* matrix is in Pascal form. *)

Transpose (NumDim, OriginalMatrix, TransMatrix);

Order := NumDim;
Accuracy := 0;

LINV1F (TransMatrix, Order, Order, TransInverse, Accuracy,
        WorkArea, Ierror);

if Ierror <> 0 then
  begin
    writeln (' Matrix inversion not complete, Ierror = ', Ierror:1)
    if Ierror = 129 then
      writeln (' The matrix is singular');
    if Ierror = 34 then
      writeln (' The accuracy test failed')
  end (* if error *)
else
  begin
    Transpose (NumDim, TransInverse, Inverse);
    MultiplyMatrices (NumDim, OriginalMatrix, Inverse, Product);
    PrintReport (OriginalMatrix, Inverse, Product)
  end
end.  (* program TestInvert *)
```

Notice that **TestInvert** transposes both the **Original** matrix before it is sent to LINV1F and the matrix **TransInverse** returned by LINV1F to obtain the **Inverse** of the original matrix. If **TestInvert** is sorted in the VMS file **testinv.pas**, it can be compiled using the VMS command **pascal**.

```
$ pascal testinv
```

The resulting object file, **testinv.obj**, must be linked with the IMSL library containing the compiled version of LINV1F as follows:

```
$ link testinv,imsls/lib
```

For the VAX-11, the single precision version of the IMSL library is used as indicated by the "s" following imsl.

Finally, **testinv.exe** can be executed using the VMS **run** command.

```
$ run testinv


=>
  The original matrix is:

      2.00     5.30
      8.60     1.00
```

```
The inverted matrix is:

  -0.02    0.12
   0.20   -0.05

The product of the matrix with its inverse is

   1.00    0.00
   0.00    1.00
```

Appendix F
Turbo Pascal

Turbo Pascal™ is designed for microcomputers running the CP/M-80, CP/M-86, or the MS-DOS operating system. Turbo Pascal is more than a compiler; it is a system, or an environment, in which you can create, edit, and execute programs. For example, when you execute a program in Turbo Pascal, if a compile-time or run-time error occurs, you will find yourself automatically in edit mode in the file at the location where the error occurred. Thus, the Turbo compiler makes tracing and correcting bugs much simpler than compilers that just flag the line where the error occurred. The editor that is used is similar to the WordStar word-processing software.

Turbo Pascal includes extensions that take advantage of the fact that the user of a microcomputer does not have to share the computer with other users and therefore, can directly allocate the computer's memory and other resources. Because many of these extensions depend on the operating system under which Turbo Pascal is running, the extensions for memory management and system-dependent features such as the graphics mode for IBM PCs will not be covered. For these enhancements, you will have to consult the *Turbo Pascal Reference Manual* (see reference 1). The topics of typed constants, untyped files, and the heap stack also will not be covered. The examples in this appendix were run on a Morrow Micro Decision MD3 using Turbo Pascal Version 2.0 under CP/M.

F.1 Implementation-dependent features

Identifiers

Identifiers in Turbo Pascal can be up to 127 characters long and all characters are

---

Turbo Pascal is a trademark of Borland International, Inc.

significant. The underscore is the only nonstandard character allowed in identifiers. An identifier may start with an underscore. The identifiers **New_Word** and **NewWord** do *not* refer to the same storage location.

Ranges

Integers range from −32768 to 32767. Overflow of integers is not detected. Real numbers range from −1.0e+38 to −1.0e−38 and from 1.0e−38 to 1.0e+38 with up to 11 significant digits in the mantissa. If a real number too large to store is created, the program halts execution; a real number too small to store is evaluated as 0.0. A **set** in Turbo Pascal can have up to 256 elements.

Comment Statements

Both { } and (* *) are valid comment delimiters. A comment set off by one set of delimiters cannot be nested inside another comment set off by the same set of delimiters. However, a comment marked by (* *) can be nested inside a comment marked by { }, and *vice versa*.

F.2   Extensions to standard Pascal

Data types

Turbo Pascal has a predefined data type called **byte** that is a subrange of integers from 0 to 255. A **byte** occupies 1 byte in memory. In addition, **mem** and **port** are arrays of data type byte. These arrays are used to access the memory and the data ports, and hence, are operating-system dependent. See the *Turbo Pascal Reference Manual* for their use in your system.

Program heading

The program heading is optional and acts almost like a comment statement, giving an identifying name to the program for the reader. Consequently, files do not have to be declared in the program heading.

Buffered input

Data entered interactively in Turbo Pascal is *buffered*, that is, it is not processed as it is entered. Instead, it is stored temporarily. This temporary storage allows the input data to be modified before the line is entered with a carriage return, making data entry easier. But, the **read** and **readln** will not execute as defined in standard Pascal. For example, the procedure **ReadString** developed in Chapter 11 uses standard Pascal to circumvent problems of nonstandard treatment of strings, but it will not work in Turbo Pascal when buffered input is active. Input statements in a Turbo program will execute as in standard Pascal if the buffered input is suppressed using a compiler option. Compiler options are discussed at the end of this section.

Additional arithmetic operators

Turbo Pascal has three operators in addition to the standard Pascal operators: **xor**, **shl**, and **shr**. **Xor** is a boolean operator called *exclusive or* that is used to compare two boolean expressions and return a value of true if only one of the boolean expressions is true; otherwise it is false. (See Problem 3 in Chapter 6.) **Xor** has the same level of precedence as +, −, **or**

```
writeln ('        A         B         C');
for A := false to true do
  for B := false to true do
    begin
      C := A xor B;
      writeln (A:10, B:10, C:10)
    end
```

=>

A	B	C
FALSE	FALSE	FALSE
FALSE	TRUE	TRUE
TRUE	FALSE	TRUE
TRUE	TRUE	FALSE

The operators **shl** (shift left) and **shr** (shift right) operate on binary words in memory and are more commonly used in assembly-level programming.

The **for** *statement*

Upon completion of the **for** statement, the control variable is equal to its final value. If the **for** statement is not executed, no assignment is made to it. For example after execution of the statements, **Index** has a value of 10.

```
for Index := 1 to 10 do
    writeln
```

After execution of the following statements,

```
for Index := 10 to 1 do
    writeln
```

Index has the same value as it had prior to the **for** statment. Since a **for** statement may not execute, do not assume that the control variable will be equal to the upper bound of the **for** statement.

The **case** *statement*

If the value of the control variable does not appear in the case label list, the statement after the **case** statement is executed. The Turbo compiler supports an **else** option; if the value of the control variable is not in the list, and an **else** follows the **case** statement, the **else** is executed. In the following example, notice that the **end** of the **case** statement follows the **else**.

```
Test := 5;
case Test of
  2, 4 : writeln (' even');
  1, 3 : writeln (' odd');
else
  writeln (Test:2, ' does not appear in the case statement')
end
```

```
=>
5 does not appear in the case statement
```

Type conversion

Turbo Pascal allows type conversion for scalar data types. Given the declaration

```
type  FuelType = (Oil, Gas, Nuclear, Coal, Hydro, Sun, Wind);

var   PlantFuel : FuelType;
```

the following statement assigns a value of **Oil** to **PlantFuel**.

```
PlantFuel := FuelType(1)
```

Character strings

Turbo Pascal has extensions to make character strings easier to work with. A nonstandard data type, **string**, is used instead of a **packed array of char**acters. (In Turbo Pascal, the word **packed** is ignored when it is encountered in a type definition.) To declare a **string** data type, the reserved word **string** is followed by the length of the string in brackets. The maximum length for a character string is 255 characters.

```
type  Word = String [10];
```

A string is compatible with an **array of char**acters of the same length and compatible with *any* other string type no matter what its length. The boolean operators can be used to compare two strings of different lengths, but if the strings are of different lengths, the shorter string is always smaller.

Control characters can be included in strings so that you can, for example, ring the bell at the terminal. The control codes and their effects are given in Table A.2, Appendix A, where you can find that ^G, or ASCII code 07, rings the bell. There are two ways to put the control character into a string. One is to type "^G". The following **writeln** statement will ring the bell and then print the message at the terminal.

```
writeln (^G^G' That was incorrect.  Please re-enter.');
```

```
=>
That was incorrect.  Please re-enter.
```

The other way to put a control code in a character string is to type the ASCII code for the control character preceded by the character #. The following statement has the same effect as the statement above.

```
writeln (#07#07' That was incorrect.  Please re-enter.')
```

Procedures for use with strings

delete (CharString, Position, NumDelete) deletes a substring of length **NumDelete** from **CharString** starting at index **Position**. After the following statements are executed, **Phrase** will have a value of 'charters':

```
Phrase := 'characters';
delete (Phrase, 5, 2)
```

insert (AddString, CharString, Position) inserts **AddString** inside **CharString** at index **Position**. For example, after the following statments have executed, **Name** has the value 'John Q. Smith':

```
Name := 'John Smith';
MiddleInitial := ' Q.';
insert (MiddleInitial, Name, 5)
```

val (CharString, Value, Code) converts **CharString** into its equivalent real or character value.

The procedure **val** is worth noting. A large part of Chapter 12 was devoted to trapping inappropriate data values by reading all input as character and converting the data to its equivalent numerical value. In Turbo Pascal, the built in procedure **val** does all the conversions for you. The **CharString** passed as an argument to **val** must correspond to a number, cannot have any leading or trailing blanks, and must be assignment compatible with the data type declared for **Value**. If the conversion cannot be made, **Code** will return with a nonzero value. So, whenever you use the **val** procedure, test the value of the return code before performing any calculations. The following fragment continues to ask for input until a real number and then an integer number are entered. Notice that the assignment to the variable **Volts** occurs within the procedure **val**.

```
writeln (' Enter a real number for the voltage in volts');
readln (CharString);
val (CharString, Volts, ReturnCode);

while ReturnCode <> 0 do
  begin
    writeln(' Please enter a real number');
    readln (CharString);
    val (CharString, Volts, ReturnCode)
  end (* while bad data *)
```

Functions for use with strings

copy (CharString, Position, NumChars) returns a character string that is a copy of **CharString**, starting from the index **Position** and that is **NumChars** long. After the following statements have been executed, **FirstName** will have a value of 'John Q.':

```
Name := 'John Q. Smith';
FirstName := copy (Name, 1, 7)
```

concat (CharString1, CharString2, . . . CharString) returns a single string that is the result of joining the strings in the argument list in the order listed. Any number of strings can be given in the argument list, but if the string created is longer than 255 characters, a run-time error results. After execution of the following statements **NewPhrase** will have a value of 'This is a test':

```
Phrase1 := 'This is ';
Phrase2 := 'a test';
NewPhrase := concat (Phrase1, Phrase2);
```

Strings can also be concatenated by using the + operator. The statement

```
NewPhrase := Phrase1 + Phrase2
```

is equivalent to the statement using concat. The plus operation has the same order of precedence as addition in expressions. If a string with a length greater than 255 characters is created, a run-time error will result.

length (CharString) returns the number of characters in **CharString**. The variable **NumChar** will have a value of 7 after the following statement is executed:

```
NumChar := length ('nothing')
```

pos (SearchFor, CharString) searches for the character string **SearchFor** within **CharString**. If the string is found, **pos** returns the index of the first character in **CharString** that matches **SearchFor**. If **SearchFor** is not found, **pos** returns a value of zero. For example,

```
writeln (' Enter a line of text');
readln (OneLine);

Yes := 'yes';
Index := pos (Yes, OneLine);
if Index > 0 then
   writeln (' The word yes occurred in that line')
else
   writeln (' The word yes did not occur in that line')
```

=>
 Enter a line of text
□well I guess maybe yes
 The word yes occurred in that line

Procedures for CRT input/output

When Turbo Pascal is installed, the control sequences, which do such things as clear the screen or change the intensity of the video, are specified. The CRT procedures depend on these control sequences having been set properly.

clreol clears all the characters from the cursor to the end of the line without moving the cursor.

clrscr clears the screen and puts the cursor in the upper left-hand corner of the screen.

crtinit sends the terminal initialization string to the screen.

crtexit sends the terminal reset string to the screen.

delline deletes the current line and moves all the lines below up one line.

insline inserts a blank line at the current line. The bottom line scrolls off the screen.

gotoxy (x,y) moves the cursor to position (x,y) on the screen, where x is the column (horizontal) and y is the row (vertical). **X** and **y** must be integers. The upper left-hand corner is (1,1) and on a screen with 80 columns and 24 lines, the bottom right-hand corner is (80,24)

lowvideo sets the screen to low video, that is, dim letters. Only text written after execution of **lowvideo** will be dim.

normvideo sets the screen to normal video intensity.

You can write a program that executes these procedures one at a time to determine their effect on your terminal.

Other procedures and functions

delay (NumMilliSecs) causes execution of the program to pause for a number of milliseconds proportional to **NumMilliSecs**, which must be an integer.

randomize initializes the random number generator with a random value. Randomize performs the same function as the function **FirstSeed** created in Chapter 10.

frac (RealNum) returns the fractional part of **RealNum**.

random returns a random number that is a real value greater than or equal to 0.0 and less than 1.0.

random (IntNum) When an integer argument is passed to the function **random**, it returns an integer value greater than or equal to 0 and less than **IntNum**.

upcase (Character) returns the upper case equivalent of **Character**. The argument **Character** must be a single character, not a **string**. If no upper case equivalent exists, the function returns **Character**.

Files

In Turbo Pascal, the extensions for files can be grouped into two categories: those that expand the operations allowed on **file** variables and those that deal with the association between Pascal files and disk files.

Subprograms for files of structural data types

As an extension to standard Pascal, Turbo Pascal includes subprograms that enable sequential **files** to be treated as random access data structures. A **file** of a structured data type can be used within a program much like an array data structure.

seek (pascalfile, ElementNumber) moves the file pointer to **ElementNumber**. After the statement

```
seek (infile, N)
```

is executed, the **file** pointer is positioned at the Nth element in **infile**. The procedure **FindAndRead** uses the seek procedure to position the read pointer at a given element, then reads the value stored there.

```
procedure FindAndRead (var realfile : RealFileType;
                      ElementNumber : integer; var ElementValue : real);
(*      This procedure finds and reads the element in realfile      *)
(*      that is in position number ElementNumber. It assigns this   *)
(*      value to ElementValue.                                      *)

begin
  seek (realfile, ElementNumber);
  read (realfile, ElementValue)
end;   (* procedure FindAndRead *)
```

filesize (pascalfile) returns the integer number of components in the file **pascalfile**.

filepos (pascalfile) returns the current position of the file pointer. The first element of a file has the position 0.

Procedures for external Pascal files

assign (pascalfile, CharString) In Turbo Pascal, an external Pascal **file** is associated with computer file by using the **assign** procedure. The following statement assigns the disk file **newstuff.dat** on drive B to the Pascal **file infile**:

```
assign (infile, 'b:newstuff.dat')
```

close (pascalfile) An external **file** created in a Turbo program must be **close**d before the program halts execution. If a **file** is not **close**d, the disk directory will not be updated and the computer file will not be saved. It is a good habit to **close** all files whether they are input or output files because in some environments it is necessary to **close** any file that has been accessed.

rename (pascalfile, CharString) renames a computer file within a Pascal program. If a computer file with the name **CharString** already exists, its contents will be replaced by the contents of the new file, so **rename** should be used with caution. **Rename** should not be used on a **file** that has been reset or rewritten but not **close**d.

erase (pascalfile) erases the computer file associated with the **pascalfile**. **Erase** should be used with care and should not be used on a **file** that has been reset or rewritten but not **close**d.

In standard Pascal, there are two standard **file**s, **input** and **output**. A standard file does not have to be declared and does not have to be reset or rewritten. Turbo Pascal has six additional standard files: **con**, **trm**, **kbd**, **lst**, **aux**, and **usr**. The **usr** file is operating-system dependent and is not described here. The standard files may be used in **read** or **write** statements to direct the computer to accept input from or send output to a local device such as the printer or modem. For example, the statement

```
writeln (lst, ' This line will be sent to the printer')
```

will print the character string on the computer's list device, which is usually the printer.

con The console is both an input and output file. When the console acts as an output file, output is directed to the computer's monitor. When the console acts as an input device, input is accepted from the keyboard and echoed on the monitor. The **con** file is the default for the **input** and **output** files. As the input file, **con** supports buffered

Appendix F

input and does not conform to standard Pascal.

trm The terminal is an input/output file that has the same properties as the console file except that it does not buffer input.

kbd This is an input file that accepts characters from the keyboard. Input from **kbd** is not echoed on the monitor. Reading from **kbd** can be useful when a password or other confidential data is requested from the user.

lst The list file is an output file usually associated with the printer.

aux The auxiliary file is an input and output file. It is associated with the punch and reader devices, which on most systems are the modem.

Compiler options

The following compiler options affect the compiled code generated by the Turbo Pascal compiler. An option is active when followed by a plus sign and inactive when followed by a minus sign. The default for each option is indicated by the sign used within the parentheses.

(*$A+*) The A option is valid only in CP/M and *must* be *inactive* when recursion is used in CP/M. When the A option is inactive, the program requires more memory and execution is slower.

(*$B+*) When the B option is active, the console is the default device so input is buffered. If the option is inactive, the terminal is the input device, input is not buffered, and **read** and **write** execute in accordance with standard Pascal.

(*$C+*) When this option is active, pressing the control and C keys simultaneously (^C) in response to a **read** or **readln** halts execution of the program and pressing ^S will freeze the screen. A second ^S will restart the screen. Once this option has been set in a program, it cannot be changed.

(*$i filename*) This option causes the file stored in the computer file **filename** to be included. An included file cannot itself contain another include statement, that is, include files cannot be nested.

(*$I+*) When the I option is active, the error code is checked after each I/O statement is executed. If an error has occurred, the error message is printed at the terminal and the program halts execution. If the option is inactive and an error occurs, I/O is suspended until the Turbo function **Ioresult** is called. **Ioresult** can be tested to determine what action should be taken. The following fragment tests whether a computer file already exists before renaming a Pascal **file**.

```
writeln (' What do you want to name the output file?');
readln (CharString);

(* use the compiler option to supress I/O error testing *)
(* while using assign and reset. *)

(*$I-*)
assign (outfile, CharString);
reset (outfile);
```

```
(* return the compiler option to its normal setting *)
(* for the call to Ioresult. *)

(*$I+*)
if Ioresult <> 0 then
  writeln (' That file already exists.  Are you sure you want',
           ' to call the file ', CharString, '?')
```

```
=>
 What do you want to name the output file?
□olddata.sav
 That file already exits.  Are you sure you want to call the file olddata.sav?
```

(*$R—*) The R option controls range checking. When the option is inactive, no checking is done to see if a scalar or subrange is out of range.

(*$U—*) The U option controls the user's ability to interrupt a program during execution. If the user-interrupt option is active, the user can interrupt execution at any time by entering ^C. If the option is passive, entering ^C during execution has no effect. If a program has any potential for an infinite loop, set the U option to active.

(*$V+*) When the option V is active, strict compatibility of arguments and variable parameters of type **string** is required. If the option is not active, a **string** may be passed as an argument into a variable parameter that is a **string** of a different length.

Additional reserved words in Turbo Pascal

absolute	**shl**	**string**
external	**shr**	**xor**
inline		

Additional standard identifiers in Turbo Pascal

assign	**constptr**	**flush**	**lo**
aux	**copy**	**frac**	**lowvideo**
auxinptr	**crtexit**	**getmem**	**lst**
auxoutptr	**crtinit**	**gotoxy**	**lstoutptr**
blockread	**delline**	**heapptr**	**mark**
blockwrite	**delay**	**hi**	**mem**
buflen	**delete**	**ioresult**	**memavail**
byte	**erase**	**insline**	**move**
chain	**execute**	**insert**	**normvideo**
con	**filepos**	**kbd**	**pi**
coninptr	**filesize**	**keypressed**	**port**
conoutptr	**fillchar**	**length**	**pos**

ptr	rename	swap	usrinptr
random	seek	trm	usroutptr
randomize	sizeof	upcase	val
release	str	usr	

F.3 Differences between Turbo Pascal and standard Pascal

The goto *statement*

In Turbo Pascal, the scope of a label is the main body of the block in which it is declared. Therefore, you cannot jump out of a procedure or a function to a label declared in an outer block. A program written in standard Pascal that uses a **goto** statement may result in a compile-time error in Turbo Pascal.

Subprogram parameter lists

The parameter list of a subprogram cannot contain another procedure or function.

Pointers

Version 2 of Turbo Pascal does not support the standard **dispose** procedure, but **dispose** has been added to Version 3. In both versions, pointers can be **mark**ed and **release**d instead of **dispose**d. See the *Turbo Pascal Reference Manual*.

Put *and* **get**

Turbo Pascal does not support **put** and **get** for files. The user must make do with **read** and **write**.

Pack *and* **unpack**

Turbo Pascal does not support the procedures **pack** and **unpack**. If an array is declared as a **packed array** the statement will compile, but the word **packed** will have no effect.

Page

The standard procedure **page** is not implemented in Turbo. For a discussion of the deviations of Turbo Pascal from standard Pascal, see Reference 2.

F.4 Creating and running a program in Turbo Pascal

When Turbo Pascal is initiated by typing **turbo** in response to the system prompt, the following menu is presented:

```
Logged drive: A

Work file:
Main file:

Edit     Compile  Run   Save

Dir      Quit  compiler Options

Text:       0 bytes
Free:   63485 bytes
```

Normally, when you start a session, the first command that you execute is the Log command which changes the default disk drive. On most systems, the A disk drive is used to store system files and the Turbo compiler, and another drive, usually the B drive, is used to store program files. After entering **L**, respond with the appropriate letter when prompted for the new drive.

Before you can create a program, you must create a work file. The work file is the default file for editing, saving, and running. To create a work file, select **W** from the main menu. Then, enter a disk file name when prompted for the work file name. You can enter a complete name, such as **newfile.dat**, or you can enter just the first name, such as **newfile**. If you enter just the first name, Turbo will assign it a file extension of **.pas**.

If the file name that you enter does not already exist, you will get a message stating that the work file is a new file. If the file should exist and you have simply made a typing mistake, enter **W** again with the correct name. If you cannot remember the name of the file, you can use the **Dir** command from the menu. This command can give you a directory of any disk in the system. When prompted for the dir mask, enter either the letter of the disk drive for which you would like the directory or a carriage return if you want a directory of the currently logged disk drive.

Suppose that you have created a new work file because you want to create a new Pascal program. To enter the Turbo editor, use the **Edit** command from the menu. Because the work file is new, your screen will clear and you will be presented with an empty file into which you can enter the program using the Turbo editor. (See the Turbo reference manual for help with the editor.) To edit an existing file, enter the file name after the **W** command, then enter the **E** command. Once the program has been edited, you can exit from the editor by typing ^**Kd**. At this point, the program resides in the computer's memory.

To save a program permanently on the disk file, you must select the **Save** command from the menu. If you attempt to begin work on another file before saving one that has been changed by the editor, you will get a prompt asking whether or not you want to save the file that you were working on.

Whether the work file has been saved or not, you can compile it using the **Compile** command or run it using the **Run** command from the menu. The compile command will cause the work file to be compiled, but not executed. If the program has compile-time errors, the result of the compilation will be an error message. When the escape key is pressed, you will be in edit in the file in which the error occurred. If there are no errors, the result will be a compiled program residing in memory. If the **Run** command is issued, after a successful compilation, the program will be executed. Again, if a run-time error occurs, when the escape key is pressed, you will be in edit in the file in which the error occurred. If the run command is issued when there is no compiled program resident in memory, the work file will be compiled and run.

If you are developing a program that has several files included using the (*$i*) compiler option, you will find the **M**ain command useful. The main command can be used to differentiate between the file that is currently being worked on (the work file) and the file that contains the program to be compiled and run (the main file). If a main file has been specified, it is the default file for compile and run, and the work file is the default file for edit and save. Otherwise, the work file is the default file for edit, save, compile, and run.

To create programs that can be executed from the operating system, you can change the compiler **O**ptions by entering the **O** command from the main menu. You will then be presented with a submenu. Among the options is the **C** option. Under this option, whenever a program is compiled, the compiled code is saved in a file with the extension **.com**. (In the submenu, the default, indicated by the arrow, is to compile to **M**emory.) Once the **C** option has been entered, you must **Q**uit the submenu by entering **Q**. You will be back in the main menu. After the **C** compiler option has been selected, whenever the compile or run command is entered from the main menu, a file containing the compiled program will be saved in a disk file on the logged drive.

After you have finished working with the Turbo compiler, enter the **Q**uit option, and you will return the operating system. Any files that you have edited and saved plus any programs compiled under the **C**om option will reside on your disk. To run a program such as **myprogram.com** simply type **myprogram** in response to the system prompt. If the program is not on the currently logged disk, either include the disk label—that is, **b:myprogram**, or change the logged disk drive.

---

1. *Turbo Pascal, Version 3.0. Reference Manual*, Borland International Inc, Scotts Valley, CA: 1985.
2. Cortesi, D.E., "Dr. Dobb's Clinic," *Dr. Dobb's Journal*, M & T Publishing Company, Palo Alto, CA: July, 1985. Responses in the November 1985 issue.

Appendix G
UCSD Pascal

The UCSD™ p-System is an operating system designed to be portable so that a program developed on one computer can be run on another computer if both computers use the p-System. The minimal requirements for running the p-System are disk storage, 64K (64000 bytes) of memory, and a video display terminal. The p-System is available for

---

UCSD is a trademark of the regents of the University of California.

UCSD Pascal A43

machines ranging from microcomputers to mainframe computers.

The p-System achieves its portability to many different computers and many different CPUs by using an intermediate language called p-code. An intermediate language is a common solution to multi-language situations. For example, at international conferences, one language is usually designated as the official language so that everyone can communicate. Similarly, p-code is the official language of the p-System; every version of the p-System can create p-code from a source program and can translate p-code into the machine language of the computer on which the version resides.

The p-System includes an operating system, file manager, editor, compiler, assembler, and many utility programs. The p-System is menu driven with the menu displayed at the top of the screen. This appendix will cover only a small part of the p-System; it will not cover untyped files, memory management, device I/O, array routines that operate on the memory location of parameters, the heap stack, or concurrent processing. The examples were tested on a Compaq portable running p-System version IV.2.1 with MS-DOS.

G.1 Implementation-dependent features

Identifiers

The number of characters in a UCSD Pascal identifier is unlimited, but only the first 8 characters are significant. The only allowable nonstandard character is the underscore, but an identifier cannot start with an underscore. Otherwise, the underscore is ignored by the compiler; the identifiers **New_Word** and **NewWord** refer to the same memory location.

Ranges

Integers range from -32768 to 32767. In most versions of the p-System a run-time error occurs if an integer outside the valid range is created. The range of real numbers depends on whether the two-word or four-word version of the p-System is installed: the largest real number is $1.0e+38$ with 6 or 7 digits of precision for the two-word version and $1.0e+308$ with 15 or 16 digits of precision for the four-word version. If a real number too large to store is created, the program halts execution. A real number too small to store is evaluated as 0.0. A **set** can have up to 4080 elements.

Comment statements

Both { } and (* *) are valid comment delimiters, but a comment must begin and end with the same type of delimiter. A comment set off by one pair of delimiters cannot be nested inside another comment that uses the same pair of delimiters. A section of code containing comments can be commented out by using the alternate symbols for comments.

Program heading

The program heading is necessary, but the list of files in the heading is optional. Consequently, the **input** and **output** files (and any other files) do not have to be declared in the program heading.

Appendix G

Case statements

If the control variable equals a value that is not in the case label list, the statement after the **case** statement is executed.

G.2 Extensions to standard Pascal

Data types

UCSD Pascal has a data type called long integer that can take values outside the range **-maxint** to **maxint**. Such a variable is created by declaring the variable to be of the type **interger[N]**, where N is an integer that is less than or equal to 36 and gives the maximum number of digits. For example,

```
var   LargeNumber : integer [25];
```

creates a variable **LargeNumber** that can have up to 25 digits. A long integer constant can be declared implicitly by setting a constant identifier equal to a number larger than **maxint**. Long integers can be used interchangeably with integers except that: the operation **mod** is undefined for long integers, a function cannot have a data type of long integer, and long integers are not treated as ordinal types and thus cannot be used for such purposes as creating subranges.

Character strings

UCSD Pascal has extensions for use with character strings. The nonstandard data type, **string**, is used instead of the standard **packed array of char**. A character string is declared by using the standard word **string** followed by the length of the string in brackets. The maximum length of a string is 255 characters.

```
type  Word = string [10];
```

A string and an array of characters that are the same length are compatible, and a string is compatible with any other string no matter what its length. Although the boolean operators can be used to compare two strings of different length, the shorter string is always smaller no matter what its contents.

Subprograms that are available in UCSD Pascal for use with strings are described in the list on page A45.

Files

In standard Pascal, there are two standard files, **input** and **output,** which do not have to be declared, reset, or rewritten prior to their use. UCSD Pascal has an additional standard file called **keyboard**. If a character is read from **input,** it is entered from the keyboard and echoed on the video display terminal; if a character is read from **keyboard** it is entered from the keyboard, but does not appear on the video display terminal. Input from **keyboard** is useful when a password or other confidential data must be entered by the user.

Interactive is a data type that can be used in UCSD Pascal to declare **files** that behave like the standard files **input** and **output.** An **interactive file** is assumed to be empty at the start of any input/output operation.

In addition there are several procedures in the list beginning on this page that can be used with files in UCSD Pascal. Most of these procedures pertain to the association between Pascal files and disk files.

Compiler options

Compiler options affect the compiled code generated by the compiler. A list of some of the compiler options available in UCSD Pascal follows; there are additional compiler options that allow sections of code to be marked so that the code will be compiled or skipped depending on the setting of a flag. Some of the options in the following list are either active or inactive: an option is active when followed by a plus sign and inactive when followed by a minus sign. The default for these options is indicated by the sign included in the list.

(*$C charstring*) The charstring will be placed in the copyright field in the codefile.

(*$I filename*) The file **filename** is included in the program. The default extension is **.text**.

(*$I+*) When option is active and an error occurs in input or output, an error message is printed at the video display terminal.

(*$L−*) The L option controls the listing of the program as it compiles. Normally, the L option is inactive and no listing file is written. If the option is active, a listing file is written to the file **lst.text** or to a file that is named, for example, (*$L+ save.text*) saves the listing in the file **save.text**. The listing file contains information about the compilation of the program, such as the offset of each line.

(*$R+*) When the R option is active, checking is performed to determine whether scalar or subrange values are within range.

(*$P+*) When the P option is active, the listing file is printed so that it skips over page breaks on continuous form paper, assuming the pages are 8½ inches long. If no sign is included (*$P*) causes a page break wherever it is placed.

(*$Q−*) When the Q option is inactive, the compiler sends information on its progress to the console; when it is active, only error messages are sent to the terminal. The default is usually minus, but depends on the value of SLOWTERMINAL in the miscellaneous information file in the p-System.

(*$T charstring*) The charstring will be printed as a title at the top of each new page in the listing file.

Subprograms available in UCSD Pascal

atan (NumVar) is an alternative form of the standard Pascal function **arctan**.

close (pascalfile, option) closes an external **file** created in a UCSD program. The **option** argument, which may be omitted, allows you control over the disk file associated with the **pascalfile**. **Option** can take the following values:

    lock   saves the current version of the disk file.
    normal   saves the previous version of the disk file.

> **crunch** saves only the elements of the disk file up to and including the last element that was accessed.
>
> **purge** deletes the disk file.

Note that if a disk file that exists prior to running a program is associated with an external file, the default option **normal** saves the *original* version of the file. To save the new version, you must use the **lock** option when closing the file, as follows:

```
close (outfile, lock)
```

concat (CharString1, CharString2, CharString3) returns the single string that results from joining the argument strings in the order in which they are listed. For example,

```
Phrase1 := 'This is ';
Phrase2 := 'a test';
NewPhrase := concat (Phrase1, Phrase2)
```

results in **NewPhrase** having the value 'This is a test'. Any number of strings can appear in the argument list, but if the resulting string length is greater than 255 characters, a run-time error results.

copy (CharString, Position, NumChars) returns a character string that is a copy of **CharString** starting from the index **Position** and is **NumChars** in length. After execution of the following statements, **FirstName** has the value 'John Q.':

```
Name := 'John Q. Smith';
FirstName := copy (Name, 1, 7)
```

delete (CharString, Position, NumDelete) removes a substring of length **NumDelete** from **CharString** starting at index **Position**. After execution of the following statements, **Phrase** has the value 'charters':

```
Phrase := 'characters';
delete (Phrase, 5, 2)
```

gotoxy (x,y) moves the cursor to position (x,y) on the screen. x is the column (horizontal) and y is the row (vertical). **x** and **y** must be integers. The upper left corner of the screen is (1,1); the lower right corner is (80,24) on a screen with 80 columns and 24 rows.

insert (AddString, CharString, Position) inserts **AddString** into **CharString** after index **Position** in **CharString**. For example, after execution of the following code, **Name** has the value 'John Q. Smith':

```
Name := 'John Smith';
MiddleInitial := ' Q.';
insert (MiddleInitial, Name, 5)
```

length (CharString) returns the number of characters in **CharString**. The variable **NumChar** has a value of 7 after execution of the following statements:

```
NumChar := length ('nothing')
```

log (NumVar) returns the real value equal to the logarithm base 10 of the argument; for example, **log (100)** is 2.0.

pos (SearchFor, CharString) searches for the character string **SearchFor** within the character string **CharString**. **Pos** returns the index of the first character in **CharString** that matches **SearchFor**. If **SearchFor** is not found, **pos** returns a value of 0. For example,

```
writeln (' Enter a line of text');
readln (OneLine);

Yes := 'yes';
Index := pos (Yes, OneLine);
if Index > 0 then
  writeln (' The word yes occurred in that line')
else
  writeln (' The word yes did not occur in that line')
```

```
=>
 Enter a line of text
□well I guess maybe yes
 The word yes occurred in that line
```

pwroften (IntVar) returns a real value equal to 10 raised to the power of the integer argument **IntVar**; for example, **pwroften** (3) is 1000.0.

reset (pascalfile, CharString) associates an external Pascal **file** with a disk file. The following statement assigns the disk file **#5:oldstuff.text** to the Pascal **file infile** and **reset**s the **file**:

```
reset (infile, '#5:oldstuff.text')
```

rewrite (pascalfile, CharString) associates an external Pascal **file** with a disk file; for example, the following statement assigns the disk file **#5:newstuff.text** to the Pascal **file outfile** and **rewrite**s the file:

```
rewrite (outfile, '#5:newstuff.text')
```

seek (pascalfile, ElementNumber) moves the pointer in the file **pascalfile** to the element at location **ElementNumber**, thus allowing a file of a structured data type to be accessed randomly. The statement

```
seek (infile, N)
```

moves the pointer to the Nth element in **infile.**

sizeof (AnyIdentifier) returns the integer number of bytes allocated for an identifier, which can be a constant, variable, or data type.

str (IntVar, CharString) returns the character representation of the integer or long integer **IntVar**. If **IntVar** has a value of 105 when **str** is executed, **CharString** will return with the value '105'.

time (HiWord, LoWord) returns the current value of the clock in sixtieths of a second as a 32-bit word. The **time** procedure in UCSD Pascal does not return the current clock time.

Appendix G

Additional reserved words in UCSD Pascal

external	process	unit
implementation	segment	uses
interface	separate	

Additional standard identifiers in UCSD Pascal

atan	insert	pos	time
attach	interactive	processid	treesearch
blockread	ioresult	pwroften	unitbusy
blockwrite	keyboard	release	unitclear
close	length	scan	unitread
concat	log	seek	unitstatus
copy	mark	semaphore	unitwait
delete	memavail	seminit	unitwrite
exit	memlock	signal	varavail
fillchar	memswap	sizeof	vardispose
gotoxy	moveleft	start	varnew
halt	moveright	str	wait
idsearch	pmachine	string	

In UCSD Pascal, **nil** is not a reserved word, but is a standard word.

G.3 Differences between UCSD Pascal and standard Pascal

Div *and* mod

In some versions of UCSD Pascal **div** and **mod** accept negative operands and return system-dependent results.

Unary negator

The unary negator is not in the standard precedence order in UCSD Pascal. Expressions such as 4*−5 and 5.6/−9.0 are illegal and must be written with parentheses to force the correct operator precedence.

Boolean *variables*

The values of **boolean** variables cannot be printed.

Records

The field list of a **record** cannot be empty.

The goto *statement*

The scope of a label is restricted to the main body of the block in which it is declared; therefore, a **goto** statement cannot refer to a label declared in an outer block.

Subprogram parameter lists

The parameter list of a subprogram cannot contain another procedure or function.

The procedures **pack** *and* **unpack**

The functions performed by **pack** and **unpack** are performed automatically by the p-System and are, therefore, not supported by UCSD Pascal.

G.4 Creating and running a UCSD Pascal program

The p-System operates with a command hierarchy that starts from the operating system menu. From the menu you have access to the major functions of the operating system: editing, filing, compiling, executing, and other functions. Options in the menu are listed with the command letter, followed by a left parenthesis, then the rest of the command name. Some options in the command menu are

```
Command: E(dit, R(un, F(ile, C(omp, L(ink, X(ecute, A(ssem,? [IV.2.1 R3.4]
```

For example, to execute a program you type "x" for X(ecute. You will then be prompted for a file name. This appendix will describe only a few of the options in the command menu. Before starting, some notation and terms used in the UCSD documentation will be explained.

The concept of a volume is used to organize the information flow within the p-System. A volume is any I/O device such as a printer, keyboard, or disk. There are two types of volumes: storage devices (or block-structured devices) such as floppy disks or hard disks that can store a directory and files, and communication devices such as printers or modems that process data as a stream rather than in a structured manner.

A volume can be referenced by its *device number* or by the *volume ID*. Each device used by the p-System has a device number that never changes. A device number consists of a number sign followed by a number, for example, the console is device #1 and the second disk drive is device #5. The volume ID is assigned by the system or by the user and is the name of the device or the ID of the disk currently in the drive. For example, the volume ID of the printer is PRINTER. Both device numbers and volume IDs are followed by a colon to separate them from file names within the volume. A floppy disk might be a volume called PROGRAM: which contains test programs and another floppy disk might be a volume called STARS: which contains programs and data for analyzing stars.

Within the p-System, the following conventions are used for naming files and volume IDs. The name of a file can be at most 15 characters long, not counting the volume ID that precedes it.

*	stands for the volume ID of the system disk.
:	stands for the volume ID of the default disk. (specified by S(et in the main menu or P(refix in the filer menu)
.Text	a text file created by the editor—usually containing source code.

Appendix G

.Code a code file containing executable code, either p-code or machine code.

.Bad an immovable file that covers a bad sector or a disk—created using B(ad-blks in the filer menu.

P-System volumes and files can be accessed only within the p-System. If you create a file on an MS-DOS disk using another editor, such as WordStar, you must execute a utility program (the DOSFILER) to transfer the file into the p-System before the file can be edited or compiled.

The edit, compile, and run functions are similar to those functions in other systems: the editor allows you to enter and modify data, the compiler allows you to translate a program from Pascal to p-code, and run allows you to execute a program. The filer, however, performs a function that is not a separate function in most operating systems. The filer allows you to query the system to find out what volumes are on-line, get a directory of the files in each volume, copy files, rename files, *etc*. For example, to find out what volumes are on-line, you first type "f" from the command menu. The filer will load and you will be presented with a new menu. If you type "v" for volumes, you will get a list similar to

```
Filer: K(rnch, M(ake, P(refix, V(ols, X(amine, Z(ero, O(n/off-line,? [6R4.0]
Vols on-line:
   1   CONSOLE:
   2   SYSTERM:
   5 # PROGRAM: [  720]
   6   PRINTER:
   7   REMIN:
   8   REMOUT:
  13 # RAMDISK: [  142]
  14 # PSYSTEM: [  545]
Root vol is - PSYSTEM:
Prefix is   - PSYSTEM:
```

To find the names of the files stored in each volume, you can use the "l" or "e" commands from the filer menu. When prompted for the volume name, you can enter either the device number (#5:) or the volume ID (PROGRAM:). In either case, you will get a listing of the files in the volume.

```
Filer: L(dir, R(em, C(hng, T(rans, D(ate, Q(uit, B(ad-blks, E(xt-dir,? [6R4.0]
PROGRAM:
TEST.TEXT          4 11-Oct-85         TEST.CODE          2 15-Oct-85
CASE.TEXT          4 29-Apr-85         CASE.CODE          2 29-Apr-85
LONG.TEXT          4 29-Apr-85         LONG.CODE          2 29-Apr-85
6/6 files<listed/in-dir>, 24 blocks used, 696 unused, 676 in largest
```

You can refer to the file LONG.TEXT either as PROGRAM:LONG.TEXT or as #5:LONG.TEXT.

The following control sequences have an equivalent effect in all of the UCSD Pascal menus.

symbol	*key*	*effect*
<esc>	escape key	aborts the last command
<cr>	return key	continues with the current command using the defaults

\<etx\>	^C	saves any changes and continues
quit	q	leaves the current menu and returns to the command menu

Note: The sequence ^C indicates that the control key and the C key should be pressed simultaneously. Quit in UCSD is the equivalent of exit in most other systems. It does not mean abandon; use \<esc\> to abandon.

To write a Pascal program in a p-System file, you must first create a new file in a new volume. The first step is to X(ecute DISKFORMAT which will request a blank disk for the B drive and then format the disk, request the size of the volume (use the maximum size given), and request a name for the volume. Once you have created the volume, enter "q" to return to the command menu. If you select the F(iler, you can verify that the volume has been created with the V(ols command and you can see that there are no files in the volume with either the L(ist-dir or E(xt-dir command. Remember that you must type "q" again to leave the filer and return to the command menu.

To enter your program into a file in the p-System volume, select "e" for E(ditor from the command menu and then select "i" for the I(nsert mode within the editor. Once you are in insert mode, you can enter text. Although you can backspace over what you have typed, you cannot move around within the file while you are in insert mode; to change a file you must exit from I(nsert by typing \<etx\>—that is ^C, which saves the information that you have entered. When the main editor menu appears, you can use the arrow keys to move around within the file and position the cursor where text is to be inserted or deleted and then type either "d" for D(elete or "i" for I(nsert. Once you have made the change, type \<etx\> again to save the change and return to the main editor menu. When your program is complete, save your file by selecting the W(rite option from the menu. You will then be prompted for a file name which must include the volume ID or device number, for example, #5:FIRSTPROG. The file will automatically be assigned an extension of .TEXT since it was created with the editor. Once the file has been saved, you can Q(uit from the editor and return to the command menu.

To compile your program, issue the C(omp command and give the name of the file that stores the program. If the program has syntax errors, the compiler will give you the options of abandoning, continuing, or editing. If you select edit, the editor will be loaded and the cursor will be positioned on the line in your program where the error occurred so that you can correct the error. If you instead elect to continue, you will have to remember all the errors that were flagged, but you will only have to edit your program once. Once the syntax errors have been corrected and the program compiles successfully, you can use the R(un command to execute the program. The output from the program will be displayed on the terminal unless you have directed it to go elsewhere. If your program has run-time errors, they will be displayed on the screen. Finally, when you are done with the p-System execute the H(alt command to stop.

Appendix H
VAX-11 Pascal

The VAX*-11 Pascal compiler is designed for the Digital Electronics Corporation (DEC) VAX computers running under the VMS operating system. The examples in this Appendix have been run on a VAX-11/780 running Pascal Version V3.0 under VMS 4.2. VAX-11 Pascal has many extensions designed to make working on a multi-user system easier, as well as extensions that are similar to assembly language commands. Because the number of extensions is large, this appendix will cover only a fraction of the extensions. Among the extensions that this appendix will *not* cover are: the binary logical operators **uand**, **unot**, **uor**, or **uxor**; the type cast operator that allows the data type of a variable or an expression to be temporarily redeclared; the allocation size functions; the low-level interlocked functions; the attributes such as **readonly** or **writeonly** that can be given to data types and that affect the way the compiler treats a variable; the function **address** for dynamic variables; keyed **files**, which are useful for databases; procedures such as **establish** that change the way errors are processed; or the extensions that allow Pascal programs to interact with the VMS operating system.

Any program written in standard Pascal will run under VAX-11 Pascal, but because VAX-11 Pascal buffers input from **textfiles**, a program written in standard Pascal may require minor adjustments to run as desired under VAX-11 Pascal.

H.1 Implementation-dependent features

Identifiers

Identifiers in VAX-11 Pascal can be up to 31 characters long. A name that is longer than 31 characters results in a warning message, and only the first 31 characters count in the identifier name. The underscore and the dollar sign are nonstandard characters that can be used in identifiers. These nonstandard characters *are* significant in a name—that is, **New_Word** and **NewWord** refer to different storage locations. The dollar sign character, however, has special meaning in the VAX/VMS operating system and should be used with caution within identifier names.

Ranges

Integers range from −2,147,483,647 to 2,147,483,647. Overflow of integers is not detected unless the subrange checking option is active during compilation. Real numbers range from −1.7e+38 to −0.29e−38 and from 0.29e−38 to 1.7e+38 with up to 7 significant digits in the mantissa. Greater accuracy is available using the extended data types **double** and **quadruple** which are described in Section H.2. If a real number is created that

---

*VAX is a trademark of Digital Equipment Corporation.

is too large to store, the program halts execution; a real number too small to store is evaluated as 0.0. A **set** in VAX-11 Pascal can have up to 256 elements.

Program heading

All files declared in the program heading are external files by default; therefore, even though there are alternate ways of declaring external files, you should list all external files, including **input** and **output**, in your program heading. The **open** command that associates an external Pascal file with a VMS file is covered in Section H.2.

Comment statements

Both { } and (* *) are valid delimiters for comments and are equivalent. A comment begun with a { can be ended with a *), and *vice versa*. Because the two types of delimiters are equivalent, comments cannot be nested.

H.2 Extensions to standard Pascal

Type compatibility

VAX-11 Pascal supports two kinds of type compatibility: *structural* compatibility and *assignment* compatibility. Data types are structurally compatible if they have the same base type, or types, and the same number of elements of each type listed in the same order. As a relaxation of the type-compatibility rules of Pascal, VAX Pascal allows a variable to be passed as an argument into a **var** parameter with which it is structurally compatible. This relaxation makes writing generalized subprograms easier because the data type identifiers do not have to be identical across all subprograms.

Representation of integers in alternative number systems

The VAX Pascal compiler allows integers to be specified in the binary, octal, or hexadecimal system as well as the decimal system. To specify an integer in an alternative number system, the number must start with a percent sign (%) followed by a special character indicating the number system and then the number enclosed in single quotation marks. If the number is negative, a minus sign must precede the percent sign; if the number is positive, the plus sign is optional. The following characters indicate the number system:

    B   binary
    O   octal
    X   hexadecimal

The characters can be in either upper or lower case. For example,

    −%b'011'   −3 given in base 2
    %X'10'   16 given in base 16
    −%O'11'   −9 given in base 8
    +%B'10'   2 given in base 2

Data types

VAX-11 Pascal has a predefined data type **unsigned** that is a subrange of integers from 0 to 2∗**maxint**. VAX-11 Pascal also has three additional predefined data types for real numbers: single, double, and quadruple. **Single** is equivalent to **real**; **double** is a double-precision real value—that is, a value with twice as many bytes as a **real** value; **quadruple** is a quadruple-precision value—that is, a value with four times as many bytes as a **real** value. Variables of type **double** and **quadruple** can be declared so that the extra bytes either extend the number of significant digits in the mantissa or extend the range of values in the exponent.

Additional arithmetic operators

In addition to the standard Pascal operators, VAX-11 Pascal has an exponentiation operator, ∗∗. The statement

```
C := A ** B
```

assigns the value of A^B to **C**. If **B** is negative, the result of the exponentiation operation varies depending on the data types and values of **A** and **B** as given in the following table.

A	B	C
real	negative	error
0	negative or zero	error
1	integer negative	1
−1	negative and odd	−1
−1	negative and even	1
any other integer	integer negative	0

The case statement

If the control variable of a **case** statement equals a value that does not appear in the case-label list, a run-time error results. The VAX-11 compiler supports an **otherwise** option. If the control variable equals a value that is not in the list and an **otherwise** follows the **case** statement, the statements in the **otherwise** clause are executed. For example,

```
Test := 5;
case Test of
  2, 4 : writeln (' even');
  1, 3 : writeln (' odd');
otherwise
  writeln (Test:2, ' does not appear in the case statement')
end

=>
 5 does not appear in the case statement
```

Notice that the end of the case statement follows the **otherwise**.

Enumerated Types

In VAX-11 Pascal, enumerated types can be parameters in **read**, **readln**, **write**, and **writeln** statements. For example,

```
writeln (' The names of the fuel types are:');
for FuelName := Oil to Wind do
  writeln (' ', FuelName);
writeln;

writeln (' Enter the name of a fuel');
readln (PlantFuel);

writeln;
writeln (' The fuel name that you entered was ', PlantFuel);
```

```
=>
The names of the fuel types are:
     OIL
     GAS
     NUCLEAR
     COAL
     HYRDO
     SUN
     WIND

Enter the name of a fuel
□nuclear
The fuel name that you entered was   NUCLEAR
```

When you enter the value for an enumerated type, you only need to enter enough characters to uniquely determine the value. In the preceding example, since each fuel type starts with a different letter, only the first letter needs to be entered. It is safer to enter as few letters as possible because the program will end in a run-time error if no match is found. So, for example, if you had entered 'nuclaer' when prompted for the fuel name, the program would have terminated.

Character strings

Like many Pascal compilers, VAX-11 Pascal has extensions to make character strings easier to work with. As in standard Pascal, a character string can be created by declaring a **packed array of char**acters. VAX-11 Pascal also has a data type called **varying of char** which allows you to create a character string in which the length varies between 0 and the specified upper bound. The following statements declare a data type **CharString** which is a character string with a length between 0 and 20:

```
type  CharString = varying [20] of char;
```

Any character string with at most **UpperBound** characters can be assigned to a variable of type **varying of char**.

A character string is compatible only with another character string of the same length. The length of a variable of type **varying of char** is the length of the string assigned to it, not its maximum value. As in standard Pascal, the boolean operators can be used to compare two strings of the same length. The comparison is based on the ASCII values of the characters in the strings.

Control characters can be included in strings, so that you can, for example, ring the bell at the terminal. The control codes and their effects are given in Table A.2 in Appendix A, where you can find that ^G (ASCII code 07) rings the bell. In VAX-11 Pascal, a control character is specified by the ASCII number code in parentheses. The string

```
'(07)'
```

is the control code that rings the bell at the terminal. The following **writeln** statement rings the bell and then prints a message at the terminal:

```
writeln (' That was incorrect.'(07)' Please re-enter.');
```

=>
 That was incorrect. Please re-enter.

Functions and operations for use with character strings

Most of the functions and procedures described in this section have options that are not covered here. You can read the documentation for VAX-11 Pascal to find out about the additional options.

bin (FixedString) converts the value of **FixedString** to a **varying** string of binary digits equivalent to the value of **FixedString**. The character string passed to **bin** cannot be a **varying** string.

hex (FixedString) is the same as the function **bin** except that **FixedString** is converted to its hexadecimal equivalent.

index (CharString, SearchFor) searches for the character string **SearchFor** within the character string **CharString**. If **SearchFor** is found, **index** returns the index of the first character in **CharString** that matches **SearchFor**. If **SearchFor** is not found, **index** returns a value of zero. For example,

```
writeln (' Enter a line of text');
readln (OneLine);

Yes := 'yes';
Position := index (OneLine, Yes);
if Position > 0 then
  writeln (' The word yes occurred in that line')
else
  writeln (' The word yes did not occur in that line')
```

=>
 Enter a line of text
□no, no, a thousand times no!
 The word yes did not occur in that line

length (CharString) returns the number of characters in **CharString**. The variable **NumChar** will have a value of 7 after the following statement is executed:

```
NumChar := length ('nothing')
```

pad (CharSring, FillChar, NumChars) adds the character stored in **FillChar** to the end of **CharString** until it is **NumChars** long. **NumChars** must be greater than the number of characters in **CharString** and at most the number of the upper bound of the string to which **pad** is assigned.

oct (FixedString) is the same as the function **bin** except that **FixedString** is converted to its octal equivalent.

substr (CharString, Position, NumChars) creates a substring of length **NumChars** from **CharString** starting at index **Position**. After the following statements are executed, **Phrase2** will have a value of 'acters':

```
Phrase1 := 'characters';
Phrase2 := substr (Phrase1, 5, 6)
```

+ Strings can be concatenated by using the + operator, as follows:

```
Phrase1 := ' This is';
Phrase2 := ' a test';
NewPhrase := Phrase1 + Phrase2;
writeln (NewPhrase)
```

```
=>
 This is a test
```

This operation has the same order of precedence as addition in expressions.

Procedures for use with character strings

readv (CharString, parameter-list) extracts data from a character string and assigns the data values to variables. The character string is treated as if it were a one line **textfile**, so the rules and errors for **readv** statements are the same as for **readln** statements. The following statements read a character string from the terminal and then assign different parts of the input string to variables.

```
writeln (' Enter a character, an integer, and a real number');
readln (OneLine);
readv (OneLine, CharVar, IntVar, RealVar);
writeln;
writeln (' The character is ', CharVar, ', the integer is ', IntVar:3,
         ', and the real number is ', RealVar:8:1)
```

```
=>
 Enter a character, an integer, and a real number
* 23 6.7
 The character is *, the integer is  23, and the real number is      6.7
```

writev (CharString, parameter-list) writes values of any data type to a character string. The character string to which the values are written is treated as a one-line **textfile**, and the rules and errors for **writev** are the same as for **writeln**. The following statements write a real, an integer, and a character variable to a character string.

```
        writev (OneLine, CharVar:1, IntVar:4, RealVar:8:1);
        writeln (' The values you entered were ', OneLine)
=>
 The values you entered were *  23     6.7
```

Transfer functions

dbl (NumVar) returns the double precision equivalent of the argument, which must be of an arithmetic data type.

int (OrdVar) returns the integer equivalent of the argument, which must be of an ordinal data type.

quad (NumVar) returns the quadruple precision equivalent of the argument, which must be of an arithmetic data type.

sngl (NumVar) returns the single precision equivalent of the argument, which must be of an arithmetic data type.

uint (OrdVar) returns the unsigned integer equivalent of the argument, which must be of an ordinal data type.

uround (RealVar) returns the rounded unsigned integer equivalent of the real argument.

utrunc (RealVar) returns the truncated unsigned integer equivalent of the real argument.

Miscellaneous functions

card (SetVar) returns the number of elements in the argument, which must be of the data type **set**.

clock returns the time in milliseconds that has been used by the current process.

expo (RealVar) returns the exponent of the real argument. For example, **expo(12.34)** is evaluated as 2 since 12.34 is represented as 0.1234e02 in standard notation (see Chapter 5).

Miscellaneous procedures

date (CharString) assigns the current date to the character string that is passed as an argument. **CharString** must be declared as a **packed array [1..11] of char**. The data is returned in the format dd-mmm-yyyy, where dd is the day given as a one or two digit number, mmm is the month given as a three character code, and yyyy is the year given as a four digit number.

time (CharString) assigns the current time to the character string that is passed as an argument. **CharString** must be declared as a **packed array [1..11] of char**. The time is returned in the format hh:mm:ss.cc, where hh is the hour on a twenty-four hour clock, mm is minutes, ss is seconds, and cc is hundredths of seconds.

halt stops execution of the program.

Files

In VAX-11 Pascal, the extensions for files can be grouped into two categories: those that deal with the association between Pascal **files** and VAX/VMS files* and those that expand the operations allowed on file variables. VAX-11 Pascal allows three types of files: sequential, random access (or relative), and keyed (or indexed). This appendix covers only sequential and random access files and only a small number of the options available for files. You can consult the documentation for VAX-11 Pascal for more information about files.

All **files** must be **open**ed prior to use within a VAX-11 Pascal program. Internal **files** can be opened implicitly by using the **reset** or **rewrite** statements, but for external files you must use the **open** statement so that the VAX/VMS computer file with which they are associated is identified.

The **open** statement gives the name of the Pascal **file**, the name of the VAX/VMS computer file to be associated with the Pascal **file**, the access method to be used for the file (sequential or random access), the print format for the file, and the original organization (sequential or random access) of the file. These parameters can be specified by assigning values to certain predeclared identifiers, for example,

```
open (infile, 'oldstuff.dat', history := old);
```

where **infile** is a **file** variable that has been declared by the user and where the predeclared identifiers can be assigned the following values:

file_name The **file_name** must be assigned the name of the VAX/VMS file to be associated with the Pascal **file**. The **file_name** must be assigned a character string (either a constant or a variable). If **file_name** is not assigned a value, the compiler will assign it the name **pascal.dat**.

history The **history** parameter indicates the current status of the VAX/VMS file and can be assigned one of the following values: **new**, **old**, **readonly**, or **unknown**. An output **file** to be created by the Pascal program is assigned a **history** of **new**. An input **file** that already exists as a VAX/VMS computer file is assigned a **history** of **old**. If the existing VAX/VMS file should not be changed, **history** can be assigned a value of **readonly**. An error will occur if a write statement is issued for a **readonly** file. Finally, **history** can be assigned a value of **unknown**, indicating that the current status of the VAX/VMS computer file assigned to **file_name** is not known: if the VAX/VMS file already exists, it will be opened as an old file; if the VAX/VMS file does not already exist, it will be opened as a new file.

access_method The **access_method** parameter indicates how the contents of a file can be accessed and can be assigned one of the following values: **sequential**, **direct**, or **keyed**. (Keyed files will not be covered here.) A **sequential** file corresponds to the standard Pascal file structure in which elements of the file can only be accessed in order. A **direct** file corresponds to a random access file. **Direct** files are covered in more detail in the next section.

record_type The **record_type** parameter indicates whether all the elements of a

---

*Properly speaking, VAX/VMS files are referred to as RMS (Record Management Services) files; however, we will refer to them as VAX/VMS files to avoid confusion. (See the vocabulary warning on page 573.)

file are of the same length and can be assigned a value of **fixed** or **variable**. **Variable** is the default for **textfile**s since each line may be a different length; **fixed** is the default for all other file types. If a file is to be a random-access file, its **record_type** must be **fixed**.

carriage_control The **carriage_control** parameter controls how output is spaced when it is printed. The term "carriage control" is used because some printers have carriages, or rollers, like typewriters that can be turned up or down to control the spacing. **Carriage_control** can be assigned any of the following values: **list**, **carriage**, **fortran**, **nocarriage**, or **none**. If **carriage_control** is assigned the value **list**, the **file** is printed single spaced with each element of the **file** separated by a carriage return. **List** is the default for all **textfile**s. If **carriage_control** is assigned the value **carriage** or **fortran**, the first character in every output line is used to control the spacing of the output. Table H.1 gives the standard carriage control characters and their effect on the printer.

Table H.1 Standard carriage control characters

character	action
+	no line feed (overprinting)
blank	single spacing (default)
0	double spacing
1	new page
$	prompting: starts new line and suppresses carriage return at the end of the line
' '(0)	prompting: suppresses new line and suppresses carriage return at the end of the line

If **carriage_control** is assigned the value **nocarriage** or **none**, no carriage control signals are sent to the printer, so the file is printed across the entire line of the output device, advancing to the next line when one line is full. **None** is the default for files other than **textfile**s.

There are several things to beware of about carriage control. One is that putting a character in column one by mistake can lead to disasters such as all of your output being printed on a single line or each line being printed on a separate page. If you occasionally print using carriage control, you should get in the habit of *always* putting a blank character at the beginning of each write statement.

When the processing of an **open** external **file** is completed, the computer file should be explicitly saved using the **close** command, as follows:

```
close (infile, disposition := save)
```

where **infile** is the name of a **file** that has previously been opened. Other values for the **disposition** parameter are; **delete**, **print**, **print_delete**, **submit**, and **submit_delete**. The default disposition for external **files** is **save**, and the default **disposition** for internal **files** is **delete**.

Subprograms for files

In addition to the standard procedures and functions for files, such as **put, get, eof**, VAX-11 Pascal has a function that allows you to test whether the buffer variable is undefined. This function can be helpful in preventing abnormal termination of programs that use **files**.

ufb (FileVar) returns a value of true if the last file operation performed on the file **FileVar** resulted in an undefined buffer variable. That is, **ufb** returns a value of false if the buffer variable for **FileVar** is currently positioned at a valid entry in the file.

Subprograms for sequential-access files

VAX-11 Pascal supports all the subprograms for standard Pascal files. In addition, it supports the following subprograms:

linelimit (pascalfile, MaxLines) limits the number of lines that can be written to the text file **pascalfile**. **MaxLines** must have an integer value. Execution of the program terminates as soon as more than **MaxLines** are written to the **textfile**. The **textfile** can be in any mode when the **linelimit** procedure is executed.

status (pascalfile) returns an integer code indicating the effect of the last operation on **pascalfile**. 0 indicates that the last operation was successful; -1 indicates that the last operation encountered the end-of-file; a positive integer indicates that an error occurred on the previous operation. A list of the positive error codes is given in the VAX-11 Pascal documentation.

truncate (pascalfile) deletes the current element plus all the elements following the current element in **pascalfile**, which must be a sequentially organized **file**. **Truncate** can only be issued for a file that is in inspection mode. After the statement has been executed, the buffer variable is positioned at the end-of-file and the **file** is in write mode.

If you are writing interactive programs that require more than simple answers from the user, you should read the sections in the VAX-11 Pascal documentation on **textfile** buffering, **files of char**, prompting of **textfiles**, and delayed device access also called lazy lookahead.

Subprograms for random-access files

As an extension to standard Pascal, VAX-11 Pascal supports random-access, or direct-access, **files**. After a file is **open**ed with the **access_method** set to **direct**, as follows, the subprograms **find, locate,** and **update** can be used.

```
open (realfile, 'realnums.dat', history := old, access_method := direct);
```

find (pascalfile, ElementNumber) allows a **file** to be accessed randomly just as arrays are. The statement

Appendix H

```
find (realfile, N)
```

moves the **file** pointer to the Nth element in **infile**. After the **find** procedure is executed, the **file** is in inspection mode. The procedure **FindAndRead** uses the **find** procedure to position the **file** pointer at a given location and then reads the value stored there.

```
procedure FindAndRead (var realfile : RealFileType;
                      ElementNumber : integer; var ElementValue : real);
(*      This procedure finds and reads the element in realfile    *)
(*      that is in position number ElementNumber.  It assigns this *)
(*      value to ElementValue.                                    *)

begin
  find (realfile, ElementNumber);
  read (realfile, ElementValue)
end;   (* procedure FindAndRead *)
```

locate (pascalfile, ElementNumber) is similar to the **find** procedure in that it moves the file pointer to a particular element number. The difference is that after the **locate** procedure is executed, the **file** is in write mode and the pointer is positioned so that the next element is written at the location given by **ElementNumber**.

update (pascalfile) writes the contents of the buffer variable at the current **file** position. A **file** that is **update**d must be in *inspection mode*, and it remains in inspection mode after being **update**d.

Additional reserved words in VAX-11 Pascal

module	**value**	**%immed**	**%stdescr**
otherwise	**varying**	**%include**	
rem	**%descr**	**%ref**	

Additional standard words in VAX-11 Pascal

add_interlocked	**double**	**lower**	**size**
address	**establish**	**next**	**status**
bin	**expo**	**oct**	**substr**
bitnext	**find**	**open**	**time**
bitsize	**findk**	**pad**	**truncate**
card	**halt**	**quad**	**uand**
clear_interlocked	**hex**	**quadruple**	**ufb**
clock	**index**	**readv**	**uint**
close	**int**	**resetk**	**undefined**
date	**length**	**revert**	**unlock**
dble	**linelimit**	**set_interlocked**	**unot**
delete	**locate**	**single**	**unsigned**

uor	upper	utrunc	writev
update	uround	uxor	

H.3 Creating and running a program in VAX-11 Pascal

After you have entered your Pascal program into a VAX/VMS file, getting the program to execute is a three step process. Suppose that you have entered your program into a VAX/VMS file called **myprogram.pas**. First, the Pascal compiler must be invoked to check the program for syntax errors. To compile **myprogram.pas**, type the VMS command **pascal** followed by the file name in response to the VMS $ prompt. The extension **.pas** can be omitted because it is the default extension for the command **pascal**.

```
$ pascal myprogram
```

If there are errors, they will be printed at the terminal. If there are no errors, the program will be translated into binary, or object, code and will be stored in a file called **myprogram.obj**.

The next step is to **link** the program. In this step, references to subprograms that are external to the compiled program are resolved. These references may be to system libraries containing subprograms for error handling or for arithmetic function calculations, to libraries that you have created, or to other VAX/VMS files that contain object code. Assuming that the original program stored in **myprogram.pas** is a simple Pascal program, **myprogram.obj** can be linked as follows:

```
$ link myprogram
```

When the program has been linked successfully, an executable file with extension **.exe** is created. To execute the file **myprogram.exe**, type

```
$ run myprogram
```

and the program will execute. If there are any run-time errors in the program, you must go back, edit the file **myprogram.pas**, and repeat the compile, link, and run steps. Although this sequence is tedious for short programs, it is efficient for large programs because subprograms can be compiled as individual modules and a change in one module does not require that the entire program be recompiled. This system of individually compiled modules also makes it easier to include subprograms that are not written in Pascal (see Appendix E).

Compiler directives and qualifiers

To include a subprogram that is stored in a separate VAX/VMS file, the **%include** directive can be used. This directive is analogous to the compiler option (*$i*) that was used in the main part of the text. The directive has the following format:

```
%include 'filename'
```

where **filename** is the complete name of the VAX/VMS computer file that contains the subprogram to be inserted.

The VAX-11 Pascal compiler has many options that can be activated at the time the

Appendix H

program is compiled. Among the options that can be specified when the compiler is invoked are the following:

/**check=all** This option enables all the checks on ranges. When **all** is specified the following checks are made: **bounds**, checks that array indices and character string lengths are within bounds; **case_selectors**, checks that the case control value is in the case-label list; **overflow**, checks that integers too large to store are not created; **pointers**, checks that pointer values are not **nil**; **subrange**, checks that subrange variables are assigned valid values and that set elements are compatible. The default is /**check=bounds**. To turn off all checking, use /**nocheck** or /**check=none**.

/**standard** This option will flag all lines in the program that are not standard Pascal. The errors will be stored in a file called **sys$error**. The default is /**nonstandard**, that is, nonstandard Pascal is not flagged.

/**list** This option produces a listing of the source file that contains information about the program compilation. The listing is stored in a file with the extension of **.lis**. In interactive mode, the default is /**nolist**, that is, no listing is produced.

To compile the program stored in **myprogram.pas**, to have all the checks performed and the nonstandard Pascal flagged, and to get the listing file, you would type

```
$ pascal myprogram/check=all/standard/list
```

Answers to Odd-Numbered Exercises

Chapter 2

1. (a) valid (b) invalid: # is not a standard alphanumeric character (c) invalid: ? is not a standard alphanumeric character (d) valid (e) invalid: **program** is a reserved word (f) valid (g) invalid: $ is not a standard alphanumeric character (h) valid (i) valid (j) invalid: does not start with a letter

3. For the Turbo Pascal and UCSD compilers, the underscore is the only nonstandard character allowed in identifiers. For the VAX-11 Pascal compiler, the underscore and the dollar sign are nonstandard characters that can be used in identifiers.

5. line 5: the **const** statement requires an equal sign, not an assignment operator: **const Ten = 10;**

   line 6: the positions of the semicolon and the colon are reversed: **var Number : Integer;** The initial capital letter in **Integer** does not matter.

   line 8: The character string does not have a closing single quotation mark. **writeln (' The number is, ');**

   line 9: The equal sign should be an assignment operator: **Number : = 378;**

   line 11: The double quotation mark at the end of the string should be a single quotation mark. **writeln (' Ten times the number is ');**

   line 13: **Number** is misspelled. **writeln (Number);**

   line 14: The period after the **end** is missing: **end.**

7. (a) $A = 3, B = 9$ (b) $X = 1.0, Y = 0.0$ (c) $A = -5, B = 5, Y = 2.236$ (d) $X = 1.5, A = 2, B = 1$

9. Because integer and character variables do not have decimal points, the problem of where to put the decimal point arises only when formatting real numbers.

Chapter 3

1. (a) 1.2, real (b) 9.8, real (c) 3.2, real (d) 3, integer (e) 1, integer (f) 3, integer (g) 3, integer (h) 0.055, real (i) 25.0, real (j) 81, integer (k) 1.00, real (l) 1, integer

3. (a) $3.0 + ((5.5 * 6.0) / 7.2) = 7.6$ (b) $(-3) * (-4) = 12$ (c) $(((3.0 / 3.0) / 3.0) / 3.0) = 0.1$

5. (a) The result is   6.61
   (b) The number is   2 and Next is   2
   (c) Num1 =   2.33 Num2 =   2.33 and Num1and2 =   4.66

7. (a) 1 centimeter $= 6.2137 \times 10^{-5}$ miles   (b) 1 pound $= 2.2481 \times 10^{-6}$ dynes
   (c) 1 gigawatt $= 10^{21}$ picowatts   (d) 1 gram per centimeter^2 $= 10$ kilograms per meter^2 (e) 1 horsepower $= 2545.1$ btus per hour   (f) 1 liter $= 61.025$ cubic inches
   (g) 1 foot-pound $= 1.356$ joules

Chapter 4

1. For a tube with a radius of   5.40 inches
                 and a length of   10.30 inches

   the surface area is   349.47 inches squared
   and the volume is   943.57 inches cubed

3.
CharVar	RealVar1	RealVar2	IntVar1	IntVar2
—	—	—	106	—
—	—	—	106	109
—	10.44	—	106	109
—	10.44	10.44	106	109
—	10.44	10.44	10	109
—	10.44	5.2	10	109
'1'	10.44	5.2	10	109

5. The procedure heading for **ReadNumbers** is missing its semicolon. The **begin** of the block of **ReadNumbers** should not have a semicolon. The second writeln statement in **ReadNumbers** is missing its closing single quotation mark. **Number** is local to **ReadNumbers** so the value entered cannot be accessed in any other part of the program. The identifier **Next** is declared twice: once as a global integer variable and once as a procedure identifier. The **end** of procedure **Next** should be followed by a semicolon, not a period. The declaration of **Number** that follows procedure **Next** is not in the declarative part of the program. (If it is moved to the top of the program, will the value assigned to **Number** in procedure **ReadNumbers** be assigned to the global variable or the local variable?)

Chapter 5

1. (a) There should not be a semicolon after **then**.   (b) The expressions compared using a boolean operator must be assignment compatible.   (c) The control variable of a **case** statement cannot be a **real** expression. The **begin** for the second case label is missing its **end;**. Because a single statement follows the **if**, instead of adding **end;**, the **begin** can be deleted. A **case** statement must have an **end**.   (d) The **case** statement should not have a **begin**.   (e) The character string in the **writeln** statement is missing its initial single quotation mark.

3. with **Num1** = 1, **Num2** = 4, and **Num3** = 0

   (a) (* no output *)
   (b) True
   (c) They're less
   (d) Done
   (e) Num1 is   1 and Num2 is   4
       Num3 =   2
   (f) More or less

   with **Num1** = 4, **Num 2** = −1, and **Num3** = 2

   (a) Logically that's: 1
   (b) False
   (c) They're less
   (d) Num1 is greater than Num2
   (e) Num1 =   4
   (f) More or less

5. Values of **Number** less than 1 or greater than 5 make the **case** statement invalid. For each valid value of **Number**, the output is:

Number	Output
1	Num1 = 4.00
2	Num2 is 1
3	Num1 = 12.00
4	Control is now 8
5	Done

7. The error band is the difference between the largest and smallest measurements. The average is the sum of the measurements divided by 6. You cannot determine the true length of the string, although you could reduce the error considerably by using a more accurate measuring device. Without knowing the true value, you cannot determine the percentage error.

9. The largest shaft that is still within tolerance has a diameter of 7.4337 cm — that is, 7.43 + 0.5% or 7.43 * (1.0005). The difference between the nominal hole size and the largest shaft is 0.0663 cm. Since 0.663 is 0.88% of 7.50, the hole must be 7.50 ± 0.88%.

Chapter 6

1. (a) The data type of the variable must be given. For example, **procedure (var NewVar : real);**   (b) The procedure identifier has been omitted. For example, **procedure ReadData (var Data : real);**. A variable cannot be declared both as a parameter and as a local variable. The procedure assigns a value to the parameter, so it should be a variable parameter, not a local variable.   (c) This procedure is syntactically correct, but is essentially a null procedure. The only executable statement changes the value of **Top**; but because **Top** is a value parameter the new value is not returned in the argument. **Top** should be a variable parameter.
   (d) The data type of the function **AddIt** has been omitted. The heading is **function AddIt (X : real): real;**. Also, the function identifier, **AddIt** must be assigned a value. So, the statement within the procedure is **AddIt := X + 1.0**.

Chapter 7

1. (a) The equal sign should be the assignment operator.   (b) The **begin-end** pair is

missing. The **begin** belongs after the **do** and the **end** after the **writeln** statement. (c) **End** cannot be used as an identifier because it is a reserved word. The last statement is incorrect because the **for** statement control variable **Counter** cannot be assigned a value. (d) These statements will compile, but the semicolon after the word **do** must be removed for the statements to execute correctly. (e) These statements will compile, but to run correctly, the initialization of **Sum** must be moved above the **for** statement.

3. (a) things
   nonsense
   or not
   could be worse nonsense
   or not
   the end of all that

   (b) Counter = 1 Index = 2
       Counter = 2
       Index = 6 Counter = 3

5. **AddToTotal** is a procedure because it returns a new value for the running total **TotalAbs** plus the value for either **TotalPos** or **TotalNeg**. It also returns new values for either **CounterPos** or **CounterNeg**. A function should return only a single value. **Average** is a function because it uses all the information in the array **Total**, plus the value of **Counter**, to return a single value, the average of the array.

7. (a) 0.50, 2.00, 0.50, 2.00, 0.50

   (b) 4.00, 16.00, 256.00, 65536.00, 4294967296.00

   (c) 9.21, 4.44, 2.98, 2.18, 1.56

   (d) 6.00, 18.00, 108.00, 1944.00, 209952.00

9. 6(a) 1.00, 0.25, 0.11, 0.06, 0.04 The partial sum is 1.46.
   6(b) 0.00, 1.00, 0.87, 0.71, 0.59 The partial sum is 3.16.
   6(c) 3.00, 1.67, 1.40, 1.29, 1.22 The partial sum is 8.57.
   6(d) 1.00, 1.33, 2.00, 3.20, 5.33 The partial sum is 12.87.
   7(a) 5.50 7(b) 4295033108.00 7(c) 20.38 7(d) 212028.00

Chapter 8

1. (a) The semicolon after the **do** must be removed. The real value 2.5 cannot be added to the integer variable **IntVar**. Alternatively, if **IntVar** is real, it cannot be the argument of the standard function **odd**.

   (b) The character variable **Digit** cannot be the argument of the standard function **chr**, which returns the character stored at the integer argument location. If the **chr** function is replaced by the **ord** function, the boolean expression is logically incorrect because every integer is greater than x and/or less than $x + 9$ for any x.

   (c) This **while-do** statement is infinite because **RealNum** will oscillate between 5.0 and 0.2, both of which are less than 10.0. The expression 1/**RealNum** uses mixed-mode arithmetic, but it is acceptable in standard Pascal.

   (d) This **repeat-until** loop is infinite because the boolean expression **A and C**, where **A** is equal to **not C**, is always false.

9. (a) y (b) $5\ln(y) + 5$ (c) $z + 1096.6$ (d) $\ln(a) - \ln(b)$ (e) e^{5x} (f) $e^{x-y} + 1$

Chapter 9

1. **(a)** The type statement requires an equal sign, not a colon.
   (b) A subrange is specified with two periods, not a comma.
   (c) The index of an array must be enclosed in square brackets, or the equivalent symbols, (. .), not simple parentheses.
   (d) **Matrix** is a data type, not a variable, so the line above the **end** will not compile. Although the syntax is correct, fragment (d) contains a logical error: The control variable **Row** is incremented from 1 to 20, but the first dimension of **OldResults** is dimensioned only from 1 to 10. Even if a subrange checking option is used, the error will not be flagged until a value greater than 10 is assigned to **Row**.

5. **(a)** For A, $r = 8.16$ and $\theta = 40.53°$ or 0.71 radians. For B, $r = 11.74$ and $\theta = 105.81°$ or 1.85 radians.
   (b) $A \cdot B = |A||B|\cos\theta = (8.16)(11.75)\cos(40.53 - 105.81) = 40.05$. $A \cdot B = \Sigma\, a_i b_i = (5.3)(11.3) + (6.2)(-3.2) = 40.05$.
   (c) Assuming that A and B lie in the x–y plane, that is, both have z components of 0, then the components of $D = A \times B$ are:
   $d_1 = a_2 b_3 - a_3 b_2 = (6.2)(0.0) - (0.0)(-3.2) = 0.0$
   $d_2 = a_3 b_1 - a_1 b_3 = (0.0)(11.3) - (5.3)(0.0) = 0.0$
   $d_3 = a_1 b_2 - a_2 b_1 = (5.3)(-3.2) - (6.2)(11.3) = -87.02$
   The components of E where $E = B \times A$ are:
   $e_1 = b_2 a_3 - b_3 a_2 = (-3.2)(0.0) - (0.0)(6.2) = 0.0$
   $e_2 = b_3 a_1 - b_1 a_3 = (0.0)(5.3) - (11.3)(0.0) = 0.0$
   $e_3 = b_1 a_2 - b_2 a_1 = (11.3)(6.2) - (-3.2)(5.3) = 87.02$
   Both E and D have a magnitude of 87.02. D points straight down along the z axis and E points straight up along the z axis.

Chapter 10

1. In the following files, the end of line markers may be after the last character written on a line or in column 80 or in column 132. Check your system documentation to find whether lines have variable or fixed length as the default.
   (a) `1   4   9   16   25   36   49   64   81   100<eoln>`
   `<eof>`
   (b) `100<eoln>`
   `<eof>`

3. The following error messages were generated by the Pascal 8000/2.0 compiler.
   (a) `*** error ***`
   `     file "infile  ", get or read was not preceded by reset`
   (b) `*** error ***`
   `     undefined variable used in expression`
   (c) `*** error ***`
   `     file "infile  ", read format error - digit expected`

9. **(a)** $P(X > 5.0) = \Phi((5.0 - 5.0)/2) = 0.5$
   (b) $P(1.0 < X < 2.0) = 1.0 - \Phi((1.0-5.0)/2) - \Phi((2.0 - 5.0)/2) = 0.44$

(c) $P(X > -1.5) = 1.0 - \Phi((-1.5 - 5.0)/2) = 0.9993$
(d) $P(X < -1.5) = \Phi((-1.5 - 5.0)/2) = 0.0006$
(e) $P(-2.0 < X < -1.0) = 1 - \Phi((-2.0 - 5.0)/2) - \Phi((-1.0 - 5.0)/2) = 0.001$
(f) $P(X < 25) = \Phi((25 - 5)/2) = 1.0$

Chapter 11

1. (a) **Hydrogen** appears in both the scalar type **Gases** and the type **GroupI**. It can appear in only one type. (b) A subrange cannot be of type **real**, so the values for **Speeds** must be changed to integers. (c) Scalar values cannot be written. (d) Scalar values cannot be read.

Chapter 12

1. (a) [1, 3, 5, 7, 9, 2, 4, 6, 8, 10] (b) [ ] (c) [5] (d) [2, 4, 6, 8, 10, 0, 5]
   (e) [1, 3, 5, 7, 9] (f) [0, 5] (g) [1, 3, 5, 7, 9, 2, 4, 6, 8, 10, 0] (h) [5]

3.
```
D := A + B;          (* Exercise 1a *)
D := A * B;          (* Exercise 1b *)
D := A * C;          (* Exercise 1c *)
D := B + C;          (* Exercise 1d *)
D := A - B;          (* Exercise 1e *)
D := C - B;          (* Exercise 1f *)
D := A * B * C;      (* Exercise 1g *)
D := A * (B + C);    (* Exercise 1h *)
```

5. (a) A thermal plant
       A thermal plant
       A thermal plant
       Not a thermal plant

   (b) It's a subset
       Thermal is a subset

Chapter 13

1. (a) **WeekRecord** is a data type not a variable. (b) **Week** is not an array.
   (c) **WordArray** is not defined. The right parenthesis in line 5 and the left parenthesis in line 6 should be deleted, the left parenthesis is missing in line 7, and the variant record does not need an **end**.

3. (a) $5.2e^{166.i}$ (b) $50.6e^{-7.2i}$ (c) $13.15e^{-135.0i}$ (d) $5e^{36.9i}$

Chapter 14

1. Statement (b) will result in a compile-time error because **allfile** is a file of an array of records and **DataPoint** is a single record. Statement (c) will result in a compile-time error because the elements of a **file** cannot be referenced like the elements of an **array**. Statement (d) will result in a compile-time error because the identifier **datafile** is a **file** of records, not a record variable.

Chapter 15

1. The answer is   8.00

Chapter 16

1. (a) Point1↑ equals 6 Point2↑ equals 26
   (b) Point1↑ equals 1 Point2↑ equals 4
   (c) Point3↑ equals Z Point4↑ equals Y

Index

Δ, 287
λ, 398
μ, 391
π, 21–22, 80, 178
σ, 392
Σ, 252
Φ, 401
$n!$ (factorial), 251
$\binom{n}{x}$ (n choose x), 396
** (exponential), 303, A54
(*$i*), 129

abs, 29, 157
Absolute errors, 182
Addition
  arithmetic, 59
  complex, 546
  matrix, 355
  set, 479–480
  vector, 344–345
Algorithm, 12–15
  development, 97–101
Alphanumeric characters, 15–16
and, 143
arctan, 29
Arctangent, 13, 29
Argument, 28, 54, 196
Arithmetic operators, 59–64
array, 313–333
  compatibility, 321
  declaration, 317–319
  element, 315
  input/output of, 316, 319, 328
  multidimensional, 324–326
  packed, 432–434
  as parameters, 321
  records, 529
  subscript (or index), 316
ASCII character code, 7, 146, 279, 486, A6–A7
Assembly language, 8–9
Assignment
  to character arrays, 435
  compatibility, 27–28, 41, 56, 61, 321
  operator, 26–28, 41
  statements, 25–28, 41

Assignment of disk files, 381–382
Average of a sample, 319, 330

Babbage, Charles, 2
Base (number conversion), 5–6, A15–A17
Batch, 4–5
begin, 31
Bernoulli, James, 394
Bernoulli distribution, 394–396
Binary
  numbers, 5, A15–17
  search, 446–447
  tree, 670
Binomial distribution, 396–397
Bit, 5
Blank character, 56
Block structure, 110–113
Body, 30, 105
Boole, George, 145
boolean, 17
  algebra, 145–151
  expressions, 145–153, 271, 275–277
  functions, 153–154, 218–219
  operators, 147
  set operations, 483–485
  variables, 151–153
Bounds, 318, 430–432
Branching statements, 139–140
Bubble sort, 453–457
Buffer, 5, A31, A38
Buffer variable, 573–575
Byron, Ada, 2
Byte, 6

Call (invoke), 106–108, 198–204, 217–220
Carriage control, A60
Cartesian coordinates
  complex numbers, 541
  vectors, 336–341
case statement, 163–166
  else, A33
  otherwise, 166, A54–A55
Central Processing Unit (CPU), 2–5

CGS units, 75
char, 17, 19–20
Character
  arrays (strings), 432–440
  ASCII and EBCDIC, 7, 146, 279, 486, A6–A7
  expression, 33, 66
  file, 370
  formatting, 67–69
  reading, 56–57, 434–435
Chemical elements, 470–517
chr, A3
Circuits, electrical
  RC, 300–301
  RL, 558
  parallel, 51
  series, 96–97
Coefficient of performance, 137
Comment statements, 24–25, 115–116
Comparing
  character arrays, 435
  characters, 146
  enumerated types, 428
  real numbers, 157
  sets, 483–484
Compatible types, 27–28, 41, 56, 61, 321
Complement of a set, 472
Compiler, 9, 39
Compiler options, A38–A39, A45, A63–A64
Compile-time errors, 39–40
Complex numbers, 540–550
  addition and subtraction, 545–548
  conjugate, 556
  coordinate systems, 541–544
  multiplication and division, 549–550
  raising to powers, 558
Compound
  growth, 269
  if statement, 159–163
  interest, 262, 288–289
  statement, 155–157, 234–235
Computational errors, 172–186
Conditional
  loops, 268–278
  statements, 154–157

I1

const, 20–22
Constants, 20–22, 27, 42
   passing as parameters, 208, 238–239
   physical, 80
   scalar types, 425–432
   standard, A2
Continuous
   numbers, 8
   distribution, 398
   time growth, 290
Control structures
   case, 163–166
   for, 226–241
   if, if-then-else, 154–159
   repeat-until, 274–278
   while-do, 271–274
   with, 526–528
Control expression
   case, 164
   for, 230, 235–239, 272–274
Convergent sequence, 250
Conversion
   characters to numbers, 279–280,
     507–517
   coordinate systems, 338–341, 541–544
   of units, 81–85
Coordinates
   cartesian, 335–338, 541
   equivalence, 545
   polar, 335–338, 541
cos, 29, 313
Covariance of a sample, 424
Critical value, 412
Cross product of vectors, 349–352

Dangling else, 159, 161–163
Data checking, 278–280
Data types, 17–20
   standard, A2
date, 583
Debugging strategies, 39–44, 68–69,
   116–125, 500
Declarative part, 30, 105
Default, 18
DeMorgan's laws, 149
Dependent variable, 624
Derivative, 225
Design of programs, 488–517
Determinant of a matrix, 367
Difference
   arithmetic, 59
   equation, 638
   set, 481–483
Digit, 15–16
Dimensional analysis, 81–85
Discrete
   numbers, 8
   probability distributions, 388–398
   time growth, 287–289

Disjoint sets, 476
Disk files, 370, 376–377, 570
dispose, 650
Distributions of random variables,
   388–406
Distributive laws, 149
div, 61–62
Divergent sequences, 251
Division
   arithmetic, 59
   by zero, 119
   complex, 549
do, 229–230, 271–273, 526–527
Documentation, 125, 501–502
Dot product of vectors, 345–349
Doubling time, 297
Doubly linked list, 654–660
downto, 229–230
Dummy procedures, 115–116
Dynamic variables, 639–666

e, 80, 178, 290–303
E (exponent), 7, 18–19
EBCDIC character code, 7, 146, 279,
   486, A7
Echo print, 59
Elements of an array, 315
Ellipsis, 52
else, 157–159, A33
   dangling, 159, 161–163
Empty
   set, 472–473
   statement, 156, 232, 273
end, 31, 105–106
End-of-file, 371
End-of-line, 371
English units, 76–78
Enumerated (scalar) type, 425–432
eof, 153–154, 378–380, 575–576
eoln, 70, 153–154, 378, 576
Equivalent resistance, 51
Error
   compile-time, 39–40
   computational, 171–186
   run-time, 42–44
   trapping, 278–280
Euler's formula, 545, 558
Exchange, 156–157, 197, 450, 456
Exchanging values, 101–104
Executable statement, 26–27, 106
Execution of a program, 39–42
exp, 292
Expected value
   random variable, 391
   sample, 410
Exponent, 7
Exponentiation, 239–241, 270, 277, 284,
   301–303
   function, 286–308

growth and decay, 295–301
   notation, 303
   operator, A54
   probability distribution, 423
   series, 291
Expressions, 32
   arithmetic, 64–66
   boolean, 145–153, 271, 275–277
   formatting, 66–69
   operator precedence, 62–64, A4–A5
   set, 484
extern, A23
External
   files, 569–573
   procedures and functions, A22–A30

Factor, 65, 150
Factorial, 251
   iterative, 396
   recursive, 602
false, 151
Fibonacci sequence, 251, 602
Field list of a record, 523
Field selector, 524–528
Field width of output, 34–38
FIFO, 669
file, Pascal, 369–385, 560–578
   buffer variable, 573–575
   components of, 370–371, 562–563
   external, 569–573
   internal, 564–569
   reading from, 372–379, 568–569
   standard procedures and functions,
     573–577
   of structured data types, 560–578
   text, 369–385
   writing to, 380–381, 564–568
File, disk, 381–382, 385, 570
   including, 129
Finite-difference equation, 287
First in/first out, 669
Floating point arithmetic, 179–180
Flowchart, 155
for statement, 226–241, 272–274,
   276–277, 288
Formatting output, 32–39, 66–69
fortran
   packages, A22
   subroutines in Pascal, A22
forward, 608–609
Fractal geometry, 618
function, 213–221
   as a parameter, 618–636
   recursive, 599–618
   standard, 28–29, A2–A3
Functional requirements, 488–489

Gauss, Karl, 399
Gaussian probability distribution, 399–402
  in hypothesis testing, 412–413
Geometric probability distribution, 423
Geometric sequence, 249
get, 575–576
Global variable, 112, 209
goto, A17–A20

Half-life, 298–299
Hardware, 2–4
Harmonic series, 253
Heading
  function, 215–216
  procedure, 105
  program, 30
Hexadecimal numbers, A15–A16
High-level languages, 8–9
Hypothesis testing, 411–413

I/O devices, 3
i (imaginary number), 540
Identical types, 321
Identifier, 15–17, 105
  global, 112, 209
  local, 112, 196, 236–237
  scope of, 110–113, 122–124
if statements, 154–159
Imaginary number, i, 540
in, 484–485
Including files, 129
Independent
  random variables, 388, 391, 580
  variable, 624
Index, array, 315–316
Infinite
  loop, 272
  recursion, 610–611
  series, 252
Inherent errors, 176–177
Initialization of variables, 42
input file, 30, 53–59
  devices, 3
Insertion sort, 457–461
integer, 6, 17–18
  formatting, 35–36
  overflow, 120–121
  maxint, 22, A31, A43, A52
Iteration versus recursion, 603, 606–607
Interactive input, 5, 54
Internal files, 564–569
Intersection of sets, 474–475, 481
Invoking a subprogram, 106–108, 198–204, 217–220

j, imaginary number, 540

Kinetic energy, 138

label, A17–A18
Label list (case), 164
Last in/first out, 604
LIFO, 667–668
Limit
  call in recursion, 604
  of a sequence, 250–251
Linear congruent equation, 581
Linked list, 641, 650–660
ln (natural log), 29, 292
Local identifiers, 112, 196, 236–237
Looping structures
  for, 226–241
  repeat-until, 274–278
  while-do, 271–274

Machine language, 8
Main program, 30–31, 105–108
Mainframe, 2
Maintenance of programs, 502
Mandelbrot, Benoit, 618
Mantissa, 7
Markov chain, 591
Matrices, 353–357
  arithmetic operations, 355–357
  determinant, 367
  transpose, 366
maxint, 22, A31, A43, A52
Mean
  of a distribution, 390
  of a sample, 410
Median, 409
Membership in a set, 484–485
Memory, 3
Menus for data entry, 440
Metric system, 76–88
Microcomputer, 2
Minicomputer, 2
Mixed mode arithmetic, 60
mod, 61–62
Mode of a sample, 409
Modular programs, 96–101, 113–115, 195
Momentum, 137
Multi-dimensional arrays, 324–326
Multiplication
  of complex numbers, 548
  cross product of vectors, 349–352
  dot product of vectors, 345–349
  of matrices, 355–357
  of reals and integers, 59–64
  scalar, 344, 354
  of sets (intersection), 474–475
Mutually exclusive
  random variables, 388, 391
  sets, 476

Natural log, 29, 292
Natural response of an RC circuit, 300

Nested
  if statements, 161–163
  procedures, 112
new, 644
Newton, Isaac, 75
nil, 647–650
Normal probability distribution, 399–402
not, 147
Null
  hypothesis, 412
  set, 472–473
  statement, 156, 232, 273
Number conversion, 5–6, A15–A17
Numerical
  constants, 80
  errors, 171–186

Octal numbers, A15–A17
odd, 153–154
Offset of character codes, 279–280
Ohm's law, 50
One-dimensional array, 315
open, 381
Operating system, 4–5
Operators
  arithmetic, 59
  boolean, 147
  precedence of, A4–A5
  relational, 145–147
  set, 477–485
or, 147
ord, 279, 428, 436, 490
Ordinal types, 142, 230, 317, 425–432
  enumerated types, 425–432
  subranges, 318, 430–432
output, 30
  carriage control, A60
  devices, 3
  formatting, 34–38, 66–69
  to a **file**, 564–568
  to a **textfile**, 380–381
Overflow, 120–121, 254

p-series, 266
pack, 432
packed array of char, 432–434
page, A21
Parallel circuit, 51
Parameters, 195–209, 618–621
  arrays, 321–323
  subprograms as, 618–621
  value, 196, 204–209
  variable, 196–204
Parentheses, 62–64
Partial sum, 252
Pascal, Blaise, 1, 80, 394

Period
  notation for records, 524–528
  of a sequence, 581
Periodic table of elements, 471
Permutations, 266
Physical constants, 80
Plotting, 622–636
Pointer-type variables, 639–666
Poisson distribution, 397–398
Polar coordinates
  for complex numbers, 541
  for vectors, 336–341
Population growth, 268, 280–285
Precedence, A4–A5
  of identifiers, 110–113, 122–124
  of operators, 62–64, 149
Precision, 177–179
pred, 428
Principal angle, 339
Probability, 386–402, 580
  Bernoulli distribution, 394–396
  binomial distribution, 396–397
  exponential distribution, 423
  Gaussian (normal) distribution, 399–402, 412–413
  geometric distribution, 423
  Poisson distribution, 397–398
  uniform distribution, 398–399
procedure, 96, 104–113, 195–209, 618–621
  recursive, 599–618
  standard, 106, A3–A4
program, 30–31, 39–45
  development, 113–116
  specifications, 489–490
Prompt, 54
Propagation of errors, 183
Pseudocode, 98
Pseudorandom numbers, 580
Punctuation, A1–A2
put, 574–576

Quadratic equation, 172
Quantification of errors, 182–183
Queue, 667–668

Radius of a circle, 11
Random
  access data structures, 572
  number generator, 579–591
  variable, 387–393
  walk, 596
Range, 430–432
RC circuit, 300–301
read, 53–59, 373–377, 568–576
Reading
  arrays of characters, 434–435
  characters, 56–57
  from a **file**, 372–379, 568–569

pointer, 373
preventing errors, 504–517
trapping errors, 278–280
readln, 53–59, 376–377, 576
ReadString, 434
real, 17–19
  comparing, 157
  formatting, 36–38
  representation, 6–7
Realization of a random variable, 393, 580
record, 522–539
  variant, 533–535
Rectangular coordinates
  complex numbers, 541
  vectors, 336–341
Recursion, 21, 252, 599–618
Relational operators
  booleans, 145–147
  sets, 484
  strings, 435–436
Relative error, 182
repeat-until, 268–270, 274–278
Representation
  errors, 181
  of numbers and characters, 6–7
Reserved words and symbols, 16–17, A1–A2
reset, 373–377, 568, 574, 576
Resistance, 51, 96
rewrite, 380, 564, 568, 574, 576
RL circuit, 558
Root finding, 172, 468
round, 29
Round-off error, 180–181
Run-time error, 42–44, 279
Running product, 239–241
Running total, 227–229

Sample, 408–410
  mean, 410
  variance, 410
Scalar (dot) product, 345–349
Scalar multiplication
  of matrices, 355
  of vectors, 344
Scalar types, 425–432
Scientific notation, 7, 18, 177–180
Scope
  of field identifiers, 527–528
  of identifiers, 122–124, 236–237
  of subprograms, 110–113
Searching, 443–447
  binary, 446–447
  sequential, 444–446
Seed, 581
Selection sort, 451–453

Semicolon
  as empty or null statement, 156, 232, 273
  between executable statements, 31
Sequences, 248–252, 602
Sequential
  data structures, 375
  files, 572
Sequential search, 444–446
Series, 252–254
Series circuit, 96–97
set, 470–487
  of char, 485–486
  of integer, 485
SI units, 75
Significant digits, 7, 177–179
sin, 29, 266
sorting
  bubble sort, 453–457
  insertion sort, 457–461
  selection sort, 451–462
Source code, 11
sqr, 29
sqrt, 29
Square wave, 559
Stack, 604–605, 668
Standard deviation, 392, 410
Standard identifiers, A2–A4
Standardized normal, 401
State space, 390
Statement
  assignment, 25–28
  compound, 155–157
  null, 156, 232, 273
  procedure call, 106–108
Statistic, 409
Statistics, 407–413
Strings, 432–437
Strongly-typed, 17, 27
Structured data types, 315
Structured programming, 12, 96–101, 113–115, 491
Stub, 496
Subprograms
  functions, 213–220
  functions vs. procedures, 220–221
  as parameters, 618–636
  procedures, 104–113, 195–209
Subrange, 318, 430–432
Subscript, array, 315–316
Subset, 472
Subtraction
  arithmetic, 59
  of complex numbers, 547
  of sets, 481–482
succ, 428
Summation, 252
Syntax, 15
  diagrams, 16

Index

errors, 39–40
Systems of measurement, 75–88

Tangent, 13
Taylor series, 266, 313
Term, 65, 150
Testing, 500–501
text, 370–371
textfile, 369–385, 560, 567, 569, 573, 576–577
  reading from, 372–379
  writing to, 380–381
then, 154
time, 583
Time constant of a circuit, 301
Time-sharing, 4
to, 229–231
Top-down design, 12, 491
Torque, 358–359
Transcendental numbers, 178
Transpose of a matrix, 366
Tree, binary, 670
Trial experimental, 580
true, 151
trunc, 29
Truncation error, 176
Truth tables, 147–149
Turbo Pascal, 433, 574, A30–A42
type
  compatible, 528–529
  declaration, 317
  identical, 321
  standard, A2

UCSD Pascal, 433, A42–A51
Unconditional loops, 226–241
Undeclared variables, 40
Undefined variables, 26, 42–43, 122, 232
Underflow, 254
Uniform probability distribution, 398–399
Union of sets, 474–475, 479–480
Units of measurement, 75–88
Universal set, 472–473
unpack, 432
until, 274
User-defined types, 315
  arrays, 313–333
  enumerated types, 425–432
  files, 369–385, 560–578
  pointers, 639–666
  records, 522–539

Value parameter, 196, 204–209
var
  in declarative part, 23–24
  in subprogram parameter list, 196–209
Variables, 23–27, 41
  global, 112, 209
  local, 112, 196, 236–237
  scope, 110–113, 122–124
  undeclared, 40
  undefined, 26, 42–43, 122, 232
Variable parameter, 196–204

Variance
  of a random variable, 391–392
  of a sample, 410
Variant **record**, 533–535
VAX-11 Pascal, A52–A64
Vectors, 335–353
  addition, 344
  cross product, 349–352
  dot (scalar) product, 345–349
  scalar multiplication, 342–344
Venn diagrams, 473

Walkthrough, 101–104
while-do, 268–275
Wirth, Niklaus, 9
with, 526–528, 532
Word, computer, 6
Work, 358–359
write, 70–71, 564–577
  to a **file**, 564–568
  to a **textfile**, 380–381
writeln, 32–39
  to a **textfile**, 381

Xor, 224
XToY, 302

YesEntered, 218–219

Zero crossing, 191
ZToPower, 240–241